The Establishment and Reconstruction of the Academician System in China

院士制度在中国
的創立与重建

王元

Jinhai Guo

The Establishment and Reconstruction of the Academician System in China

上海交通大學出版社
SHANGHAI JIAO TONG UNIVERSITY PRESS

Springer

Jinhai Guo
Institute for the History of Natural Sciences
Chinese Academy of Sciences
Beijing, China

Translated by
Xiaoxuan Zhou
East China University of Science and Technology
Shanghai, China

Weizhen Gu
East China University of Science and Technology
Shanghai, China

ISBN 978-981-15-7207-4 ISBN 978-981-15-7208-1 (eBook)
https://doi.org/10.1007/978-981-15-7208-1

Jointly published with Shanghai Jiao Tong University Press
Not for sale outside the Mainland of China (Not for sale in Hong Kong SAR, Macau SAR, and Taiwan, and all countries, except the Mainland of China)

This Springer imprint is published by the registered company Springer Nature Singapore Pte Ltd.
The registered company address is: 152 Beach Road, #21-01/04 Gateway East, Singapore 189721, Singapore

Foreword

The academician system is recognized as being one of the most important institutional arrangements of science academies worldwide. It was first established in Europe during the seventeenth century. Western science was introduced into China at the end of the Ming Dynasty, and it instilled new vitality into the declining, ancient empire. China began the difficult and tortuous process of integrating itself into this world-renowned academic system. At the beginning of the twentieth century, the imperial examinations were abolished. Around the same time, many traditional science and technology disciplines were interrupted, and then the process of systematizing Chinese science began. In 1928, Academia Sinica, the highest academic research institution in the Republic of China, was established as the first National Academy of Sciences in Chinese history. In 1948, the academician system was established at Academia Sinica. From then onward, the academician system, which had originated in Western civilization, became firmly established within China. On November 1, 1949, only 1 month after the founding of the New China, the Chinese Academy of Sciences was established. However, for some 40 years, the Chinese Academy of Sciences did not follow the academician system. There are various reasons for this. In 1955, the Chinese Academy of Sciences, influenced by the Soviet Union, established academic divisions and a membership system. Still, however, it did not appoint academicians. There were many setbacks, and it was not until 1994 that the academy's academician system was established.

In the decades that followed, scholars, both at home and abroad, investigated Academia Sinica's and the Chinese Academy of Sciences' academician system. They have published many papers and achieved notable results. However, most of these studies are somewhat scattered. Some historical facts in them still need to be verified, and some are even erroneous. At present, there is not a single monograph on the process of China founding an academician system. Since 2004, Dr. Jinhai Guo has spent a lot of time and energy studying the history of China's academician system. He visited the Second Historical Archives of China, the Documents and Archives Department of the General Office at the Chinese Academy of Sciences, the Archives of Tsinghua University and the Archives of Peking University to consult multiple pieces of archive materials. After careful sifting and researching, he compiled this book, the first monograph on China founding its own academician system. It is very gratifying.

The focus of this book is the process of China founding an academician system under the Republic of China. It shows the reader a panoramic, historical reconstruction of this milestone event in China's twentieth-century history of science and academia, and comments on the significance, impact, and shortcomings of this event. It also investigates in detail the process of rebuilding the academician system in New China. The Chinese Academy of Sciences did not establish an academician system for more than 40 years and it did not even grant anyone the title of academician until 15 years after reform and opening up. Therefore, China was some 200–300 years later than developed American/European countries in having academicians. However, it was also later than many developing countries, and this is a useful discussion point. This book pays special attention to the use of archive materials and relevant diaries, so as to show the true history of this event as much as possible. I had the honor of being one of the earliest readers of this book, and it is my view that this book is rich in historical materials, full, and accurate in narration, and objective in exposition. In particular, the background/context into how the Academia Sinica academician system was established has been carefully sifted and analyzed, showing the author's considerable skill.

Although I speak positively in Jinhai Guo's monograph, that does not mean that this book is, by any means, flawless. After the election of the first batch of members of Academia Sinica at the end of the Liberation War, when the Kuomintang was doomed to failure, it was only about 18 months away from the founding of New China. At that time, there was not much activity in China's mainland. In order to fully understand how the system became established in China, it is also necessary to write about the Chinese Academy of Sciences' and the Chinese Academy of Engineering's academician system.

I have always kept a low profile and never boasted about my students. But Jinhai Guo is well known for his hard work and rigorous study at the Institute for the History of Natural Sciences, Chinese Academy of Sciences. He is also known for his work on the history of mathematics in China. There is a trend of people "putting money above everything else" in society and this trend has had increased impetuous in academia. Consequently, Guo's colleagues say that "people who study hard like Guo Jinhai are precious." In this regard, I am both satisfied and proud; but I have also had to remind him about the importance of rest. I have even bluntly said, "you need to learn how to rest" and "take your children to the park at the weekends or during the holidays." Because in my opinion, spending time with your children and seeing the mountains and rivers might on the surface delay working hours, but in fact will give impetus to scientific research work, and will also bring inspiration. In any case, as a teacher and a senior, I advised Guo to play hard rather than to work hard. Ours might be a unique, and probably rare, the relationship between teacher and student.

I often tell Jinhai Guo that I admire my seniors Li Yan and Qian Baocong very much. We have benefited immensely from their foundation work into the history of Chinese mathematics. But I admire the academic and human qualities of Mr Qian more. During the summer of 1998, when I was in charge of compiling the *Complete Works of History of Science by Li Yan and Qian Baocong*,[1] I received an article from Prof. Qian Xu, Mr Qian's daughter, recalling her father. After reading the article, I was deeply moved and decided to include the article in the complete works as an appendix. Of course, I had to ask Mr Li's family to also write a memoir. Prof. Li Bingquan, Mr Li's son and vice-president of the Third Hospital of Peking University, wrote a manuscript about his father's contributions to the history of mathematics. I said, "I hope you can write some anecdotes about your father's family life, such as how he educated his children." Prof. Li replied, "as soon as my father got home, he went to work in his study and didn't come out until dinner. When we ate together, he seldom talked or smiled." So I showed him Prof. Qian Xu's article of recollection, and Prof. Li later rewrote his manuscript. In my opinion, as a scholar, one should not only have profound knowledge and the capacity to study rigorously; they should also demonstrate that they can build a warm family. During the Anti-Japanese War, Mr Qian's family moved westward with the staff of the University of Chekiang to Guizhou. Their life was difficult, but I came to discover that they included baked sweet potatoes in their poems. What warmth, and an optimistic attitude! I am glad to see that Jinhai Guo made great improvements, in this respect, in recent years. I wanted to write this passage to encourage and support Jinhai Guo!

Beijing, China
February 2013

Guo Shuchun

---

[1]Li Yan 李俨and Qian Baocong 钱宝琮(1998). *Liyan Qian Baoocng kexueshi quanji* 李俨钱宝琮科学史全集 (Complete Works of History of Science by Li Yan and Qian Baocong), 10 vols. Shenyang: Liaoning jiaoyu chubanshe 辽宁教育出版社.

Acknowledgments

The history of China's academician system first caught my interest in 2004. While I was browsing the archives at Tsinghua University, I accidentally discovered material on the university's 1947 nominations for the first batch of members of Academia Sinica. This got me really interested. I discussed the topic with Wang Yangzong who was a research fellow at the Institute for the History of Natural Sciences, Chinese Academy of Sciences. He told me that he thought this was a subject worth exploring. I started by looking at the election of the first batch of members of Academia Sinica and the history of China's academician system. I have spent 10 years publishing a series of papers on this topic, and I published a Chinese-language version of the book *The Establishment and Reconstruction of the Academician System in China*.[2] This book is the English version of that monograph. The framework of the original book has been slightly adjusted, and errors in the original book have been corrected. The contents of the original book have also been simplified due to space limitations. Shanghai Jiao Tong University Press organized special translators for the English version of this manuscript. Tang Zongxian and Li Ye are to thank for their efforts in ensuring the smooth publication of this English-language book.

Professor Li Zhaohua from the School of Mathematical Sciences at Tianjin Normal University was my Master's instructor and Research fellow Guo Shuchun from the Institute for the History of Natural Sciences of Chinese Academy of Sciences was my instructor during my Doctorate. They are both scholars who have made considerable academic achievements in the field of history of China's mathematics. They have both helped guide me through the research gate of the history of science and technology. Over the years, they have given me a lot of academic guidance. Guo Shuchun gladly wrote a foreword to this book. Fan Hongye, a research fellow at the Institute of Policy and Management, Chinese Academy of Sciences, made pertinent amendments to the first draft and outline of this book. He also provided precious pictures of the badge for members of

[2]Jinhai Guo 郭金海 (2014). *Yuanshi zhidu zai zhongguo de chuangli yu chongjian* 院士制度在中国的创立与重建 (The Establishment and Reconstruction of the Academician System in China). Shanghai: Shanghai Jiaotong daxue chubanshe 上海交通大学出版社.

Academia Sinica. Research fellow Wang Yangzong not only suggested that I study the election of the first batch of Members of Academia Sinica but also provided useful information and suggested amendments in several of my related papers.

Yuan Xiangdong, a research fellow at the Academy of Mathematics and Systems Science, Chinese Academy of Sciences, also made valuable suggestions, both in the writing process of this book and upon seeing the first, Chinese draft. Mr. Yuan also introduced me to Wang Yuan, an academician at the Chinese Academy of Sciences, who generously signed the Chinese version of my book. Both Ge Nengquan at the Chinese Academy of Engineering and Shan Tianlun at the Chinese Academy of Social Sciences provided valuable, historical materials based on their own experiences and personal knowledge. They carefully reviewed some of the first drafts of this book. During my first visit to Nanjing, Prof. Chen Youqing at Hohai University provided me with residence. Han Jianmin, the former President of the Shanghai Jiao Tong University Press, Dr. Zhou Leiming at the China Pharmaceutical University, Zhu Junpeng at the Archives of Tsing Hua University, Liu Chang at the Chinese Academy of Engineering, and Zou Dahai, a research fellow at the Institute for the History of Natural Sciences, Chinese Academy of Sciences, also provided help in various forms.

I am very grateful to all of the above, my teachers and friends! I would also like to thank the Second Historical Archives of China, the Archives of Chinese Academy of Sciences, the Archives of Tsinghua University, the Archives of Peking University and the Archives of Nankai University for allowing me to use the material they have collected. Thanks to the Shanghai Jiao Tong University Press and the Springer Publishing Group for their support and meticulous publishing work. The successful completion of this book is due largely to the loose and free academic environment and rich, academic atmosphere at my workplace: the Institute for the History of Natural Sciences, Chinese Academy of Sciences. Over the past 10 years, my roots in this soil have enabled my academic studies to grow, bit by bit. I will forever be eternally grateful and remember it.

Before the Chinese version of the book was published, it was selected by the National Social Sciences Fund as a later stage funded project. After publication, it won the third-term "Science, History and Philosophy Youth Works Award," in 2017. This gave me great encouragement and made me more confident about pursuing my work in the difficult field of China's History of Modern Science and Technology. Therefore, I would also like to take this opportunity to express my gratitude to the experts who participated in the review, the National Office for Philosophy and Social Sciences, the founding organizations of "Science, History and Philosophy Youth Works Award", Tsinghua University's Science, Technology and Social Research Center, and the Council of the Award Foundation.

I dedicate this book to my parents Guo Deyou and Han Qinglan, and my wife Zhao Li. Their long-term support for my work has been an important factor in my academic achievements. My son, Guo Mozhi, has brought me a lot of joy over the past 10 or so years. I would like to express my deep gratitude to them. This book is also dedicated to honor my late mother.

Introduction

Among academic circles, there is no clear, universal definition for what an academician system is. The academician system in this book refers to an institutional arrangement with various regulations, standards, forms, structures, and operational mechanisms. The system is used by a national science academy or a non-government academic society that functions as a national science academy for its members, fellows, or academicians. The establishment of academician system at least satisfies the requirements that members, fellows, or academicians of the above organizations formally begin group activities and their organizations are established and start working.

The academician system was born in Europe during the scientific revolution in the seventeenth century. It was a product of institutionalization of modern Western scientific research. In 1666, the French Royal Academy of Sciences (Académie des Sciences) was established. A group of outstanding scientists were selected as members in the name of the French King, Louis XIV. They were paid to do research there. The establishment of this institution meant the birth of the first state-owned academician system in history. However, the Royal Society, formally known as the Royal Society of London for Improving Natural Knowledge, was established in 1660. It was a loosely organized academic society, independent of the government, and its members were known as "fellows." In 1662, it was granted a royal charter and became "the Royal Society." When this happened, this group established legal, political status, and began to function as the national academy of sciences. The composition of its fellows was complicated in the early days, but by 1860, the majority of its members were scientists. By the end of the nineteenth century, all fellows had been admitted into the society, based on their scientific achievements (but the quota of members of royal family into the Royal Society is still retained to this day). Fellows of the Royal Society are regarded as academicians of the academy of sciences, and they have maintained a high reputation in the international academic community.[3]

A number of continental European countries followed in the footsteps of France, in setting up academies of science, selecting academicians, and establishing academician systems. Commonwealth countries predominantly followed the example of Britain and set up academies without research entities. Their members were similarly known as either "fellows" or "academicians." At the beginning of the twentieth century, the academician system became popular in the West, and was regarded as an important institution in promoting academic development. Although the national academies of science varied from country to country (in form, size, and functions), groups of academicians were generally regarded as their core component. They were noted for their outstanding achievements and contributions and credited with honorary titles.

After the academician system was born in Europe, the Chinese people became aware of Western science. However, it was not until around the year 1910 that an equivalent modern, scientific organization, and system emerged in the country. China's system did not evolve from its own traditional social system but was transplanted from the West. The national academy of sciences and the academician system were to thank for this. In 1928, the Nanjing national government established the Academia Sinica (literally the "Central Research Academy"). It became recognized as the highest national academic research institution. It was the first national academy of sciences in China's history and integrated the natural sciences with humanities and social sciences. Since then, and with the development of various modern sciences in the country, many scientists have made outstanding academic achievements. Groups of academic elites have emerged and gained a reputation in the international arena. In 1948, the Academia Sinica established its academician system. This meant that the academic elite could finally enjoy honorary titles for their achievements, and make their mark in Chinese history. More than 300 years after Western learning spread to the east (during the late Ming dynasty), China's modern scientific system has become fully integrated with the rest of the world.

During the Academia Sinica's First General Assembly of Members in September 1948, Acting President Chu Chia-Hua said he was deeply touched by the establishment of the academician system, and that he had great hopes for China's future scientific development. He said: "China's import of western science began in the Ming Dynasty, but it had been repeatedly influenced by traditional ideas and the environment, and was without achievement. The late Qing Dynasty reform movement advocated new means of education and introduced Western culture to us, but science had still failed to take root in our country. It has also been twenty

[3]Henry Lyons (1944). *The Royal Society, 1660–1940: A History of Its Administration under Its Charters*. Cambridge: Cambridge University Press. 232–233, 276; Fan Hongye 樊洪业 (2005). Yuanshi zhidu qibian 院士制度七辩 (Seven Debates on the Academician System). *Kexue shibao* 科学时报 (Science Times), 2005-06-06 (6).

years since the National Government established Academia Sinica, after Nanjing became the capital city of China. But as of today, members of Academia Sinica have been elected and our academician system has become truly complete. The foundation for scientific research has been laid; scientific research has entered a new stage. We hope that scientific research will be accelerated in the future under the leadership of all of you and better achievements and contributions will be made in 20 years."[4] However, subsequent political changes in China have changed the track of the Academia Sinica and Chinese science as a whole. On April 23, 1949, the People's Liberation Army liberated Nanjing, the Academia Sinica closed on the Chinese mainland, and the first batch of members of Academia Sinica parted ways.

On November 1, 1949, the Chinese Academy of Sciences (CAS) was formally established in the People's Republic of China. The Academia Sinica's academician system was completely abandoned with this regime change. Taking lessons from the Soviet Union, in 1955, the CAS set up academic division memberships. It did not, however, set up an academician system. This came 39 years later, in 1994—both the CAS and the Chinese Academy of Engineering (CAE) established academician systems after all members of academic divisions of CAS became academicians. The academician system was, therefore, completely reconstructed in China. There were, however, constant disputes over the Chinese Academy of Sciences' system, how it should select and increase its academic division members and establish its academician system, and the relations between its members and its academicians. After the regime change and the Thought Reform Movement, the relationship between the government, the national academies of science, and academic elites has also changed. The reconstruction of the academician system in China was not only tortuous but also unique.

This book attempts to systematically study the process of China establishing and reconstructing its academician system. Based on archive material it is focused first on the Academia Sinica during the Kuomintang rule on the mainland of China, and discusses the establishing process of the academician system in China revolving around the brewing, preparations and establishment of the system by the Council of Academia Sinica. It then looks at the Chinese Academy of Sciences after the founding of the People's Republic of China, and explores the reconstrcting process of academician system in China combined with the situation of the Chinese Academy of Engineering and the Chinese Academy of Social Sciences.

In my opinion, the establishment and reconstruction of the academician systems in China is a topic that is worthy of serious study. First of all, it was a milestone event in China's scientific institutionalization and was closely linked to the process of Chinese modernization. It was an important chapter in China's twentieth-century science, technology, and academic history. Secondly, it also reflected the different

[4]Zhongyang yanjiuyuan zuzhi banfa yuanshi huiyi jilu jiqi xuanju guicheng deng youguan wenshu 中央研究院组织办法、院士会议纪录及其选举规程等有关文书 (Relevant Documents such as the Organization Methods of Academia Sinica, Minutes of the General Assembly of Members and the Regulations for Election). Nanjing: Zhongguo dier lishi danganguan 中国第二历史档案馆 (The Second Historical Archives of China), Total No. 393, File No. 2824.

processes for establishing core academic systems, based on the different social conditions for the Republic of China, and the new China. Studying this will help people gain a better understanding of China's scientific institutionalization processes from the Republic of China to the founding of the People's Republic of China in 1949, how academic systems changed around 1949, and the relationships between the government, the national academy of sciences and academic elites.

Moreover, focusing on the Academia Sinica and studying the establishing process of academician system in China will help people deeply understand how China transplanted a modern, Western academic system under the contact and collision of Chinese and Western civilization. It will also help people understand what kind of changes and patterns took place during the process of localization of Western academic system in China. The Council of Academia Sinica was an important, scientific community in the Republic of China. Studying the establishing process of academician system in China will help us deeply understand the council's academic appraisal activities as well.

Since the academician system of Academia Sinica was established in 1948, academic circles have published many achievements related to the theme of this book. For example, in 1948, Shiah Nae published the article "The Analysis of the First Batch of Members of Academia Sinica," which investigates data on the ages, birthplaces, educational backgrounds, and service institutions of the first batch of members of Academia Sinica.[5] Since the 1990s, Fan Hongye has also published a number of relevant articles, including "The Introduction of the Concept of Academy from the West and the Entablement of Academies in China," "Seven Debates on the Academician System," and "The Establishment of the Former Academia Sinica and its Election of the First Batch of Members."[6] In 1998, Chen Shiwei of Harvard University completed his doctoral thesis titled "Government and Academy in Republican China: History of Academia Sinica, 1927–1949," which was the first historical work on the Academia Sinica written in English. This paper

[5]Shiah Nae 夏鼐 (1948). Zhongyang yanjiuyuan diyijie yuanshi de fenxi 中央研究院第一届院士的分析 (The Analysis of the First Batch of Members of Academia Sinica). *Guancha* 观察 (Observation) 5 (14): 3–5.

[6]Fan Hongye 樊洪业 (1990). Yanjiuyuan dongjian kao "研究院"东渐考 (The Introduction of the Concept of Academy from the West and the Establishment of Academies in China). *Ziran bianzhengfa tongxun* 自然辩证法通讯 (Journal of Dialectics of Nature) 12 (4) :47–54; Fan Hongye 樊洪业 (2005).Yuanshi zhidu qibian 院士制度七辩 (Seven Debates on the Academician System). *Kexue shibao* 科学时报 (Science Times)06–06 (6); Fan Hongye 樊洪业 (1990). Qian zhongyang yanjiuyuan de chuangli jiqi shoujie yuanshi xuanju 前中央研究院的创立及其首届院士选举 (The Establishment of the Former Academia Sinica and its Election of the First Batch of Members). *Jindaishi yanjiu* 近代史研究 (Modern Chinese History Studies) (3): 218–231.

used archive material to explore the history of Academia Sinica, including the academician system. It was influential in promoting research on the history of Academia Sinica.[7] Chen Shiwei also published "A Review on the Election of Academicians at Academia Sinica in 1948" and "From Nanjing to Nangang: On the Two Academia Sinica Elections of 1949" in 2009 and 2011, respectively.[8] Zhou Leiming has also used archive material to study the election of the first batch of members of Academia Sinica.[9] In 2010, Li Lairong of Nankai University completed her doctoral dissertation titled "Academician System in Republic of China: Establishment and Operation of the Academician System in 1948."[10] She used archive material to investigate the election of the first batch of members of

[7]Chen Shiwei (1998). Government and Academy in Republican China: History of Academia Sinica, 1927–1949. Cambridge: Doctoral Thesis of Harvard University.

[8]Chen Shiwei 陈时伟. Zhongyang yanjiuyuan 1948 nian yuanshi xuanju shulun 中央研究院1948 年院士选举述论 (A Review on the Election of Academicians at Academia Sinica in 1948), In Zhongguo shehui kexueyuan jindaishi yanjiusuo minguoshi yanjiushi, Sichuan shifan daxue lishi wenhua xueyuan 中国社会科学院近代史研究所民国史研究室, 四川师范大学历史文化学院 (Research Office of History of the Republic of China of Institute of Modern History, Chinese Academy of Social Sciences, School of History and Culture of Sichuan Normal University) (Eds.) (2009). *1940 niandai de Zhongguo*—九四O年代的中国 (China in the 1940s), Vol. 2. Beijing: Shehui kexue wenxian chubanshe 社会科学文献出版社. 1025–1043; Chen Shiwei 陈时伟 (2011). Cong Nanjing dao Nangang: Zhongyang yanjiuyuan 1949 nian qianhou liangci yuanshi xuanju shulun 从南京到南港: 中央研究院1949年前后两次院士选举述论(From Nanjing to Nangang: On the Two Academia Sinica Elections of 1949). *Zhongguo xueshu* 中国学术 (China Scholarship) 8 (2), Total 28: 30–73.

[9]Zhou Leiming 周雷鸣 (2006). 1948 nian zhongyang yanjiuyuan yuanshi xuanju—九四八年中央研究院院士选举 (The Election of Academicians at Academia Sinica in 1948). *Nanjing shehui kexue* 南京社会科学 (Social Sciences in Nanjing) (2): 78–86.

[10]Li Lairong 李来容 (2010). Yuanshi zhidu yu minguo xueshu: 1948 nian yuanshi zhidu de queli yu yunzuo 院士制度与民国学术——1948年院士制度的确立与运作 (Academician System in Republic of China: Establishment and Operation of the Academician System in 1948). Tianjin: Doctoral Dissertation of Nankai University.

Academia Sinica. Wang Yangzong explored the origin and evolution of the academician system in China, as well as the processes for establishing academic divisions and membership system in the early days of the CAS.[11] Xie Yong,[12] Li Zhenzhen,[13] Cao Cong,[14] and Hu Zonggang[15] have also published relevant research on the academic divisions of the Chinese Academy of Sciences and its membership system. Liu Fengsong has done research on the contemporary Chinese academician system.[16]

These papers have enriched research on the history of the academician system in China, and they are helpful and enlightening for people seeking an understanding of the relevant historical facts in this field. However, research to date is still incomplete, and there has yet to be an academic monograph published on the establishment and reconstruction of the academician system in China. There was an important prelude to the establishment of Academia Sinica's academician system: the election of elected members of the Second Council. From this followed the nomination of the first batch of member candidates of Academia Sinica, the response regarding the official list of the first batch of member candidates of Academia Sinica, the evolution of CAS academic division members to academicians, the establishment of the CAE's membership system, and other important details and historical facts that have yet to be fully discussed. Moreover, some of the existing research results include historical errors.

In view of the above, I have self-published a series of research findings and collected further original documents, such as archives and diaries of relevant

---

[11]Wang Yangzong 王扬宗 (2005). Zhongguo yuanshi zhidu de jianli jiqi wenti 中国院士制度的建立及其问题 (The Academician System in China and Its Defect).Kexue wenhua pinglun 科学文化评论 (Science & Culture Review)2 (6): 5–22; Wang Yangzong 王扬宗. Cong yuanshi dao xuebu weiyuan: Zhongguo kexueyuan xueshu tizhi jianli de kunjing 从院士到学部委员——中国科学院学术体制建立的困境 (From Academicians to Academic Division Members: Difficulties in Establishing Academic System of the CAS). In Yu Miin-ling 余敏玲 (Ed.) (2012). Liangan fenzhi: Xueshu jianzhi tuxiang xuanchuan yu zuqun zhengzhi (1945–2000) 两岸分治：学术建制、图像宣传与族群政治 (1945–2000) (Divided Rule across the Taiwan Straits: Educational Reorganization, Visual Propaganda, and Ethnic Politics, 1945–2000). Taipei: Institute of Modern History, Academia Sinica. 65–135.

[12]Xie Yong 谢泳 (2005). 1949 nian hou zhishi jingying yu guojia de guanxi: Cong yuanshi dao xuebu weiyuan 1949 年后知识精英与国家的关系——从院士到学部委员 (The Relationship Between the Intellectual Elite and the Nation after 1949 : From Academicians to Academic Division Members). *Kaifang shidai* 开放时代 (Open Times)(6): 51–64.

[13]Li Zhenzhen 李真真 (1992). Zhongguo kexueyuan xuebu de choubei yu jianli 中国科学院学部的筹备与建立 (The Preparation and Establishment of the Divisions of Academia Sinica). *Ziran bianzhengfa tongxun* 自然辩证法通讯 (Journal of Dialectics of Nature)14 (4): 40–50, 80.

[14]Cao Cong(2004). *China's Scientific Elite*. London and New York: Routledge Curzon.

[15]Hu Zong-gang胡宗刚 (2005). Hu Hsen-hsu luoxuan xuebu weiyuan kao 胡先骕落选学部委员考 (Verification on Losing the Academician Election of Hsen-Hsu Hu). *Ziran bianzhengfa tongxun* 自然辩证法通讯 (Journal of Dialectics of Nature) 27(5): 67–72, 111.

[16]Liu Fengsong刘峰松 (2012). Zhongguo dangdai yuanshi zhidu yanjiu: Lishi yu wenti 中国当代院士制度研究: 历史与问题 (A Study of the Contemporary Membership System in China: History and Its Challenges). Beijing: Doctoral Dissertation of Institute for the History of Natural Sciences, Chinese Academy of Sciences.

figures. I have drawn on the most recent research findings in the writing of this book. In the process of historical research, archive material from original documents proved the most reliable, and this book has paid special attention to excavating and utilizing this resource. The archives used in the book are mainly from the Second Historical Archives of China, the Archives of Chinese Academy of Sciences, the Archives of Tsinghua University, and the Archives of Peking University. At the same time, this book pays attention to the completeness and systematicness and also focuses on sorting, researching, and analyzing the important links and details of basic historical facts. In order to give wider scope to this field of research, it will also factor in how the academician system was brewing in the Council of Academia Sinica, and draw on how this system had already proven popular in the West. This book is divided into 10 chapters:

Chapter 1 outlines the international and domestic context that Academia Sinica established its academician system. I begin by focusing on the birth and development of the academician system in the West to introduce the international background that the academician system had already proven popular in the West, and subsequently, was regarded as an important institution for promoting academic development when the Second Council of Academia Sinica ruminated on an academician system. I then discuss China's domestic, academic background: the institutionalization of Chinese sciences since the founding of the Republic of China; the establishment of Academia Sinica; the rise of Chinese academia, and the common ideals of pursuing academic independence in Chinese society.

In Chapter 2, I discuss what happened prior to the establishment of Academia Sinica's academician system: how the Council of Academia Sinica was established and its members were elected. First, I examine how the council was arranged and established. I then look at how the Second Council members were elected and distinguish this election from that of the first batch of members of Academia Sinica. I also analyze the similarities and differences between this election of council members and the elections of the Ministry of Education's Academic Review Committee members in 1940.

Chapter 3 examines how Academia Sinica's Second Council members met, and how they structured the academician system and relevant regulations. I begin by discussing the Second Council's deliberation of the academician system of Academia Sinica and the progress made by the council during the Anti-Japanese War. I then analyze how the council discussed and decided to establish an academician system during its Third Annual Meeting, and how the council amended *The Organization Act of Academia Sinica* and *The Regulations of the Council of Academia Sinica*. Finally, I study how the council formulated *The Regulations for the Election of Members of Academia Sinica* and *The Regulations for the General Assembly of Members of Academia Sinica*.

Chapter 4 explores how the first batch of member candidates of Academia Sinica was nominated and how they qualified to become Academia Sinica's official member candidates. First, I look at how the Second Council members worked with the Preparatory Committee for the Election of the First Batch of Members of Academia Sinica to draft a reference list of member candidates. I then study the official nominations, and their many twists and turns. Finally, I discuss the preliminary examinations of nominees by the Preparatory Committee for the Election of the First Batch of Members of Academia Sinica, and how the Academia Sinica's Second Council determined who would qualify as a member candidate. I analyze the nominations' gains and losses, and the impact of social relation resources on academic evaluation of member candidates.

Chapter 5 examines the response to the official list of the first batch of member candidates of Academia Sinica, and the subsequent official election of members. First, I discuss the responses made by Bo Shang, Wang Hefu, Yuan Hanqing, and Fu Ssu-Nien. I then look at the official election, the reasons for its success, the members themselves, and how academic circles responded to the election results.

Chapter 6 discusses the First General Assembly of Members of Academia Sinica. First, I introduce the symposium of members of Academia Sinica received prior to the assembly, the design of the certificate and badge issued to them, the opening ceremony, and preparatory meeting of the assembly, and subsequent meetings held by the subcommittee. Second, I examine how the assembly elected its members of the Third Council, and the issues they resolved. I then analyze how the first batch of members of Academia Sinica implemented solutions to these issues. Finally, I study the choices made by the first batch of members of Academia Sinica about leaving or staying in mainland China before and after 1949.

Chapter 7 starts from the preparations for the establishment of Chinese Academy of Sciences, and discusses the process that the Academia Sinica's academician system was completely abandoned in mainland China, and the whole story that Chinese Academy of Sciences prepared and established the academic divisions and the membership system of academic divisions. The purpose of the chapter is to reveal the origin and prelude of the reconstruction of the academician system in China. First, I examine the preparatory work involved and the reasons why Academia Sinica's academician system was completely abandoned in mainland China. Second, I explore the backgrounds and reasons for establishing the academic divisions and the membership system of academic divisions of Chinese Academy of Sciences, and then elaborate the relevant process.

Chapter 8 studies the 1957 supplement of members of academic divisions and the 1980 co-optation of members of the academic divisions of Chinese Academy of Sciences. The chapter will reflect the improvement of the selection system of members of academic divisions and the evolution of the nature of academic division members of Chinese Academy of Sciences from a job title to an academically honorary title. First, I examine the backgrounds, selection criteria, scope and processes of the 1957 supplement of members of academic divisions of Chinese Academy of Sciences. Then, I discuss the situation prior to the 1980 co-optation of members of the academic divisions of Chinese Academy of Sciences, and how the co-optation was prepared and processed.

Chapter 9 examines how a joint academician system nearly came about in 1980 involving the Chinese Academy of Sciences, Chinese Academy of Social Sciences and the State Science and Technology Commission. I look at why this plan was suspended, and how China's academician system was reconstructed between 1980 and 1981. I explore the initial work that went into formulating regulations for a joint group of academicians, and how the three units submitted relevant documentation to the State Council and obtained approval. I look at the opposition raised among the Chinese Academy of Sciences towards this system, and lay out in the summary the reasons for why this plan was suspended.

Chapter 10 studies the Chinese academic circles' calls after 1981 for an academician system. It explores how the nature of academic division members of Chinese Academy of Sciences has continuously evolved, and how members have attained status as academicians. It examines the establishment of the membership system of Chinese Academy of Engineering, and the preparation work involved. It also looks at the Chinese Academy of Social Sciences' effort for establishing its membership system. The chapter will show the tortuous and unique course of the reconstruction of the academician system in China after 1981.

This book concludes by exploring the relationship between the government, the national academy of sciences and the academic elite around 1949 in China, and how these relationships influenced the establishment and reconstruction of China's academician systems.

Contents

1 The Chinese Academic Circle in the Era of Academician System Prevailing in the West 1
1.1 The Birth and Rise of the Western Academician System 1
1.2 The Development of China's Scientific Institutionalization Process and the Establishment of Academia Sinica 3
1.3 The Rise of Chinese Academia and the Ideal of Academic Independence 8
1.4 Summary 16

2 Establishment of the Council of Academia Sinica and Election of Elected Council Members 17
2.1 Preparations and Establishment of the Council 17
2.2 Preparations for the Election of Elected Members of the Second Council 24
2.3 Operation Process of Election: From Primary Election to Final Election 28
2.4 Summary 34

3 The Council Structured the Academician System and Relevant Regulations 37
3.1 The Brewing of the Academician System and the Beginning of Legislation 37
3.2 The Legislation of Election Law: The Formulation of Regulations for the Election of Members 50
3.3 The Legislation of Assembly: The Formulation of Regulations for the General Assembly of Members 59
3.4 Summary 63

4 **How the First Batch of Member Candidates Were Nominated and Their Qualification Examinations** . . . 65
4.1 Formulating a Member Candidates Reference List . . . 65
4.2 Twists and Turns During the Official Nomination . . . 67
4.3 The Preliminary Examinations of Nominees by the PCE . . . 82
4.4 The Qualificaton Examinations of Member Candidates by the Council . . . 87
4.5 Summary . . . 95

5 **From the Response to the Official List of Member Candidates to the Official Election** . . . 97
5.1 Reactions from Bo Shang, Wang Hefu, Yuan Hanqing and Others . . . 97
5.2 Response from Fu Ssu-Nien . . . 101
5.3 Official Election of the First Batch of Members . . . 112
5.4 Summary . . . 131

6 **The First General Assembly of Members of Academia Sinica** . . . 133
6.1 From the Members' Symposium to the Subcommittee's Meeting . . . 134
6.2 The Implementation of Member Authority: The Election of the Third Council Members . . . 146
6.3 Other Agendas and Resolutions of the Assembly . . . 158
6.4 The Implementation of the Resolution After the Assembly and the Scattering of the Members . . . 168
6.5 Summary . . . 176

7 **From the Abolition of Academician System to the Establishment of the Membership System of Academic Divisions** . . . 179
7.1 Abandonment of Academician System of Academia Sinica and System Selection of the CAS . . . 180
7.2 Preparations and Establishment of Academic Divisions and the Membership System . . . 189
7.3 Summary . . . 213

8 **The Expansion of the Members of Academic Divisions of the Chinese Academy of Sciences Before and After the Cultural Revolution** . . . 215
8.1 The 1957 Supplement of Members of the Academic Divisions . . . 215
8.2 Co-optation of Academic Division Members in 1980 . . . 241
8.3 Summary . . . 255

9 Preparation and Suspension of Academician System Between 1980 and 1981 257
9.1 Preparatory Work and Formulation of Academician Regulations 257
9.2 From Reporting the "Request for Instructions" to the Emergence of Differences 271
9.3 The Fourth General Assembly of the CASAD Members and the Plan to Establish the Academician System Ran Aground 276
9.4 Summary 282

10 The Establishment of the Academician System in New China 285
10.1 From Academic Division Members to CAS Members: The Membership System Established by the CAS 285
10.2 The Preparations and Establishment of the CAE and Its Membership System 309
10.3 The Efforts of the CASS to Establish a Membership System 344
10.4 Summary 357

Conclusion 359

Glossary 367

Select Bibliography 389

Index 399

About the Author

Jinhai Guo born in 1974 in Tianjin, is a research fellow at the Institute for the History of Natural Sciences, Chinese Academy of Sciences. He graduated from the Chinese Academy of Sciences in 2003 and received a Ph.D. in Science. He specializes in the history of mathematics in China, history of modern science and technology in China, and the history of the Chinese Academy of Sciences. He has published more than 40 academic papers and authored *The Foundation of Modern Mathematics in China: The Department of Mathematics and Mathematical Communications Activities in Universities Before the Anti-Japanese War (Xiandai shuxue zai zhongguo de dianji: Quanmian kangzhan qian de daxue shuxuexi jiqi shuxue chuanbo huodong* 现代数学在中国的奠基：全面抗战前的大学数学系及其数学传播活动). He is currently a member of the Council of the Chinese Society of Mathematical History and a member of the Chinese Academy of Sciences' expert advisory group.

Abbreviations

AD	Academic Division (*Xuebu* 学部)
CAE	Chinese Academy of Engineering (*Zhongguo gongchengyuan* 中国工程院)
CAS	Chinese Academy of Sciences (*Zhongguo kexueyuan* 中国科学院)
CASS	Chinese Academy of Social Sciences (*Zhongguo shehui kexueyuan* 中国社会科学院)
CPC	Communist Party of China (*Zhongguo gongchandang* 中国共产党)
PCE	Preparatory Committee for the Election of the First Batch of Members (*Diyici yuanshi xuanju choubei weiyuanhui* 第一次院士选举筹备委员会)

List of Figures

Fig. 1.1 Tsai Yuan-Pei, president of Academia Sinica 7
Fig. 2.1 Tsai Yuan-Pei posed with Director-General Yang Quan at Academia Sinica 18
Fig. 2.2 After Yang Quan was assassinated, V. K. Ting succeeded as director-general of Academia Sinica 20
Fig. 2.3 Photo taken at the Inaugural Meeting of the Council of Academia Sinica on September 7, 1935 (part) 23
Fig. 2.4 Photo taken at the Fifth Annual Meeting of the First Council of Academia Sinica on March 22, 1940 30
Fig. 3.1 Chu Chia-Hua, acting president of Academia Sinica and president of the council 39
Fig. 3.2 Wong Wen-hao, honorary secretary of the council 40
Fig. 3.3 The Proposal "Please Stipulate the Name, Establishment, Election and Related Matters of Member of Academia Sinica" 45
Fig. 3.4 The Proposal "Please Set up Yuan Lü (院侣) of Academia Sinica and Amend the Organization and Tasks of the Council" (part). 46
Fig. 3.5 Fu Ssu-Nien, director of the Institute of History and Philology of Academia Sinica. 51
Fig. 3.6 Adam Pen-Tung Sah, director-general of Academia Sinica. 52
Fig. 3.7 Original manuscript of *The Regulations for the Election of Members of Academia Sinica*, drafted by Fu Ssu-Nien and revised by the group on March 17, 1947 (part) 53
Fig. 4.1 The Nomination Form for Wu Ta-You proposed by Chinese Physical Society 72
Fig. 4.2 Nomination Form for Hua Loo-Keng proposed by National Tsing Hua University 79
Fig. 4.3 Nomination form for Shiing-Shen Chern proposed by the National University of Peking 80

Fig. 4.4 Nomination form for Chiang Li-Fu proposed by Academia Sinica . . . 81
Fig. 5.1 Photo taken at the Fifth Annual Meeting of the Second Council of Academia Sinica . . . 114
Fig. 5.2 Recommendation List of Members of the Humanities Group circled by Hu Shih, convener of the group . . . 119
Fig. 5.3 Election ticket for the first batch of members of Academia Sinica . . . 121
Fig. 5.4 The elected by general election of the council in *the Announcement of Academia Sinica* . . . 123
Fig. 5.5 *The Announcement of Academia Sinica* on April 1, 1948. . . 127
Fig. 6.1 The badge for members of Academia Sinica in 1948 . . . 138
Fig. 6.2 Chiang Kai-shek delivered a speech at the First General Assembly of Members of Academia Sinica on September 23, 1948 . . . 141
Fig. 6.3 Photo of the First General Assembly of Members of Academia Sinica on September 23, 1948 . . . 145
Fig. 7.1 *Draft for the Establishment of the People's Academy of Sciences* (part) . . . 182
Fig. 7.2 Some leaders of the CAS took photos with members of the delegation visiting the Soviet Union in 1953 . . . 185
Fig. 7.3 Kuo Mo-jo, president of the CAS delivering a speech at the Inaugural Assembly of the Academic Divisions . . . 211
Fig. 8.1 Venue of the Second General Assembly of the CASAD Members . . . 238
Fig. 8.2 The result of the supplement announced by *Guangming Daily* on May 31, 1957 . . . 240
Fig. 9.1 Hu Qiaomu, president of the CASS . . . 258
Fig. 9.2 *Academician Regulations of the Chinese Academy of Sciences (Draft)* (Part) drafted by Tsien San-Tsiang . . . 263
Fig. 9.3 *Request for Instructions on the Establishment of Academician System* (Page 1) . . . 272
Fig. 9.4 Letter to Fang Yi and Li Chang by Hu Qiaomu on May 19, 1981 . . . 282

List of Tables

Table 2.1 Final election result of the elected members of the Second Council of Academia Sinica . . . 32
Table 4.1 A reference list of member candidates and the quota for some disciplines proposed by the PCE . . . 68
Table 4.2 List of member candidates drawn up by each group of the sub-group examination committee . . . 89
Table 5.1 Summary results of votes obtained by three groups from correspondence voting: mathematics and physics, biology, humanities . . . 115
Table 5.2 Results of correspondence voting and group voting of mathematics and physics group . . . 116
Table 5.3 Correspondence voting results of biology group . . . 117
Table 5.4 Correspondence voting results of humanities group . . . 118
Table 5.5 General election results of the election of the first batch of members of Academia Sinica . . . 122
Table 6.1 The results of the first vote for the third council members of Academia Sinica . . . 152
Table 6.2 The results of the second vote for the third council members of Academia Sinica . . . 154
Table 6.3 The results of the third vote for the third council members of Academia Sinica . . . 155
Table 6.4 The proposals of members and resolutions of the First General Assembly of Members of Academia Sinica . . . 161
Table 7.1 List of first academic division members of CAS (1955) . . . 205
Table 8.1 Results of the Division of Mathematics, Physics and Chemistry's solicitation of comments on initial referees at the end of November 1956 . . . 222

Table 9.1 Comparison of the Draft by Yu Guang-yuan and finalized *Academician Regulations (Draft)* 268
Table 10.1 Sub-group list of selection group of the CAE's preparatory leading group 329
Table 10.2 A list of the first batch of CAE members (1994) 339

Chapter 1
The Chinese Academic Circle in the Era of Academician System Prevailing in the West

As early as the 17th century, the academician system was born in Europe as a product of the institutionalization of western modern scientific research. In 1666, the Académie des Sciences was established. In the name of the royal family, it recruited outstanding scientists as academicians to work in the academy. This marks the birth of the first academician system, which was established by the state, in world history. In the process of transplanting modern western scientific organizations, systems and systematizing science, China established the first national academy of sciences, Academia Sinica, in 1928. Twenty years later, in 1948, Academia Sinica established the academician system. This chapter outlines the international and domestic context that Academia Sinica established its academician system in.

1.1 The Birth and Rise of the Western Academician System

"Academician" is the honorary title of the members of the national academy of sciences in the world. Its cultural connotation is generally consistent with the "fellows" of foreign non-governmental professional academic organizations represented by the Royal Society of London for Improving of Natural Knowledge. This means that the academic achievements and contributions of the candidates are recognized by the academic community and also represent candidates who have acquired a high academic status.[1] In English, members of the Royal Society of London for Improving of Natural Knowledge are called Fellows. Other countries generally call "academicians" as members. Some countries, including Russia, call the members of

[1]Chen Shiwei陈时伟 (2011). Cong Nanjing dao Nangang: Zhongyang yanjiuyuan 1949 nian qianhou liangci yuanshi xuanju shulun从南京到南港: 中央研究院1949年前后两次院士选举述论 (From Nanjing to Nangang: On the Two Academia Sinica Elections of 1949). *Zhongguo xueshu* 中国学术 (China Scholarship) 8 (2), Total 28: 33.

J. Guo, *The Establishment and Reconstruction of the Academician System in China*,
https://doi.org/10.1007/978-981-15-7208-1_1

the national academy of sciences as academicians. The latter two translations both indicate that academicians are members of the academy.

The Royal Society of London for Improving of Natural Knowledge, also known as the Royal Society of London, was founded in 1660 and is still one of the most important academic groups in Great Britain. It has no research entity and is a loosely organized academic group. Many years after the foundation of the society, its members were not limited to scientists. This is mainly related to its society's decision to absorb new members of a certain proportion from officials, intellectuals and the elite class in order to obtain sufficient funds for development. According to statistics, in the early days only one-third of its society members were made up of outstanding and contributing scientists, while aristocrats, barons and knights, gentlemen and members who did not understand science accounted for two-thirds.[2] At the beginning of the nineteenth century, the membership system of the Royal Society of London has undergone profound changes. Its membership has changed from mostly non-scientific members and only a small number of scientists to an outstanding number of prominent scientists. With this change, the institute's scientific research has continuously made important progress and earned a high reputation, which has consolidated its unique position in British scientific institutions. Its members are regarded as academicians of the national academy of sciences and have always maintained a high reputation in the international academic community.

The Académie des Sciences, also known as the Paris Academy of Sciences, was established in 1666. It is the first national academic research institution in Europe that is directly operated by the nation and provides salaries for scientists. When it was founded, a group of outstanding scientists were selected as academicians in the name of the French King Louis XIV, and they were paid to do research there. This marks the birth of the first academician system of the Academy of Science, which was established by the state, in world history. After the first batch of academicians materialized, the Académie des Sciences has continuously selected outstanding scientists as academicians. The first 30 years of the Académie des Sciences were not very formal in terms of system construction and it did not draw up a statute. In 1699, Louis XIV personally promulgated the charter of the academy, and under the patronage of the royal family reorganized the academy and moved the royal academy to the Louvre. By the eighteenth century, the members of the Academy had reached 70. This group of academicians actively engaged in scientific research. By publishing academic works, publications and providing advisory services to the government, they had effectively promoted the development of French science and made France the scientific epicenter in Europe.

Judging from the history of the world, the Académie des Sciences has a great influence on the establishment of academies and academicians in other countries around the world. After it was established, other countries, especially continental European countries, followed the example of France in setting up academies of science, selecting academicians and establishing academician systems. This is true of the

[2]Henry Lyons (1944). *The Royal Society, 1660–1940: A History of Its Administration under Its Charters*. Cambridge: Cambridge University Press.52–53.

Russian Petersburg Royal Academy of Sciences, the Prussian Royal Academy of Sciences and the Swedish Academy. Among them, the Russian Petersburg Royal Academy of Sciences was officially established in 1725 by order of Peter the Great.[3] It has research entities with academicians selected from among the most outstanding scientists as its constituent entities. At the beginning of the establishment of the Petersburg Royal Academy of Sciences, the scientific research level was in a leading position in Europe. It was renamed Russian Academy of Sciences and Soviet Academy of Sciences in 1917 and 1925 respectively.[4] Some academies, represented by the Russian Petersburg Royal Academy of Sciences, have since become models for other countries outside Europe to follow. Commonwealth countries predominantly followed the example of Britain in setting up academies of science without research entities. Their members were similarly known as either "fellows" or "academicians".

More than 200 years after the establishment of the Académie des Sciences, especially in the beginning of twentieth century, the national academy of sciences, a form of academic research institutions, had been widely accepted by the entire world. Academician system had been widely used in the West as an important institution of the national academy of sciences to promote academic development. Before the Academician System was established by Academia Sinica in 1948, many countries in the world had national academies of science and academician systems. In particular, the European countries comprised of Britain, France, Germany, Russia, Spain, Sweden, Norway, Ireland, Denmark, Switzerland, Belgium and Portugal have the most national academies and societies, followed by the United States. Among the countries that have established national academies of science in Asia, Japan established the Japanese Academy of Sciences in 1879, being the first to establish an academician system. National academies of science in different countries have different organizational forms, scales and functions, but academicians are generally regarded as the main body, outstanding academic achievements and contributions as the conditions for academicians, and academicians are regarded as high academic honorary titles.

1.2 The Development of China's Scientific Institutionalization Process and the Establishment of Academia Sinica

China's first national academy of sciences, the Academia Sinica, was founded in 1928, 262 years later than the Académie des Sciences. In fact, Emperor Kangxi already knew something about the Académie des Sciences through French Jesuits

---

[3]Alexander Vucinich (1984). *Empire of Knowledge: The Academy of Sciences of the USSR (1917–1970)*. Berkeley and Los Angeles: University of California Press.6.

[4]Xiong Shuming熊树明 and Liu Yongwei刘勇卫 (Eds.) (2009). *Shijie zhuyao guojia yuanshi zhidu zhinan*世界主要国家院士制度指南 (A Guide to the Academician System of Major Countries in the World). Jinan: Shandong jiaoyu chubanshe山东教育出版社.248.

who came to China in the 1880s and 1890s, and was said to be planning to set up an academy in China.[5] But by the end of the Qing Dynasty, China had not been able to establish an academy of sciences. However, education and science have undergone changes in China and the institutionalization of science has since begun.

After the founding of the Republic of China in 1912, the institutionalization of science began to advance to a higher level in China. In 1913, the Ministry of Industry and Commerce established the National Geological Survey of China, headed by V. K. Ting, bringing together outstanding geologists such as Zhang Hongzhao and Wong Wen-hao. In 1914, the Science Society of China was organized by Hu M. T., Chao Yuen Ren, Chow Jen, Bingzhi, Ren Hongjun, Zhang Yuanshan, Guo Tanxian, King P. C., Yang Quan and other Chinese students from Cornell University in the United States. In the continuous expansion of work and affairs, the Society has made outstanding contributions to the institutionalization of science in China, including the construction of social organizations, scientific research institutions, academic exchange systems and the formation of scientists' social roles. Its social status has also been continuously promoted, gradually becoming a symbol of folk science during the Republic of China.[6] In 1922, the Science Society of China established the Institute of Biology, which trained a large number of talents and achieved great scientific research results.[7] The Society and the National Geological Survey of China have made biology and geology a pioneer in promoting scientific localization in China.

In addition to the Science Society of China, by 1927, national academic organizations such as the Science and Art Society of China, Chinese Astronomical Society, the Geological Society of China, Chinese Meteorological Society, Chinese Physiological Society and the Natural Science Society of China had been established. Some data shows that from 1916 to 1917, there were 12 academic organizations affiliated with the Ministry of Education of the Republic of China's government in Beijing and on record. From 1918 to 1921, the number increased to 14. And it increased to 18 in 1922. From 1923 to 1926, the number increased from 20 to 31.[8] From 1912 to

[5]Translated by Zhao Chen赵晨, collated by Liu Yaowu刘耀武 (1981). *Kangxi huangdi*康熙皇帝 (Emperor Kangxi). Harbin: Heilongjiang renmin chubanshe.61.

[6]Zhang Jian张剑 (2008). *Zhongguo jindai kexue yu kexue tizhihua*中国近代科学与科学体制化 (Modern Chinese Science and Its Institutionalization). Chengdu: Sichuan renmin chubanshe四川人民出版社.298.

[7]Zhang Jian张剑 (2005). *Kexue shetuan zai jindai Zhongguo de mingyun: Yi Zhongguo kexueshe wei zhongxin*科学社团在近代中国的命运—以中国科学社为中心 (The Science Association and the Change of Society in Modern China: A Study on the Science Society of China). Jinan: Shandong jioayu chubanshe山东教育出版社.197–233.

[8]The Education Yearbook Compilation Committee of the Ministry of Education 教育部教育年鉴编纂委员会(Eds.) (1971). *Diyici Zhongguo jiaoyu nianjian*第一次中国教育年鉴 (The First China's Education Yearbook). In Wu Xiangxiang吴相湘and Liu Shaotang刘绍唐(Eds.). *Minguo Shiliao congkan*民国史料丛刊 (Historical Materials Series of the Republic of China), Vol. 4, Category 1. Taipei: Zhuanji wenxue chubanshe传记文学出版社.1489.

1925, there were a total of 44 academic organizations nationwide.[9] These reflect the rise of modern Chinese academic organizations after 1912.

Universities are institutions that train high-level talents and carry out academic research activities. They play an important role in the institutionalization of science. In 1912, when the Republic of China was established, Tsai Yuan-Pei became the education chief and urged the reform of the university academic system. As a result, the Ministry of Education promulgated the Ren-Zi-Gui-Chou educational system and gradually established a relatively complete new educational system[10] In the process of establishing the new education system, public universities have gradually become popular and developed their academic system as well. The emergence of private universities and church universities has also become an important force in China's higher education.

With the systematization of Chinese science moving to a higher level, some overseas Chinese students have made positive introductions and publicity in foreign academies, societies and academician systems.[11] A group of politicians and academics represented by Sun Yat-sen and Tsai Yuan-Pei also advocated for the establishment of academic institutes or research institutes in China. In the winter of 1924, when Sun Yat-sen left Guangdong for the north, he proposed to set up "the central academy as the highest academic research institution in the country to lay the foundation for revolutionary construction" and appointed Wang Ching-wei, Yang Quan and Huang Changgu to be responsible for drafting the academy plan.[12] However, Sun Yat-sen was soon unable to recover from his illness, and this proposal failed to materialize. Tsai Yuan-Pei studied in Germany in his early years and had many contacts with Li Shizeng who studied in France to understand the academic systems of Germany, France and other countries. On April 10, 1924, Tsai Yuan-Pei attended the China Society in London and delivered a paper entitled "*The Development of Chinese Education*". In this article, he solemnly suggested that the boxer indemnity from Britain should be used to "set up a large-scale research institute" in China. Referring to the "Science Museum and Natural History Museum in South Kensington", this research institute "not only undertakes the task of carrying out higher education and encouraging scientific development, but also will become the center

[9]The Education Yearbook Compilation Committee of the Ministry of Education 教育部教育年鉴编纂委员会(Eds.) (1986). *Dierci Zhongguo jiaoyu nianjian*第二次中国教育年鉴 (The Second China's Education Yearbook). Shanghai: Shangwu yinshuguan商务印书馆.795–796.

[10]Jin Yilin金以林 (2000). *Jindai Zhongguo daxue yanjiu*近代中国大学研究 (Study on University in Modern China). Beijing: Zhongyang wenxian chubanshe中央文献出版社.36–42.

[11]Yang Quan杨铨 (1915). Xuehui yu kexue学会与科学 (Society and Science). *Kexue*科学 (Science) 1 (7): 707; Ren Hongjun任鸿隽 (1917). Waiguo kexueshe ji benshe zhi lishi外国科学社及本社之历史 (Foreign Science Societies and Our Scociety's History). *Kexue*科学 (Science) 3(1): 2–18.

[12]Guoli Zhongyang yanjiuyuan choubei jingguo ji zuzhi国立中央研究院筹备经过及组织 (The Preparatory Process and Organization of Academia Sinica). In Liu Guiyun刘桂云 and Sun Chengrui孙承蕊(Eds.) (2008). *Guojia tushuguan cang guoli zhongyang yanjiuyuan shiliao congbian*国家图书馆藏国立中央研究院史料丛编 (Collection of Historical Materials of Academia Sinica Collected by the National Library), Vol. 1. Beijing: Guojia tuchuguan chubanshe国家图书馆出版社.49.

of information and research".[13] At the beginning of 1923, the congressman Huang Gongsu put forward "the case of establishing the National Academy of Sciences" and drafted regulations and instructions. In the proposition, he proposed that it was urgent to create a National Academy of Sciences and an academician system and following the example of the French Academy of Sciences and the Royal Society of London for Improving of Natural Knowledge was key.[14]

All these show that as China's science systematization develops to a higher level, the establishment of a national academy of sciences following the western scientific system, the improvement of China's academic research standards, and the promotion of the development of China's science are not only the ideals pursued by a group of scholars, but also the efforts of some politicians. Such ideals and efforts still needed to be mature in China, a country which was divided by warlords and changed the political situation at that time.

On April 17, 1927, the day before the Nanjing National Government was established, the Kuomintang Central Executive Committee held the 74th meeting of the Central Political Conference in Nanjing. At the meeting, Li Shizeng put forward a motion to establish the Academia Sinica, which was unanimously approved by the delegates attending the meeting. It was then decided that Li Shizeng, Tsai Yuan-Pei and Zhang Jingjiang would be responsible for drafting *The Organization Act of Academia Sinica*.[15] In May, the 90th meeting of the Central Political Conference decided to set up a preparatory office for the Academia Sinica and presumed Tsai Yuan-Pei, Li Shizeng and Zhang Jingjiang as preparatory committee members.[16] On June 9, 1928, Tsai Yuan-Pei, who had been appointed president of Academia Sinica, convened the First General Assembly in Shanghai East Asia restaurant, announcing the formal establishment of Academia Sinica (Fig. 1.1).

There is no doubt that the establishment of Academia Sinica is the product of China's systematization of science. It is closely related to the active advice and organization of Tsai Yuan-Pei and Li Shizeng, two important figures in the field of politics and academics, and cannot be separated from Chiang Kai-shek's support. Chiang Kai-shek. His support for the establishment of the Academia Sinica is related to his understanding of the importance of science and technology to national development. After all, after the baptism of the May 4th New Culture Movement and the publicity

---

[13]Guo Pingshu高平叔(Eds.) (1984). *Tsai Yuan-Pei Quanji*蔡元培全集 (The Complete Works of Tsai Yuan-Pei), Vol. 4. Beijing: Zhonghua shuju.475–476.

[14]Yiyuan jianyi chuangshe guoli kexueyuan议员建议创设国立科学院 (Congressman Suggested the Establishment of a National Academy of Sciences). *Kexue*科学 (Science) 1923, 8(2): 199–201.

[15]Chen Shiwei陈时伟 (2003). Zhongyang yanjiuyuan yu Zhongguo jindai xueshu tizhi de zhiyehua, 1927–1937 nian中央研究院与中国近代学术体制的职业化, 1927–1937年 (Academia Sinica and the Professionalization of Modern Chinese Academic System, 1927–1937). *Zhongguo xueshu*中国学术 (China Scholarship) 4 (3), Total 15: 185.

[16]Guoli Zhongyang yanjiuyuan choubei jingguo ji zuzhi国立中央研究院筹备经过及组织 (The Preparatory Process and Organization of Academia Sinica). In Liu Guiyun刘桂云 and Sun Chengrui孙承蕊(Eds.) (2008).*Guojia tushuguan cang guoli zhongyang yanjiuyuan shiliao congbian* 国家图书馆藏国立中央研究院史料丛编 (Collection of Historical Materials of Academia Sinica Collected by the National Library), Vol. 1. Beijing: Guojia tuchuguan chubanshe国家图书馆出版社.49.

Fig. 1.1 Tsai Yuan-Pei, president of Academia Sinica

and promotion of the Science Society of China and the vast number of overseas students, China "had known the importance of science among the masses" and "both the government and the people" advocated science at that time.[17] At the same time, the establishment of the country's highest academic research institution was the legacy of Sun Yat-sen, the founder of the Kuomintang, and was later proposed by Li Shizeng, a veteran of the Kuomintang. Chiang Kai-shek, as a follower of Sun Yat-sen and leader of the Kuomintang's impending new regime, was also logical to support the establishment of the Academia Sinica. During the preparatory period of the Academia Sinica, the scientific research activities to be carried out by Japanese academic institutions in China were obviously aggressive[18] and accelerated the establishment of the Academia Sinica.

However, the establishment of Academia Sinica ended the history of China without a national academy of sciences and was a milestone in the process of institutionalization of modern science in China. As a result, China's modern science entered a new stage of development.

---

[17]Benshe qing bo peikuan guanshui shang zhengfu shuotie bing jihuashu本社请拨赔款关税上政府说贴并计划书 (This society asks the government to pay the compensation tariff and submits the plan). *Kexue*科学 (Science) 1923, 8(2): 192.

[18]Guoli Zhongyang yanjiuyuan 17 niandu zongbaogao (II) 国立中央研究院十七年度总报告(下) (The 17th Annual Report of Academia Sinica (II)). In Liu Guiyun刘桂云 and Sun Chengrui孙承蕊(Eds.) (2008). *Guojia tushuguan cang guoli zhongyang yanjiuyuan shiliao congbian*国家图书馆藏国立中央研究院史料丛编 (Collection of Historical Materials of Academia Sinica Collected by the National Library), Vol. 2. Beijing: Guojia tuchuguan chubanshe国家图书馆出版社.71.

1.3 The Rise of Chinese Academia and the Ideal of Academic Independence

Academia Sinica belongs to the centralized national academy of sciences with research entities funded by the government. This is quite similar to the French Academy of Sciences and the Soviet Academy of Sciences. Tsai Yuan-Pei made reference to "the nature and form of national academies of science" and considered "the political system and needs of our country" when he set up the Academia Sinica.[19] During the preparatory period, Tsai Yuan-Pei and others formulated the system between "honorary member" and "corresponding research fellow". The "honorary member" system is actually the academician system that changed its name, but in the end it remained on paper and failed to be implemented. One of the factors restricting its implementation was that there were still very few outstanding scientists in China back then. It was very short of them and did not have a sufficient talent base. At that time, Chinese society and Japanese scholars generally believed that "there is no science in China", which reflected this predicament.[20]

However, since the establishment of the Academia Sinica, the political situation in the country has stabilized. Government or private academic institutions and state or private universities have successively contested to establish research institutes. Domestic academic research and scientific experiments have become extremely popular.[21] By 1937, before the outbreak of the Anti-Japanese War, national professional societies in the field of science and technology had basically been established. The specialized publications created by many professional societies are gradually recognized by the world scientific community. As Woo Y. H., dean of the Faculty of Science of National Southwest Associated University and the member of the Council of Academia Sinica, said in March 1944, "Many specialized scientific societies have been established in China in the past 15 years, each with its own specialized publications. These publications have gained considerable status in the world. Its materials and contents are often quoted by foreign experts."[22] In addition, research institutions such as the Academia Sinica, the National Academy of Peiping, the National Geological Survey of China and the Institute of Biology of the Science Society of China, as well as universities with solid teaching and research foundations such as National

---

[19]Chu Chia-Hua朱家骅 (1972). Guoli Zhongyang yanjiuyuan jianshuo国立中央研究院简说 (A Brief Introduction to Academia Sinica). In Du Yuanzai杜元载(Eds.). *Geming wenxian*革命文献 (Revolutionary Documents), series 59. Taipei: Zhongguo guomindang zhongyang weiyuanhui dangshi weiyuanhui中国国民党中央委员会党史委员会.219.

[20]Wong Wen-hao翁文灏(1933). Zhongguo de kexue gongzuo中国的科学工作(China's Scientific Work). *Dili pinglun*独立评论(Independent Review) (34): 6–9.

[21]Zhou Xiangeng周先庚 (1934). Xueshu yanjiu de tujing学术研究的途径 (Approaches to Academic Research). *Dili pinglun*独立评论 (Independent Review) (126): 6.

[22]Zhongyang yanjiuyuan pingyihui dierjie dierci nianhui jilu中央研究院评议会第二届第二次年会纪录 (Minutes of the Second Annual Meeting of the Second Council of Academia Sinica). Nanjing: Zhongguo dier lishi danganguan中国第二历史档案馆 (The Second Historical Archives of China), Total No. 393, File No. 1556.

Tsing Hua University, the National University of Peking, National Central University, the University of Nanking and Yenching University, are gradually recognized by the international academic community.

More importantly, by the first half of the 1940′s, after a series of efforts and activities, such as scientific education, academic research and organization of academic groups, scientific pioneers, mainly returning overseas students, had laid the foundation for basic disciplines of modern science in China. Some disciplines even reached the forefront of world science, and important progress had been made in the humanities and social sciences. China's academic circles are gradually rising and a number of scientists have emerged in the international academic circles.

For example, in mathematics, Hua Loo-Keng made world-class work in number theory and has made pioneering work with his own characteristics in the theory of automorphic function, matrix geometry, classical group and function of several complex variables.[23]Shiing-Shen Chern proved Gauss-Bonnet formula in 1943 and made important achievements in the field of differential geometry.[24] In physics, Yao Y. T. made important contributions to the research of gas conduction and has made achievements in atomic and molecular spectroscopy.[25] Chao Chung-Yoa was the first to observe the generation and annihilation of positron and electron pairs in studying absorption coefficient and scattering experiments of hard γ rays, which played an important role in the discovery of positrons and the acceptance of quantum electrodynamics theory in physics.[26] Wu Ta-You predicted the existence of the uranium system through his doctoral thesis, and then proposed the excited state and free dissociation theory of the helium atom.[27] In theory of turbulence, Chou Pei-yuan first proposed the pulsation equation, established the general theory of turbulence, and proposed two methods to solve the turbulent motion. Because of its influence, the "turbulent model theory" has formed internationally.[28] In chemistry, Chuang

---

[23]Wang Yuan王元 (1999). *Hua Loo-Keng*华罗庚. Nanchang: Jiangxi jiaoyu chubanshe江西教育出版社.79–132.

[24]Shiing-Shen Chern陈省身. Xuesuan 60 nian学算60年 (60 Years of Studing Mathematics). In Shiing-Shen Chern (2011). *Shiing-Shen Chern wenxuan: Zhuanji tongsu yanjiang ji qita*陈省身文选: 传记、通俗演讲及其它 (Selected Works of Cheng Xingshen: Biographies, Popular Speeches and Others) Beijing: Kexue chubanshe科学出版社.41.

[25]Song Zengfu宋增福 (1996). Yao Y. T饶毓泰. In China Association for Science and Technology 中国科学技术协会(Eds.) (1996). *Zhongguo kexue jishu zhuanjia zhuanlue*中国科学技术专家传略 (Biographies of Chinese Science and Technology Experts), Science Part, Volume 1 of Physics Book. Shijiazhuang: Hebei jioayu chubanshe河北教育出版社.53–54.

[26]Zheng Wenli郑文莉 and Mao Zhenlin毛振麟. Chao Chung-Yoa赵忠尧. In China Association for Science and Technology中国科学技术协会(Eds.) (1996). *Zhongguo kexue jishu zhuanjia zhuanlue* 中国科学技术专家传略(Biographies of Chinese Science and Technology Experts), Science Part, Volume 1 of Physics Book. Shijiazhuang: Hebei jioayu chubanshe河北教育出版社.240–241.

[27]Han Ruqi韩汝琦. Wu Ta-You吴大猷. In China Association for Science and Technology中国科学技术协会(Eds.) (1996). *Zhongguo kexue jishu zhuanjia zhuanlue*中国科学技术专家传略 (Biographies of Chinese Science and Technology Experts), Science Part, Volume 1 of Physics Book. Shijiazhuang: Hebei jioayu chubanshe河北教育出版社.429.

[28]Wang Xueying汪雪瑛. Chou Pei-yuan周培源. In China Association for Science and Technology 中国科学技术协会(Eds.) (1996). *Zhongguo kexue jishu zhuanjia zhuanlue*中国科学技术专家传

Chang-Kong made outstanding contributions to the studies on synthesis of steroids and structure of natural organic compounds.[29] Wu Hsien has achieved world-leading results in clinical biochemistry, balance between gas and electrolyte, denaturation of proteins and immunochemistry.[30] Woo Sho-Chow did pioneering research on the electronic spectrum and molecular structure of polyatomic molecules and made world-class contributions.[31] In terms of geology and geography, Lee J. S.'s creation of geological mechanics, the discovery of Quaternary glaciers in Lushan Mountain and southern China, and the study of Chinese insect fossils have important influences on the international geological circle.[32] Wong Wen-hao made important achievements in structural geology, seismic geology, sedimentology and other fields.[33] Huang T. K. provided the method to subdivide the Permian in China.[34] Young Chung-Chien made outstanding achievements in Chinese vertebrate paleontology as well as Mesozoic and Cenozoic stratigraphic research.[35] In meteorology, Chu Coching made valuable research achievements in monsoon, typhoon, natural zoning, phenology and other

略 (Biographies of Chinese Science and Technology Experts), Science Part, Volume 1 of Physics Book. Shijiazhuang: Hebei jioayu chubanshe河北教育出版社.262–263.

[29]Huang Yaozeng黄耀曾. Chuang Chang-Kong庄长恭. In China Association for Science and Technology中国科学技术协会(Eds.) (1993). *Zhongguo kexue jishu zhuanjia zhuanlue*中国科学技术专家传略 (Biographies of Chinese Science and Technology Experts), Science Part, Volume 1 of Chemistry Book. Beijing: Zhongguo kexue jishu chubanshe中国科学技术出版社.123–130.

[30]Zheng Bocheng郑伯承 and Ray Wu吴瑞. Wu Hsien吴宪. In China Association for Science and Technology中国科学技术协会(Eds.) (2005). *Zhongguo kexue jishu zhuanjia zhuanlue*中国科学技术专家传略(Biographies of Chinese Science and Technology Experts), Medicine Part, Volume 1 of Basic Medicine Book. Beijing: Renmin weisheng chubanshe人民卫生出版社.40–45.

[31]Zeng Guangfu曾广赋, Wu Yanan吴雅南 and Xi Shiquan席时权. Woo Sho-Chow吴学周. In China Association for Science and Technology中国科学技术协会(Eds.) (1993). *Zhongguo kexue jishu zhuanjia zhuanlue*中国科学技术专家传略(Biographies of Chinese Science and Technology Experts), Science Part, Volume 1 of Chemistry Book. Beijing: Zhongguo kexue jishu chubanshe中国科学技术出版社.322–325.

[32]Zeng Wenqu曾问渠. Lee J. S.李四光. In the Editorial Team of the Complete Dictionary of Scientific Biography«科学家传记大辞典» 编辑组(Eds.) (1991). *Zhongguo Xiandai kexuejai zhuanji* 中国现代科学家传记(Biographies of Modern Chinese Scientists), Vol. 1. Beijing: 科学出版社.304–309.

[33]Shi Baoheng石宝珩 and Pan Yuntang潘云唐. Wong Wen-hao翁文灏. In the Editorial Team of the Complete Dictionary of Scientific Biography«科学家传记大辞典»编辑组(Eds.) (1994). *Zhongguo Xiandai kexuejai zhuanji*中国现代科学家传记 (Biographies of Modern Chinese Scientists), Vol. 5. Beijing: 科学出版社.334–345.

[34]Jiang Chunfa姜春发. Huang T. K. 黄汲清. In the Editorial Team of the Complete Dictionary of Scientific Biography«科学家传记大辞典»编辑组(Eds.) (1991). *Zhongguo Xiandai kexuejai zhuanji*中国现代科学家传记(Biographies of Modern Chinese Scientists), Vol. 1. Beijing: 科学出版社.348–349.

[35]Ye Xiangkui叶祥奎. Young Chung-Chien杨钟健. In the Editorial Team of the Complete Dictionary of Scientific Biography«科学家传记大辞典»编辑组(Eds.) (1994). *Zhongguo Xiandai kexuejai zhuanji*中国现代科学家传记 (Biographies of Modern Chinese Scientists), Vol. 5. Beijing: 科学出版社.381–383.

fields.[36] In terms of engineering technology, Hou T. P. made outstanding achievements in the research and development of soda ash manufacturing.[37] Adam Pen-Tung Sah creatively applied dyadic method and complex vector in mathematics to solve the problem of three-phase circuit, which was highly praised by the international electrician community.[38]

In biology and agronomy, Ping Chi conducted a large number of pioneering researches in vertebrate morphology, neurophysiology, fauna taxonomy, and paleontology.[39] His research on Cretaceous insect fossils in China is particularly remarkable and enjoys an important position in the world.[40] Hu Hsen-hsu carried out large-scale field collection and investigation of plant resources, predicted that there must be a distribution of Carex plants in China, and discovered the world's rare living fossil metasequoia, which shocked China and the rest of world. In medicine, pharmacology and physiology, Robert K. S. Lim, a scientist who could carry out high-level research in early China, mainly studied gastrointestinal physiology and neurophysiology. His research on the gastric secretion mechanism is recognized as a classic work by the international medical community.[41] Chen Ko Kuei discovered the pharmacological effects of ephedrine and further studied with colleagues the pharmacological effects of many compounds with structures similar to ephedrine, thus promoting the synthesis of numerous sympathetic amines.[42] Tsai Chiao discovered the anterior

---

[36]Shen Wenxiong沈文雄. Chu Coching竺可桢. In the Editorial Team of the Complete Dictionary of Scientific Biography«科学家传记大辞典»编辑组(Eds.) (1994). *Zhongguo Xiandai kexuejai zhuanji*中国现代科学家传记(Biographies of Modern Chinese Scientists), Vol. 5. Beijing: 科学出版社.349–353.

[37]Li Zhichuan李祉川. Hou Debang (Hou T. P.) 侯德榜. In the Editorial Team of the Complete Dictionary of Scientific Biography《科学家传记大辞典》编辑组(Eds.) (1994). *Zhongguo Xiandai kexuejai zhuanji*中国现代科学家传记 (Biographies of Modern Chinese Scientists), Vol. 5. Beijing: 科学出版社.774–776.

[38]Xu Qiaozhen许乔蓁. Sa Bendong (Adam Pen-Tung Sah)萨本栋. In China Association for Science and Technology中国科学技术协会(Eds.) (1996). *Zhongguo kexue jishu zhuanjia zhuanlue*中国科学技术专家传略 (Biographies of Chinese Science and Technology Experts), Science Part, Volume 1 of Physics Book. Shijiazhuang: Hebei jioayu chubanshe河北教育出版社.249–251.

[39]Zhai Qihui翟启慧. Ping Chi秉志. In China Association for Science and Technology中国科学技术协会(Eds.) (1996). *Zhongguo kexue jishu zhuanjia zhuanlue*中国科学技术专家传略 (Biographies of Chinese Science and Technology Experts), Science Part, Volume 1 of Biology Book. Shijiazhuang: Hebei jioayu chubanshe河北教育出版社.21–29.

[40]Shi Hu施浒. Hu Hsen-hsu胡先骕. In China Association for Science and Technology中国科学技术协会(Eds.) (1996). *Zhongguo kexue jishu zhuanjia zhuanlue*中国科学技术专家传略 (Biographies of Chinese Science and Technology Experts), Science Part, Volume 1 of Biology Book. Shijiazhuang: Hebei jioayu chubanshe河北教育出版社.59–68.

[41]Wang Zhijun王志均. Li Kesheng (Robert K. S. Lim) 林可胜. In the Editorial Team of the Complete Dictionary of Scientific Biography《科学家传记大辞典》编辑组(Eds.) (1991). *Zhongguo Xiandai kexuejai zhuanji*中国现代科学家传记 (Biographies of Modern Chinese Scientists), Vol. 1. Beijing: 科学出版社.599–601.

[42]Song Zhenyu宋振玉. Chen Ko Kuei陈克恢. In China Association for Science and Technology中国科学技术协会(Eds.) (1996). *Zhongguo kexue jishu zhuanjia zhuanlue*中国科学技术专家传略 (Biographies of Chinese Science and Technology Experts), Medicine Part, Volume 1 of Pharmacy Book. Beijing: Zhongguo kexue jishu chubanshe中国科学技术出版社.91–93.

nucleus of the tectum, the central part of vision and eye movement function, later called Tsai's area in 1920s. He also made important discoveries in the fields of glucose metabolism and blood physiology.[43] Feng Teh-Pei discovered the "stretch response" in the study of neurology and muscular abilitsy, which is what Polish muscle chemist Parnas called the "Feng Effect".[44] In the field of neuromuscular junction physiology, he discovered new features of neuromuscular junction electrophysiology. From 1936 to 1941, his laboratory published 26 papers. Part of their work provided evidence for the theory of chemical transfer that was being formed at that time. Some experimental results directly compensated for and promoted the theory of British scientist H. H. Dale.[45]

In terms of history, literature, and philosophy, Chen Yin-k'o made important discoveries in the translation, collation and interpretation of Buddhist scriptures, as well as in the study of phonology, Mongolian origin, Li clan in the Tang Dynasty, government military system, and Sino-Indian cultural exchange.[46] Chen Yuan made outstanding achievements in many fields such as history of the Yuan Dynasty, Chinese and Western transportation, Chinese religions, and Chinese historical philology.[47] Hu Shih wrote many books and has great influence on history, literature and philosophy.[48] Feng You-lan had profound attainments in the history of Chinese philosophy

[43]Cai Xueli蔡雪丽.Tsai Chiao蔡翘. In China Association for Science and Technology中国科学技术协会(Eds.) (2005). *Zhongguo kexue jishu zhuanjia zhuanlue*中国科学技术专家传略(Biographies of Chinese Science and Technology Experts), Medicine Part, Volume 1 of Basic Medicine Book. Beijing: Renmin weisheng chubanshe人民卫生出版社.95–109.

[44]Tan Depei谭德培 and Cai Kunhua蔡坤华. Feng Teh-Pei冯德培. In China Association for Science and Technology中国科学技术协会(Eds.) (1996). *Zhongguo kexue jishu zhuanjia zhuanlue*中国科学技术专家传略 (Biographies of Chinese Science and Technology Experts), Science Part, Volume 1 of Biology Book. Shijiazhuang: Hebei jioayu chubanshe河北教育出版社.353–355.

[45]Rao Yi饶毅(1996). Jinian shijie zhuming shenjing shengwuxuejia Feng Teh-Pei纪念世界著名神经生物学家冯德培 (Commemorates World-renowned Neurobiologist Feng Teh-Pei). *21 shiji*二十一世纪(The 21st Century), the April Issue (Total34): 102–107.

[46]Zhao Zhongwen赵忠文(Eds.) (1992). *Zhongguo lishixue dacidian*中国历史学大辞典 (Dictionary of Chinese History). Jilin: Yanbian daxue chubanshe延边大学出版社.320; Gao Guokang高国抗 and Rao Zhanxiong饶展雄. Chen Yin-k' o. In Zhang Shunhui张舜徽(Eds.) (1984). *Zhongguo shixuejia zhuan*中国史学家传 (Biography of Chinese Historians). Shenyang: Liaoning renmin chubanshe辽宁人民出版社.338–341.

[47]Zhao Zhongwen赵忠文(Eds.) (1992). *Zhongguo lishixue dacidian*中国历史学大辞典 (Dictionary of Chinese History). Jilin: Yanbian daxue chubanshe延边大学出版社.313.

[48]Zhao Zhongwen赵忠文(Eds.) (1992). *Zhongguo lishixue dacidian*中国历史学大辞典(Dictionary of Chinese History). Jilin: Yanbian daxue chubanshe延边大学出版社.374–375; Geng Yunzhi耿云志. Hu Shih胡适. In Chen Qingquan陈清泉, Su Shuangbi苏双碧, Li Guihai李桂海, Xiao Li萧黎, Ge Zengfu葛增福(Eds.) (1985). *Zhongguo shixuejia pinghzuan*中国史学家评传 (Commentaries on Chinese Historians). Henan: Zhongzhou guji chubanshe中州古籍出版社.1380–1407.

and had written classic works such as *History of Chinese Philosophy*[49] and *Neo-Confucianism.*[50] In archaeology, the excavation of the Yin Ruins in Anyang presided over or attended by Li Chi, Tung Tso-pin and Liang Ssu-yung shocked the world.[51] In terms of architecture and art history, Liang Ssu-cheng has carried out extensive investigation, research, illustrating and writing of Chinese ancient buildings.[52] Chao Yuen Ren is engaged in the study of phonology theory, Chinese phonology, Chinese grammar, semantics and dialects, mediaeval sounds, etc. Many of his academic achievements have attracted the attention of international linguists.[53]

In addition to the above scholars, there are also a number of scholars in the fields of natural sciences and humanities and social sciences who have achieved higher academic attainments and standards in the research fields they specialize in. I will not introduce them one by one.

With the emergence of this group of scholars, the academician system established by Academia Sinica already has a relatively sufficient talent base. At the same time, the high voice of Chinese academia for national academic independence and Chiang Kai-shek's clear instructions on academic independence also prompted the Council of Academia Sinica to put the establishment of the academician system on the agenda. As early as the May 4th New Culture Movement raised the banner of "democracy" and "science", striving for national academic independence has become the direction of Chinese academic circles. As Liang Qichao pointed out in 1925, "every independent country has the possibility and necessity of independence in its learning."[54] After the outbreak of the Anti-Japanese War in 1937, the voice of Chinese academic circles for national academic independence was even higher. Hua Loo-Keng, who had become one of the world leaders and scholars in number theory that year, returned to China from Cambridge University with the first theme of "promoting academic

---

[49]Feng You-lan冯友兰 (1931). *Zhongguo zhexueshi*中国哲学史 (History of Chinese Philosophy). Shanghai: Shenzhou guoguangshe神州国光社; Feng You-lan冯友兰 (1934). *Zhongguo zhexueshi*中国哲学史 (History of Chinese Philosophy). Shanghai: Shangwu yinshuguan商务印书馆.

[50]Feng You-lan冯友兰 (1939). *Xin lixue*新理学 (Neo-Confucianism). Changsha: Shangwu yinshuguan商务印书馆.

[51]Zhao Zhongwen赵忠文(Eds.) (1992). *Zhongguo lishixue dacidian*中国历史学大辞典 (Dictionary of Chinese History). Jilin: Yanbian daxue chubanshe延边大学出版社.259.

[52]Huang Yanfu黄延复. Liang Ssu-cheng梁思成. In Li Xin李新(Editor-in-Chief), Zhongguo shehui kexueyuan jindaishi yanjiusuo Zhoghua minguoshi yanjiushi中国社会科学院近代史研究所中华民国史研究室 (Research Office of History of the Republic of China of Institute of Modern History, Chinese Academy of Social Sciences)(Eds.) (2011). *Zhonghua minguoshi renwuzhuan*中华民国史·人物传 (History of the Republic of China, Biographies of Characters), Vol. 3. Beijing: Zhonghua shuju中华书局.1975–1982.

[53]Xiong Shanghou熊尚厚.Chao Yuen Ren赵元任. In Li Xin李新(Editor-in-Chief), Zhongguo shehui kexueyuan jindaishi yanjiusuo Zhoghua minguoshi yanjiushi中国社会科学院近代史研究所中华民国史研究室 (Research Office of History of the Republic of China of Institute of Modern History, Chinese Academy of Social Sciences)(Eds.) (2011). *Zhonghua minguoshi renwuzhuan*中华民国史·人物传(History of the Republic of China, Biographies of Characters), Vol. 8. Beijing: Zhonghua shuju中华书局.5303–5306.

[54]Liang Qichao梁启超 (1925). Xuewen duli yu qinghua dierqi shiye学问独立与清华第二期事业 (Academic Independence and Tsinghua's Second Career). *Qinghua zhoukan*清华周刊 (Tsinghua Weekly) (350): 5.

independence", but it failed. Despite his failure, 12 years later he told his student Hsu L. C.: "My current theme is the same."[55] It can be said that striving for national academic independence has become an ideal pursued by a group of scientists represented by Hua Loo-Keng. The situation at home and abroad improved during China's Anti-Japanese War in 1943, and Japan was bound to be defeated. Due to this situation, the Chinese academic circles began to plan for the revival of Chinese academic circles after the war. On July 18, a joint annual meeting was held by six academic societies: the Science Society of China, the Geographical Society of China, Chinese Meteorological Society, Chinese Mathematical Society, China Zoological Society and the Botanical Society of China. In his written address, Chiang Kai-shek, president of the Kuomintang and chairman of the National Defense Supreme Council, also specifically instructed: "Applied science must be based on pure science. There must be a large group of people studying pure science together. With constant inventions, China's science has a deep foundation. It gradually removed its dependence and build up its own independent academic."[56]

As the symbol and center of the Kuomintang regime, Chiang Kai-shek's instructions pushed the efforts of Chinese academia to strive for national academic independence to a climax. From March 8 to the 10 of 1944, the Second Annual Meeting of the Second Council of Academia Sinica was held in Chongqing. At the meeting, Acting President Chu Chia-Hua reiterated Chiang Kai-shek's instructions and pointed out: "A country's academic independence cannot be achieved, that is, the conditions for a country's independence are not yet complete."[57] It should be said that the ideal of Chinese academic circles striving for national academic independence at that time mainly meant that China should establish an independent and complete modern academic system and independently cultivate high-level academic talents, so that Chinese academic could catch up with the world's advanced level and achieve the goal of Chinese academic independence from the west and stand erect in the world's academic forest. However, Chu Chia-Hua also linked academic independence with national independence, which gave "academic independence" a higher significance for the future of the country. In this way, how to realize the academic independence of the country has become one of the key issues to be discussed in this council. As

---

[55]Dictated by Xu Lizhi (Hsu L. C.)徐利治, Interviewed by Yuan Xiangdong袁向东 and Guo Jinhai 郭金海 (2017). *Xu Lizhi fangtanlu*徐利治访谈录 (An Interview with Xu Lizhi). Changsha: Hunan jiaoyu chubanshe湖南教育出版社.232.

[56]Chiang Kai-shek蒋介石. Dui 6 xueshu tuanti nianhui shumian zhici对六学术团体年会书面致词 (Written Speech to the Annual Meetings of Six Academic Societies). In Qin Xiaoyi秦孝仪(Eds.) (1984). *Xian zongtong jianggong sixiang yanlun zongji*先"总统"蒋公思想言论总集(The Late "President" Mr. Chiang's Thoughts and Comments), Vol. 40. Taipei: Zhongguo guomindang zhongyang weiyuanhui dangshi weiyuanhui中国国民党中央委员会党史委员会.3.

[57]Chu Chia-Hua朱家骅. Guoli zhongyang yanjiuyuan pingyihui dierjie dierci nianhui kaimuci (minguo 33 nian 3 yue 8 ri)国立中央研究院评议会第二届第二次年会开幕词 (民国三十三年三月八日) (Opening Speech at the Second Annual Meeting of the Second Council of Academia Sinica, March 8, 33, Republic of China). In Wang Yujun王聿均 and Sun Bin孙斌(Eds.) (1977). *Chu Chia-Hua xiansheng yanlun ji*朱家骅先生言论集 (Speech by Mr. Chu Chia-Hua). Taipei: Zhongyang yanjiuyuan jinaishi yanjiusuo中央研究院近代史研究所 (Institute of Modern History of Academia Sinica).85—86.

a result, the issue of establishing the academician system was formally put on the agenda of the council.

It is also worth noting that by the 1940s, Academia Sinica had enjoyed a certain reputation in the world after more than ten years of development and was recognized by some famous academic institutions and organizations. In March 1943, Eric Robertson Dodds, a professor at Oxford University in England, Joseph Needham, a chemical embryologist at Cambridge University and a member of the Royal Society of London, visited China. They brought to Academia Sinica letters from the Royal Society of London, the British Association for the Advancement of Science, the Universities Bureau of the British Empire, the British Academy, Oxford University, Cambridge University and Mumbai University of India to "express admiration".[58] This prompted Academia Sinica to decide in the future to "engage in the communication between the academic cultures of the two countries, to bring honor to the academic culture and to benefit mankind."[59] Later, America sent George B. Cressey to China in November 1943, carrying letters from the National Academy of Sciences and the Council of Learned Societies in the United States to Academia Sinica, and also expressing admiration for Academia Sinica.[60] In order to carry out in-depth international academic exchanges and have an equal dialogue with international academic circles, it is a trend to establish the academician system. As the Academia Sinica said when introducing the process of handling the election of the first batch of members: "after more than ten years of efforts, the academy becomes China's national institute, which has gradually been recognized by foreign academic organizations. In order to seek academic independence in our country and to cooperate with international academia, it is necessary to strengthen the foundation of the academy and complete the academic organization in our country."[61]

---

[58]Zhongyang yanjiuyuan pingyihui dierjie dierci nianhui jilu中央研究院评议会第二届第二次年会纪录(Minutes of the Second Annual Meeting of the Second Council of Academia Sinica). Nanjing: Zhongguo dier lishi danganguan中国第二历史档案馆 (The Second Historical Archives of China), Total No. 393, File No. 1556.

[59]Chu Chia-Hua朱家骅. Huanying yingguo Eric Robertson Dodds, Joseph Needham liang jiaoshou zhici(minguo 32 nian 4 yue 10 ri)欢迎英国陶德斯、尼德汉两教授致词 (民国三十二年四月十日) (Welcome speeches to British professors Eric Robertson Dodds and Joseph Needham, April 10, 32, Republic of China). In Wang Yujun王聿均 and Sun Bin孙斌(Eds.) (1977). *Chu Chia-Hua xiansheng yanlun ji*朱家骅先生言论集 (Speech by Mr. Chu Chia-Hua). Taipei: Zhongyang yanjiuyuan jinaishi yanjiusuo中央研究院近代史研究所 (Institute of Modern History of Academia Sinica).84–85.

[60]Zhongyang yanjiuyuan pingyihui dierjie dierci nianhui jilu中央研究院评议会第二届第二次年会纪录(Minutes of the Second Annual Meeting of the Second Council of Academia Sinica). Nanjing: Zhongguo dier lishi danganguan中国第二历史档案馆 (The Second Historical Archives of China), Total No. 393, File No. 1556; Zhongyang yanjiuyuan dierjie pingyihui tiaoli jilu ti'an ji baogao中央研究院第二届评议会条例、记录、提案及报告 (Regulations, Records, Proposals and Reports of the Second Council of Academia Sinica). Nanjing: Zhongguo dier lishi danganguan 中国第二历史档案馆 (The Second Historical Archives of China), Total No. 393, File No. 2926.

[61]Guoli zhongyang yanjiuyuan banli diyici yuanshi xuanju jingguo qingxing jielue国立中央研究院办理第一次院士选举经过情形节略(Summary of Academia Sinica's Election of Its First Batch of Members). Nanjing: Zhongguo dier lishi danganguan中国第二历史档案馆 (The Second Historical Archives of China), Total No. 393, File No. 1085.

1.4 Summary

As a product of the institutionalization of modern western scientific research, the academician system was born in Europe in the seventeenth century, which was in the period of scientific revolution. After the Académie des Sciences established the academician system in 1666, continental European countries followed suit and established national academies, selecting academicians and establishing the academician system. Commonwealth countries predominantly followed the example of Britain, and set up academies without research entities. Their members were similarly known as either "fellows" or "academicians". At the beginning of the twentieth century, the academician system became popular in the west, and was regarded as an important institution in promoting academic development.

In the twenty years since the establishment of Academia Sinica, the Chinese academic community has grown rapidly, the process of scientific institutionalization has accelerated, the foundation of basic disciplines of modern science have been laid one after another, and many excellent research results have emerged. A number of scientists have emerged in the fields of natural science, humanities and social sciences, and those individuals have also emerged in the international academic circle. Academician system established by Academia Sinica already has relatively mature talent conditions. At the same time, Academia Sinica has been recognized by some famous foreign academic institutions and organizations and enjoys a certain international reputation. It hopes to strengthen international academic exchanges and perfect its own academic organizations by establishing the academician system. Moreover, in the Chinese academic circles during the special period of Anti-Japanese War, the atmosphere of nationalism was quite strong, and the call for national academic independence was very high.

It is against this international and domestic background that Academia Sinica established the academician system in 1948 and the academician system was established in China.

Chapter 2
Establishment of the Council of Academia Sinica and Election of Elected Council Members

"Carrying out scientific research" and "guiding, contacting and rewarding academic research" were the two major tasks undertaken by Academia Sinica. In the original system design, the second task was undertaken by the council. Although Academia Sinica had been actively preparing since its founding for the council to set up, it had encountered setbacks. In 1935, under the direct impetus of Director-General V. K. Ting, the council was finally established. During the Anti-Japanese War, the First Council of Academia Sinica successfully presided over the election of the elected council members for the Second Council in the rear area.[1] After the war, the Second Council successfully organized the election of the first batch of members of Academia Sinica and the First General Assembly of Members of Academia Sinica. The establishment of the council and the election of elected council members laid an organizational foundation and provided useful experience and reference for the establishment of the academician system of Academia Sinica, which was the prelude to the establishment of the system.

2.1 Preparations and Establishment of the Council

Prior to the formal establishment of Academia Sinica, the National Government had already provided for the council to set up in the *Revised Organizational Regulations for Academia Sinica* announced on April 10, 1928. According to the regulations, the council "is the highest scientific appraisal organ in the country and is made up of 30 domestic experts appointed by the president".[2] Soon after its formal establishment,

[1]There are two types of council members in Academia Sinica: ex-officio members and elected council members. The former is held by the president, directors and director-general of Academia Sinica (since 1943, the director-general has been added as an ex-officio member), while the latter is elected among scholars nationwide.

[2]Guoli zhongyang yanjiuyuan 17 niandu zongbaogao (I) 国立中央研究院十七年度总报告 (上) (The 17th Annual Report of Academia Sinica (I)). In Liu Guiyun刘桂云 and Sun Chengrui孙承

J. Guo, *The Establishment and Reconstruction of the Academician System in China*,
https://doi.org/10.1007/978-981-15-7208-1_2

Fig. 2.1 Tsai Yuan-Pei posed with Director-General Yang Quan at Academia Sinica

Academia Sinica began its preparations for the council. On June 30, 1928, Tsai Yuan-Pei presided over the Second General Assembly (Fig. 2.1). At the meeting, it was decided that the English name for the council was the "National Research Council" which would be held once a year.[3] The council chose this translated name solely for the consideration that "the nature of the council is equal to that of the National Research Council in European and American countries".[4] On August 13, the Third General Assembly of Academia Sinica discussed the composition and candidates of the council. The final candidates were distributed among 11 groups in accordance with the *Revised Organizational Regulations for Academia Sinica*

蕊(Eds.) (2008). *Guojia tushuguan cang guoli zhongyang yanjiuyuan shiliao congbian*国家图书馆藏国立中央研究院史料丛编(Collection of Historical Materials of Academia Sinica Collected by the National Library), Vol. 1. Beijing: Guojia tuchuguan chubanshe国家图书馆出版社.447.

[3]Guoli zhongyang yanjiuyuan 17 niandu zongbaogao (I) 国立中央研究院十七年度总报告(上) (The 17th Annual Report of Academia Sinica (I)). In Liu Guiyun刘桂云 and Sun Chengrui孙承蕊(Eds.) (2008). *Guojia tushuguan cang guoli zhongyang yanjiuyuan shiliao congbian*国家图书馆藏国立中央研究院史料丛编(Collection of Historical Materials of Academia Sinica Collected by the National Library), Vol. 1. Beijing: Guojia tuchuguan chubanshe国家图书馆出版社.497.

[4]Guoli zhongyang yanjiuyuan choubei jingguo ji zuzhi国立中央研究院筹备经过及组织(The Preparatory Process and Organization of Academia Sinica). In Liu Guiyun刘桂云 and Sun Chengrui孙承蕊(Eds.) (2008). *Guojia tushuguan cang guoli zhongyang yanjiuyuan shiliao congbian* 国家图书馆藏国立中央研究院史料丛编(Collection of Historical Materials of Academia Sinica Collected by the National Library), Vol. 1. Beijing: Guojia tuchuguan chubanshe国家图书馆出版社.51.

mentioned above.[5] On January 13, 1929, Tsai Yuan-Pei presided over the Fourth General Assembly, which discussed the selection of the council members, but there was no result. Thus, Wang Shih-Chieh, Chu Coching and Lee J. S. were presumed to draft first the organizational regulations of the council.[6] On January 4, 1930, when the Academia Sinica held its Ninth General Assembly, the drafting of the regulations was completed, but during the meeting, a decision was made to "reserve" the *Draft Regulations of the Council of Academia Sinica*.[7] Then, preparations for the council were shelved.

However, in May 1934, after becoming the director-general of Academia Sinica, V. K. Ting made great efforts to prepare for the establishment of the council.[8] Under his direct impetus, the council was set up the following year. V. K. Ting, courtesy name Zaijun, was a geologist with an international reputation (Fig. 2.2). He was once the director of National Geological Survey of China, and was a research professor in the Department of Geology of the National University of Peking before he took office in Academia Sinica. He was not only good at science, but also had the ability to handle affairs, "he is really a rare figure in modern China", said by Tsai Yuan-Pei.[9] After taking up the post of director-general, V. K. Ting and Tsai Yuan-Pei agreed that Academia Sinica "has a rough scale and the establishment of a council is actually needed".[10] Moreover, V. K. Ting believed that the matter "cannot be put off any

[5]Guoli zhongyang yanjiuyuan 17 niandu zongbaogao (I) 国立中央研究院十七年度总报告(上) (The 17th Annual Report of Academia Sinica (I)). In Liu Guiyun刘桂云 and Sun Chengrui孙承蕊(Eds.) (2008). *Guojia tushuguan cang guoli zhongyang yanjiuyuan shiliao congbian*国家图书馆藏国立中央研究院史料丛编(Collection of Historical Materials of Academia Sinica Collected by the National Library), Vol. 1. Beijing: Guojia tuchuguan chubanshe国家图书馆出版社.502-503.

[6]Guoli zhongyang yanjiuyuan 17 niandu zongbaogao (I) 国立中央研究院十七年度总报告(上) (The 17th Annual Report of Academia Sinica (I)). In Liu Guiyun刘桂云 and Sun Chengrui孙承蕊(Eds.) (2008). *Guojia tushuguan cang guoli zhongyang yanjiuyuan shiliao congbian*国家图书馆藏国立中央研究院史料丛编(Collection of Historical Materials of Academia Sinica Collected by the National Library), Vol. 1. Beijing: Guojia tuchuguan chubanshe国家图书馆出版社.507.

[7]Guoli zhongyang yanjiuyuan 18 niandu zongbaogao国立中央研究院十八年度总报告(The 18th Annual Report of Academi a Sinica). In Liu Guiyun刘桂云 and Sun Chengrui孙承蕊(Eds.) (2008). *Guojia tushuguan cang guoli zhongyang yanjiuyuan shiliao congbian*国家图书馆藏国立中央研究院史料丛编(Collection of Historical Materials of Academia Sinica Collected by the National Library), Vol. 2. Beijing: Guojia tuchuguan chubanshe国家图书馆出版社. 294.

[8]The first director-general of Academia Sinica was Yang Quan. On June 18, 1933, Chiang Kai-shek sent special agents to assassinate Yang Quan. On April 24, 1934, Academia Sinica formally hired V. K. Ting as its director-general. On May 18, V. K. Ting took up the post. See Song Guangbo宋广波(2009). *Ding Wenjiang(V. K. Ting) nianpu*丁文江年谱(The Chronicle of V. K. Ting). Harbin: Heilongjiang jiaoyu chubanshe黑龙江教育出版社410-412.

[9]Tsai Yuan-Pei蔡元培(2011). V. K. Ting duiyu zhongyang yanjiuyuan de gongxian丁文江对于中央研究院的贡献(V. K. Ting's Contributions to Academia Sinica). In Tsai Yuan-Pei蔡元培, Wen Mingguo文明国(Eds.). *Tsai Yuan-Pei zishu*蔡元培自述(Tsai Yuan-Pei's Self-Statement). Beijing: Renmin ribao chubanshe人民日报出版社. 260.

[10]Guoli zhongyang yanjiuyuan shoujie pingyihui diyici baogao国立中央研究院首届评议会第一次报告(The First Report by the First Council of Academia Sinica). In Liu Guiyun刘桂云 and Sun Chengrui孙承蕊(Eds.) (2008). *Guojia tushuguan cang guoli zhongyang yanjiuyuan shiliao congbian*国家图书馆藏国立中央研究院史料丛编(Collection of Historical Materials of Academia

Fig. 2.2 After Yang Quan was assassinated, V. K. Ting succeeded as director-general of Academia Sinica

longer".[11] In the spring of 1935, Tsai Yuan-Pei and V. K. Ting suggested to the National Government to set up the council in the shortest possible time and to make "relevant amendments" to Article 5 of *The Organization Act of Academia Sinica*.[12] The National Government finally adopted their suggestions.

On May 27, 1935, the National Government amended and passed *The Organization Act of Academia Sinica* and promulgated *The Regulations of the Council of Academia Sinica* formulated by V. K. Ting and others. The regulations were considered as a programmatic constitution of the council, which provided for the election of the elected members of the First Council, the number of the elected council members and their disciplines, the functions and powers of the council, the qualifications of council members, the term duration of office, by-elections, etc. With regard to the qualifications of council members, article 3 of the regulations stipulated two items: one was "the person to be elected should have some special works or inventions in

---

Sinica Collected by the National Library), Vol. 7. Beijing: Guojia tuchuguan chubanshe国家图书馆出版社.133.

[11]Tsai Yuan-Pei蔡元培(2011). V. K. Ting duiyu zhongyang yanjiuyuan de gongxian丁文江对于中央研究院的贡献(V. K. Ting's Contributions to Academia Sinica). In Tsai Yuan-Pei蔡元培, Wen Mingguo文明国(Eds.). *Tsai Yuan-Pei zishu*蔡元培自述(Tsai Yuan-Pei's Self-Statement). Beijing: Renmin ribao chubanshe人民日报出版社. 260.

[12]Guoli zhongyang yanjiuyuan shoujie pingyihui diyici baogao国立中央研究院首届评议会第一次报告(The First Report by the First Council of Academia Sinica). In Liu Guiyun刘桂云 and Sun Chengrui孙承蕊(Eds.) (2008). *Guojia tushuguan cang guoli zhongyang yanjiuyuan shiliao congbian*国家图书馆藏国立中央研究院史料丛编(Collection of Historical Materials of Academia Sinica Collected by the National Library), Vol. 7. Beijing: Guojia tuchuguan chubanshe国家图书馆出版社.133.

his/her academic field", and the other was "those who are from the academic institutions should have led or presided over their institutions for more than five years with outstanding achievements". The elected must be at least one of them.[13] When V. K. Ting and others worked out the qualifications for the members of the council, they not only emphasized academic contributions, but also included the achievements of presiding over academic institutions. It should be the result of considering that scientists with leadership ability were required to perform the functions and powers of an evaluation committee, and scientists who are purely engaged in academic research may not be fully competent. Later, the Preparatory Committee for the Election of the Second Council Members of Academia Sinica stressed that the candidates should have "considerable scientific qualifications and at the same time have leadership skills",[14] which directly related to this idea. What is interesting is that when the Second Council of Academia Sinica formulated *The Regulations for the Election of Members of Academia Sinica* in 1946, those two qualifications were basically transferred to the qualifications of members of Academia Sinica. This will be described in detail later.

After the promulgation of the regulations, Academia Sinica held the selection meeting for the elected members of the First Council from June 19 to 20, 1935, at the general office in Chengxian Street, Nanjing. The meeting elected 30 members. Their list is as follows:

> Physics: Li Shu-Hua (Vice President of the National Academy of Peiping), Chiang Li-Fu (Dean of Department of Mathematics, Nankai University), Yeh Chi-Sun (Dean of the School of Science, National Tsing Hua University);
> Chemistry: Wu Hsien (Dean of Department of Biochemistry, Peiping Union Medical College), Hou T. P. (Chief Engineer of Yongli Soda Plant), Chou T. Q. (Directorof Institute of Materia Medica National Academy of Peiping);
> Engineering: Li Xie (Chairman of the Yellow River Water Conservancy Committee), Ling Hung-hsun (Director of Zhushao Railway Administration) and Tang Bingyuan (General Manager of Qingfeng Cotton Mill in Wuxi);
> Zoology: Ping Chi (Directorof the Institute of Biology of the Science Society of China), Robert K. S. Lim (Dean of Department of Physiology, Peiping Union Medical College), Hu Jingfu (Dean of Department of Biology, Yenching University);

---

[13]Guoli zhongyang yanjiuyuan shoujie pingyihui diyici baogao国立中央研究院首届评议会第一次报告(The First Report by the First Council of Academia Sinica). In Liu Guiyun刘桂云 and Sun Chengrui孙承蕊(Eds.) (2008). *Guojia tushuguan cang guoli zhongyang yanjiuyuan shiliao congbian*国家图书馆藏国立中央研究院史料丛编(Collection of Historical Materials the Academia Sinica Collected by the National Library), Vol. 7. Beijing: Guojia tuchuguan chubanshe国家图书馆出版社.120.

[14]Zhongyang yanjiuyuan pingyihui dierjie pingyiyuan xuanju choubei weiyuanhui baogao中央研究院评议会第二届评议员选举筹备委员会报告(Report of the Preparatory Committee for the Election of the Second Council Members of Academia Sinica). Nanjing: Zhongguo dier lishi danganguan中国第二历史档案馆(The Second Historical Archives of China), Total No. 393, File No. 1549.

Botany: Xie Jiasheng (Director of National Agricultural Research Bureau, Ministry of Industry), Hu Hsen-hsu (Director of Fan Memorial Institute of Biology), and Chun W. Y. (Director of the Institute of Agroforestry and Botany, National Sun Yatsen University);
Geology: V. K. Ting (Director-General of Academia Sinica), Wong Wen-hao (Director of National Geological Survey of China, Ministry of Industry), Chu Chia-Hua (former Director of the Geological Survey of Kwangtung and Kwangsi, and Chairman of the Geological Society of China);
Astronomy: Zhang Yun (Professor of Astronomy and Director of the Observatory, National Sun Yatsen University);
Meteorology: Zhang Qiyun (Professor of Department of Geography, National Central University);
Psychology: Kuo Zing-Yang (President of the University of Chekiang);
Social Sciences: Wang Shih-Chieh (former Professor of Law School and former President, National Wu-han University), Ho L. (Dean of School of Economics, Nankai University), Chou Keng-Shang (Professor of Law School, National Wu-han University);
History: Hu Shih (Dean of School of Arts, the National University of Peking), Chen Yuan (President of the Catholic University of Peking), Chen Yin-k'o (Professor of History, National Tsing Hua University);
Linguistics: Chao Yuen Ren (Director of Linguistic Section of Institute of History and Philology, Academia Sinica);
Archaeology: Li Chi (Director of Archaeological Section of Institute of History and Philology, Academia Sinica);
Anthropology: Woo Ting-liang (Director of Anthropological Section of Institute of History and Philology, Academia Sinica).[15]

In the month after the election, the magazine *Science* (*Kexue*科学) of the Science Society of China published an editorial praising the elected council members as "representatives of all parties in the scientific circle of our country, except for a few non-scientists. They are our best choice".[16] That statement was relatively objective.

Judging from the discipline distribution of the elected, it had the characteristics of comprehensive discipline distribution. Those 30 people were nominally distributed in 14 disciplines including physics, chemistry, engineering, zoology, botany, geology, astronomy, meteorology, psychology, social sciences, history, linguistics, archaeology, and anthropology, but actually they were also involved in mathematics, physiology, and agriculture. Those 17 disciplines almost covered all fields of academic

[15]Guoli zhongyang yanjiuyuan shoujie pingyihui diyici baogao国立中央研究院首届评议会第一次报告(The First Report by the First Council of Academia Sinica). In Liu Guiyun刘桂云 and Sun Chengrui孙承蕊(Eds.) (2008). *Guojia tushuguan cang guoli zhongyang yanjiuyuan shiliao congbian*国家图书馆藏国立中央研究院史料丛编(Collection of Historical Materials of Academia Sinica Collected by the National Library), Vol. 7. Beijing: Guojia tuchuguan chubanshe国家图书馆出版社.138–140.

[16]Guan Hua观化(1935). Guoli zhongyang yanjiuyuan pingyihui chengli国立中央研究院评议会成立(Establishment of the Council of Academia Sinica). *Kexue*科学(Science) 19 (6): 826.

Fig. 2.3 Photo taken at the Inaugural Meeting of the Council of Academia Sinica on September 7, 1935 (part)

research in China at that time. Judging from the positions of the elected, 24 of those 30 people were presiding officers of academic institutions or their subordinate departments. It could be seen that election also emphasized the two qualifications of Council Members. Judging from the work units of the elected, they were widely distributed and mainly distributed in important academic institutions.

The 30 elected council members and 11 ex officio members including Academia Sinica's President Tsai Yuan-Pei and Directors Ting S. L. (Institute of Physics), Chuang Chang-Kong (Institute of Chemistry), Chow Jen (Institute of Engineering), Lee J. S. (Institute of Geology), Yu C. S. (Institute of Astronomy), Chu Coching (Institute of Meteorology), Fu Ssu-Nien (Institute of History and Philology), Wang Ging-Hsi (Institute of Psychology), Tao L. K. (Institute of Social Sciences) and Wang Chia-Chi (Institute of Zoology and Botany) form the First Council, with Tsai Yuan-Pei as the president.

On September 7, 1935, the First Council of Academia Sinica held its Inaugural Meeting at the Institute of History and Philology of Academia Sinica, No. 1, Jiming Temple, Nanjing. It marked the establishment of the council seven years after the establishment of Academia Sinica. The establishment of the council was an important achievement of Academia Sinica and China's academic institutionalization. Tsai Yuan-Pei believed that "this is one of the events in the history of Academia Sinica that can 'be written in a special style'".[17] In Chu Chia-Hua's view: "The establishment of the council was based on Mr. Zaijun's long-term plan for Academia Sinica. With the council, there was a subsequent General Assembly of Members of Academia Sinica. With the assembly, the structure of Academia Sinica was formally completed" (Fig. 2.3).[18]

[17]Tsai Yuan-Pei蔡元培(2011). Zhongyang yanjiuyuan yu Zhongguo kexue yanjiu gaikuang中央研究院与中国科学研究概况(A Survey of Academia Sinica and Scientific Research in China). In Tsai Yuan-Pei蔡元培, Wen Mingguo文明国(Eds.). *Tsai Yuan-Pei zishu*蔡元培自述(Tsai Yuan-Pei's Self-Statement). Beijing: Renmin ribao chubanshe人民日报出版社391.

[18]Chu Chia-Hua朱家骅. V. K. Ting yu zhongyang yanjiuyuan丁文江与中央研究院(V. K. Ting and Academia Sinica). In Hu Shih胡适 et al. (1979). *V. K. Ting zhegeren*丁文江这个人(V. K. Ting). Taipei: Zhuanji wenxue chubanshe传记文学出版社.158.

2.2 Preparations for the Election of Elected Members of the Second Council

On September 7, 1935, the First Annual Meeting of the council was held immediately after its inaugural meeting. The council elected V. K. Ting as its honorary secretary,[19] Li Shu-Hua, V. K. Ting, Hu Shih, Chou Keng-Shang, Xie Jiasheng, Li Chi and Ho L. as its drafting members for the regulations and Li Shu-Hua as the chairman of the Drafting Committee. In addition to drafting *The Regulations of Procedure for the Council of Academia Sinica* and *The Regulations of Handling Affairs for the Council of Academia Sinica*, the committee also drafted *The Regulations for the Election of the Council of Academia Sinica*. On September 8, after Chairman Li Shu-Hua let "all the draft rules and regulations be voted on one by one" and the council members' "careful discussion", the council amended and approved those draft rules and regulations.[20] Among them, *The Regulations for the Election of the Council of Academia Sinica* systematically stipulated the election procedures for elected council members and candidates for the president of Academia Sinica.[21]

After the First Annual Meeting, the first Council of Academia Sinica held two annual meetings in April 1936 and May 1937. After the Anti-Japanese War broke out in July 1937, Academia Sinica was ordered to move westward in the autumn of that year. The general office and institutes moved again and again. In addition, the limited material conditions, transportation difficulties and soaring prices seriously affected the normal working order and development. The council changed the Fourth Annual Meeting originally scheduled for April 1938 to be held at Yunnan University in Kunming from March 13 to 14, 1939. As the First Council would expire in July of the following year, discussion on the election of the members of the Second Council would become a major agenda of that annual meeting.

On March 14, the council held a detailed discussion on the agenda and decided that the divisions and quota of the election should be handled in accordance with the provisions of the election for members of the First Council. Of course, ex officio

[19]On January 5, 1936, V. K. Ting died tragically. In April 1936, the Council of Academia Sinica re-elected Wong Wen-hao as its honorary secretary, and by-elected Ye Liangfu as its elected council member for geological discipline.

[20]Guoli zhongyang yanjiuyuan shoujie pingyihui diyici baogao国立中央研究院首届评议会第一次报告(The First Report by the First Council of Academia Sinica). In Liu Guiyun刘桂云 and Sun Chengrui孙承蕊(Eds.) (2008).*Guojia tushuguan cang guoli zhongyang yanjiuyuan shiliao congbian*国家图书馆藏国立中央研究院史料丛编(Collection of Historical Materials of Academia Sinica Collected by the National Library), Vol. 7. Beijing: Guojia tuchuguan chubanshe国家图书馆出版社. 142–144.

[21]Guoli zhongyang yanjiuyuan shoujie pingyihui diyici baogao国立中央研究院首届评议会第一次报告(The First Report by the First Council of Academia Sinica). In Liu Guiyun刘桂云 and Sun Chengrui孙承蕊(Eds.) (2008). *Guojia tushuguan cang guoli zhongyang yanjiuyuan shiliao congbian*国家图书馆藏国立中央研究院史料丛编(Collection of Historical Materials of Academia Sinica Collected by the National Library), Vol. 7. Beijing: Guojia tuchuguan chubanshe国家图书馆出版社.126–128, 156.

members could not be candidates for elected council members.[22] At the same time, the council decided to set up the Preparatory Committee for the Election of the Second Council Members (hereinafter referred to as the "Preparatory Committee") to handle all election matters. On the spot, Wong Wen-hao, honorary secretary of the council, was presumed to be concurrently the chairman, with Ren Hongjun, director-general of Academia Sinica, Wang Shih-Chieh, Chu Chia-Hua, Fu Ssu-Nien, Yeh Chi-Sun and Tao L. K. serving as members.[23] These members of the Preparatory Committee had not only good scientific literacy, but also considerable administrative experience and organizational ability. Wong Wen-hao pointed out: "This organization should always maintain its academic spirit. It should carefully consider the election, candidates and methods, and expect a fair and impartial result."[24]

On March 16, 1939, the Preparatory Committee held its first meeting in Kunming. Preparations for the election officially began. The first task of the Preparatory Committee was to investigate the professors of national universities and independent colleges, and then draw up a list of voters to participate in the primary election. In order to do this work well, the Preparatory Committee first investigated the list of national colleges and universities to the Ministry of Education on April 7 that year. Then, on the basis of the list, questionnaires were sent to those national colleges and universities on June 2 and November 28 respectively to investigate the situation of serving professors in the first half of the year 1939 and newly recruited professors in the second half of the year 1939 (until December 1). In order to avoid omissions, on February 12, 1940, the Ministry of Education was also investigated about the newly established national universities. After that, the professors of relevant disciplines in Guangxi University, which had been newly converted into a national university, were reviewed in order to draw up a list of voters for the university.

During the investigation, the Preparatory Committee held its third meeting in Chongqing on November 26, 1939. The meeting decided to vote separately for a national university or a national independent college as a constituency. All colleges and universities are required to complete the primary election by March 1940.[25]

---

[22]Zhongyang yanjiuyuan pingyihui dierjie pingyiyuan xuanju choubei weiyuanhui baogao中央研究院评议会第二届评议员选举筹备委员会报告(Report of the Preparatory Committee for the Election of the Second Council Members of Academia Sinica). Nanjing: Zhongguo dier lishi danganguan中国第二历史档案馆(The Second Historical Archives of China), Total No. 393, File No. 1549.

[23]Zhongyang yanjiuyuan pingyihui dierjie pingyiyuan xuanju choubei weiyuanhui baogao中央研究院评议会第二届评议员选举筹备委员会报告(Report of the Preparatory Committee for the Election of the Second Council Members of Academia Sinica). Nanjing: Zhongguo dier lishi danganguan中国第二历史档案馆(The Second Historical Archives of China), Total No. 393, File No. 1549.

[24]Li Xuetong李学通(2005). *Wong Wen-hao nianpu*翁文灏年谱(The Chronicle of Dr. Wong Wen-hao). Jinan: Shandong jioayu chubanshe山东教育出版社.183.

[25]Zhongyang yanjiuyuan pingyihui pingyiyuan xuanju wendian ji houxuanren cankao mingdan中央研究院评议会评议员选举文电及候选人参考名单(The Telegram and the Candidate Reference List for the Election of Council Members of the Council of Academia Sinica). Nanjing: Zhongguo dier lishi danganguan中国第二历史档案馆(The Second Historical Archives of China), Total No. 393, File No. 1547.

In January 1940, the Preparatory Committee received questionnaires sent back by most national colleges and universities. Then, according to the way each institution as a unit, a list of voters for the primary election was drawn up. Moreover, the Preparatory Committee determined the members of the Selection Committee for the disciplines of astronomy, archaeology and anthropology because the professors of national universities and colleges in each of the three disciplines were less than 5 people.

During the preparations of the election, another important task for the Preparatory Committee was to draw up a reference list of candidates for Council Members. On March 16, 1939, at its first meeting, the Preparatory Committee decided that, for the convenience of voters, a candidate reference list should be prepared with emphasis on experts other than university professors, and designated persons in groups to be responsible for drawing up the list.[26] When the third meeting was held in Chongqing on November 26, 1939, it was also decided to request the directors of Academia Sinica to draw up the reference list by division, and then to request the President to approve it, which would then be sent to the voters by the Preparatory Committee for reference.[27]

In order to make the reference list of Council Members more valuable, the Preparatory Committee decided to send the list made by the designated personnel to the directors of Academia Sinica for reference. If the original list is drawn up by the director in the division, it would also be sent back for revision. At the same time, the Preparatory Committee stressed to the directors: "The list should seek to meet the legal qualifications. Those who are qualified for the election should be properly selected and listed to represent the relevant subjects."[28] With regard to the reference list, the Preparatory Committee, at its fourth meeting held in Kunming on December 23, 1939, decided to send copies of the reference list drawn up by the directors to all members of the Committee for deliberation, and formulated two strict deliberation rules: (1)" If two or more members of the Committee think there are unqualified

[26]Zhongyang yanjiuyuan pingyihui dierjie pingyiyuan xuanju choubei weiyuanhui baogao中央研究院评议会第二届评议员选举筹备委员会报告(Report of the Preparatory Committee for the Election of the Second Council Members of Academia Sinica). Nanjing: Zhongguo dier lishi danganguan中国第二历史档案馆(The Second Historical Archives of China), Total No. 393, File No. 1549; Zhongyang yanjiuyuan pingyihui pingyiyuan xuanju wendian ji houxuanren cankao mingdan中央研究院评议会评议员选举文电及候选人参考名单(The Telegram and the Candidate Reference List for the Election of Council Members of the Council of Academia Sinica). Nanjing: Zhongguo dier lishi danganguan中国第二历史档案馆(The Second Historical Archives of China), Total No. 393, File No. 1547.

[27]Zhongyang yanjiuyuan pingyihui pingyiyuan xuanju wendian ji houxuanren cankao mingdan中央研究院评议会评议员选举文电及候选人参考名单(The Telegram and the Candidate Reference List for the Election of Council Members of the Council of Academia Sinica). Nanjing: Zhongguo dier lishi danganguan中国第二历史档案馆(The Second Historical Archives of China), Total No. 393, File No. 1547.

[28]Zhongyang yanjiuyuan pingyihui pingyiyuan xuanju wendian ji houxuanren cankao mingdan中央研究院评议会评议员选举文电及候选人参考名单(The Telegram and the Candidate Reference List for the Election of Council Members of the Council of Academia Sinica). Nanjing: Zhongguo dier lishi danganguan中国第二历史档案馆(The Second Historical Archives of China), Total No. 393, File No. 1547.

persons in the list, they may suggest the President to delete them "; (2) "If there are four or more members in the Committee who think that there are qualified persons who have been omitted, they may suggest the President to add them."[29] At the same time, Tsai Yuan-Pei was also very cautious about drawing up the reference list. He did not make a final decision after each director-branch drew up a list. In January 1940, Tsai Yuan-Pei entrusted the matter to Wong Wen-hao. The reason was that he lived in Hong Kong then and felt "it is very difficult to decide here with no one to consult if there are any doubts."[30]

In the same month Tsai Yuan-Pei entrusted Wong Wen-hao to handle the matter, Wong Wen-hao and other members of the Preparatory Committee drew up a general reference list after reviewing the reference list of candidates proposed by the directors of Academia Sinica. As to whether the first-term elected Council Members should be included in the reference list, the opinions of the directors of Academia Sinica were not completely consistent. After discussions at the Preparatory Committee meeting, it was decided to include all of them. The total reference list was of 121 persons.[31] It included many well-known scholars with high academic status in the country at that time. In addition to the vast majority of the first-term elected Council Members, it also included Woo Y. H., Ny Tsai-ze, Kiang Tsai-han, Sun Dan, Su Buchin, Peter P. T. Sah, Tseng Chao-lun, Sun Xuewu, Woo Sho-Chow, Chen Ko Kuei, Wang C. Y., Chen Shisan C., Tai Fon-Land, Hsieh Chia-yung, John Lee, Tang Yueh, Ku Chieh-kan, Li Fang Kuei, Luo Changpei, Tung Tso-pin, Liang Ssu-cheng, etc. There were about 80 or 90 people on the list who had achieved certain academic achievements and presided over academic institutions for more than 5 years, accounting for more than 66% of the total number on the list. It showed that scholars who met the two qualifications of the council members were the candidates most valued by the Preparatory Committee and the directors of the Academia Sinica. Moreover, on September 20, 1939, when the second meeting of the Preparatory Committee was held, article 1, which stipulated five criteria for drawing up a reference list and selecting candidates for the council, clearly stated: "the qualifications stipulated in article 3 of the regulations[32] shall be

[29]Zhongyang yanjiuyuan pingyihui dierjie pingyiyuan xuanju choubei weiyuanhui baogao中央研究院评议会第二届评议员选举筹备委员会报告(Report of the Preparatory Committee for the Election of the Second Council Members of Academia Sinica). Nanjing: Zhongguo dier lishi danganguan中国第二历史档案馆(The Second Historical Archives of China), Total No. 393, File No. 1549.

[30]Tsai Yuan-Pei蔡元培(1995). Zhi Wong Wen-hao han(1940 nian 1 yue 11 ri) 致翁文灏函(1940年1月11日)(Letter to Wong Wen-hao, January 11, 1940). In Tsai Yuan-Pei蔡元培. *Tsai Yuan-Pei wenji*蔡元培文集(Tsai Yuan-Pei's Collected Works), Vol. 12, Letters II. Taipei: Jinxiu chuban shiye gufen youxian gongsi锦绣出版事业股份有限公司.755.

[31]Zhongyang yanjiuyuan pingyihui dierjie pingyiyuan xuanju choubei weiyuanhui baogao中央研究院评议会第二届评议员选举筹备委员会报告(Report of the Preparatory Committee for the Election of the Second Council Members of Academia Sinica). Nanjing: Zhongguo dier lishi danganguan中国第二历史档案馆(The Second Historical Archives of China), Total No. 393, File No. 1549.

[32]"Regulations" means *The Regulations of the Council of Academia Sinica*.

paid due attention to, and the intention is to enable the elected to have considerable scientific qualifications and leadership ability as well."[33]

In addition to drawing up a list of voters for primary elections and a reference list of candidates for council members, the Preparatory Committee had also drawn up a voting method for primary elections. As for the specific measures, the Preparatory Committee held careful discussions at the third and fourth meetings. The final decision for voting would be made at national universities and national independent colleges, and the presidents of the universities and colleges would be appointed as supervisors to preside over the distribution of votes, supervision and collection of votes for their constituencies. At the same time, "the general list of voters of the school", "matters needing attention to for supervisors" and "instructions for voters" had been prepared for the supervisor in each constituency, and documents and materials such as "reference list of candidates" and "election votes" had also been prepared for the electors. From the end of January to the beginning of February 1940, the Preparatory Committee sent the above documents and materials to the presidents of those national universities and national independent colleges, and sent the relevant documents and votes to the Selection Committee.[34] At this point, preparations for the election were on the whole in place.

2.3 Operation Process of Election: From Primary Election to Final Election

The primary election for the elected members of the Second Council of Academia Sinica was held on February 20, 1940. There were 13 national universities and 8 national independent colleges participating. The national universities were National Central University, National Southwest Associated University, National Northwest University, National Sun Yatsen University, National Jiao Tong University, National Tongji University, National Jinan University, National Wu-han University, the University of Chekiang, Sichuan University, Hunan University, Amoy University and Yunnan University. Independent colleges were Zhongzheng Medical College, Jiangsu Medical College, Shanghai Medical College, Guiyang Medical College, Northwest Agricultural College, National Northwest Engineering College, Shanghai Business School and Tangshan Engineering College of Jiao Tong University. These

---

[33]Zhongyang yanjiuyuan pingyihui dierjie pingyiyuan xuanju choubei weiyuanhui baogao中央研究院评议会第二届评议员选举筹备委员会报告(Report of the Preparatory Committee for the Election of the Second Council Members of Academia Sinica). Nanjing: Zhongguo dier lishi danganguan中国第二历史档案馆(The Second Historical Archives of China), Total No. 393, File No. 1549.

[34]Zhongyang yanjiuyuan pingyihui dierjie pingyiyuan xuanju choubei weiyuanhui baogao中央研究院评议会第二届评议员选举筹备委员会报告(Report of the Preparatory Committee for the Election of the Second Council Members of Academia Sinica). Nanjing: Zhongguo dier lishi danganguan中国第二历史档案馆(The Second Historical Archives of China), Total No. 393, File No. 1549.

universities and colleges were located in Sichuan, Yunnan, Shaanxi, Fujian, Guizhou, Hunan and war zone Shanghai. Northeastern University, Guangxi University, Sino-French Industrial College and other national universities failed to take part in the election.

There were about 560 voters in the primary election. In the primary election, the voters cast 538 valid votes, and 548 candidates obtained the votes. After the primary election, the council held a preparatory meeting for the Fifth Annual Meeting in Chongqing Niujiaotuo Resources Committee on March 14, 1940. As Nanjing had fallen into enemy hands, ballot boxes from all constituencies had been assembled and were opened at the meeting. According to *The Regulations for the Election of the Council of Academia Sinica*, candidates to be the elected from all disciplines must obtain one-fourth of the valid votes of their disciplines. The number of candidates who obtained the votes of all disciplines in the primary election was 72. Among them, those who obtained the highest votes in their disciplines were Woo Y. H. 38, Ny Tsai-ze 37, Tseng Chao-lun 31, Hsieh Chia-yung 19, Tang Yueh 9, Ma Yin-chu 38, Chen Yin-k'o 21, Hu Shih 19 and Ling Chunsheng 5.[35]

After the primary election, the First Council of Academia Sinica held its Fifth Annual Meeting from March 22 to 23, 1940, and completed the election in strict accordance with the election procedures. According to *The Regulations for the Election of the Council of Academia Sinica*, the number of candidates elected for each discipline should be twice the number of elected Council Members for that discipline. The primary elections in chemistry and botany reached a total of 6 votes per candidate. Chou T. Q. and Xie Jiasheng were eliminated due to insufficient votes. Although Wu Hsien's votes reached the number of candidates for election in chemistry, he was unable to enter the run-off process because he was ranked 7th in the number of votes in chemistry. Chen Ko Kuei, Ouyang Zhu, Young Chung-Chien, Zhu Sen, Chiang Ting-fu, Jin Yufu, Li Fang Kuei, Liang Ssu-cheng, Liang Ssu-yung, Kuo Mo-jo and Xu Yitang, who got the votes for their discipline's candidate in the primary election, all lost the election due to their lower ranking.

Judging from the results of the primary election, those who reached the number of votes for the candidates in engineering and social sciences were both short to the quota, of which engineering was short of 3 candidates and social sciences was short of 1. That required a by-election by the council to select from among the other candidates in the two disciplines. Those who reached the number of votes for the candidates in the primary elections of physics, zoology and psychology all exceeded the quota limit. However, Yeh Chi-Sun and Hu K. F. in physics, Wu Hsien-Wen, Tsai Burg and Liu Chung-lo in zoology, and Kuo Zing-Yang and Lu Zhiwei in psychology all had the same number of votes and all ranked just within the quota limit of candidates elected in their discipline. That required the council to vote for one candidate each (Fig. 2.4).

---

[35]Zhongyang yanjiuyuan pingyihui pingyiyuan xuanju wendian ji houxuanren cankao mingdan中央研究院评议会评议员选举文电及候选人参考名单(The Telegram and the Candidate Reference List for the Election of Council Members of the Council of Academia Sinica). Nanjing: Zhongguo

Fig. 2.4 Photo taken at the Fifth Annual Meeting of the First Council of Academia Sinica on March 22, 1940

At 8: 00 am on March 22, the annual meeting was opened in Jialing Hotel in Chongqing. Because President Tsai Yuan-Pei died of illness in Hong Kong on March 5, Wang Shih-Chieh was elected as temporary chairman by the council members present at the meeting. On the afternoon of that day, the Preparatory Committee made a report on the preparations for the election. The next day, the council held two agendas, one was to elect the candidate for the presidency, and the other was to complete the election of council members. There were 30 council members at the meeting, including Ping Chi, Chun W. Y., Chen Yin-k'o, Fu Ssu-Nien, Kuo Zing-Yang, Ho L., Hu Hsen-hsu, Chiang Li-Fu, Li Chi, Li Shu-Hua, Lee J. S., Robert K. S. Lim, Ling Hung-hsun, Mao T. E., Ren Hongjun, Tao L. K., Wang Chia-Chi, Wang Shih-Chieh, Wang Ging-Hsi, Wong Wen-hao, Woo Ting-liang, Xie Jiasheng, Ye Liangfu, Yeh Chi-Sun, Yu C. S., Chow Jen, Chu Chia-Hua, Chu Coching, Zhang Qiyun and Zhang Yun.[36] The election of candidate for the President was added to the council's provisional agenda after Tsai Yuan-Pei suddenly died of illness. Hu Shih, then ambassador to the United States, and Wong Wen-hao and Chu Chia-Hua

dier lishi danganguan中国第二历史档案馆(The Second Historical Archives of China), Total No. 393, File No. 1547.

[36]Zhongyang yanjiuyuan dierjie pingyiyuan juexuan xuanjupiao中央研究院第二届评议员决选选举票(Tickets of the Final Election of the Second Council Members of Academia Sinica). Nanjing: Zhongguo dier lishi danganguan中国第二历史档案馆(The Second Historical Archives of China), Total No. 393, File No. 1551.

were very popular for election. Two days before the annual meeting, Chiang Kai-shek suddenly "issued a note, citing Gu Mengyu".[37] This caused a strong backlash among Council Members. In the election, council members resisted Chiang Kai-shek's note and finally voted to elect Wong Wen-hao, Chu Chia-Hua and Hu Shih as alternate Presidents. In September of the same year, the National Government officially announced that Chu Chia-Hua was acting president of Academia Sinica.

After selecting the candidate for the presidency, the council began to decide on the candidates for elected members of the Second Council. First of all, in engineering and social sciences, respectively, among those who did not get the votes of the candidates, by-elections would be held for the candidates of each discipline. From Yeh Chi-Sun and Hu K. F., Wu Hsien-Wen, Tsai Burg and Liu Chung-lo, Kuo Zing-Yang and Lu Zhiwei, respectively, one person was elected as the candidate in physics, zoology and psychology. The by-election resulted for the engineering discipline were: Ling Hung-hsun 18, Shen Yi 8, Ren Shangwu 7, Wang C. Y. 6, Li Chenggan 4, Yun Zhen 4, Yan Enyu 3, Qian Changzuo 3, Hou T. P. 2, Du Dianying 2, Xu Kai 2, Wang S. C. 2, Shi Jiayang, Ku Yu-hsiu, Zhi Bingyuan, Peng Jiqun, Zhuang Qianding, Guan Dongbo, Li Shutian, Xiao Guanying, Adam Pen-Tung Sah, He Jie, Cheng Ganyun all got one vote each. The results of the social sciences by-election were: Ho L. 17, Hsiao Kung-chuan 4, Yang Duanliu 3, Huang Yuanbin, Fong H. D. and Hsiao Chu all got one vote each. The voting results for Yeh Chi-Sun and Hu K. F. etc. were: Yeh Chi-Sun 25, Hu K. F. 3, Wu Hsien-Wen 13, Tsai Burg 7, Liu Chung-lo 3, Kuo Zing-Yang 15 and Lu Zhiwei 9.[38] After that, according to the voting results and the number of candidates in each discipline, the council elected Ho L., and Ling Hung-hsun, Shen Yi and Ren Shangwu as candidates in social sciences and engineering respectively, and decided that Yeh Chi-Sun, Wu Hsien-Wen and Kuo Zing-Yang were candidates in physics, zoology and psychology respectively.

Then, the council examined the qualifications of the candidates who had been elected as elected council members in various disciplines. The examination was divided into three groups: sky, earth, and man. The five disciplines of physics, chemistry, engineering, astronomy and meteorology belonged to sky section. Zoology, botany, geology, social sciences, history, psychology, linguistics, archaeology and Anthology belonged to the groups of earth and man. Lai Lian, acting dean of National Northwest Engineering College, won 38 votes in the primary election, ranking third in the engineering discipline, but was disqualified because he had no works. According to Chu Coching, Lai Lian and Xiao Guanying of National Sun Yatsen University

[37] Fu Ssu-Nien zhi Hu Shih (1940 nian 8 yue 14 ri) 傅斯年致胡适(1940年8月14日)(Fu Ssu-Nien's Letter to Hu Shih, August 14, 1940). In Zhongguo shehui kexueyuan jindaishi yanjiusuo Zhoghua minguoshi zu中国社会科学院近代史研究所中华民国史组 (Section of History of the Republic of China of Institute of Modern History, Chinese Academy of Social Sciences)(Eds.) (1979). *Hu Shih wanglai shuxinxuan*胡适往来书信选(Selected Correspondence of Hu Shih), Middle Volume. Beijing: Zhonghua shuju中华书局.475.

[38] Zhongyang yanjiuyuan dierjie pingyiyuan juexuan xuanjupiao中央研究院第二届评议员决选选举票(Tickets of the Final Election of the Second Council Members of Academia Sinica). Nanjing: Zhongguo dier lishi danganguan中国第二历史档案馆(The Second Historical Archives of China), Total No. 393, File No. 1551.

Table 2.1 Final election result of the elected members of the Second Council of Academia Sinica

Discipline	Names of the elected and number of votes obtained[a]	Quota number
Physics	Chiang Li-Fu (23), Woo Y. H. (20 or 19)[b], Li Shu-Hua (16)	3
Chemistry	Hou T. P. (23), Tseng Chao-lun (19), Chuang Chang-Kong (14)	3
Engineering	Ling Hung-hsun (24), Mao T. E. (22), Wang C. Y. (14)	3
Zoology	Ping Chi (22), Robert K. S. Lim (20), Chen Shisan C. (15)	3
Botany	Tai Fon-Land (24), Chun W. Y. (21), Hu Hsen-hsu (16)	3
Geology	Wong Wen-hao (25), Chu Chia-Hua (21), Hsieh Chia-yung (13)	3
Astronomy	Zhang Yun (17)	1
Meteorology	John Lee (15 or 14)	1
Psychology	Tang Yueh (17)	1
Social sciences	Wang Shih-Chieh (19), Ho L. (17 or 16), Chou Keng-Shang (13)	3
History	Hu Shih (23), Chen Yin-k'o (22), Chen Yuan (13)	3
Linguistics	Chao Yuen Ren (24)	1
Archaeology	Li Chi (23)	1
Anthropology	Woo Ting-liang (20)	1

Sources Zhongyang yanjiuyuan dierjie pingyiyuan juexuan xuanjupiao中央研究院第二届评议员决选选举票(Tickets of the Final Election of the Second Council Members of Academia Sinica). Nanjing: Zhongguo dier lishi danganguan中国第二历史档案馆(The Second Historical Archives of China), Total No. 393, File No. 1551

[a]In order of the number of votes obtained, the numbers in brackets are the number of votes won

[b]After reviewing the voting result of council members in the archives, the author found that there was a difference of 1 vote between the votes of Woo Y. H., John Lee and Ho L. counted by the First Council of Academia Sinica and the result of the review. However, whether they had one more or less vote would not affect the election results

were still suspected of running for election, and "neither of them was really eligible for election".[39] After Lai Lian's qualification was cancelled, the council nominated Wang C. Y., who came fourth in the engineering by-election, as the candidate. As a result, the number of candidates for elected council members reached 60 in full.

On the afternoon of March 23, the council held a run-off election for elected council members. Of the 27 members present at the meeting, Wang Shih-Chieh asked for leave for work and Wong Wen-hao was elected as temporary chairman. After voting on the spot, the council successfully elected 30 of the 60 candidates. The result of the run-off is shown in Table 2.1.

According to the profiles of these 30 elected, 20 of them including Woo Y. H., Tseng Chao-lun, Chuang Chang-Kong, Mao T. E., Ping Chi, Robert K. S. Lim, Chen Shisan C., Tai Fon-Land, Chun W. Y., Hu Hsen-hsu, Wong Wen-hao, Zhang

[39]Chu Coching竺可桢(2005). *Chu Coching quanji*竺可桢全集(The Complete Works of Coching Chu), Vol. 7. Shanghai: Shanghai keji jiaoyu chubanshe上海科技教育出版社.316, 322.

Yun, Wang Shih-Chieh, Chou Keng-Shang, Hu Shih, Chen Yin-k'o, Chen Yuan, Chao Yuen Ren, Li Chi and Woo Ting-liang all had a certain number of academic achievements, some of which had reached international standards, and had presided over academic institutions for at least 5 years. They all had or basically had the two qualifications of the council members, while the other 10 people all had one qualification: Hou T. P., Ling Hung-hsun, Wang C. Y., Hsieh Chia-yung, John Lee and Tang Yueh had made either outstanding academic research achievements or remarkable contributions to the establishment of industries, and Chiang Li-Fu, Li Shu-Hua, Chu Chia-Hua and Ho L. had presided over academic institutions for more than 5 years and had achieved notable results. Only 4 of the 30 candidates were elected mainly for hosting academic institutions, reflecting the fact that the council did not attach importance to purely academic leaders.

The discipline distribution of the elected members of the Second Council was exactly the same as that of the first-term, maintaining the feature of comprehensive discipline distribution. Judging from the academic level, the overall level of the elected members of the Second Council was almost equal to that of the First Council. Therefore, there is no obvious difference in academic authority between the two councils. However, the representativeness of the Second Council is smaller than that of the first, since National Geological Survey of China, National Agricultural Research Bureau, the Institute of Biology of the Science Society of China, the University of Chekiang, Yenching University and other important academic institutions that had the elected members of the First Council of Academia Sinica had no representatives.

Judging from the period of birth, the second-term and first-term elected Council Members were mostly born from 1891 to 1900, of whom 18 were appointed for the first-term and 19 for the second-term. In 1940, they were between 40 and 49 years old and were in the prime of their life. From the perspective of academic qualifications, there were, in the first-term, about 18 people who had stayed in the United States, 9 people who had stayed in Europe, 1 person who had stayed in both the United States and Europe; 22 people who obtained doctoral degrees and 3 people who obtained master's degrees. In the second-term, 19 elected Council Members were employed to stay in the United States, 9 to stay in Europe, 1 to stay in both the United States and Europe; 21 obtained doctoral degrees and 5 obtained master's degrees. Generally speaking, most of the second-term and first-term elected Council Members were in the prime of their life when they were elected, and almost all of them studied in Europe and/or America, especially in the United States, and most received doctorates. After their return to China, they mostly devoted themselves to the scientific cause of China and belonged to a generation that laid the foundation for Chinese science in the 20th century.

2.4 Summary

At the time of the election of elected members of the Second Council of Academia Sinica, China was suffering from Japan's full-scale invasion. China's politics was out of order and society was out of order. Many academic institutions were scattered in the rear area, which was often invaded by war, and some were still being evacuated. On the eve of the election, President Tsai Yuan-Pei died suddenly, and Academia Sinica was not presided over for a while. Under these unfavorable conditions, there were two keys to the successful holding of that election. First, the first Council had established a more thorough and democratic election procedure based on *The Regulations of the Council of Academia Sinica* drafted by V. K. Ting and others, which had laid an effective institutional foundation for the election. Second, the Preparatory Committee had made sufficient preparations for the election by drawing up a list of primary voters, a reference list of candidates for the council members and a primary voting method.

This session of election for elected council members was conducted on the basis of the first election, but it was obviously improved over the latter. First of all, it is shown that the selection method of candidates for the council members is more democratic and strict. Electing candidates for council members was a basic and important part of the election. When the first election was held in Academia Sinica, the candidates were selected by Jiang Menglin, Mei Yi-chi and Chou Keng-Shang who were temporarily appointed by Tsai Yuan-Pei, and were not selected by experts from the national universities and the Selection Committee. It is doomed to be less democratic in the first selection of candidates than in the second one. In the second election for the council members, the council strictly examined the qualifications of the candidates elected. There was no such link in the election for the first elected council members. These links are very important to ensure the academic standard of council members and the objectivity and fairness of the election.

Secondly, it is manifested that the preparations for election was more careful. Before Academia Sinica held the election of elected members of the First Council, it did not set up a special preparatory organization, it only held a preparatory meeting the day before the formal election, agreed on the quota allocation and election principles for elected council members in various disciplines, and presumed the candidates for them. It was different for the election of elected members of the Second Council. About a year before the election, the Preparatory Committee headed by Wong Wen-hao was set up. The Preparatory Committee had carefully considered and arranged the formulation of the primary election voter list, the reference list of the candidates and the primary election voting method. For the purpose of drawing up the reference list of candidates, a special person had originally been appointed to draw up the list, and then, in order to avoid suspicion, the directors of Academia Sinica were asked to draw up the list by division. In order to ensure the quality of the reference list, the lists drawn up by the directors were copied to the members of the Preparatory Committee for deliberation, and strict deliberation methods were formulated. As a result, the general reference list played an important reference role in the election. When the election of elected members of the First Council was held in Academia Sinica, the list drawn up by Jiang Menglin, Mei Yi-chi and Chou Keng-Shang was both a candidate list and a candidate reference list, because in the formal election, candidates could be elected separately from the list. However, that practice of drawing up a reference list is obviously less prudent than that of having the Preparatory Committee.

The fundamental reason for these improvements and changes in the election of elected members of the Second Council is that when the election of elected members of the First Council was organized by Academia Sinica, a lot of work was run by V. K. Ting alone, and the time was too short to do it. However, the election of elected members of the Second Council was carefully prepared by a specialized committee and all links were strictly handled in accordance with regulations, which had greatly improved the election method.

Judging from the process and results, the election of elected members of the Second Council was relatively fair. After the national university professors and the Selection Committee voted to select candidates for council members, the First Council of Academia Sinica conducted qualification examination and run-off of the candidates in strict accordance with the election procedures and qualifications. As for the selection of council members, the council attached the greatest importance to both outstanding academic achievements and academic leadership ability, and preferred the former between pure scholars with outstanding academic achievements and academic leaders with less academic achievements. These had reduced the influence and interference of non-academic factors on the election, reflecting the spirit of the council to pursue academic independence and academic freedom, and possessing the behavioral characteristics of western advanced scientific organizations to abide by scientific norms and attach importance to academic standards. It can be seen from the following article that the Second Council elected this time also abided by scientific norms and attached importance to academic standards when preparing for the election of the first batch of members of Academia Sinica. To some extent, it means that this kind of behavior characteristic, as an academic tradition, was inherited by the Second Council and had a positive impact on the establishment of academician system in Academia Sinica.

The election of elected council members also provided useful experience and reference for the establishment of academician system of Academia Sinica. After the Anti-Japanese War, when the Second Council organized the election of the first batch of members of Academia Sinica, it also designed the link of qualification examinations of candidates in *The Regulations for the Election of Members of Academia Sinica*. Before the election, drawing on the experience of the First Council in presiding over the election of elected members of the Second Council, a preparatory committee for the election was established and a reference list of member candidates was drawn up. The preparatory committee effectively promoted the nomination of member candidates by utilizing the resources of social relations with the National University of Peking and National Tsing Hua University. Moreover, by drawing up of a reference list of member candidates, a better frame of reference was set up for the selection of members of Academia Sinica by the council.

Chapter 3
The Council Structured the Academician System and Relevant Regulations

After the establishment of the Second Council of Academia Sinica, the academician system began to be brewed and was formally put on the agenda at the Second Annual Meeting in March 1944. After the victory of the Anti-Japanese War, Academia Sinica demobilized and returned to the east, speeding up the pace of establishing the academician system. In October 1946, the council held its Third Annual Meeting in Nanjing. As Chinese society was still in violent turbulence, Fu Ssu-Nien held a pessimistic attitude and believed that "the world is in chaos," and that it was not appropriate to elect members of Academia Sinica; he said, "if it is not done well, Academia Sinica can be toppled." However, most council members were optimistic and supported the election of members of Academia Sinica.[1] In this way, the annual meeting decided to set up members in Academia Sinica to improve the academy's system. Subsequently, the legislative work on the academician system began. This chapter examines how the council structured the academician system and relevant regulations.

3.1 The Brewing of the Academician System and the Beginning of Legislation

The Second Council of Academia Sinica was established on July 3, 1940, after being approved by the National Government. The president and the honorary secretary of the council were Chu Chia-Hua and Wong Wen-hao, respectively. The First Annual Meeting of this term's council was held in Chongqing from March 13 to March 15, 1941; the theme of the meeting was how to effectively coordinate and plan

[1] 1947 nian 6 yue 20 ri Fu Ssu-Nien zhi Hu Shih xin 1947年6月20日傅斯年致胡适信 (A letter from Fu Ssu-Nien to Hu Shih on June 20, 1947). In Geng Yunzhi耿云志(Eds.) (1994). *Hu Shih yigao ji micang shuxin*胡适遗稿及秘藏书信(Hu Shih's Posthumous Manuscripts and Secret Collection of Letters), Vol. 37. Hefei: Huangshan shushe黄山书社. 524.

J. Guo, *The Establishment and Reconstruction of the Academician System in China*,
https://doi.org/10.1007/978-981-15-7208-1_3

national academic research and strengthen international academic exchanges. The meeting decided that Academia Sinica and the Ministry of Education should jointly convene a national academic conference to promote national academic development. At the same time, Wong Wen-hao temporarily proposed the proposal of "whether the honorary members stipulated in Article 8 of *The Organization Act of Academia Sinica* can be established or not."[2] After the discussion, the meeting decided that "it should first be considered by each group of council members, and then be discussed at the next annual meeting."[3] This meant that the establishment of the academician system would be formally put on the council's agenda one year later.

However, after this annual meeting, the Second Annual Meeting was not held until 1944, during which the council failed to function properly. Academia Sinica was also in a dilemma; before Tsai Yuan-Pei's death in 1940, there were calls in the society, because of the promotion of the Ministry of Education, for Academia Sinica to be changed from being directly attached to the National Government to being attached to Executive Yuan. Later, in order to compete with Academia Sinica for power, the Minister of Education Chen Lifu suggested that Chiang Kai-shek merge Academia Sinica with either the Ministry of Education or Examination Yuan, before the Ninth Plenary Session of the Fifth Kuomintang Central Committee in December 1941.[4] After being unblocked by Academia Sinica, this proposal was "dismissed" before March 9, 1942.[5] Furthermore, the Academic Review Committee of the Ministry of Education did all the academic review work that should have been handled by the Council of Academia Sinica, which embarrassed the council. Although these situations were unfavorable to Academia Sinica, they prompted the council to resolve

[2]Zhongyang yanjiuyuan diyijie dierjie diyici nianhui ti'an ji youguan wenshu中央研究院评议会第一届第二届第一次年会提案及有关文书(Proposals and Related Documents of the First Annual Meeting of the First and the Second Council of Academia Sinica). Nanjing: Zhongguo dier lishi danganguan中国第二历史档案馆(The Second Historical Archives of China), Total No. 393, File No. 1563.

[3]Zhongyang yanjiuyuan diyijie dierjie diyici nianhui ti'an ji youguan wenshu中央研究院评议会第一届第二届第一次年会提案及有关文书(Proposals and Related Documents of the First Annual Meeting of the First and the Second Council of Academia Sinica). Nanjing: Zhongguo dier lishi danganguan中国第二历史档案馆 (The Second Historical Archives of China), Total No. 393, File No. 1563.

[4]Chen Shiwei陈时伟. Zhongyang yanjiuyuan 1948 nian yuanshi xuanju shulun中央研究院1948年院士选举述论 (A Review on the Election of Academicians at Academia Sinica in 1948), In Zhongguo shehui kexueyuan jindaishi yanjiusuo minguoshi yanjiushi, Sichuan shifan daxue lishi wenhua xueyuan中国社会科学院近代史研究所民国史研究室: 四川师范大学历史文化学院(Research Office of History of the Republic of China of Institute of Modern History, Chinese Academy of Social Sciences, School of History and Culture of Sichuan Normal University) (Eds.) (2009). *1940 niandai de Zhongguo*一九四O年代的中国 (China in the 1940s), Vol. 2. Beijing: Shehui kexue wenxian chubanshe社会科学文献出版社.1029; Li Lairong李来容 (2010). Yuanshi zhidu yu minguo xueshu: 1948 nian yuanshi zhidu de queli yu yunzuo院士制度与民国学术—1948年院士制度的确立与运作 (Academician System in Republic of China: Establishment and Operation of the Academician System in 1948). Tianjin: Doctoral Dissertation of Nankai University. 171–172.

[5]Chu Coching竺可桢 (2006). *Chu Coching quanji*竺可桢全集 (The Complete Works of Coching Chu), Vol. 8. Shanghai: Shanghai keji jiaoyu chubanshe上海科技教育出版社. 306.

Fig. 3.1 Chu Chia-Hua, acting president of Academia Sinica and president of the council

to enhance its function of "guiding, contacting, and rewarding academic research," and to further discuss the establishment of an academician system. On September 30, 1943, the council called for a meeting of council members in Chongqing at Niujiaotuo 22. The acting president of Academia Sinica and the president of the council, Chu Chia-Hua, and honorary secretary of the council, Wong Wen-hao as well as the council members Wang Shih-Chieh, Ting S. L., and Fu Ssu-Nien attended the meeting. The meeting decided to change the date of the Second Annual Meeting to the first half of March 1944, in order to enrich the preparations. The meeting discussed whether the honorary members of Academia Sinica should be elected at this annual meeting, but failed to do so in the end (Figs. 3.1 and 3.2).[6]

Going into 1944, China's victory in the Anti-Japanese War was just around the corner. The Second Council of Academia Sinica began to prepare for the Second Annual Meeting in earnest. On March 6, Chu Chia-Hua and Wong Wen-hao invited council members in Chongqing to hold a symposium at the Chongqing Sino-British Cultural Association. At this meeting, Wong Wen-hao raised the issue of "whether to elect honorary members at the Second Annual Meeting." After the discussion, the

[6]Pingyihui erjie zai yu pingyiyuan tanhuahui评议会二届在渝评议员谈话会 (The Symposium of Members of the Second Council in Chongqing). Nanjing: Zhongguo dier lishi danganguan中国第二历史档案馆 (The Second Historical Archives of China), Total No. 393, File No. 577.

Fig. 3.2 Wong Wen-hao, honorary secretary of the council

meeting decided to "organize an 'Honorary Member Candidate Selection Committee' at the meeting."[7]

The Second Annual Meeting was held from March 8 to March 10. There were 27 council members present: Chu Chia-Hua, Wong Wen-hao, Li Shu-Hua, Chu Coching, Fu Ssu-Nien, Hsieh Chia-yung, Mao T. E., Ting S. L., Woo Y. H., Zhang Yun, Tao L. K., Chen Shisan C., Lee J. S., Li Chi, Woo Sho-Chow, Chang Yu-Che, Wang Ging-Hsi, Tai Fon-Land, Chiang Li-Fu, Ho L., Chuang Chang-Kong, John Lee, Tseng Chao-lun, Chow Jen, Woo Ting-liang, Wang Chia-Chi, and Tang Yueh. With the victory of the Anti-Japanese War in sight, China's academic career would develop greatly, and Chu Chia-Hua was excited about it. In his opening speech, he expressed his excitement and pointed out the tasks of the meeting, in the light of the post-war national construction work:

> The Second Annual Meeting of the Second Council of Academia Sinica was held in the eighth year of the Anti-Japanese War. We look over the fact that China's academic career will follow the victory of the Anti-Japanese War; there is infinite excitement in looking forward to the future. Every gentlemen, it is a great pleasure to have come all the way here. This Council is organized by national academic circles to select its council members. Each member is an authority in the national academic circle. The resolution made by the Council concerns the academic development of the whole country, not just Academia Sinica. … Post-war China must carry out grand construction according to the industrial plan of Father of the State. Therefore, post-war demobilization and revival require more contributions and

[7]Zhongyang yanjiuyuan zai yu pingyiyuan tanhuahui jilu中央研究院在渝评议员谈话会纪录 (Minutes of the Symposium of the Council Members of Academia Sinica in Chongqing). Nanjing: Zhongguo dier lishi danganguan中国第二历史档案馆(The Second Historical Archives of China), Total No. 393, File No. 546.

> efforts from our academic circle. The demobilization of academic undertakings and their revival should also be planned in advance to match the demobilization and revival of the country. All these are to be discussed at this annual meeting.[8]

He also reiterated the above-mentioned instructions on academic independence, contained in Chiang Kai-shek's written address to the joint annual meeting of the six academic organizations, the Science Society of China, the Geographical Society of China, Chinese Meteorological Society, Chinese Mathematical Society, China Zoological Society, and the Botanical Society of China, on July 18, 1943, and linked academic independence with national independence. Moreover, he stressed: "If a country's academic independence cannot be achieved, then the conditions for a country's independence are not yet complete. How we should plan our country's academic independence so as to lay its foundation at an early date remains to be discussed at this annual meeting. The demobilization and revival of academic undertakings after the war, as well as the pursuit of academic independence in our country and academic cooperation with friendly countries, must start with general progress and full cooperation in domestic academic research."[9]

After the opening ceremony, the annual meeting discussed a total of 31 proposals, as well as a number of provisional proposals and motions in the subsequent agenda around the work of Academia Sinica itself and the grand plans for post-war academic revival. The council attached great importance to the issue of how to realize national academic independence. The proposals put forth to the government included this one: "Academic independence is related to the future of the country. In order to achieve academic independence and establish a scientific research foundation for the country, it is necessary to enrich the equipment of domestic research institutions and universities."[10] Echoing this, thecouncil formally put the establishment of the academician system on the agenda, according to the decision of the March 6 symposium on "whether to elect honorary members or not."

On the afternoon of March 8, the first meeting of the annual meeting assumed that Wong Wen-hao would be the convener, and an honorary member candidate selection committee composed of Wong Wen-hao, Wang Ging-Hsi, Chuang Chang-Kong, Chiang Li-Fu, Li Shu-Hua, and Lee J. S. was formed. The second meeting was held on the afternoon of March 9, and the selection committee presented its comments in the review comment: "It is suggested that the selection committee not

[8]Zhongyang yanjiuyuan pingyihui dierjie dierci nianhui jilu中央研究院评议会第二届第二次年会纪录 (Minutes of the Second Annual Meeting of the Second Council of Academia Sinica). Nanjing: Zhongguo dier lishi danganguan中国第二历史档案馆 (The Second Historical Archives of China), Total No. 393, File No. 1556.

[9]Zhongyang yanjiuyuan pingyihui dierjie dierci nianhui jilu中央研究院评议会第二届第二次年会纪录 (Minutes of the Second Annual Meeting of the Second Council of Academia Sinica). Nanjing: Zhongguo dier lishi danganguan中国第二历史档案馆 (The Second Historical Archives of China), Total No. 393, File No. 1556.

[10]Zhongyang yanjiuyuan pingyihui dierjie dierci nianhui jilu中央研究院评议会第二届第二次年会纪录 (Minutes of the Second Annual Meeting of the Second Council of Academia Sinica). Nanjing: Zhongguo dier lishi danganguan中国第二历史档案馆 (The Second Historical Archives of China), Total No. 393, File No. 1556.

hold the election for time being." After the discussion, the meeting decided to "act according to the review comment."[11]

Later, at the third meeting held on the morning of March 10, Wong Wen-hao temporarily proposed that "all translated names of Academia Sinica should be formally stipulated." The case included the translated names of Academia Sinica, the Council, council members, and the Member of Academia Sinica. Member of Academia Sinica was later translated as the Academician of Academia Sinica; at the meeting, however, the Council did not confirm the translated name of the Member of Academia Sinica. Instead, it decided that "the name, establishment, election, and any related matters shall be examined and approved by the organizing committee and discussed at the next annual meeting. Woo Y. H., Wong Wen-hao, Fu Ssu-Nien, Lee J. S., and Li Shu-Hua shall be selected as members to be convened by Woo Y. H."[12] On the afternoon of March 10, the fourth meeting of the annual meeting discussed "the possibility of electing honorary members of the Council." It was reviewed and approved at this meeting, but they decided "not to elect this year."[13]

After the closing ceremony of this annual meeting, Chu Chia-Hua hosted a symposium of the council members in Chongqing on March 16, at the Sino-Indian Society of Chongqing's Liangfu Branch Road. John Lee, Chang Yu-Che, Tao L. K., Wang Ging-Hsi, Woo Ting-liang, Lee J. S., Woo Y. H., Wang Chia-Chi, Woo Sho-Chow, Zhang Yun, Mao T. E., Fu Ssu-Nien, Hsieh Chia-yung, Chu Coching, Ting S. L., Li Shu-Hua, Chuang Chang-Kong, Chiang Li-Fu, and Tai Fon-Land attended. At the meeting, Woo Y. H. made a review report on the case concerning the name, establishment, election, and related matters of the Member of Academia Sinica:

(1) The translation of the Member of Academia Sinica includes the options of "Yuan Yuan (院员)", "Yuan Shi (院士)", "Yuan Zheng (院正)", "Yuan Lü (院侣)", and "Xue Lü (学侣)". It should be submitted to a referendum (hereinafter temporarily called "Yuan Yuan" to facilitate the discussion).
(2) Academia Sinica (within the meaning of Articles 1 and 2 of *The Organization Act of Academia Sinica*) sets up an academic association, which is divided into two groups: natural science and humanities. The members of the academic association are called "Yuan Yuan of Academia Sinica" (see the English translation above). The organization of the academic association shall be filed with the government, and its regulations shall be stipulated separately.

[11]Zhongyang yanjiuyuan pingyihui dierjie dierci nianhui jilu中央研究院评议会第二届第二次年会纪录 (Minutes of the Second Annual Meeting of the Second Council of Academia Sinica). Nanjing: Zhongguo dier lishi danganguan中国第二历史档案馆 (The Second Historical Archives of China), Total No. 393, File No. 1556.

[12]Zhongyang yanjiuyuan pingyihui dierjie dierci nianhui jilu中央研究院评议会第二届第二次年会纪录 (Minutes of the Second Annual Meeting of the Second Council of Academia Sinica). Nanjing: Zhongguo dier lishi danganguan中国第二历史档案馆 (The Second Historical Archives of China), Total No. 393, File No. 1556.

[13]Zhongyang yanjiuyuan pingyihui dierjie dierci nianhui jilu中央研究院评议会第二届第二次年会纪录 (Minutes of the Second Annual Meeting of the Second Council of Academia Sinica). Nanjing: Zhongguo dier lishi danganguan中国第二历史档案馆 (The Second Historical Archives of China), Total No. 393, File No. 1556.

(3) The qualifications of the candidates for "Yuan Yuan" are the same as those of the council members. For time being, the disciplines are limited to those of the established institutes of Academia Sinica.
(4) "Yuan Yuan" shall be elected by the council and shall be permanent members of Academia Sinica after being elected. The number and disciplines distribution shall be determined by the council. (After the academic association is established, how to elect "Yuan Yuan" should be discussed.)
(5) The relationship between the council members and "Yuan Yuan" should be discussed."[14]

The word "Yuan Shi" had already been used in the first point of this report; according to Chu Chia-Hua's memory, this was proposed by Fu Ssu-Nien.[15] The "academic association" mentioned in the second point referred to the General Assembly of Members, which will be touched upon later. Although this report was not yet very detailed, it had already touched on important issues such as the translation of the name of the member of Academia Sinica, the qualifications of members of Academia Sinica, disciplines, elections, and the relationship between council members and members of Academia Sinica, which all must be resolved in order to establish an academician system. Moreover, the third and fourth points later became the two basic guiding principles for Academia Sinica to establish the academician system. The symposium was very cautious about the report and decided to "invite all council members to study it and bring it up for discussion at the next annual meeting."[16] In this way, the establishment of the academician system in Academia Sinica became one of the major plans for China's academic revival after the war and was included on the agenda of the council.

On August 31, 1945, the eve of the formal victory of the Anti-Japanese War, Academia Sinica held a symposium in its general office in Chongqing between the council members and directors. The meeting decided that the Third Annual Meeting of the Second Council, originally scheduled to be held on November 1, could not be convened due to restoration problems and was scheduled to be held in Nanjing on March 5 of the following year. At that time, "the proposal to set up 'Member of Academia Sinica' would be discussed again; if it was passed, it would be set up

[14]Zhongyang yanjiuyuan zai yu pingyiyuan tanhuahui jilu中央研究院在渝评议员谈话会纪录 (Minutes of the Symposium of the Council Members of Academia Sinica in Chongqing). Nanjing: Zhongguo dier lishi danganguan中国第二历史档案馆 (The Second Historical Archives of China), Total No. 393, File No. 546.

[15]Chu Chia-Hua朱家骅. 30 nian lai de zhongyang yanjiuyuan (minguo 47 nian 6 yue 9 ri)三十年来的中央研究院 (民国四十七年六月九日) (Academia Sinica for 30 Years, June 9, 47, Republic of China). In Wang Yujun王聿均 and Sun Bin孙斌(Eds.) (1977). *Chu Chia-Hua xiansheng yanlun ji*朱家骅先生言论集 (Speech by Mr. Chu Chia-Hua). Taipei: Zhongyang yanjiuyuan jinaishi yanjiusuo 中央研究院近代史研究所 (Institute of Modern History of Academia Sinica). 111.

[16]Zhongyang yanjiuyuan zai yu pingyiyuan tanhuahui jilu中央研究院在渝评议员谈话会纪录 (Minutes of the Symposium of the Council Members of Academia Sinica in Chongqing). Nanjing: Zhongguo dier lishi danganguan中国第二历史档案馆(The Second Historical Archives of China), Total No. 393, File No. 546.

immediately."[17] Since then, Academia Sinica began to demobilize and returned to the east, which accelerated the pace of establishing the academician system.

From October 20 to 24, 1946, Academia Sinica held the Third Annual Meeting of the Second Council in Nanjing, before the demobilization work was completely finished. One of the important issues was the establishment of the academician system. Chu Chia-Hua, Wong Wen-hao, the new Director-General of Academia Sinica, Adam Pen-Tung Sah, and 28 council members, Wang Shih-Chieh, Wang Jiabo, Ho L., Woo Y. H., Woo Ting-liang, Woo Sho-Chow, John Lee, Li Shu-Hua, Ping Chi, Wang Ging-Hsi, Chow Jen, Chou Keng-Shang, Robert K. S. Lim, Chu Coching, Mao T. E., Hu Shih, Hu Hsen-hsu, Tang Yueh, Chen Yuan, Ling Hung-hsun, Fu Ssu-Nien, Zhang Yun, ChienShung-shu, Hsieh Chia-yung, and Loo Tsung-Le attended.[18]

At the opening ceremony, Chu Chia-Hua delivered a speech explaining the significance of this annual meeting: "Since the Second Annual Meeting of the council was held in Chongqing, in March of the 33rd year of the Republic of China, academic research has encountered unprecedentedly serious difficulties in these two or three years, even beginning from the Anti-Japanese War to these present nine years. In the future, academic research in the whole country and the promotion of the work of Academia Sinica all need to have a common policy and plan, in order to follow and strengthen the research. Therefore, this annual meeting is indeed of historic significance." He also said: "It has been less than 20 years since the establishment of Academia Sinica. The council has an even shorter history. Not long after the end of the Anti-Japanese War, when construction is about to begin, it faced drastic changes in all circumstances how to perfect the organization of Academia Sinica itself, and how to make the representative nature of Academia Sinica more reasonable, so as to enhance the efficiency of work, and so that national academics could have the opportunity to develop as far as possible. We need your advice."[19] Among them, "how to perfect the organization of the Academia Sinica itself" mainly referred to the establishment of an academician system.

At 9:00 am on October 22, the Third General Meeting of the annual meeting began. According to the pre-war decision, they discussed the proposal "Please Stipulate the Name, Establishment, Election and Related Matters of Member of Academia Sinica." At the same time, the proposal "Please Set up Yuan Lü (院侣) of Academia Sinica

[17]Zhongyang yanjiuyuan 1945 niandu 5 yue 8 yue 12 yue pingyiyuan tanhuahui jilu中央研究院一九四五年度五月八月十二月评议员谈话会纪录 (Minutes of the Symposiums of the Council Members of Academia Sinica in May, August and December, 1945). Nanjing: Zhongguo dier lishi danganguan中国第二历史档案馆 (The Second Historical Archives of China), Total No. 393, File No. 647.

[18]Zhongyang yanjiuyuan pingyihui dierjie disanci nianhui jilu中央研究院评议会第二届第三次年会纪录 (Minutes of the Third Annual Meeting of the Second Council of Academia Sinica). Nanjing: Zhongguo dier lishi danganguan中国第二历史档案馆 (The Second Historical Archives of China), Total No. 393, File No. 1557.

[19]Zhongyang yanjiuyuan pingyihui dierjie disanci nianhui jilu中央研究院评议会第二届第三次年会纪录 (Minutes of the Third Annual Meeting of the Second Council of Academia Sinica). Nanjing: Zhongguo dier lishi danganguan中国第二历史档案馆 (The Second Historical Archives of China), Total No. 393, File No. 1557.

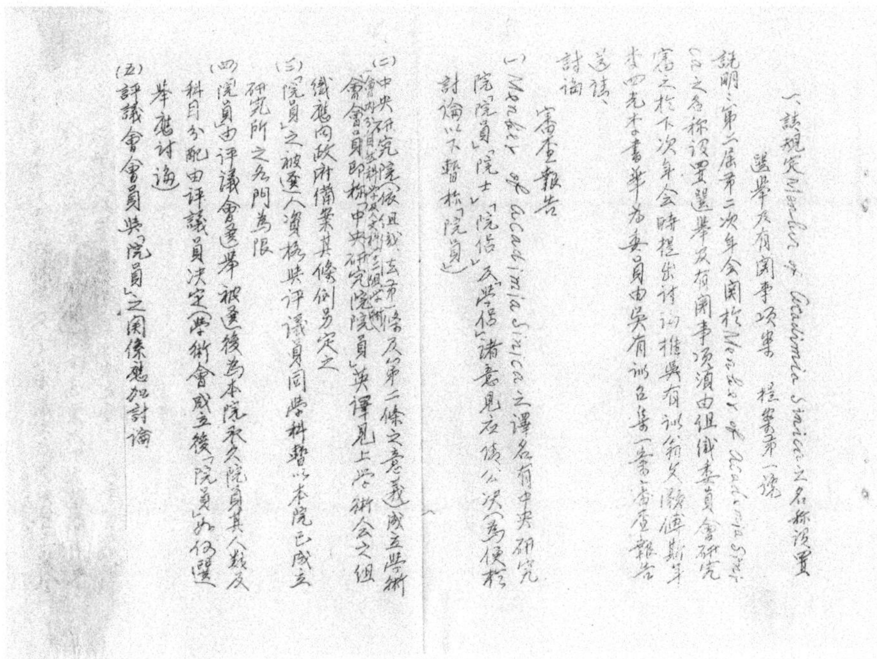

Fig. 3.3 The Proposal "Please Stipulate the Name, Establishment, Election and Related Matters of Member of Academia Sinica"

and Amend the Organization and Tasks of the Council," put forth by Wong Wen-hao and Adam Pen-Tung Sah and countersigned by Woo Y. H., Fu Ssu-Nien, Wang Ging-Hsi, and Chu Coching, was discussed. The meeting decided to organize the first examination committee, with Woo Y. H., Wong Wen-hao, Fu Ssu-Nien, Li Shu-Hua, Hu Shih, Adam Pen-Tung Sah, Chou Keng-Shang, Ping Chi, and Chen Yuan presumed to be members; Hu Shih convened the meeting (Figs. 3.3 and 3.4).[20]

The latter proposal referred to "Member of Academia Sinica" as "Yuan Lü of Academia Sinica," stating that "in order to complete the system of the Academy of Science and to award the national academy, it is proposed to set up Yuan Lü in Academia Sinica and amend any relevant laws and regulations of the council as appropriate." This proposal included the qualification, election, grouping, and disciplines of each group, task, term of office, the treatment of Yuan Lü, the main meaning of setting up Yuan Lü, the election and term of office of council members, the nature

[20]Zhongyang yanjiuyuan pingyihui dierjie disanci nianhui jilu中央研究院评议会第二届第三次年会纪录 (Minutes of the Third Annual Meeting of the Second Council of Academia Sinica). Nanjing: Zhongguo dier lishi danganguan中国第二历史档案馆(The Second Historical Archives of China), Total No.393, File No. 1557; Zhongyang yanjiuyuan pingyihui dierjie disanci nianhui ti'an中央研究院评议会第二届第三次年会提案 (Proposals of the Third Annual Meeting of the Second Council of Academia Sinica). Nanjing: Zhongguo dier lishi danganguan中国第二历史档案馆 (The Second Historical Archives of China), Total No. 393, File No. 1565.

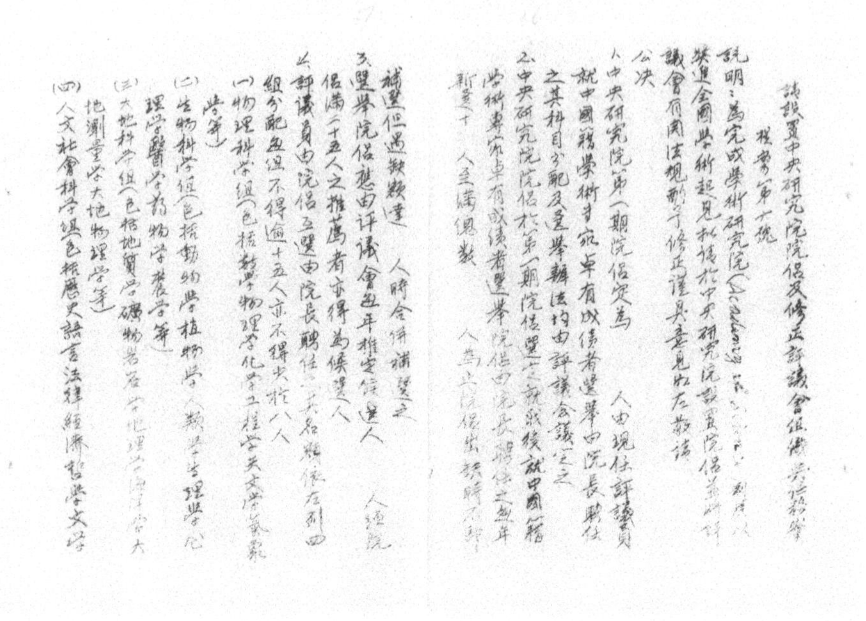

Fig. 3.4 The Proposal "Please Set up Yuan Lü (院侣) of Academia Sinica and Amend the Organization and Tasks of the Council" (part)

and task of the council, etc. Some of them referred to the academician system as commonly used in the world, but still not fully in line with the world. The most obvious one was the stipulation on the qualification of Yuan Lü, which required that the candidates be "Chinese academic experts and outstanding achievers." By "outstanding achievements," they meant outstanding academic research achievements or outstanding academic leadership achievements. First of all, this was because Article 5 of *The Organization Act of Academia Sinica*, which was amended and passed at the Fourth General Meeting of the annual meeting, stipulated that members of Academia Sinica should be "elected on the basis of outstanding academic achievements in the whole country." Among them, "achievement" referred to academic research achievements and academic leadership achievements.[21] The amendment referred to this proposal. Secondly, Adam Pen-Tung Sah, who jointly put forth the proposal with Wong Wen-hao, divided the members into the groups of "contributors to science" and "influencers of science" at the first examination committee meeting of the annual meeting.[22]

---

[21]Zhongyang yanjiuyuan pingyihui dierjie disanci nianhui jilu中央研究院评议会第二届第三次年会纪录 (Minutes of the Third Annual Meeting of the Second Council of Academia Sinica). Nanjing: Zhongguo dier lishi danganguan中国第二历史档案馆(The Second Historical Archives of China), Total No. 393, File No. 1557.

[22]Zhongyang yanjiuyuan pingyihui dierjie disici nianhui jilu中央研究院评议会第二届第四次年会纪录 (Minutes of the Fourth Annual Meeting of the Second Council of Academia Sinica).

At the same time, some aspects of this proposal were drawn up on separately without referring to the academician system that prevailed in the world. As stipulated in Article 11 of the proposal, the director-general and the directors of the institutes were ex officio Yuan Lü. Since academicians were generally elected internationally, there had been no such thing as ex officio Yuan Lü. Even at the beginning of the establishment of the Royal Society of London for the Improving of Natural Knowledge, aristocrats enjoyed the benefits of joining the society, they also had to join through election. Article 13 stipulated that "Yuan Lü may receive special research fees on a monthly basis"—this was also the case, because academicians generally did not have any special economic treatment outside of their functions and duties worldwide.

At the Third General Meeting, after discussing, the participating council members unanimously approved of the establishment of the "Member" in Academia Sinica. Thus, the title of "Member" was decided. Hu Shih proposed the titile "Yuan Yuan"; Hu Hsen-hsu suggested "Yuan Ru(院儒)" and "Yuan Fu(院傅)." Later, Hu Shih proposed adoping "Yuan Shi," "Xue Zheng(学正)," or "Xue Shi(学士)." Both Woo Y. H. and Chou Keng-Shang advocated for the use of "Yuan Yuan"; Wong Wen-hao suggested using "Yuan Shi" (meaning "academicians"), "Xue Shi," "Yuan Zheng," or "Xue Zheng." Subsequently, ChienShung-shu suggested adopting "Hui Yuan(会员)" or "Hui Yuan of the Academic Association(xue shu hui学术会) of Academia Sinica." However, Chu Chia-Hua thought that there should not be any other title under Academia Sinica, so he did not agree to the addition of the "Academic Association." Due to different opinions, a voting was adopted in the end. The results of the first vote were: nine votes for "Yuan Shi," eight votes for "Yuan Yuan," five votes for "Xue Zheng," three votes for "Hui Yuan," four votes for "Xue Shi," and one vote for "Yuan Zheng." Since then, participating council members voted for the second time for "Yuan Shi," "Yuan Yuan," and "Xue Zheng," which all ranked in the top three. The results were: thirteen votes for "Yuan Shi," nine votes for "Yuan Yuan," and five votes for "Xue Zheng." The Council therefore decided to adopt the title "Yuan Shi" when referring to "Member."[23]

After the title of "Member" was determined, the participating Council members discussed the quota, General Assembly of Members, and composing of the first batch of members of Academia Sinica. Hu Shih suggested 60 members for the first session, Fu Ssu-Nien suggested 80, and Wong Wen-hao suggested 100; he argued that the General Assembly of Members of Academia Sinica must be held in the capital city. Adam Pen-Tung Sah advocated that "the Director-General and the directors of all institutes should be selected among the members as fully as possible." This opinion is basically reiterated the Article 11of his and Wong Wen-hao's proposal "Please Set up Yuan Lü (院侣) of Academia Sinica and Amend the Organization and Tasks of the

Nanjing: Zhongguo dier lishi danganguan中国第二历史档案馆(The Second Historical Archives of China), Total No. 393, File No. 1558.

[23]Zhongyang yanjiuyuan pingyihui dierjie disici nianhui jilu中央研究院评议会第二届第四次年会纪录 (Minutes of the Fourth Annual Meeting of the Second Council of Academia Sinica). Nanjing: Zhongguo dier lishi danganguan中国第二历史档案馆 (The Second Historical Archives of China), Total No. 393, File No. 1558.

Council." Chu Chia-Hua believed that members were not necessarily the directors of each institute; "otherwise, they are too aged to be chosen." Therefore, Adam Pen-Tung Sah also suggested that "if the director-general and the directors of all institutes must be members of Academia Sinica, there must be a time limit as well." Subsequently, Articles 11 & 12 of this proposal, put forth by Chu Chia-Hua, were discussed by the group.[24]

On the morning of October 23, the first examination committee held a review conference; at the meeting, the examination committee discussed *The Organization Act of Academia Sinica* and *The Regulations of the Council of Academia Sinica.* They proposed to elect 80 to100 members of Academia Sinica for the first term and 15 members every year thereafter. Council members were to be selected by members of Academia Sinica.[25] In the afternoon, the Fourth General Meeting of this annual meeting discussed the review report of the first examination committee. The committee's review comments were: to review the proposal of "Please Stipulate the Name, Establishment, Election, and Related Matters of Member of Academia Sinica," and the proposal of "Please Set up Yuan Lü (院侣) of Academia Sinica and Amend the Organization and Tasks of the Council" jointly, and "Amend *The Organization Act of Academia Sinica* and *The Regulations of the Council of Academia Sinica.*" The meeting amended and approved the two regulations according to the review comment.[26]

The amended and approved *The Organization Act of Academia Sinica* added new regulations on Academia Sinica's member qualifications, quota, elections, terms of office, functions and powers, and grouping and disciplines of each group, and revised regulations of the council and its honorary members.[27] Article 5 stipulated the qualification of members, which read as follows: "Academia Sinica shall have a number of members, who shall be elected according to one of the following qualifications, for outstanding academic achievements throughout the country. First, they should have special works, inventions, or contribution to academic research. Second, they should have more than five years' experience in leading or presiding over academic institutions with outstanding achievements.[28] These two member qualifications were

[24]Zhongyang yanjiuyuan pingyihui dierjie disici nianhui jilu中央研究院评议会第二届第四次年会纪录 (Minutes of the Fourth Annual Meeting of the Second Council of Academia Sinica). Nanjing: Zhongguo dier lishi danganguan中国第二历史档案馆 (The Second Historical Archives of China), Total No. 393, File No. 1558.

[25]Chu Coching竺可桢 (2006). *Chu Coching quanji*竺可桢全集 (The Complete Works of Coching Chu), Vol. 10. Shanghai: Shanghai keji jiaoyu chubanshe上海科技教育出版社. 234.

[26]Zhongyang yanjiuyuan pingyihui dierjie disanci nianhui jilu中央研究院评议会第二届第三次年会纪录 (Minutes of the Third Annual Meeting of the Second Council of Academia Sinica). Nanjing: Zhongguo dier lishi danganguan中国第二历史档案馆 (The Second Historical Archives of China), Total No. 393, File No. 1557.

[27]Zhongyang yanjiuyuan pingyihui dierjie disanci nianhui jilu中央研究院评议会第二届第三次年会纪录 (Minutes of the Third Annual Meeting of the Second Council of Academia Sinica). Nanjing: Zhongguo dier lishi danganguan中国第二历史档案馆 (The Second Historical Archives of China), Total No. 393, File No. 1557.

[28]Zhongyang yanjiuyuan pingyihui dierjie disanci nianhui jilu中央研究院评议会第二届第三次年会纪录 (Minutes of the Third Annual Meeting of the Second Council of Academia Sinica).

basically transferred from the qualifications of the council members of Academia Sinica.[29]

At that time, other countries in the world had generally regarded those with outstanding academic achievements and contributions as academicians. In addition to these, the council had also regarded the achievements of leadership and presiding over academic institutions as Academia Sinica's member qualifications, which was obviously not fully in line with the rest of the world. This was a result of the particularity of China's modern scientific development. In the west, there was a certain number of academic elites before the modern scientific system appeared; after the establishment of the Royal Society of London and the Académie des Sciences, some of them became members or academicians. Although it also absorbed many non-scientific members many years after the establishment of the Royal Society, in order to obtain sufficient funds and develop further, by the end of the nineteenth century this situation had completely changed. First, there was either a group of returned students or a few scientific pioneers who did not have the experience of studying abroad, and through their establishment of academic institutions, the modern scientific system was able to emerge in China. This was reflected in the birth of some academic institutions; most of these scientific pioneers had since become leaders of these institutions. It was only after training and having students studying abroad that a group of scientists emerged in the international academic circle. However, many leaders of these institutions had not attained important academic achievements in China, mainly because they had devoted their energy to academic undertakings in the golden period of their lives. However, they did make important contributions to the development of modern science in China. To a certain extent, the transfer of the qualifications of council members by the Council of Academia Sinica resulted from the consideration of this situation.

Judging by Article 11 of the amended *The Organization Act of Academia Sinica*, the council was retained. In response, the council stated clearly: "The establishment of the council is of great importance. There are more than 100 members of Academia Sinica and it is not easy to hold regular meetings. Hence, the responsibility of contacting and guiding the academic circles of the whole country must be confined to a selected, smaller organization or it cannot push forward. Therefore, the original system of the council will be retained."[30]

Nanjing: Zhongguo dier lishi danganguan中国第二历史档案馆 (The Second Historical Archives of China), Total No. 393, File No. 1557.

[29]The only difference is that the first qualification of members of Academia Sinica added the word "contribution" to the first qualification of the council members: the former "there should be special works, inventions or contributions for the academic research", the latter "there should be special works or inventions for the academic research".

[30]Zhongyang yanjiuyuan zuzhifa pingyihui tiaoli pingyihui guicheng xianxing fagui zhangcheng mulu deng中央研究院组织法、评议会条例、评议会规程、现行法规章程目录等 (The Organization Act, the Regulations of the Council, the Aritcles for the Council, the Catalogue of Current Regulations and Articles of Academia Sinica, etc.). Nanjing: Zhongguo dier lishi danganguan中国第二历史档案馆 (The Second Historical Archives of China), Total No. 393, File No. 2539.

One of the major differences between the amended *The Regulations of the Council of Academia Sinica* passed by this meeting and the original regulations was that Article 4 of the original regulations, "election of honorary members of Academia Sinica" had been deleted. In addition, Article 8 had been completely removed and the original Articles 1, 2, 3, and 6 were slightly revised. These amendments "all came from the supplement of members of Academia Sinica in *The Organization Act of Academia Sinica* and there were no other principles."[31]

3.2 The Legislation of Election Law: The Formulation of Regulations for the Election of Members

On January 27, 1947, Academia Sinica submitted the amended draft of *The Organization Act of Academia Sinica* and *The Regulations of the Council of Academia Sinica*, which were approved at the fourth meeting of the council's Third Annual Meeting, to the National Government and to Legislative Yuan for deliberation. After that, it was revised and published by the National Government on March 13.[32] On March 15, Academia Sinica held a symposium of Nanjing and Shanghai council members. Chu Chia-Hua, Wong Wen-hao, Adam Pen-Tung Sah, Fu Ssu-Nien, Woo Y. H., Ling Hung-hsun, Hsieh Chia-yung, Loo Tsung-Le, Li Chi, Wang Chia-Chi, Jaw Jeou-Jang, Woo Sho-Chow, John Lee, Mao T. E., and Hu Shih, who was in Nanjing, attended the meeting. The meeting was chaired by Wong Wen-hao, Hu Shih, and Chu Chia-Hua in turns. The main topic was to discuss the drafting of *The Regulations for the Election of Members of Academia Sinica* and *The Regulations for the General Assembly of Members of Academia Sinica* (Figs. 3.5 and 3.6).

Before the conference, Adam Pen-Tung Sah had already drawn up a draft of each of the two regulations. Fu Ssu-Nien read Adam Pen-Tung Sah's draft once and "added his own opinions and also drew up the draft for reference."[33] Fu Ssu-Nien and Adam Pen-Tung Sah's drafts were read out loud at the talks. Later, everyone

---

[31]Zhongyang yanjiuyuan zuzhifa pingyihui tiaoli pingyihui guicheng xianxing fagui zhangcheng mulu deng中央研究院组织法、评议会条例、评议会规程、现行法规章程目录等 (The Organization Act, the Regulations of the Council, the Aritcles for the Council, the Catalogue of Current Regulations and Articles of Academia Sinica, etc.). Nanjing: Zhongguo dier lishi danganguan中国第二历史档案馆 (The Second Historical Archives of China), Total No. 393, File No. 2539.

[32]Zhongyang yanjiuyuan yuanshi xuanju guicheng ji jinghu pingyiyuan tanhuahui jilu中央研究院院士选举规程及京沪评议员谈话会纪录 (*The Regulations for the Election of Members of Academia Sinica* and Minutes of the Symposiums of the Council Members in Nanjing and Shanghai). Nanjing: Zhongguo dier lishi danganguan中国第二历史档案馆 (The Second Historical Archives of China), Total No. 393, File No. 1034.

[33]Zhongyang yanjiuyuan yuanshi xuanju guicheng ji jinghu pingyiyuan tanhuahui jilu中央研究院院士选举规程及京沪评议员谈话会纪录(*The Regulations for the Election of Members of Academia Sinica* and Minutes of the Symposiums of the Council Members in Nanjing and Shanghai). Nanjing: Zhongguodierlishidanganguan中国第二历史档案馆(The Second Historical Archives of China), Total No. 393, File No. 1034.

Fig. 3.5 Fu Ssu-Nien, director of the Institute of History and Philology of Academia Sinica

compared the points of attention in the two drafts in the member election as the basis for drafting, and agreed on the following three points:

(a) The generation of the member candidates is important, so nomination and examination procedures should be carefully stipulated.
(b) Preliminary work for the election of members is extremely difficult and important, and must be organized by an organization. The "election committee" drawn up in the two drafts can be renamed "the Preparatory Committee for the Election of Members" and its functions and powers will be clearly defined.
(c) There are various difficulties involved in holding a General Assembly of Members of Academia Sinica, such as the huge amount of funds required, the difficulty in preparations, the difficulty in setting a quorum due to the large number of participants, etc. However, there should be meetings for existing members of Academia Sinica, such as reading papers and electing members of Academia Sinica, so it is necessary to hold a plenary meeting. Once the state is stable, the above difficulties can be easily solved. Therefore, how to respect the system, consider the actual situation, and set the rules for the General Assembly

Fig. 3.6 Adam Pen-Tung Sah, director-general of Academia Sinica

of Members of Academia Sinica should be carefully considered and decided upon."[34]

Finally, the meeting decided to set up a group to draft *The Regulations for the Election of Members of Academia Sinica* and *The Regulations for the General Assembly of Members of Academia Sinica*, and held a plenary meeting on March 26 to review them. Fu Ssu-Nien, Wong Wen-hao, Adam Pen-Tung Sah, Hu Shih, Mao T. E., Woo Y. H., and Li Chi were selected as team members, with Adam Pen-Tung Sah as the convener.[35]

On March 17, the group held a meeting in the conference room of the general office of Academia Sinica; Wong Wen-hao was unable to attend the meeting because he was in Shanghai. Hu Shih and Adam Pen-Tung Sah presided over the meeting. At the meeting, Fu Ssu-Nien put forth a draft regulation for the election of members of Academia Sinica. Hu Shih spoke highly of Fu Ssu-Nien's draft; he wrote in his diary on the same day: "the second session of the symposium of the Council of Academia

[34]Zhongyang yanjiuyuan yuanshi xuanju guicheng ji jinghu pingyiyuan tanhuahui jilu中央研究院院士选举规程及京沪评议员谈话会纪录 (*The Regulations for the Election of Members of Academia Sinica* and Minutes of the Symposiums of the Council Members in Nanjing and Shanghai). Nanjing: Zhongguo dier lishi danganguan中国第二历史档案馆 (The Second Historical Archives of China), Total No. 393, File No. 1034.

[35]Zhongyang yanjiuyuan yuanshi xuanju guicheng ji jinghu pingyiyuan tanhuahui jilu中央研究院院士选举规程及京沪评议员谈话会纪录 (*The Regulations for the Election of Members of Academia Sinica* and Minutes of the Symposiums of the Council Members in Nanjing and Shanghai). Nanjing: Zhongguo dier lishi danganguan中国第二历史档案馆 (The Second Historical Archives of China), Total No. 393, File No. 1034.

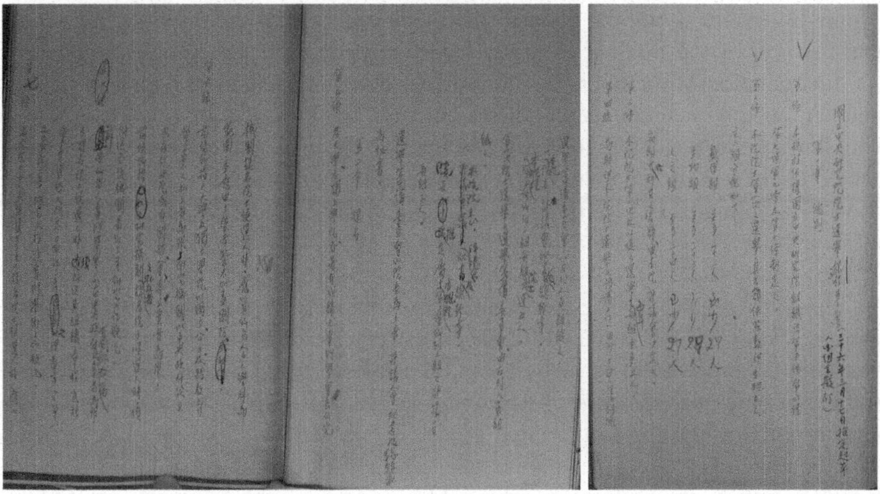

Fig. 3.7 Original manuscript of *The Regulations for the Election of Members of Academia Sinica*, drafted by Fu Ssu-Nien and revised by the group on March 17, 1947 (part)

Sinica will continue to discuss the election law for members of Academia Sinica. The second draft proposed by Mengzhen[36] is very good; that is, it serves as a good basis for discussion (Fig. 3.7)."[37]

Fu Ssu-Nien's draft was divided into 5 chapters—"General Provisions," "Nomination," "Qualification Examinations of member Candidates," "Election of Members," and "Supplementary Provisions"—with 19 articles (the draft regulations were uniformly sorted)."General Provisions" contained four articles, stating that the regulations were formulated in accordance with Articles 5, 6, 7, 9, and 10 of the revised and published *The Organization Act of Academia Sinica*. They stipulated the maximum number of members of Academia Sinica in the three groups of mathematics and physics, biology, and humanities (33 in mathematics and physics, biology and 34 in humanities), the maximum number of members in the second and subsequent elections (5), and the composition of the Preparatory Committee for the Election of Members. "Nomination" contained five articles, which stipulated the qualifications of nominating institutions (universities, independent colleges, specialized institutes, and research institutions with outstanding achievements), requirements and points of attention for nomination, and requirements for members or council members of Academia Sinica to nominate member candidates. "A Qualification Examination of Member Candidates" contained five articles, which stipulated the procedure for the qualification examination, the announcement content and time of the list of member candidates, and the treatment of opinions from all walks of life. "Election of Members" contained three articles, which stipulated the election method of members

36 Mengzhen(孟真) is Fu Ssu-Nien's style name.

37 Hu Shih胡适, Ji Xianlin季羡林(Eds.) (2003).*Hu Shih Quanji*胡适全集 (Complete Works of Hu Shih), Vol. 33. Hefei: Anhui jiaoyu chubanshe安徽教育出版社. 626.

(group discussion should be adopted first, then each group should vote; the person who had 4/5 of the number of votes from the people present should be elected. The first election of members would not be carried out among groups); the list of members would be announced after the election, the President of Academia Sinica would inform the candidate to start his/her post, and the candidate who had not been elected could remain a candidate for future elections ofmembers. "Supplementary Provisions" contained two articles which stipulated the conditions for amending and implementing *The Regulations for the Election of Members of Academia Sinica*.[38]

The group would "discuss and amend Fu Ssu-Nien's draft chapter by chapter and article by article."[39] The amendment was divided into the above five chapters, totaling 17 articles.[40] The main differences between the final amendment and Fu Ssu-Nien's draft were as follows: Article 2 of Fu Ssu-Nien's draft read, "For the election of the first batch of members of Academia Sinica, the number of members should be distributed according to the three groups of mathematics and physics, biology, and humanities as follows: the mathematics and physics group should have no more than 33 members, the biology group should have no more than 33 members, and the humanities group should have no more than 34 members. The number of subjects in each group should be determined by the Council." On this basis, the amendment increased the minimum number of members in the three groups of mathematics and physics, biology, and humanities (all 27). Article 13 of the group's amendment stipulated that the council should exert the corresponding functions and powers of the General Assembly of Members of Academia Sinica during the election of the first batch of members of Academia Sinica. The corresponding provision on this point had not been proposed in Fu Ssu-Nien's draft. Article 15 of Fu Ssu-Nien's draft read: "After receiving the full list and the documents submitted by the Preparatory Committee for the Election of Members, each candidate should be discussed in groups at the General Assembly of Members of Academia Sinica. Members should be elected in groups, with 4/5 of the attendees indicating a winning vote. If the first vote is insufficient, a second vote might be adopted, and the quorum for election shall be as the first vote. In the election of the first batch of members of Academia Sinica, the Council should exert the functions and powers of this article and should not be

[38]Zhongyang yanjiuyuan yuanshi xuanju guicheng ji jinghu pingyiyuan tanhuahui jilu中央研究院院士选举规程及京沪评议员谈话会纪录 (*The Regulations for the Election of Members of Academia Sinica* and Minutes of the Symposiums of the Council Members in Nanjing and Shanghai). Nanjing: Zhongguo dier lishi danganguan中国第二历史档案馆 (The Second Historical Archives of China), Total No. 393, File No. 1034.

[39]Zhongyang yanjiuyuan yuanshi xuanju guicheng ji jinghu pingyiyuan tanhuahui jilu中央研究院院士选举规程及京沪评议员谈话会纪录 (*The Regulations for the Election of Members of Academia Sinica* and Minutes of the Symposiums of the Council Members in Nanjing and Shanghai). Nanjing: Zhongguo dier lishi danganguan中国第二历史档案馆 (The Second Historical Archives of China), Total No. 393, File No. 1034.

[40]Zhongyang yanjiuyuan yuanshi xuanju guicheng ji jinghu pingyiyuan tanhuahui jilu中央研究院院士选举规程及京沪评议员谈话会纪录 (*The Regulations for the Election of Members of Academia Sinica* and Minutes of the Symposiums of the Council Members in Nanjing and Shanghai). Nanjing: Zhongguo dier lishi danganguan中国第二历史档案馆 (The Second Historical Archives of China), Total No. 393, File No. 1034.

divided into groups." The group deleted the phrase "if the first vote is insufficient, a second vote might be adopted, and the quorum for election shall be as the first vote," and changed the phrase "not be divided into groups" in the last sentence of this article to "but not be divided into groups." The group deleted all the contents of Article 17 of Fu Ssu-Nien's draft, namely: "those who have been included in the full list of member candidates and have not been elected as members shall remain candidates for future member elections"; Article 19 of Fu Ssu-Nien's draft read: "This regulation shall come into force after it is decided by the conference of the council (submitted to the National Government for approval and filing)." The group deleted the words "submitted to the National Government for approval and filing" from this article. In addition, the group revised the wording of some regulations in Fu Ssu-Nien's draft and combined Articles 6 & 7 into one article, but most of the amendments did not change from the original intent of Fu Ssu-Nien's draft. On the whole, the amendment did not differ greatly from the basic principles and ideas of Fu Ssu-Nien's draft. Not only were the election procedures stipulated by both of them consistent, but their requirements for each link of the election procedure were basically the same.

When discussing Article 15 of Fu Ssu-Nien's draft, the panel discussed this point in great detail: "if the first vote is insufficient, a second vote might be adopted, and the quorum for election shall be as the first vote." The group originally intended to change it to "if the number of elected members in each group is less than the minimum number stipulated in Article 2 of this regulation, the group may continue to vote until the full number is reached." In the end, the group decided to "delete all of them, and this point will be noticed by the Council (for the first time) or the General Assembly of Members of Academia Sinica, and it shall be decided at that time."[41] As expected by Fu Ssu-Nien, in March 1948, when the Council held a final election for the first batch of members of Academia Sinica at its Fifth Annual Meeting, it happened that the number of candidates elected in the first ballot was less than the minimum of 80. As a result, the Council decided to continue voting and held four by-elections.[42]

In addition to the text of this draft, Fu Ssu-Nien had attached an explanation on the groups of mathematics and physics, biology, and humanities and the distribution of the number of people in each discipline. After discussion in the group meeting, Fu Ssu-Nien's plan for the distribution of the number of people in different branches and disciplines was not adopted. The meeting held that "the division of each group and the allocation of seats in each group shall be decided by the council once each election

---

[41]Zhongyang yanjiuyuan yuanshi xuanju guicheng ji jinghu pingyiyuan tanhuahui jilu中央研究院院士选举规程及京沪评议员谈话会纪录 (*The Regulations for the Election of Members of Academia Sinica* and Minutes of the Symposiums of the Council Members in Nanjing and Shanghai). Nanjing: Zhongguo dier lishi danganguan中国第二历史档案馆 (The Second Historical Archives of China), Total No. 393, File No. 1034.

[42]Zhongyang yanjiuyuan yuanshi houxuanren mingdan ji xuanjupiao中央研究院院士候选人名单及选举票 (The List of Candidates of Members of Academia Sinica and the Ballots). Nanjing: Zhongguo dier lishi danganguan中国第二历史档案馆 (The Second Historical Archives of China), Total No. 393, File No. 1620.

is held." Since the development level of each discipline in each group may not be the same, if one discipline was especially developed, the number of seats in that group may be increased or decreased as appropriate. However, the members of the group agreed with Fu Ssu-Nien's comments on the discipline of Chinese literature in this draft: "the selection is limited to those who lead the literary trend and study classics, not including literary and artistic writers."[43] In addition, Fu Ssu-Nien believed, in this draft, that when the council handled the election of the first batch of members of Academia Sinica, "there must be a general statement" and wrote the following general statement:

> The election of members of Academia Sinica must be in accordance with Article 5 of *The Organization Act of Academia Sinica*. It must be based on one of the two qualifications listed, and it must also be a person with outstanding academic achievements in the national academic circle. If those who have taught for many years, have cultivated many talents, and have indeed achieved outstanding results in the academic circle are not fully met the two qualifications, it is not appropriate to have them listed. Those who have indeed achieved outstanding results in setting up specialized undertakings, but haven't academic achievement, are also not suitable for inclusion. Academia Sinica is a research institution, not an educational or enterprise institution, and its scope is limited, so it does not completely list first-class talents. In addition, the scope of Academia Sinica is interpreted as science and organized learning according to Article 2 of *The Organization Act of Academia Sinica*. Therefore, all of those who engaged in fine arts and literary creation cannot be included in the election. It's not biased.[44]

On March 26, 1947, Academia Sinica held a second meeting with the council members from Nanjing and Shanghai in the auditorium of the general office. First of all, Adam Pen-Tung Sah, as the convener of the drafting group of seven council members of *The Regulations for the Election of Members of Academia Sinica* and *The Regulations for the General Assembly of Members of Academia Sinica*, reported on the drafting process of the drafting on March 17. Later, the symposium would discuss *The Draft Regulations for the Election of of Members of Academia Sinica* revised and approved by the group one by one, and then revise and approve them. As "the matters of the *General Assembly of Members* are of great importance," it was decided that when soliciting comments from all council members in correspondence, Articles 13 & 14 of the draft, passed through this amendment, should be stated as assumptions. If the council members had any comments, they could put forth discussions and make amendments until "the assembly is held." In addition, it was decided that the draft nomination list for member candidates, proposed by Fu Ssu-Nien, would be revised by Wong Wen-hao and Adam Pen-Tung Sah and attached to *The Draft Regulations*

[43]Zhongyang yanjiuyuan yuanshi xuanju guicheng ji jinghu pingyiyuan tanhuahui jilu中央研究院院士选举规程及京沪评议员谈话会纪录 (*The Regulations for the Election of Members of Academia Sinica* and Minutes of the Symposiums of the Council Members in Nanjing and Shanghai). Nanjing: Zhongguo dier lishi danganguan中国第二历史档案馆 (The Second Historical Archives of China), Total No. 393, File No.1034.

[44]Zhongyang yanjiuyuan yuanshi xuanju guicheng ji jinghu pingyiyuan tanhuahui jilu中央研究院院士选举规程及京沪评议员谈话会纪录 (*The Regulations for the Election of Members of Academia Sinica* and Minutes of the Symposiums of the Council Members in Nanjing and Shanghai). Nanjing: Zhongguo dier lishi danganguan中国第二历史档案馆 (The Second Historical Archives of China), Total No. 393, File No. 1034.

for the Election of Members of Academia Sinica, and comments from all the council members would be solicited through correspondence.[45] Among them, the framework of *The Draft Regulations for the Election of Members of Academia Sinica* and the basic principles and ideas of most of its regulations had no essential differences from Fu Ssu-Nien's draft.

When the nomination of the first batch of Members started in May 1947, the Preparatory Committee for the Election of the First Batch of Members of Academia Sinica sent notices of nomination, as well as *The Regulationss for the Election of Members of Academia Sinica* and *The Organization Act of Academia Sinica*, to academic institutions such as National Tsing Hua University and National University of Peking.[46] Among them, *The Regulations for the Election of Members of Academia Sinica* was the same as its draft revised and approved by this meeting. From the articles of *The Regulations for the Election of Members of Academia Sinica*, we can see the philosophy of the council in organizing the election of the first batch of members of Academia Sinica.

First, the election was entirely presided over and operated by the council. This was mainly reflected in the fact that those with election qualifications—members of the Preparatory Committee for the Election of the First Batch of Members of Academia Sinica who were responsible for preparatory work such as nomination and preliminary examination of member candidates as well as those deciding the list of member candidates and those with the right to amend *The Regulations for the Election of Members of Academia Sinica* were all held by the council or its members. These were clearly stipulated in Articles 4, 9, 10, 13, 14, and 16 of the amended *Regulations for the Election of Members of Academia Sinica* and passed by the second meeting of council members from Nanjing and Shanghai on March 26, 1947. For example, Article 4 stipulated the composition of the preparatory committee for the election of members: "First, the president of Academia Sinica, the honorary secretary of the council, and the director-general. Second, the council members who were elected by the council in the three groups listed in Article 2 of this regulation. Five in each group. The preparatory committee for the election of members was chaired by the president of Academia Sinica, with the honorary secretary of the council and the director-general as secretaries." At that time, there had not been a new president for Academia

[45]Zhongyang yanjiuyuan yuanshi xuanju guicheng ji jinghu pingyiyuan tanhuahui jilu中央研究院院士选举规程及京沪评议员谈话会纪录(*The Regulations for the Election of Members of Academia Sinica* and Minutes of the Symposiums of the Council Members in Nanjing and Shanghai). Nanjing: Zhongguo dier lishi danganguan中国第二历史档案馆 (The Second Historical Archives of China), Total No. 393, File No. 1034.

[46]Zhongyang yanjiuyuan yuanshi xuanju guicheng ji yuanshi houxuanren timingbiao中央研究院院士选举规程及院士候选人提名表 (The Regulations for the Election of Members of Academia Sinica and Nomination Form for Member Candidates). Beijing: Beijing daxue danganguan北京大学档案馆 (Archives of Peking University), File No. BD1947058; Zhongyang yanjiuyuan juban diyici yuanshi xuanju tuijian houxanren mingdan ji youguan guizhang banfa he laiwang wenshu中央研究院举办第一次院士选举推荐候选人名单及有关规章办法和来往文书(List of Recommended Candidates for the Election of the First Batch of Members of Academia Sinica and Relevant Regulations, Methods and Correspondence). Beijing: Qinghua daxue danganguan清华大学档案馆 (Archives of Tsinghua University), Total No.1, Cat. No. 4–2, File No. 191.

Sinica. The chairperson of the preparatory committee for the election of members was Acting President Chu Chia-Hua, and the secretaries were the Honorary Secretary Wong Wen-hao and the Director-General Adam Pen-Tung Sah, respectively. Article 9 stipulated that the preparatory committee should be responsible for the preliminary examination of the candidates' qualifications. Article 10 stipulated that the council should be responsible for deciding the official list of member candidates. Article 13 stipulated that the council should be responsible for examining criticisms from all sides when the election of first batch of members of Academia Sinica. Article 14 stipulated that the first batch of members of Academia Sinica should be elected by the council. Article 16, which belonged to Chap. 5, "Supplementary Provisions," stipulated that the council had the right to amend *The Regulations for the Election of Members of Academia Sinica*: "*The Regulations* may be amended by a decision of 2/3 in the council, on the proposal of more than five members of the council or on the proposal of more than 10members of Academia Sinica." The members of the Second Council of Academia Sinica had a high academic status and expectation in domestic academia. The authority of the election that presided over by them and was run by them independently was said to be beyond doubt.

Second, democratic elections. This can be seen from the articles of Chaps. 3, 4, and 5 of *The Regulations for the Election of Members of Academia Sinica*. These three chapters were mainly about the regulations on the qualification examinations of member candidates, the election, and how to deal with problems arising from the election. The qualification examinations of member candidates must first go through a preliminary examination by the preparatory committee for the election of members before submitting the preliminary list. And the preparatory committee submits the foundation for each member candidate's qualifications and any relevant documents to the council. When reviewing the preliminary list, the council should examine the candidates in groups, according to their respective categories, and discuss them throughout the whole experience. Only when more than half of the members present passed the examination could they then be included in the official list of member candidates. If 10 council members made a written proposal, the nominees not included in the preliminary list could be included in the final list, if approved by more than half of the members present. The Council would then make an official list and announce it immediately. The announcement should indicate the foundation on which each candidate met a certain qualification and should notify the council members. The announcement period should last at least 4 months. In the meantime, people from all walks of life were allowed to bring their criticisms and suggestions to the preparatory committee. After reviewing the criticisms and suggestions, the preparatory committee would make a selection of candidates for reference and submit it to the council together with the full volume. During the election, the council would first examine the official list of member candidates and documents submitted by the preparatory committee in groups, according to each group that member candidates were belonged to. Finally, the plenary meeting of the council would vote, and the member candidate must obtain at least 4/5 votes of the total number of people present before being elected. If more than five council members proposed to amend *The Regulations for the Election of Members of Academia Sinica* after the election had gone wrong, the amendment must be approved by at least 2/3 of the assessors.

According to the articles of *The Regulations for the Election of Members of Academia Sinica*, the entire procedure for the election ofmembers drawn up by the council can also be seen as: defining the qualification of member candidates, allocating the three groups of mathematics and physics, biology, and humanities → establishing a preparatory committee for the election of members → nominating qualified universities, independent colleges, specialized institutes, and research institutions and sending the nomination form and related documents to the preparatory committee → conducting a preliminary examination of the nominations by the preparatory committee → deciding the preliminary examination results through the council → announcing the list of member candidates → listening to opinions from all sides → the preparatory committee preparing a section of the named criticisms that could be used for reference and selection to submit to the council→ holding a formal election of members (group discussion first, then impromptu collective voting by the council members) → the President announcing the list of the elected members and notifying the elected members to start their work → revising *The Regulations for the Election of Members of Academia Sinica* if the election had problems. The whole procedure included not only preparatory work for the election and all the important links of the election, but also specific measures to deal with problems after the election, which was relatively strict.

3.3 The Legislation of Assembly: The Formulation of Regulations for the General Assembly of Members

From the aforementioned, it can be seen that on March 15, 1947, the main goal of the symposium of the council members from Nanjing and Shanghai was to discuss the draft of *The Regulations for the Election of Members of Academia Sinica* and *The Regulations for the General Assembly of Members of Academia Sinica.* Adam Pen-Tung Sah had already drawn up a draft of each of the two regulations before the conference. Based on the decision of setting up a group of seven council members at the symposium to "prudently draft" the two regulations, the drafts had not been generally accepted.[47] Moreover, due to many difficulties in holding the General Assembly of Members of Academia Sinica, such as "the huge amount of funds required, difficulty in preparation, and difficulty in setting a quorum due to the large number of people attending the meeting," some council members did not approve of holding the assembly. Therefore, this symposium would discuss whether the assembly should

[47]Zhongyang yanjiuyuan yuanshi xuanju guicheng ji jinghu pingyiyuan tanhuahui jilu中央研究院院士选举规程及京沪评议员谈话会纪录(*The Regulations for the Election of Members of Academia Sinica* and Minutes of the Symposiums of the Council Members in Nanjing and Shanghai). Nanjing: Zhongguo dier lishi danganguan中国第二历史档案馆 (The Second Historical Archives of China), Total No. 393, File No. 1034.

be held or not. They finally decided to hold the assembly, and stressed that "the assembly should be held since members of Academia Sinica did exist."[48]

On March 17, 1947, when the seven council members of the panel (Fu Ssu-Nien, Adam Pen-Tung Sah, Hu Shih,Mao T. E., Woo Y. H., and Li Chi) met, "whether the General Assembly of Members of Academia Sinica should be held or not was still the main point of discussion."[49] Not only that, but everyone was quite divided on how to hold the assembly. To sum up, there were five opinions: (1) In order to respect the system of setting up members of Academia Sinica, a assembly must be held. (2) It seemed that there was no need to hold a plenary meeting due to reasons such as large expenses, difficulty in handling, difficulty in entertaining, difficulty in meeting a quorum due to large population, etc. (3) Members of Academia Sinica could assemble in different districts; the reading of papers or elections could be held by district meetings. Finally, the opinions of all the district meetings would be integrated into the opinions of the meeting of all the members. (4) A plenary meeting should be held for the first time. In the future, it would be stipulated that the quorum for attending the plenary meeting would be small, not more than half, so that the quorum could be easily reached, and the plenary meeting would be held in the future. In case there was still not a quorum, voting by correspondence would be used to handle the election. (5) The member's assembly could be classified according to its items, the quorum of the assembly should be determined according to the nature of the items, and a proxy authorization or voting method might be adopted for certain items. After a discussion, it was decided to report the above opinions at the plenary session. In the end, "whether the General Assembly of Members of Academia Sinica will be held or not remained to be discussed," and the meeting decided that "*The Draft Regulations for the General Assembly of Members of Academia Sinica* will be drafted after discussion in the plenary."[50]

As the topic focused on *The Draft Regulations for the Election of Members of Academia Sinica*, the second symposium of the council members from Nanjing and Shanghai held by Academia Sinica on March 26 failed to fomulate *The Draft Regulations for the General Assembly of Members of Academia Sinica* on the agenda. In the following six months, the council would focus on the nomination of the first

---

[48]Zhongyang yanjiuyuan yuanshi xuanju guicheng ji jinghu pingyiyuan tanhuahui jilu中央研究院院士选举规程及京沪评议员谈话会纪录 (*The Regulations for the Election of Members of Academia Sinica* and Minutes of the Symposiums of the Council Members in Nanjing and Shanghai). Nanjing: Zhongguo dier lishi danganguan中国第二历史档案馆 (The Second Historical Archives of China), Total No. 393, File No. 1034.

[49]Zhongyang yanjiuyuan yuanshi xuanju guicheng ji jinghu pingyiyuan tanhuahui jilu中央研究院院士选举规程及京沪评议员谈话会纪录 (*The Regulations for the Election of Members of Academia Sinica* and Minutes of the Symposiums of the Council Members in Nanjing and Shanghai). Nanjing: Zhongguo dier lishi danganguan中国第二历史档案馆 (The Second Historical Archives of China), Total No. 393, File No. 1034.

[50]Zhongyang yanjiuyuan yuanshi xuanju guicheng ji jinghu pingyiyuan tanhuahui jilu中央研究院院士选举规程及京沪评议员谈话会纪录 (*The Regulations for the Election of Members of Academia Sinica* and Minutes of the Symposiums of the Council Members in Nanjing and Shanghai). Nanjing: Zhongguo dier lishi danganguan中国第二历史档案馆 (The Second Historical Archives of China), Total No. 393, File No. 1034.

batch of member candidates of Academia Sinica. It was not until the nomination was completed that the Second Council held its Fourth Annual Meeting in Nanjing on October 17, where it discussed the proposal of "Please Stipulate *The Regulations for the General Assembly of Members of Academia Sinica*." After discussion, it was decided that they would "authorize the council members in Nanjing to draw up a draft, send it to each of them for comments, and propose the next council for discussion."[51] As a result, the formulation of *The Regulations for the General Assembly of Members of Academia Sinica* was transferred to the council members in Nanjing. At this time, Fu Ssu-Nien, who played a key role in stipulating *The Regulations for the Election of Members of Academia Sinica*, went to the United States for medical treatment due to illness in June of that year. Because of this, he would no longer take part in concrete preparations for the academician system of Academia Sinica.

Judging from the situation after the meeting, the stipulation of *The Regulations for the General Assembly of Members of Academia Sinica* was a relatively slow progress. On February 16, 1948, due to the approaching election of members of Academia Sinica, *The Regulations for the General Assembly of Members of Academia Sinica* "needed to be stipulated urgently." Chu Chia-Hua and Wong Wen-hao invited the council members in Nanjing to attend the symposium at Academia Sinica on February 19. The invitation letter was accompanied by *The Draft Regulations for the General Assembly of Members of Academia Sinica*, drafted by Wong Wen-hao. On February 19, Chu Chia-Hua, Wong Wen-hao, Adam Pen-Tung Sah, John Lee, Woo Ting-liang, Li Chi, Ling Hung-hsun, and Hsieh Chia-yung attended the symposium, and Chu Chia-Hua presided over the meeting. At the meeting, Wong Wen-hao explained *The Draft Regulations for the General Assembly of Members of Academia Sinica* one article by one article, and said in particular: "The draft includes plans to divide the General Assembly of Members of Academia Sinica into two categories: one is the plenary meeting and the other is the regional meeting. Moreover, the General Assembly of Members of Academia Sinica should be closely linked with the council, so the council was designated as the permanent appraisal organ of the General Assembly of Members of Academia Sinica. No voting system shall be adopted for the regional meetings."[52] This classification of the General Assembly of Members of Academia Sinica might have been influenced by the method of Assembly of the Academicians of the National Academy of Sciences, United States. As Adam Pen-Tung Sah said at the symposium:

> I have recently received a report from the National Academy of Sciences of United States for the fiscal year 1944–1945. The academy meets every spring and autumn, once in Washington

---

[51]Zhongyang yanjiuyuan pingyihui dierjie disici nianhui jilu中央研究院评议会第二届第四次年会纪录 (Minutes of the Fourth Annual Meeting of the Second Council of Academia Sinica). Nanjing: Zhongguo dier lishi danganguan中国第二历史档案馆 (The Second Historical Archives of China), Total No. 393, File No. 1558.

[52]Zhongyang yanjiuyuan zai jing pingyiyuan niding yuanshi huiyi guicheng caoan tanhuahui jilu 中央研究院在京评议员拟订院士会议规程草案谈话会纪录 (Minutes of the Symposium of the Council Members in Nanjing to Draw up the Draft of Regulations for the General Assembly of Members of Academia Sinica). Nanjing: Zhongguo dier lishi danganguan中国第二历史档案馆 (The Second Historical Archives of China), Total No. 393, File No. 547.

and once elsewhere. Although the number of its members even reaches 450, the quorum for the meeting is only 20, which is not proportional to the total number of its members. It acts as a reference.[53]

The council members that attended the meeting had some opinions on this classification of the General Assembly of Members of Academia Sinica and the stipulation in Wong Wen-hao's draft that "themember's plenary meetings shall be held every three years." For example, Chu Chia-Hua pointed out: "If the plenary meeting of members is held every 3 years, that seems to be inconsistent with Article 6 of *The Organization Act of Academia Sinica* and Article 14 of Chapter 4 of *The Regulations for the Election of Members of Academia Sinica* that say every year thereafter."[54] He suggested that the "plenary meeting ofmembers should be held once a year, with a quorum of 1/3 of the total members. The method of grouping elections has yet to be studied. Can we make it flexible and not divide into groups?"[55] The contradiction pointed out by Chu Chia-Hua between Wong Wen-hao's draft and *The Organization Act of National Academia Sinica* and *The Regulations for the Election of Members of the National Academia Sinica* did exist. After discussing, the council adopted most of Chu Chia-Hua's suggestions and revised Wong Wen-hao's draft. After the amendment, the regulations in Articles 2, 4, and 6 of Wong Wen-hao's draft, concerning the regional meetings, and in Articles 8 to 10, concerning the specific operation of the member's functions and powers and the funds for member meetings, were deleted. In accordance to the functions and powers of members stipulated in Article 9 of *The Organization Act of Academia Sinica*, the four tasks of the General Assembly of Members were added after the amendment: "to elect members and honorary members," "to elect council members," "to discuss and stipulate national academic policies," and "to discuss matters entrusted by the government." At the same time, the article that "important academic papers must be read out loud during the General Assembly of Members" was added. This reflected the council's emphasis on the academic nature of the General Assembly of Members itself. The quorum for the General Assembly of Members was "half of all members."

---

[53]Zhongyang yanjiuyuan zai jing pingyiyuan niding yuanshi huiyi guicheng caoan tanhuahui jilu 中央研究院在京评议员拟订院士会议规程草案谈话会纪录 (Minutes of the Symposium of the Council Members in Nanjing to Draw up the Draft of Regulations for the General Assembly of Members of Academia Sinica). Nanjing: Zhongguo dier lishi danganguan中国第二历史档案馆 (The Second Historical Archives of China), Total No. 393, File No. 547.

[54]Article 14 of *The Regulations for the Election of Members of Academia Sinica* reads: "When the General Assembly of Members of Academia Sinica elects members, the list of candidates for members and the documents submitted by the Preparatory Committee for the Election shall be divided into groups to discuss each candidate; The candidate will be elected after getting the approval of 4/5 of the attendees in his or her group and is reported to the plenary meeting. " These cannot be achieved by district meetings.

[55]Zhongyang yanjiuyuan zai jing pingyiyuan niding yuanshi huiyi guicheng caoan tanhuahui jilu 中央研究院在京评议员拟订院士会议规程草案谈话会纪录 (Minutes of the Symposium of the Council Members in Nanjing to Draw up the Draft of Regulations for the General Assembly of Members of Academia Sinica). Nanjing: Zhongguo dier lishi danganguan中国第二历史档案馆 (The Second Historical Archives of China), Total No. 393, File No. 547.

After the meeting, on February 25, Chu Chia-Hua and Wong Wen-hao again solicited comments from the council members on the amendment, hoping to "respond before March 20 so as to bring together the deliberations to be held on March 25."[56] Before March 20, Li Shu-Hua wrote back "with great approval," and ChienShung-shu also said "with complete approval." Loo Tsung-Le agreed, but thought that " to discuss and stipulate national academic policies" seemed to be misunderstood as the possibility of ruling academia, and should be amended" and suggested that it be changed to "to discuss and stipulate the policy of national academic administration." Moreover, he pointed out: "academic research should be absolutely free and not restricted."[57] This reflected the fact that Loo Tsung-Le attached great importance to and advocated for academic freedom. However, the council did not accept Loo Tsung-Le's proposal, probably because most of the council members did not think this would cause such misunderstanding. On March 25, 1948, the Fifth Annual Meeting of the Second Council passed *The Draft Regulations for the General Assembly of Members of Academia Sinica*, which was revised and approved at the symposium of council members in Nanjing on February 19. At the same time, the meeting named it as *The Regulations for the General Assembly of Members of Academia Sinica*. The difference between *The Regulations* and *The Draft Regulations for the General Assembly of Members of Academia Sinica* was that Article 6 of the latter was changed to "1/3 of all members shall be present at the General Assembly of Members as a quorum." In addition, Article 7 of the latter was changed to: "the council shall be the permanent organ of the General Assembly of Members." Thus far, the council had completed primary legislative work on the academician system of Academia Sinica.

3.4 Summary

Although Academia Sinica failed to establish an academician system during the Anti-Japanese War, the brewing of the academician system by the Second Council made progress in that time. After the Anti-Japanese War, the Second Council focused on the establishment of an academician system in the Third Annual Meeting, and decided to build up members in Academia Sinica. Subsequently, the name of "Member" was determined; the quota of the first batch of members, the General Assembly of Members, the composition of members and other important issues were discussed,

56Zhongyang yanjiuyuan zai jing pingyiyuan niding yuanshi huiyi guicheng caoan tanhuahui jilu 中央研究院在京评议员拟订院士会议规程草案谈话会纪录 (Minutes of the Symposium of the Council Members in Nanjing to Draw up the Draft of Regulations for the General Assembly of Members of Academia Sinica). Nanjing: Zhongguo dier lishi danganguan中国第二历史档案馆 (The Second Historical Archives of China), Total No. 393, File No. 547.

57Zhongyang yanjiuyuan zai jing pingyiyuan niding yuanshi huiyi guicheng caoan tanhuahui jilu 中央研究院在京评议员拟订院士会议规程草案谈话会纪录 (Minutes of the Symposium of the Council Members in Nanjing to Draw up the Draft of Regulations for the General Assembly of Members of Academia Sinica). Nanjing: Zhongguo dier lishi danganguan中国第二历史档案馆 (The Second Historical Archives of China), Total No. 393, File No. 547.

and *The Organization Act of Academia Sinica* and the *The Regulations of the Council of Academia Sinica* were revised. After the meeting, the seven members panel stipulated *The Regulations for the Election of Members of Academia Sinica*, in which Fu Ssu-Nien played a key role. *The Regulations for the Election of Members of Academia Sinica* stipulated a relatively strict member election system and democratic election procedures that were completely presided over by the council and independently operated. The stipulation of the *The Regulations for the General Assembly of Members of Academia Sinica* was mainly the responsibility of the council members in Nanjing, and Wong Wen-hao had made important contributions to it. The legislative work, especially the stipulation of the last two aforementioned regulations, laid an important legal foundation for the establishment of the academician system of Academia Sinica.

Chapter 4
How the First Batch of Member Candidates Were Nominated and Their Qualification Examinations

After the council drew up *The Regulations for the Election of Members of Academia Sinica* in April 1947, they turned their focus to the election of the first batch of members of Academia Sinica. The selection of member candidates was critical to the election of this batch of members. Without a certain quantity of outstanding scholars as member candidates, it would be difficult for the council to continue selecting members of Academia Sinica that could generally be accepted by the academic community. The nomination and qualification examinations of member candidates were the two basic links in the generation of member candidates, which played a decisive role in determining whether there was a certain quantity of outstanding scholars as member candidates. On March 15, 1947, there was a symposium of council members from Nanjing and Shanghai. They collectively agreed that "The generation of the member candidates is important, so nomination and examination procedures should be carefully stipulated."[1] This chapter explores these processes, and how they worked with the first batch of member candidates of Academia Sinica. It analyzes the formal nomination losses and gains, and looks at the impact of social relation resources on academic evaluation of member candidates.

4.1 Formulating a Member Candidates Reference List

In order to elect the first batch of members of Academia Sinica as soon as possible, the council arranged "the Preparatory Committee for the Election of the First Batch of Members" (hereinafter referred to as "PCE"). This took place after the Second

[1]Zhongyang yanjiuyuan yuanshi xuanju guicheng ji jinghu pingyiyuan tanhuahui jilu中央研究院院士选举规程及京沪评议员谈话会纪录(*The Regulations for the Election of Members of Academia Sinica* and Minutes of the Symposiums of the Council Members in Nanjing and Shanghai). Nanjing: Zhongguo dier lishi danganguan中国第二历史档案馆(The Second Historical Archives of China), Total No. 393, File No. 1034.

J. Guo, *The Establishment and Reconstruction of the Academician System in China*,
https://doi.org/10.1007/978-981-15-7208-1_4

symposium of Nanjing and Shanghai council members, which was on March 26, 1947. According to Article 4 in *The Draft Regulations for the Election of Members of Academia Sinica* (revised and approved by the symposium), the PCE would include the president of Academia Sinica, the honorary secretary of the Council and the Director-General, and five members each from the mathematics and physics, biology, and humanities groups. The PCE would be chaired by the president of Academia Sinica, and the honorary secretary of council and the director-general would serve as secretaries. Chu Chia-Hua drew up a reference list of PCE members and solicited comments from the council members. The following 15 people were elected: The mathematics and physics group's Woo Y. H., Woo Sho-Chow, Hsieh Chia-yung, Ling Hung-hsun and Mao T. E.; the Biology group's Ping Chi, Wang Chia-Chi, Loo Tsung-Le, Robert K. S. Lim, Wang Ging-Hsi, and the Humanities group's Hu Shih, Fu Ssu-Nien, Wang Shih-Chieh, Tao L. K. and Li Chi.[2]

At 3 pm on May 9, 1947, the first PCE meeting was held in the Academia Sinica General Office. Chu Chia-Hua was the host, and Wong Wen-hao, Adam Pen-Tung Sah, Fu Ssu-Nien, Wang Chia-Chi, Woo Sho-Chow, Tao L. K., Robert K. S. Lim, Loo Tsung-Le, Li Chiand Mao T. E. attended the meeting. Members discussed which disciplines to include in each of the three groups, and they elected council members to draft a reference list of potential member candidates in various disciplines. They discussed the general situation of qualified members in various disciplines, and how to nominate further members. As a result, 26 disciplines were decided, i.e. mathematics and physics group: mathematics, physics (including geographical physics), chemistry, geology (including paleontology), natural geography, astronomy, meteorology, engineering (including civil engineering, machinery, electrical machinery, chemical engineering, mining and metallurgy); biological group: zoology, botany, physical anthropology, psychology, physiology, medical science (including pathological science, physiological science, pharmacology, medical science, veterinary science), agronomy; humanities group: philosophy, Chinese literature, history, linguistics, archaeology and art history, ethnonymics, human geography, jurisprudence, politics, economics, sociology. A total of 24 members were elected. The specific candidates and the disciplines they would be responsible for were: Chiang Li-Fu (mathematics), Woo Y. H. and Li Shu-Hua (physics), Hou T. P. and Woo Sho-Chow (chemistry), Wong Wen-hao and Hsieh Chia-yung (geology, natural geography), Jaw Jeou-Jang (meteorology), Mao T. E. and Chow Jen and Adam Pen-Tung Sah (engineering), Ping Chi and Wang Chia-Chi (zoology), Chien Shung-shu and Loo Tsung-Le (botany), Woo Ting-liang (physical anthropology), Wang Ging-Hsi (psychology), Robert K. S. Lim (physiology), Robert K. S. Lim and Wang Ging-Hsi (medical science), Tai Fon-Land and Loo Tsung-Le (agronomy), Hu Shih and Fu Ssu-Nien and Li Chi (philosophy, Chinese literature, history, linguistics, archaeology and art history, ethnonymics, human geography), Tao L. K. and Wang Shih-Chieh (jurisprudence,

[2]Zhongyang yanjiuyuan pingyihui diyici yuanshi xuanju choubei weiyuanhui zuzhi ji huiyi jilu中央研究院评议会第一次院士选举筹备委员会组织及会议记录(The Organization and Minutes of the Preparatory Committee for the Election of the First Batch of Members of Academia Sinica). Nanjing: Zhongguo dier lishi danganguan中国第二历史档案馆(The Second Historical Archives of China), Total No. 393, File No. 134.

politics, economics, sociology).[3] The PCE had no limit on the number of nominees, but it required the heads of various disciplines to convene their council members, and after consideration, draw up a list of nominees and submit it within four weeks.[4]

The decision of the PCE to draw up a reference list of candidates before an election was based on the experiences of the second term council members. Later, the council members responsible for drafting the reference list became separate from those elected by the PCE. For example, Ling Hung-hsun was not elected, but he was still responsible for drafting the reference list of engineering candidates.

In early June 1947, the council members finished drawing up their reference list. Academia Sinica's General Office analyzed the general situation of their domestic experts and "obtained a lot of material."[5] The PCE drew up a reference list of member candidates covering almost all disciplines, and recommended them for election. They also drew up a quota for some disciplines, as is shown in Table 4.1.

Table 4.1 shows that there were 181 people in total, and 78 people were elected. Therefore, the election had a success rate of 43.1%. Hsu Pao-Lu, Li Hsien-wen and Liang Ssu-yung were not included in the reference list, but they were elected later. The reference list formulated by the PCE can therefore be said to have been relatively strict. Before the council elected the first batch of members of Academia Sinica, the PCE had first established a better framework of reference for them. The PCE proposed that the quota for most disciplines be different to the actual number of members elected. However, these quotas only served as a reference point for the council, and had no decisive influence on the election results themselves.

4.2 Twists and Turns During the Official Nomination

According to *The Regulations for the Election of Members of Academia Sinica*, there were two ways to nominate the first batch of member candidates of Academia Sinica. The first was via their universities, independent colleges, specialized institutes and research institutes, as a result of their outstanding achievements. Alternatively,

[3] Zhongyang yanjiuyuan pingyihui diyici yuanshi xuanju choubei weiyuanhui zuzhi ji huiyi jilu中央研究院评议会第一次院士选举筹备委员会组织及会议记录(The Organization and Minutes of the Preparatory Committee for the Election of the First Batch of Members of Academia Sinica). Nanjing: Zhongguo dier lishi danganguan中国第二历史档案馆(The Second Historical Archives of China), Total No. 393, File No. 134.

[4] Zhongyang yanjiuyuan pingyihui diyici yuanshi xuanju choubei weiyuanhui zuzhi ji huiyi jilu中央研究院评议会第一次院士选举筹备委员会组织及会议记录(The Organization and Minutes of the Preparatory Committee for the Election of the First Batch of Members of Academia Sinica). Nanjing: Zhongguo dier lishi danganguan中国第二历史档案馆(The Second Historical Archives of China), Total No. 393, File No. 134.

[5] Zhongyang yanjiuyuan pingyihui diyici yuanshi xuanju choubei weiyuanhui zuzhi ji huiyi jilu中央研究院评议会第一次院士选举筹备委员会组织及会议记录(The Organization and Minutes of the Preparatory Committee for the Election of the First Batch of Members of Academia Sinica). Nanjing: Zhongguo dier lishi danganguan中国第二历史档案馆(The Second Historical Archives of China), Total No. 393, File No. 134.

Table 4.1 A reference list of member candidates and the quota for some disciplines proposed by the PCE

Discipline	Proposed quota	Number of proposed nominees	Name list
Mathematics	5	8	Chiang Li-Fu, Shiing-Shen Chern, Hua Loo-Keng, Kiang Tsai-han, Su Buchin, Chen Kien Kwong, Hiong King-lai, Sun Guangyuan[6]
Astronomy	1	3	Yu C. S., Chang Yu-Che, Zhang Yun
Physics	6	9	Li Shu-Hua, Yeh Chi-Sun, Yao Y. T. , Woo Y. H. , Ny Tsai-ze, Chao Chung-Yoa, Wu Ta-You, Chou Pei-yuan, Wang Ganchang
Chemistry	6	10	Wu Hsien, Chuang Chang-Kong, Peter P. T. Sah, Tseng Chao-lun, Huang Minlon, Woo Sho-Chow, Huang T. C., Li F. H., Kao C. H., Sun Xuewu
Geology	8	14	Young Chung-Chien, Si Xingjian, Hsieh Chia-yung, Huang T. K., Lee J. S. , Sun Y. C., Wong Wen-hao, Chu Chia-Hua, Zhang Hongzhao, Yin T. H., Meng H. M., Wang C. C., Sun C. C. and Nan Y. T.
Meteorology	3	3	Chu Coching, Jaw Jeou-Jang, John Lee
Engineering	6	10	Hou T. P., Ling Hung-hsun, Mao T. E., Chow Jen, Wang C. Y., Adam Pen-Tung Sah, Zhi Bingyuan, Tsai Fang-yin, Yap Chu-Phay, Cheng Xiaogang
Zoology	6	10	Ping Chi,Cheney Zhen, Hu Jingfu, Wangj Jiaji, Wu Hsien-Wen, Tung Ti-Cho, Pai Sitsan, Zhu Xi, Liu C. C., Tan Chia-chen

(continued)

Table 4.1 (continued)

Discipline	Proposed quota	Number of proposed nominees	Name list
Botany	6	10	Chang Chin-yueh, Chien Shung-shu, Hu Hsen-hsu, Loo Tsung-Le, Liou Tchen-Ngo, Tai Fon-Land, Pei Chien, Jao Chin-chih, Li Tsi-tung, Yin Hung-chang
Medical Science		31	(biochemistry, nutriology) Wu Hsien, Lu Gwei-djen, Wang Ying-Lai; (pathology) Hu C. H., Kang Xirong, Liang Boqiang; (bacteriology) Xie Shaowen, Tang F. F., Yan Chunhui, Liu Weitong; (internal medicine) Liu S. H., Chung H. L., Chang Hsiao-chien, Li Zongen, Qi Shounan, Dong Chenglang; (surgery) Guan Songtao, James K. Shen, Zhang Xianlin, Huang C. S.; (radiology) Hsieh C. K. and Rong Dushan; (public health) Yuan I-Chin, Chen Zhiqian, Liu Ruiheng; (nnatomy) Shen Shizhang and Ma Wen-chao; (obstetrics and gynecology) Sun Keji and Li Siwei; (ophthalmology) Lin Wenbing, Guo Bingkuan
Pharmacology		4	Chen Ko Kuei, Chou T. Q., Zhang Changshao, Zhang Yi
Physical Anthropology	1	1	Woo Ting-liang
Psychology	3	3	Wang Ging-Hsi, Lu Zhiwei, Tang Yueh
Physiology		7	Robert K. S. Lim, Feng Teh-Pei, Zhang Xijun, Tsai Chiao, Tang P. S., Wang Shijun, Xu Fengyan

(continued)

Table 4.1 (continued)

Discipline	Proposed quota	Number of proposed nominees	Name list
Agronomy	5	8	Shen Zonghan, Zhao Lianfang, Yu T. F., Teng S. C., Liu Chung-lo, Chen Rong, Tsai Pang-hwa, Wei Jingchao
Philosophy		4	Tang Yung-tung, Feng You-lan, Chin Yuch-Lin, Lü Cheng
Chinese literature		7	Woo Tsin-Hang, Zhang Yuanji, Hu Shih, Yang Shu-Tah, Yu Cha-Shih, Zhu Qifeng, Shen Jianshi
History		6	Chen Yin-k'o, Chen Yuan, Ku Chieh-kan, Fu Ssu-Nien, Liu Yei-Tsen and Chiang Ting-fu
Linguistics		4	Chao Yuen Ren, Li Fang Kuei, Luo Changpei, Wang Li
Archaeology and Art History		5	Li Chi, Tung Tso-pin, Kuo Mo-jo, Liang Ssu-cheng, Xiang Da
Jurisprodence	4	6	Wang Chung Hwei, Wang Shih-Chieh, Yen Shu-Tang, Li Haopei, Mei Ju-Ao, Kuo Yun-kuan
Politics	3	5	Chou Keng-Shang, Tsien T. S., Hsiao Kung-chuan, Chang C. F. and Hsu Shuhsi
Economics	6	9	Chen Zong, Ma Yin-chu, Ho L., Yang Duan Liu, Fong H. D., Wu Chi-Yuen, Yang Hsi-meng, Wu Bao San, Lieu D. K.
Sociology	3	5	Tao L. K. , Sun Pen-wen, Chen Ta, Pan Kuang-tan, Wu Ching-chao

Sources Hu Shih Loo Tsung-Le dengren suo ti de yuanshi houxuanren mingdan ji youguan hanjian胡适、罗宗洛等人所提的院士候选人名单及有关函件(The Nomination Lists of Member Candidates Proposed by Hu Shih, Loo Tsung-Le, etc. and Related Correspondence). Nanjing: Zhongguo dier lishi danganguan中国第二历史档案馆(The Second Historical Archives of China), Total No. 393, File No. 1615

five members from each of the three groups (mathematics and physics, biology and humanities) could nominate a candidate jointly. The latter method was rarely used, although Mao T. E., along with four members of the mathematics and physics group, nominated Liang Pu.[7] Chu Chia-Hua explained the reason this method was rare as follows: "According to *The Regulations for the Election of Members Academia Sinica*, five council members could jointly nominate member candidates of Academia Sinica, but all members deemed that the election of the first batch of members should be carried out by the council. Therefore, they were very cautious and did not exercise this power."[8]

Candidates were officially nominated in mid-May, 1947. On May 16, the PCE sent *The Announcement of Academia Sinica's PCE* to the *Central Daily* in Nanjing, Chongqing, Hankou, Guangzhou and other places. It also sent an announcement to the *Ta Kung Pao* in Shanghai and Tianjin, and to the *North China Daily* and *Zhongxing Daily* in Peking. The announcement stated that between 80 and 100 candidates were up for election, in the mathematics and physics, biology and humanities groups. The deadline for selecting them was July 20, 1947.[9] Around the same time, the PCE sent a number of documents to National University of Peking, National Tsing Hua University, National Wu-han University, and other academic institutions. Documents included *The Announcement*, *The Organization Act of Academia Sinica*, *The Regulations for the Election of Members of Academia Sinica* and *The Nomination Form for Member Candidates.*

By July 20, more than 50 institutions, including Northeastern University, Yunnan University, Shandong University, Anhui University, Hunan University, National Jinan University, Yingshi University, National Pei-yang University, Cheeloo University, Amoy University, Yenching University, Tangshan Engineering College, the Geological Society of China, Chinese Physical Society, National Agricultural Research Bureau, and Fan Memorial Institute of Biology, had all completed their nominations (Fig. 4.1).

The formal nominations were not plain sailing. National University of Peking, National Tsing Hua University, National Wu-han University, National Central University and many other important universities had many well-known scholars,

---

[7]Guoli zhongyang yanjiuyuan diyici yuanshi xuanju houxuanren timingce国立中央研究院第一次院士选举候选人提名册(Nomination Lists of the Member Candidates for the Election of the First Batch of Members of Academia Sinica). Nanjing: Zhongguo dier lishi danganguan中国第二历史档案馆(The Second Historical Archives of China), Total No. 393, File No. 1597.

[8]Chu Chia-Hua朱家骅. Guoli zhongyang yanjiuyuan pingyihui dierjie diwuci nianhui kaihuici(minguo 37 nian 3 yue 25 ri) 国立中央研究院评议会第二届第五次年会开会词(民国三十七年三月二十五日)(Opening Speech at the Fifth Annual Meeting of the Second Council of Academia Sinica, March 25, 37, Republic of China). In Wang Yujun王聿均 and Sun Bin孙斌(Eds.) (1977). *Chu Chia-Hua xiansheng yanlun ji*朱家骅先生言论集(Speech by Mr. Chu Chia-Hua). Taipei: Zhongyang yanjiuyuan jinaishi yanjiusuo中央研究院近代史研究所(Institute of Modern History of Academia Sinica).90.

[9]Zhongyang yanjiuyuan guanyu yuanshi houxuanren timing wenti de tonggao中央研究院关于院士候选人提名问题的通告(The Announcement of Nomination problems of Member Candidates of Academia Sinica). Nanjing: Zhongguo dier lishi danganguan中国第二历史档案馆(The Second Historical Archives of China), Total No. 393, File No. 1617.

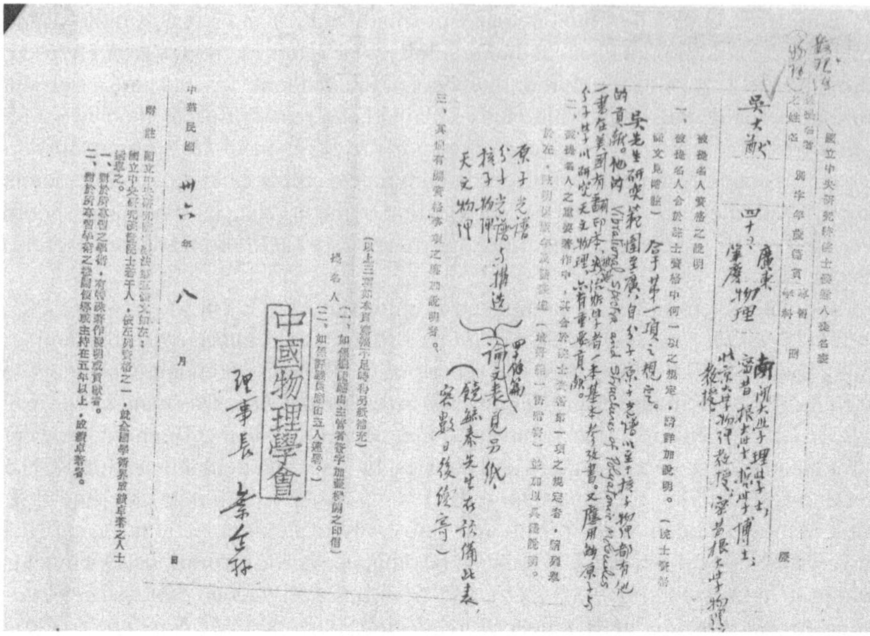

Fig. 4.1 The Nomination Form for Wu Ta-You proposed by Chinese Physical Society

and these were the main candidates that were put forward. However, after Adam Pen-Tung Sah sent a letter to the presidents of these universities on July 12, 1947, to urge them to submit candidates, the PCE did not receive the nomination lists of these universities until July 17.[10] Therefore, Adam Pen-Tung Sah wrote to Wong Wen-hao that day, to discuss a solution.

> Since our preparation work for the election of a first batch of members, the council has received more than 50 nomination forms. However, National University of Peking, National Tsing Hua University, National Wu-han University, and the National Central University have not yet submitted nominations. I have already sent letters urging them to do so. The difficulty lies in the fact that many nominees' previous works have either become scattered or lost during the war, and they are not easy to recover. However, the deadline for nomination is July 20, as was shown in the previous notice, which is not far away… by that time, if these famous universities and colleges can submit their nominations, that is, if they can send them within the time limit, there will be no additional examination difficulties. Otherwise, the deadline will potentially have to be extended by one month, until August 20. Is this feasible?"[11]

[10] Dongbei daxue deng yuanxiao hansong yuanshi houxuanren timingdan ji youguan hanjian东北大学等院校函送院士候选人提名单及有关函件(The Nomination Lists of Member Candidates Sent by Northeastern University and Other Universities and Colleges and Related Correspondence). Nanjing: Zhongguo dier lishi danganguan中国第二历史档案馆(The Second Historical Archives of China), Total No. 393, File No. 1613.

[11] Zhongyang yanjiuyuan guanyu yuanshi houxuanren timing wenti de tonggao中央研究院关于院士候选人提名问题的通告(The Announcement of Nomination problems of Member Candidates

Wong Wen-hao was deeply touched by Adam Pen-Tung Sah's words, since his own works had been destroyed during the war. "Every time I hear of publications being lost during the Anti-Japanese War, I have felt it. Before, I (humbly calling himself as "younger brother" in his response) had printed more than 300 scientific papers. Most of them were stored in National Geological Survey of China, however, only 20 reprints of them have been found. It pains me that the number is so far from the original figure; it is inconceivable. It is very difficult for our nominees to send their entire works to Academia Sinica."[12] He "fully agreed" with Adam Pen-Tung Sah's extension plan, and suggested that "private correspondence should be sent to urge schools including National University of Peking, National Tsing Hua University, National Wu-han University, and the National Central University." If they could not send their relevant works, he proposed "they may also list the titles of their journals, and the year and place of publication." He said that he would personally send a letter to Hu Shih, the President of National University of Peking.[13] Both National Wu-han University and the National Central University sent late nominations, and so the PCE decided to extend the deadline to August 20 "for the sake of actual difficulties".[14]

By August 20, 67 institutions had submitted candidates for nomination to Academia Sinica. Besides these, councilmember Mao T. E. had nominated Liang Pu, and Tian Sanqing and Sun Haozhan had both nominated themselves. The National University of Peking nominated people from all over the country and proved an important influence on the PCE. Hu Shih made it clear prior to the National University of Peking nominating candidates that although "the number of members in each discipline is still uncertain and there is no clear stipulation as to whether the nomination scope should be limited to those who currently hold or have held positions in the university, I think the nominations from the National University of Peking should be nationwide, in order to include those who have contributed the most to various disciplines as members."[15] Later, he disclosed his plan to Adam Pen-Tung Sah.

---

of Academia Sinica). Nanjing: Zhongguo dier lishi danganguan中国第二历史档案馆(The Second Historical Archives of China), Total No. 393, File No. 1617.

[12]Zhongyang yanjiuyuan guanyu yuanshi houxuanren timing wenti de tonggao中央研究院关于院士候选人提名问题的通告(The Announcement of Nomination problems of Member Candidates of Academia Sinica). Nanjing: Zhongguo dier lishi danganguan中国第二历史档案馆(The Second Historical Archives of China), Total No. 393, File No. 1617.

[13]Zhongyang yanjiuyuan guanyu yuanshi houxuanren timing wenti de tonggao中央研究院关于院士候选人提名问题的通告(The Announcement of Nomination problems of Member Candidates of Academia Sinica). Nanjing: Zhongguo dier lishi danganguan中国第二历史档案馆(The Second Historical Archives of China), Total No. 393, File No. 1617.

[14]Zhongyang yanjiuyuan guanyu yuanshi houxuanren timing wenti de tonggao中央研究院关于院士候选人提名问题的通告(The Announcement of Nomination problems of Member Candidates of Academia Sinica). Nanjing: Zhongguo dier lishi danganguan中国第二历史档案馆(The Second Historical Archives of China), Total No. 393, File No. 1617.

[15]Dongbei daxue deng yuanxiao hansong yuanshi houxuanren timingdan ji youguan hanjian东北大学等院校函送院士候选人提名单及有关函件(The Nomination Lists of Member Candidates Sent by Northeastern University and Other Universities and Colleges and Related Correspondence). Nanjing: Zhongguo dier lishi danganguan中国第二历史档案馆(The Second Historical Archives of China), Total No. 393, File No. 1613.

"The National University of Peking is now nominating members. In principle, these nominations should not be limited to university personnel, but should include people from all over the country. They can be mailed on July 15; is that too late?[16] Adam Pen-Tung Sah expressed deep admiration for the National University of Peking's move, and he wrote back the following: "This matter was actually discussed in the preparatory meeting. The electoral regulations do not restrict units to nominating their own personnel; however, as there is no clear guideline on this, we will have to review the nominating unit's opinions. I deeply admire National University of Peking's 'magnanimity' for putting forward this opinion."[17] We can assert from this that Adam Pen-Tung Sah hoped academic institutions would be able to nominate people from all over the country.

Hu Shih hoped that National Tsing Hua University would follow suit and also nominate people from outside the institution. On July 4, 1947, National Tsing Hua University's Professor Shiing-Shen Chern, who was acting director of Institute of Mathematics of Academia Sinica, conveyed Hu Shih's thoughts to the president of the university, Mei Yi-chi. "I've just met Shizhi (Hu Shih) and talked about the election of members at Academia Sinica. Mr Hu wants National Tsing Hua University and the National University of Peking to select candidates from across the whole country, and nominate these in future elections so the elections are fair. I hereby present my report. I hope that you may talk with Mr Hu in person at your convenience."[18] But this did not work. Mei Yi-chi thought: "There won't be any omissions if universities, research institutes and societies all recommend their own candidates. Besides, the affairs of the election should be decided by council members. If the nomination scope is enlarged beyond our university, there may be unintentional omissions, and these would arouse the grievances of our counterparts!"[19] Because of this concern,

[16] Hu Shih Loo Tsung-Le dengren suo ti de yuanshi houxuanren mingdan ji youguan hanjian胡适、罗宗洛等人所提的院士候选人名单及有关函件(The Nomination Lists of Member Candidates Proposed by Hu Shih, Loo Tsung-Le, etc. and Related Correspondence). Nanjing: Zhongguo dier lishi danganguan中国第二历史档案馆(The Second Historical Archives of China), Total No. 393, File No. 1615.

[17] Hu Shih Loo Tsung-Le dengren suo ti de yuanshi houxuanren mingdan ji youguan hanjian胡适、罗宗洛等人所提的院士候选人名单及有关函件(The Nomination Lists of Member Candidates Proposed by Hu Shih, Loo Tsung-Le, etc. and Related Correspondence). Nanjing: Zhongguo dier lishi danganguan中国第二历史档案馆(The Second Historical Archives of China), Total No. 393, File No. 1615.

[18] Zhongyang yanjiuyuan juban diyici yuanshi xuanju tuijian houxanren mingdan ji youguan guizhang banfa he laiwang wenshu中央研究院举办第一次院士选举推荐候选人名单及有关规章办法和来往文书(List of Recommended Candidates for the Election of the First Batch of Members of Academia Sinica and Relevant Regulations, Methods and Correspondence). Beijing: Qinghua daxue danganguan清华大学档案馆(Archives of Tsinghua University), Total No. 1, Cat. No. 4-2, File No. 191.

[19] Zhongyang yanjiuyuan juban diyici yuanshi xuanju tuijian houxanren mingdan ji youguan guizhang banfa he laiwang wenshu中央研究院举办第一次院士选举推荐候选人名单及有关规章办法和来往文书(List of Recommended Candidates for the Election of the First Batch of Members of Academia Sinica and Relevant Regulations, Methods and Correspondence). Beijing: Qinghua daxue danganguan清华大学档案馆(Archives of Tsinghua University), Total No. 1, Cat. No. 4-2, File No. 191.

National Tsing Hua University initially only nominated its own personnel. However, things changed after the National University of Peking's nominations. The university not only explained the contributions of its nominees and the reasons for nominating them, but it also attached the university's official seal to each nomination form. Hu Shih executed this in person, and this greatly touched Adam Pen-Tung Sah. Shortly after the nomination deadline was postponed, Adam Pen-Tung Sah wrote a very sincere, personal letter to Mei Yi-chi. He said:

> There are many reasons why Academia Sinica is holding a member election. As a student of yours, I respect you and your reputation, but I do hope that every university can come to include scholars from outside the institute, so that the council has an appropriate and sufficient mix of candidates on which to vote. The list sent by Tsinghua only includes professors working for Tsinghua. Now that the deadline for nomination has been extended to August 20, I wonder if National Tsing Hua University can make another list and include scholars from all over the country. The National University of Peking is nominating scholars across the nation, as this will greatly help with the future voting of all members (especially "the laymen"). When National Tsing Hua University is willing to reconsider its nominations, we will try our best to give help, for example, if there is insufficient reference material (such as the university being unable to collect papers).[20]

Adam Pen-Tung Sah added in his letter: "There are a number of references (in the nomination paperwork), such as the title of papers, etc. The relevant columns can be left blank for time being, and we can figure it out later. However, I still hope that each list can be stamped with the university's seal, as well as your own."[21] In his early years, Adam Pen-Tung Sah studied at National Tsing Hua University and he taught in its Department of Physics prior to 1937. He had a special relationship with National Tsing Hua University. Mr Sa intended to impress Mei Yi-chi with these special policies.

On July 25, 1947, the day after Adam Pen-Tung Sah wrote this letter, Shiing-Shen Chern spoke with the head of Academia Sinica and made a joint proposal to Mei Yi-chi with Chou Pei-yuan, a physics professor of National Tsing Hua University, regarding the nomination of people outside of the school.

> During a recent talk with the head of Academia Sinica in Nanjing, we learned that all academic institutions have been requested to submit a list of member candidates. We hope that they will draw up a list of scholars from all over the country, which will serve as a reference point for future elections. So far, only one school, National University of Peking,

[20]Zhongyang yanjiuyuan juban diyici yuanshi xuanju tuijian houxanren mingdan ji youguan guizhang banfa he laiwang wenshu中央研究院举办第一次院士选举推荐候选人名单及有关规章办法和来往文书(List of Recommended Candidates for the Election of the First Batch of Members of Academia Sinica and Relevant Regulations, Methods and Correspondence). Beijing: Qinghua daxue danganguan清华大学档案馆(Archives of Tsinghua University), Total No. 1, Cat. No. 4-2, File No. 191.

[21]Zhongyang yanjiuyuan juban diyici yuanshi xuanju tuijian houxanren mingdan ji youguan guizhang banfa he laiwang wenshu中央研究院举办第一次院士选举推荐候选人名单及有关规章办法和来往文书(List of Recommended Candidates for the Election of the First Batch of Members of Academia Sinica and Relevant Regulations, Methods and Correspondence). Beijing: Qinghua daxue danganguan清华大学档案馆(Archives of Tsinghua University), Total No. 1, Cat. No. 4-2, File No. 191.

> intends to do this. It has nominated more than 100 people. Each list of nominees should indicate the important contributions of that candidate and the reasons for nominating them. It is extremely important that this list then be sealed by the university and signed in person by Mr Shizhi. The election of members of Academia Sinica is a national academic event. Our university has always attached great importance to research and is a leading Chinese academic institution. The right to nominate members should not be taken for granted. Recently, the deadline for nominating candidates has been extended to August 20. Can you draw up a separate list of candidates who are not working at Tsinghua to account for the shortcomings of the previous list? You only need to list the nominees, the reasons for nominating them and their important contributions. The works of the nominees can be validated at this end, so there is no need to search for them. Do let me know whether this is feasible.[22]

Following Adam Pen-Tung Sah's call, Shiing-Shen Chern and Chou Pei-yuan jointly made their proposal, and Mei Yi-chi finally decided to nominate people outside of National Tsing Hua University. On July 30, 1947, Mei Yi-chi wrote back to Adam Pen-Tung Sah, saying, "I am writing to you today to nominate several colleagues in each group who are not at the university. I am sure it can be sent mid-August, since there is no limit for the number of nominees. Although we do not know those scholars very well, we will recommend them based on the best of what we know."[23] On August 18, 1947, National Tsing Hua University nominated 92 people to the PCE from outside of the school. It had previously nominated 43 members of university staff. The latter included a number of scholars who had once worked at National Tsing Hua University, such as Peter P. T. Sah and Jaw Jeou-Jang. Mei Yi-chi also spoke to Adam Pen-Tung Sah about his views on the nomination of "foreign researchers".

> Currently, there are many famous overseas researchers, who have spent a short period of time overseas. There is no problem with this; but is it appropriate to elect those who went abroad in their early years and have accomplished much in their studies but have never returned/perhaps will not return to China? My opinion is this: such scholars should be taken into consideration once they return to China (there will always be opportunities to nominate them in the future). With regards those who refuse to return to China, although they have attained a high academic status, they should only be regarded as experts. They do not deserve the opportunity to become members based on their spirit and outlook on life. Even if they were elected, they would not benefit Academia Sinica and China's academic circles.[24]

---

[22]Zhongyang yanjiuyuan juban diyici yuanshi xuanju tuijian houxanren mingdan ji youguan guizhang banfa he laiwang wenshu中央研究院举办第一次院士选举推荐候选人名单及有关规章办法和来往文书(List of Recommended Candidates for the Election of the First Batch of Members of Academia Sinica and Relevant Regulations, Methods and Correspondence). Beijing: Qinghua daxue danganguan清华大学档案馆(Archives of Tsinghua University), Total No. 1, Cat. No. 4-2, File No. 191.

[23]Zhongyang yanjiuyuan juban diyici yuanshi xuanju tuijian houxanren mingdan ji youguan guizhang banfa he laiwang wenshu中央研究院举办第一次院士选举推荐候选人名单及有关规章办法和来往文书(List of Recommended Candidates for the Election of the First Batch of Members of Academia Sinica and Relevant Regulations, Methods and Correspondence). Beijing: Qinghua daxue danganguan清华大学档案馆(Archives of Tsinghua University), Total No. 1, Cat. No. 4-2, File No. 191.

[24]Zhongyang yanjiuyuan juban diyici yuanshi xuanju tuijian houxanren mingdan ji youguan guizhang banfa he laiwang wenshu中央研究院举办第一次院士选举推荐候选人名单及有关规章办法和来往文书(List of Recommended Candidates for the Election of the First Batch of Members of Academia Sinica and Relevant Regulations, Methods and Correspondence). Beijing:

Adam Pen-Tung Sah agreed with this. In his reply, he said: "Some people expressed views in the preparatory meeting that 'since they have abandoned the motherland, the motherland should naturally give them up, too'. I also felt that this was the right thing to do."[25] Wang Chia-Chi and Ping Chi shared the same opinion. They were in charge of drawing up the list of candidates within the zoology discipline, and said: "Some scholars have made a lot of contributions abroad, but they have not yet returned to China, or only returned briefly, for one or two years. Therefore, for the time being, we say that they have not made any contributions."[26] Although Mei Yi-chi strongly opposed the nomination of anyone who had not returned to China for an extended period of time, it is significant to note that Tsien H. S., an aviation engineer and professor at the Massachusetts Institute of Technology, was among the 92 people nominated by National Tsing Hua University. Mr Qian had not been back to China for a long period of time. Mei Yi-chi failed to impose his will there, and his nomination may have been based on other professors at the university.

From the existing archives, we can surmise that the nomination procedure at National Tsing Hua University was more democratic than at many other institutions, but also much stricter. Professors and associate professors from various departments were divided into three groups: mathematics and physics, biology and humanities. This was in line with the member groups stipulated in *The Organization Act of Academia Sinica* and *The Regulations for the Election of Members of Academia Sinica*. Members of the university's professors' association would vote on each other's preliminary candidates and send these by registered post. The school council would then draw up a list based on the results of the nominations (candidates could only qualify if they were nominated by five or more people) and send the list of these candidates to the PCE of Academia Sinica. The maximum number of nominees from each member of the professors' association was 11 and the minimum number was three.[27] The school council was strict in insisting that candidates needed five members

Qinghua daxue danganguan清华大学档案馆(Archives of Tsinghua University), Total No. 1, Cat. No. 4-2, File No. 191.

[25]Zhongyang yanjiuyuan juban diyici yuanshi xuanju tuijian houxanren mingdan ji youguan guizhang banfa he laiwang wenshu中央研究院举办第一次院士选举推荐候选人名单及有关规章办法和来往文书(List of Recommended Candidates for the Election of the First Batch of Members of Academia Sinica and Relevant Regulations, Methods and Correspondence). Beijing: Qinghua daxue danganguan清华大学档案馆(Archives of Tsinghua University), Total No. 1, Cat. No. 4-2, File No. 191.

[26]Hu Shih Loo Tsung-Le dengren suo ti de yuanshi houxuanren mingdan ji youguan hanjian胡适、罗宗洛等人所提的院士候选人名单及有关函件(The Nomination Lists of Member Candidates Proposed by Hu Shih, Loo Tsung-Le, etc. and Related Correspondence). Nanjing: Zhongguo dier lishi danganguan中国第二历史档案馆(The Second Historical Archives of China), Total No. 393, File No. 1615.

[27]Zhongyang yanjiuyuan juban diyici yuanshi xuanju tuijian houxanren mingdan ji youguan guizhang banfa he laiwang wenshu中央研究院举办第一次院士选举推荐候选人名单及有关规章办法和来往文书(List of Recommended Candidates for the Election of the First Batch of Members of Academia Sinica and Relevant Regulations, Methods and Correspondence). Beijing: Qinghua daxue danganguan清华大学档案馆(Archives of Tsinghua University), Total No. 1, Cat. No. 4-2, File No. 191.

of the professors' association to nominate them if they were to be included in the final, submitted list. The final nominees also had to be nominated "by five members of the same school".[28] But even then, some of these candidates were not necessarily approved. Both Li Tsi-tung and Tang Yueh were elected by five and six experts respectively, but neither was elected when they were considered by the school council.[29] Later, when the school began nominating people outside of the school, it "consulted domestic scholars extensively"[30] and candidates went through the school's "separate recommendations by professors within the same research field of the school".[31] The National University of Peking also adopted a democratic method of election or at least went on the basis of recommendations made by professors at the university. It submitted its nomination materials to the PCE on 4 July, 1947. Hu Shih's sample letter to the school's nomination experts was also included.[32] (Figs. 4.2 and 4.3).

As the highest academic research institution in the country, the host of this member election, Academia Sinica, also participated in the nomination. It was specifically nominated by its research institutes and preparatory offices, but the director was nominated by Academia Sinica.[33] At that time, Academia Sinica had 12 research institutes

[28]Zhongyang yanjiuyuan juban diyici yuanshi xuanju tuijian houxanren mingdan ji youguan guizhang banfa he laiwang wenshu中央研究院举办第一次院士选举推荐候选人名单及有关规章办法和来往文书(List of Recommended Candidates for the Election of the First Batch of Members of Academia Sinica and Relevant Regulations, Methods and Correspondence). Beijing: Qinghua daxue danganguan清华大学档案馆(Archives of Tsinghua University), Total No. 1, Cat. No. 4-2, File No. 191.

[29]Zhongyang yanjiuyuan juban diyici yuanshi xuanju tuijian houxanren mingdan ji youguan guizhang banfa he laiwang wenshu中央研究院举办第一次院士选举推荐候选人名单及有关规章办法和来往文书(List of Recommended Candidates for the Election of the First Batch of Members of Academia Sinica and Relevant Regulations, Methods and Correspondence). Beijing: Qinghua daxue danganguan清华大学档案馆(Archives of Tsinghua University), Total No. 1, Cat. No. 4-2, File No. 191.

[30]Dongbei daxue deng yuanxiao hansong yuanshi houxuanren timingdan ji youguan hanjian东北大学等院校函送院士候选人提名单及有关函件(The Nomination Lists of Member Candidates Sent by Northeastern University and Other Universities and Colleges and Related Correspondence). Nanjing: Zhongguo dier lishi danganguan中国第二历史档案馆(The Second Historical Archives of China), Total No. 393, File No. 1613.

[31]Dongbei daxue deng yuanxiao hansong yuanshi houxuanren timingdan ji youguan hanjian东北大学等院校函送院士候选人提名单及有关函件(The Nomination Lists of Member Candidates Sent by Northeastern University and Other Universities and Colleges and Related Correspondence). Nanjing: Zhongguo dier lishi danganguan中国第二历史档案馆(The Second Historical Archives of China), Total No. 393, File No. 1613.

[32]Dongbei daxue deng yuanxiao hansong yuanshi houxuanren timingdan ji youguan hanjian东北大学等院校函送院士候选人提名单及有关函件(The Nomination Lists of Member Candidates Sent by Northeastern University and Other Universities and Colleges and Related Correspondence). Nanjing: Zhongguo dier lishi danganguan中国第二历史档案馆(The Second Historical Archives of China), Total No. 393, File No. 1613.

[33]Beiping yanjiuyuan deng yanjiu jiguan suoti de yuanshi houxuanren mingdan ji youguan hanjian 北平研究院等研究机关所提的院士候选人名单及有关函件(The Nomination Lists of Member Candidates Sent by the National Academy of Peiping and Other Research Institutions and Related Correspondence). Nanjing: Zhongguo dier lishi danganguan中国第二历史档案馆(The Second Historical Archives of China), Total No. 393, File No. 1614.

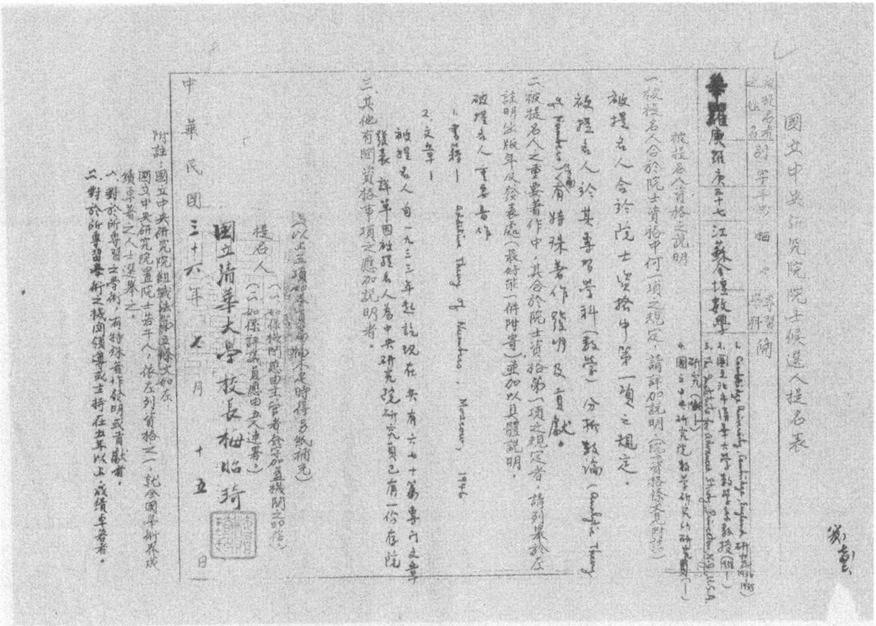
國立中央研究院院士候選人提名表

Fig. 4.2 Nomination Form for Hua Loo-Keng proposed by National Tsing Hua University

in mathematics, astronomy, physics, chemistry, geology, meteorology, engineering, zoology, botany, psychology, history and philology, sociology, and a preparatory office for medical research institutes. The nomination of some research institutes went through the process of discussion by the General Assembly of the institutes or democratic election. For example, the nomination of the Institute of History and Philology was approved by the Sixth General Assembly of the Second Session from 1946 to 1947.[34] Xu Zhongshu, Chen Shouyi, Shen Jianshi, Xu Hongbao, Liang Ssu-yung, Wang Li and Ling Chunsheng nominated by the Institute were not on the nomination list drafted by Director Fu Ssu-Nien in his capacity as a council member. Nominations for institutes in mathematics, chemistry, zoology, botany, meteorology, psychology and sociology were also different from the nomination lists drawn up by the directors of Chiang Li-Fu, Woo Sho-Chow, Wang Chia-Chi, Loo Tsung-Le, Jaw Jeou-Jang, Wang Ging-Hsi, and Tao L. K. as council members. For example, Sun Guangyuan, Yang K. C. and Chu Kun-ching who were nominated by Chiang Li-Fu were not in the list of the Institute of Mathematics, while Hsu Pao-Lu, who was not nominated by him, was in the list. Chou T. Q. and Chen Ko Kuei, nominated by Woo Sho-Chow and Hou T. P. were not nominated by the Institute of Chemistry. John C.

[34]Beiping yanjiuyuan deng yanjiu jiguan suoti de yuanshi houxuanren mingdan ji youguan hanjian 北平研究院等研究机关所提的院士候选人名单及有关函件(The Nomination Lists of Member Candidates Sent by the National Academy of Peiping and Other Research Institutions and Related Correspondence). Nanjing: Zhongguo dier lishi danganguan中国第二历史档案馆(The Second Historical Archives of China), Total No. 393, File No. 1614.

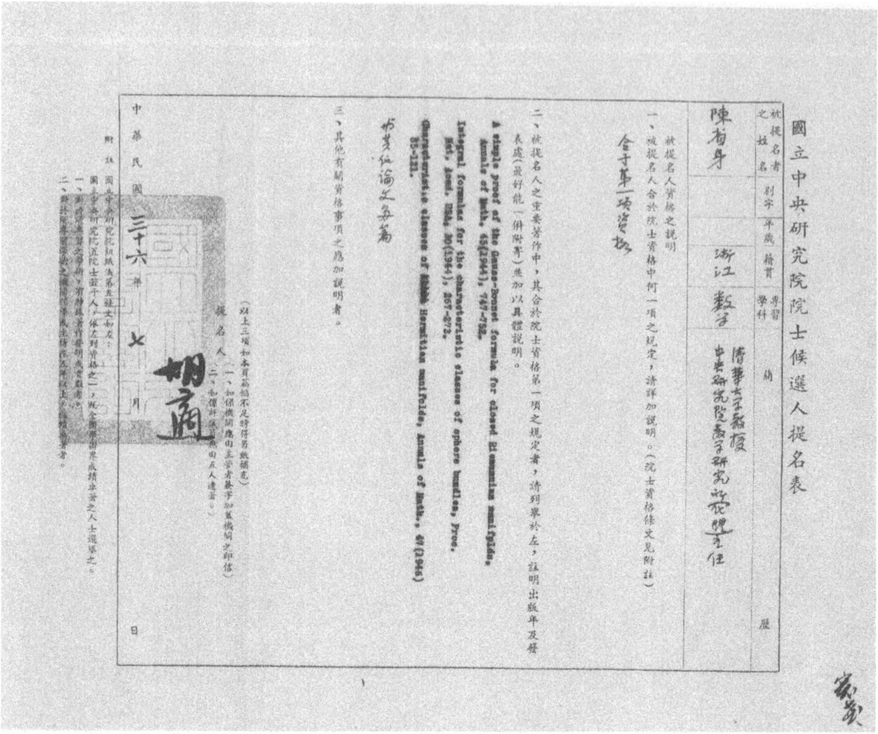

國立中央研究院院士候選人提名表

被提名者之姓名：陳省身　別字　年歲　籍貫：浙江　專習學科：數學　職：清華大學教授 中央研究院數學研究所代理主任　履歷

被提名人資格之說明

一、被提名人合於院士資格中何一項之規定，請詳加說明。（院士資格條文見附註）

合於第一項資格

二、被提名人之重要著作中，其合於院士資格第一項之規定者，請列舉於左，註明出版年及發表處（最好能一併附寄）並加以具體說明。

A simple proof of the Gauss-Bonnet formula for closed Riemannian manifolds, Annals of Math. 45(1944), 747-752.

Integral formulas for the characteristic classes of sphere bundles, Proc. Nat. Acad. USA, 30(1944), 267-272.

Characteristic classes of Hermitian manifolds, Annals of Math., 47(1946) 85-121.

共發表論文多篇

三、其他有關資格事項之應加說明者。

（以上三項如本頁篇幅不足時得另紙補充）

提名人（一、如係機關應由主管者簽字加蓋機關之印信 二、如係評議員應由五人連署。）胡適

中華民國三十六年七月　日

附註 國立中央研究院組織法第五條文如左：
國立中央研究院置院士若干人，依左列資格之一，就全國學術界成績卓著之人士選舉之。
一、對於所專習之學術，有特殊著作發明或貢獻者。
二、對於所專習學術之機關領導或主持在五年以上，成績卓著者。

Fig. 4.3 Nomination form for Shiing-Shen Chern proposed by the National University of Peking

H. Wu, Li Haopei, Mei Ju-Ao, Zhang Zhirang, Shao Xunke, Liu Naicheng, Chen Su-ching, Wang Ganyu, Wang Tieh-yai, Zhu Bingnan and Sun Pen-wen nominated by Tao L. K. were not nominated by the Institute of Social Sciences (Fig. 4.4).[35]

Academia Sinica could not prevent people from nominating candidates based on their personal connections to them. On June 17, 1947, Adam Pen-Tung Sah asked Chow Jen, the director of the Engineering Institute, to add Yu Ta-wei to the nomination list.[36] Yu Ta-wei was then Minister of Communications. In his early years,

[35]Beiping yanjiuyuan deng yanjiu jiguan suoti de yuanshi houxuanren mingdan ji youguan hanjian 北平研究院等研究机关所提的院士候选人名单及有关函件(The Nomination Lists of Member Candidates Sent by the National Academy of Peiping and Other Research Institutions and Related Correspondence). Nanjing: Zhongguo dier lishi danganguan中国第二历史档案馆(The Second Historical Archives of China), Total No. 393, File No. 1614; Guo li zhongyang yanjiuyuan diyici yuanshi xuanju houxuanren timingce国立中央研究院第一次院士选举候选人提名册(Nomination Lists of the Member Candidates for the Election of the First Batch of Members of Academia Sinica). Nanjing: Zhongguo dier lishi danganguan中国第二历史档案馆(The Second Historical Archives of China), Total No. 393, File No. 1597.

[36]Beiping yanjiuyuan deng yanjiu jiguan suoti de yuanshi houxuanren mingdan ji youguan hanjian 北平研究院等研究机关所提的院士候选人名单及有关函件(The Nomination Lists of Member Candidates Sent by the National Academy of Peiping and Other Research Institutions and Related

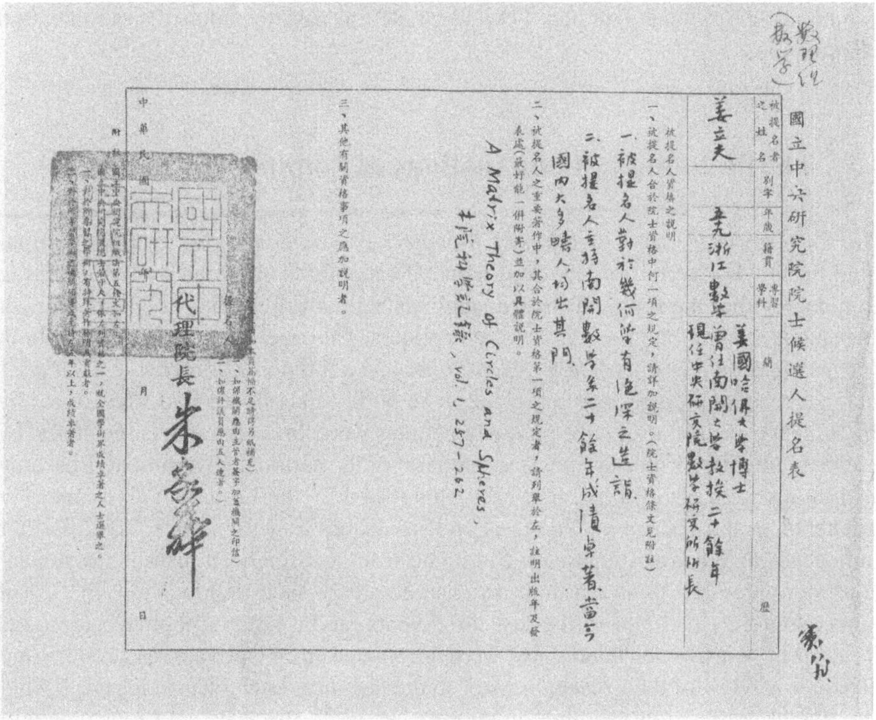

Fig. 4.4 Nomination form for Chiang Li-Fu proposed by Academia Sinica

he studied mathematics in the United States and received a doctorate from Harvard University in 1922. After that, he went to Germany where he studied ballistics. After returning to the country, he turned to politics and took part in the Northern Expedition. He was a senior official in the Ordnance Department for many years.[37] Even though Yu Ta-wei had not worked in the scientific research field, Chow Jen complied with Adam Pen-Tung Sah and asked Yu Ta-wei to fill in the member "Qualification Questionnaire Form". Related documents were prepared, so that they could nominate him,[38] and later, Yu Ta-wei was nominated by the Institute of Engineering

Correspondence). Nanjing: Zhongguo dier lishi danganguan中国第二历史档案馆(The Second Historical Archives of China), Total No. 393, File No. 1614.

[37]Zhongyang yanjiuyuan shulizu yuanshi houxuanren timingbiao (2) 中央研究院数理组院士候选人提名表(二)(The Nomination Forms of Member Candidates of the Mathematics and Physics Group of Academia Sinica(II)). Nanjing: Zhongguo dier lishi danganguan中国第二历史档案馆(The Second Historical Archives of China), Total No. 393, File No. 1602.

[38]Beiping yanjiuyuan deng yanjiu jiguan suoti de yuanshi houxuanren mingdan ji youguan hanjian 北平研究院等研究机关所提的院士候选人名单及有关函件(The Nomination Lists of Member Candidates Sent by the National Academy of Peiping and Other Research Institutions and Related Correspondence). Nanjing: Zhongguo dier lishi danganguan中国第二历史档案馆(The Second Historical Archives of China), Total No. 393, File No. 1614.

as a member candidate. He had, however, failed to pass the council's qualification examinations.

4.3 The Preliminary Examinations of Nominees by the PCE

The Regulations for the Election of Members of Academia Sinica clearly stipulated those who were formally nominated. Academic institutions were required to nominate within the scope of their disciplines, and attach to these the signatures of relevant executives and the seals of the institutes. They were also required to submit *The Nomination Form of Member Candidates* and a full publication list. Universities and independent colleges could include state, public and private works registered by the Ministry of Education. Specialized institutes were limited to works registered by relevant ministries, commissions and agencies of the national government. Research institutions were also limited to works established by the National Government or registered in the Executive Yuan, as well as in relevant ministries, commissions and agencies. When nominating a candidate from a specialized institute, organizations were required to attach their articles, details of their membership, the list of supervisors they had had in the last three years, and a note on their research and specialized work. When candidates were nominated from a private research institution, an overview of their research work in the last three years was required.[39] After the official nomination deadline on August 20, 1947, the PCE held four meetings between August 27 and October 13. They conducted a preliminary review into the nominations of all parties in accordance with *The Regulations for the Election of Members of Academia Sinica*.

On August 27, the PCE reviewed the qualifications of candidates from the nominating institutes. They looked at the specialties of candidates, and discussed how to rank various institutions and nominees. The council decided to adopt of all the lists proposed by various universities. There were some doubts about the submissions by independent colleges, but they corresponded with the Ministry of Education and sought advice. This was also the case with some specialized societies; in the end, the qualifications of candidates from Chinese Frontier Research Association and Chinese Association of Frontier Political Affairs were retained. The nomination of every institute of Academia Sinica was then changed to the nomination of Academia Sinica. The National Institute for Compilation and Translation was not able to nominate any candidates; the member candidate nominated by Mao T. E. needed to be supplemented by another four members. China's Muslim Youth Service Agency for Building Country and the Xiehe Agency did not conform to the relevant regulations. Chinese Chemical Society submitted its nomination forms too late and did

[39]Zhongyang yanjiuyuan yuanshi xuanju guicheng ji yuanshi houxuanren timingbiao中央研究院院士选举规程及院士候选人提名表(The Regulations for the Election of Members of Academia Sinica and Nomination Form for Member Candidates). Beijing: Beijing daxue danganguan北京大学档案馆(Archives of Peking University), File No. BD1947058.

not specify its nominees' qualifications. Some candidates could not be nominated because Academia Sinica had no records on them. Those who were eligible for nomination but were not nominated were not included in the member election. When the news on the final candidates was issued, the following must be stated: The PCE found that some experts and scholars were not included in the list of member candidates during the examination of nominations. In fact, the reason is that they were not nominated by any institution. Academia Sinica has limited qualifications and cannot nominate these candidates on its behalf. The council has drawn up a list of members for each discipline within three groups.[40]

On August 28, Wong Wen-hao reported at the PCE's plenary meeting that on the previous day, members had drawn up a list of candidates for a number of disciplines. Some of them, however, exceeded their quorum. "It is urgent we draw up another revised nomination list based on academic excellence and seniority." Later, participants exchanged views on the number of sub-divisions within each group. The council also discussed "how much scope to give to the subject of Chinese literature" and "how to determine a preliminary nomination list." They decided to allow—within a limited capacity—those who had done intensive research into ancient books and were academic leaders. They excluded literary and artistic writers for the time being, and the subject was renamed "Chinese Literature and History" (hereinafter referred to as "literature and history" or "literary history"). After this, the PCE deliberated "how to decide on a preliminary list" of member candidates. The council decided to examine nominees according to the simplest criterion, and to notify candidates and their institutions if they needed to supplement works that had not yet been submitted.[41]

On October 12, ahead of the Second Council's Fourth Annual Meeting, the PCE held its plenary meeting. The main purpose of the meeting was to re-examine *The Nomination Form of Member Candidates*, so that during the annual meeting, the council could proceed smoothly. Members decided to nominate Loo Tsung-Le, Woo Y. H. , Ping Chi, Wang Chia-Chi, Woo Sho-Chow, and Adam Pen-Tung Sah to review the nomination form and related correspondence. They discussed "how to publish a list of member candidates after it had been drawn up", "whether nominating institutions should be listed once their preliminary list has been put forward for review" and "how someone qualifies as a member candidate." They decided that only the name of the candidate and the basis for them qualifying should be published. Once the preliminary list was submitted to the council, the institutions

---

[40]Zhongyang yanjiuyuan pingyihui diyici yuanshi xuanju choubei weiyuanhui zuzhi ji huiyi jilu中央研究院评议会第一次院士选举筹备委员会组织及会议记录(The Organization and Minutes of the Preparatory Committee for the Election of the First Batch of Members of Academia Sinica). Nanjing: Zhongguo dier lishi danganguan中国第二历史档案馆(The Second Historical Archives of China), Total No. 393, File No. 134.

[41]Zhongyang yanjiuyuan pingyihui diyici yuanshi xuanju choubei weiyuanhui zuzhi ji huiyi jilu中央研究院评议会第一次院士选举筹备委员会组织及会议记录(The Organization and Minutes of the Preparatory Committee for the Election of the First Batch of Members of Academia Sinica). Nanjing: Zhongguo dier lishi danganguan中国第二历史档案馆(The Second Historical Archives of China), Total No. 393, File No. 134.

that had nominated them would not be listed. If candidates were heads of department, deans or principals, and had led or presided over work for more than five years, this counted as their having obtained outstanding achievements, i.e. if they had led or presided over an academic institution for more than five years and that institution had made outstanding achievements.[42] The PCE finally examined and approved the nominations from 57 institutions. They decided to preliminarily nominate 510 member candidates.[43] Following a panel review, 108 people were removed from this list, meaning 402 candidates remained.[44]

On October 13, the PCE held its plenary meeting again. Adam Pen-Tung Sah reported the results of the previous day's panel meeting and why some of those on the original roster had been excluded, including those who had not submitted nomination forms. The council decided that "it is quite difficult to draw a line between the nominees who are suspected of being traitors to education and those who have been forced to join pseudo-cultural organizations. Even though there were candidates that we wanted to include, we are eliminating those people who have been executed because of previous convictions, but these nominees will still be sent to the council as part of the preliminary list."[45] By the end of this meeting, a full list of member candidates had been nominated.

On the whole, there were losses and gains during the formal nomination process. The "gains" were the large number of primary candidates and the large number of nominees with either a primary or secondary member qualification. Their nomination was relatively concentrated; however, as we've seen previously, Academia Sinica members generally had both primary and secondary qualifications. Primary qualifications were some form of special work, i.e. an inventions or contribution to

[42]Zhongyang yanjiuyuan pingyihui diyici yuanshi xuanju choubei weiyuanhui zuzhi ji huiyi jilu中央研究院评议会第一次院士选举筹备委员会组织及会议记录(The Organization and Minutes of the Preparatory Committee for the Election of the First Batch of Members of Academia Sinica). Nanjing: Zhongguo dier lishi danganguan中国第二历史档案馆(The Second Historical Archives of China), Total No. 393, File No. 134.

[43]Zhongyang yanjiuyuan diyici yuanshi xuanju timing jigou tongjibiao中央研究院第一次院士选举提名机构统计表(Statistical Table of Nominating Institutes during the Election of the First Batch of Members of Academia Sinica). Nanjing: Zhongguo dier lishi danganguan中国第二历史档案馆(The Second Historical Archives of China), Total No.393, File No.1599; Guo li zhongyang yanjiuyuan diyici yuanshi xuanju houxuanren timingce国立中央研究院第一次院士选举候选人提名册(Nomination Lists of the Member Candidates for the Election of the First Batch of Members of Academia Sinica). Nanjing: Zhongguo dier lishi danganguan中国第二历史档案馆(The Second Historical Archives of China), Total No. 393, File No. 1597.

[44]Zhongyang yanjiuyuan pingyihui diyici yuanshi xuanju choubei weiyuanhui zuzhi ji huiyi jilu中央研究院评议会第一次院士选举筹备委员会组织及会议记录(The Organization and Minutes of the Preparatory Committee for the Election of the First Batch of Members of Academia Sinica). Nanjing: Zhongguo dier lishi danganguan中国第二历史档案馆(The Second Historical Archives of China), Total No.393, File No.134.

[45]Zhongyang yanjiuyuan pingyihui diyici yuanshi xuanju choubei weiyuanhui zuzhi ji huiyi jilu中央研究院评议会第一次院士选举筹备委员会组织及会议记录(The Organization and Minutes of the Preparatory Committee for the Election of the First Batch of Members of Academia Sinica). Nanjing: Zhongguo dier lishi danganguan中国第二历史档案馆(The Second Historical Archives of China), Total No. 393, File No. 134.

a piece of specific academic research. A secondary qualification was recognition for being a leader or presiding over an academic institution for more than five years, and therefore gaining outstanding achievements. Of the 510 candidates nominated in the very first election, 274 had a primary qualification. They accounted for 53.73% of the total number of candidates. There were 114 nominees who had both primary and secondary qualifications, and they accounted for 22.35% of the total number of candidates. 110 nominees had a secondary qualification, and they accounted for 21.57% of the total number of candidates. 388 candidates had either primary or primary and secondary qualifications, and they accounted for 76.08% of the total number of candidates. Six institutions nominated candidates nationwide: Academia Sinica, National University of Peking, National Tsing Hua University, National Wu-han University, the National Central University and Henan University. Out of the 81 people who were eventually elected, the majority received nominations from representatives at several of these institutions. 54 people were nominated by three or more institutions, and they accounted for 66.7% of those who were elected. The more institutions that nominated a preliminary candidate, the higher that candidate's chance was at winning the election. Therefore, we can say that the nominations were relatively concentrated.

Another "gain" obviously lay in the fact that the lists submitted by Academia Sinica, National University of Peking, and National Tsing Hua University were of great importance and provided the council with important candidates and reference points. The success rates of the candidates submitted by these institutions were among the highest in the country, even though they nominated higher numbers of candidates than other nominating institutions. Academia Sinica nominated 155 people, 72 of whom were elected; therefore, they had a success rate of 46.5%. National University of Peking nominated 119 people, 63 of whom were elected, so they had a success rate of 52.9%. National Tsing Hua University nominated 134 people, 47 of whom were elected, so their election success rate was 35.1%. National Wu-han University nominated 99 people, 33 of whom were elected, so their success rate was 33.3%. The National Central University nominated 37 people, five of whom were elected, and their success rate was 13.5%. Out of the 28 nominees from Henan University, 12 were elected, so the success rate was 42.9%. Candidates nominated by Academia Sinica and the National University of Peking had higher chances of reaching the election stage than the latter three institutions. Although National Tsing Hua University's success rate was slightly lower than Henan University's, it obviously nominated many more people.

The National University of Peking was able to be so successful because, as we have seen, it took the election very seriously. Hu Shih oversaw the university's nomination work, and he made it his mission to ensure that members were recognized for their contributions across all disciplines. Although National Tsing Hua University implemented a more democratic and a stricter selection method for nominating people inside the school, and it "separately tasked professors outside the school with recommending people across different specialties,"[46] it was also very successful because of its resources, and its strong relationship with both Academia Sinica, and the National University of Peking. National Tsing Hua University's selection method for people outside of the school originated from the model implemented by the National University of Peking, which in turn depended on the influences of Adam Pen-Tung Sah, Shiing-Shen Chern and Chou Pei-yuan. Their influence was spread across Academia Sinica, the National University of Peking, and National Tsing Hua University, and therefore it fed into the special policies implemented by Academia Sinica.

One of the "losses" during the formal nomination process was the fact that formally nominating candidates meant "leaving behind some pearls". However, this was only a small problem, due to the nominations from Academia Sinica, National University of Peking, National Tsing Hua University, National Wu-han University, the National Central University and Henan University. Chu Chia-Hua, Shiah Nae and Young Chung-Chien all recognized that some excellent people would get left behind. On March 25, 1948, Chu Chia-Hua said in his opening speech at the Fifth Annual Meeting of the Second Council of Academia Sinica: "there are still some famous members who are not listed as member candidates. This is somewhat regrettable to Academia Sinica."[47] During the same year, after the first batch of members of Academia Sinica were elected, Shiah Nae wrote: "the nomination method is weak, and so some scholars who should have been elected have not been nominated. It has made impossible for them to be elected, and this means that people feel like they have left some pearls behind."[48] Young Chung-Chien wrote in his memoirs: "At the time, the election of members was considered a special honor in China. The way that members were generated, it was all done very carefully; there were hardly any drawbacks, and even ordinary people paid attention to this event. But there was also the phenomenon of some truly learned scholars being left out and not elected. One example was Pei W. C. He made great academic achievements in recent years, especially during the Anti-Japanese War. In 1947, while I was in Peking and various agencies were selecting their preliminary candidates, Hu Shizhi (Hu Shih) asked me, 'Why haven't China's geological agencies recommended Pei W. C.? Don't related

[46]Zhongyang yanjiuyuan juban diyici yuanshi xuanju tuijian houxanren mingdan ji youguan guizhang banfa he laiwang wenshu中央研究院举办第一次院士选举推荐候选人名单及有关规章办法和来往文书(List of Recommended Candidates for the Election of the First Batch of Members of Academia Sinica and Relevant Regulations, Methods and Correspondence). Beijing: Qinghua daxue danganguan清华大学档案馆(Archives of Tsinghua University), Total No. 1, Cat. No. 4-2, File No. 191.

[47]Zhongyang yanjiuyuan dierjie pingyihui tiaoli jilu ti'an ji baogao中央研究院第二届评议会条例、记录、提案及报告(Regulations, Records, Proposals and Reports of the Second Council of Academia Sinica). Nanjing: Zhongguo dier lishi danganguan中国第二历史档案馆(The Second Historical Archives of China), Total No. 393, File No. 2926.

[48]Shiah Nae夏鼐(1948). Zhongyang yanjiuyuan diyijie yuanshi de fenxi中央研究院第一届院士的分析(The Analysis of the First Batch of Members of Academia Sinica). *Guancha*观察(Observation) 5 (14): 3.

institutes select candidates from their archaeological circles or is he about to be recommended?' It turned out that, in the end, he was not recommended."[49]

4.4 The Qualificaton Examinations of Member Candidates by the Council

Between October 15 and 17, 1947, Academia Sinica's Second Council held its Fourth Annual Meeting. The main agenda of the meeting was to further review the member candidates approved by the PCE after they had been preliminarily examined, and draw up a formal list of member candidates that they could agree to publicize. The meeting began at 9 am on October 15 in the Academia Sinica auditorium at 1 Jiming Temple Road, Nanjing. Among the attendees were Chu Chia-Hua, Wong Wen-hao, Adam Pen-Tung Sah, Wang Chia-Chi, Woo Y. H. , Woo Ting-liang, Woo Sho-Chow, Li Chi, Li Shu-Hua, Ping Chi, Chow Jen, Chou Keng-Shang, Robert K. S. Lim, Hu Shih, Hu Hsen-hsu, Mao T. E., Chen Yuan, Chen Shisan C., Tao L. K. , Ling Hung-hsun, Chuang Chang-Kong, Jaw Jeou-Jang, Chien Shung-shu, Hsieh Chia-yung, and Loo Tsung-Le.[50] In his opening speech, Chu Chia-Hua stressed that the establishment of an academician system was a "particular priority" ahead of the council's Third Annual Meeting in 1946. He explained that Academia Sinica had some institutional defects, and gave some general background to why establishing an academician system was so important. He also detailed some of the preparatory work that had been carried out and said "the main task of the council is to examine and appoint member candidates. We wish to hold a special, formal council election next spring and set up an association of members. More members will be elected each year. With members, Academia Sinica can become a complete system and it can become the academic foundation of our country. It can be of great significance in facilitating scientific research in the future."[51]

The First General Meeting of the annual meeting began at 3 pm on October 15. All of the council members and attendees who had attended the morning's opening ceremony attended the meeting, with the exception of Tao L. K. The general meeting passed a proposal to "ratify *The Regulations for the Election of Members of the National Academia Sinica*." After that, they discussed "the preliminary nomination

[49]Young Chung-Chien杨钟健(1983). *Young Chung-Chien huiyilu*杨钟健回忆录(Memoirs of Young Chung-Chien). Beijing: Dizhi chubanshe地质出版社.167.

[50]Zhongyang yanjiuyuan pingyihui dierjie disici nianhui jilu中央研究院评议会第二届第四次年会纪录(Minutes of the Fourth Annual Meeting of the Second Council of Academia Sinica). Nanjing: Zhongguo dier lishi danganguan中国第二历史档案馆(The Second Historical Archives of China), Total No. 393, File No. 1558.

[51]Zhongyang yanjiuyuan dierjie diyi san si wuci nianhui jilu中央研究院评议会第二届第一、三、四、五次年会纪录(Minutes of the First, Third, Fourth and Fifth Annual Meetings of the Second Council of Academia Sinica). Nanjing: Zhongguo dier lishi danganguan中国第二历史档案馆(The Second Historical Archives of China), Total No.393, File No.2551.

list of member candidates submitted by Academia Sinica's PCE and relevant documents". The council decided that less than 50% of the member nominees should be elected. At 9 am the following day, the council conducted a group review. Li Shu-Hua, Ping Chi and Hu Shih convened over the mathematics and physics, biology, and humanities groups respectively.[52] None of them worked full-time at Academia Sinica. The conference chose them specifically because of their high prestige in academic circles, but also to avert any suspicion of interference.

On October 16, there was a meeting between a subsidiary group in the morning. They met to assess the list of member candidates drawn up within the three groups. There were 49 members of the mathematics and physics group, 48 members of the biology group, and 56 members of the humanities group, meaning a total of 152 members spread across three groups. Wu Hsien was included in the first two groups. This is shown in Table 4.2.

The member candidates were examined and approved by each group, and then entered into a list. 402 candidates were approved during the PCE's preliminary examination. However, according to Table 4.2, most of the people on this list were later removed. The remaining 152 were almost all celebrities in academic circles. Most of them had made important academic achievements and contributions, but a few others had scraped through the selection process. Shiah Nae was acting director of the Institute of History and Philology of Academia Sinica at that time. He acted on behalf of the institute as a nonvoting participant to attend the examination meeting of the humanities group. He thought that out of the 31 people elected in the literature and history field, some of them were not worthy of nomination. In the social sciences, few suitable candidates in economics and jurisprudence, especially in economics. Because "Mr Ou" was present, seven people (Ma Yin-chu, Lieu D. K., Ho L., Fong H. D., Yang Hsi-meng, Ou Pao-san and Wu Daye) were still proposed. In fact, it could be temporarily omitted. In jurisprudence, Wang Chung Hwei and Wang Shih-Chieh were proposed. But later, Yen Shu-Tang, Kuo Yun-kuan and Li Haopei were also reluctantly added to the list."[53] "Mr Ou" (Ou Pao-san) was a full-time research fellow at Academia Sinica's Institute of Social Sciences, and he attended the meeting with Shiah Nae. Although the economics candidates mentioned by Shiah Nae and put forward by the humanities group did not conform to the correct process of recording,[54] Mr Shiah evaluation was generally objective.

According to Shiah Nae's diary, the humanities group's review team discussed whether to remove pseudo-members of the National University of Peking from the list. "Rong Geng was the only one, so it was decided not to include him."[55] Therefore,

[52]Zhongyang yanjiuyuan pingyihui dierjie disici nianhui jilu中央研究院评议会第二届第四次年会纪录(Minutes of the Fourth Annual Meeting of the Second Council of Academia Sinica). Nanjing: Zhongguo dier lishi danganguan中国第二历史档案馆(The Second Historical Archives of China), Total No. 393, File No. 1558.

[53]Shiah Nae夏鼐(2011). *Shiah Nae riji*夏鼐日记(Shiah Nae's Diary), Vol. 4. Shanghai: Huadong shifan daxue chubanshe华东师范大学出版社.150.

[54]As shown in Table 4.2, there were a total of eight member candidates in the field of economics proposed by the humanities group. These including Yang Duanliu and Chen Zong, but not Wu Daye.

[55]Shiah Nae夏鼐(2011). *Shiah Nae riji*夏鼐日记(Shiah Nae's Diary), Vol. 4. Shanghai: Huadong shifan daxue chubanshe华东师范大学出版社.150.

Table 4.2 List of member candidates drawn up by each group of the sub-group examination committee

Group	Discipline and number	Member candidates examined and approved by the group
Mathematics and physics group	Mathematics (8)	Kiang Tsai-han, Chiang Li-Fu, Hsu Pao-Lu, Shiing-Shen Chern, Chen Kien Kwong, Hua Loo-Keng, Hiong King-lai, Su Buchin
	Astronomy (2)	Yu C. S., Chang Yu-Che
	Physics (8)	Wu Ta-You, Woo Y. H. , Li Shu-Hua, Chou Pei-yuan, Yeh Chi-Sun, Chao Chung-Yoa, Ny Tsai-ze, Yao Y. T.
	Chemistry (6)	Wu Hsien, Woo Sho-Chow, Chuang Chang-Kong, Tseng Chao-lun, Peter P. T. Sah, Chou T. Q.
	Geology (12)	Yin T. H., Wang C. C., Chu Chia-Hua, Lee J. S. , Lee S. P., Meng H. M., Yu C. C., Sun Y. C., Wong Wen-hao, Huang T. K., Young Chung-Chien, Hsieh Chia-yung
	Meteorology (1)	Chu Coching
	Engineering (12)	Wang C. Y., Wang H. C., Chow Jen, Shi Jiayang, Hou T. P., Mao T. E., Ling Hung-hsun, Cheng Xiaogang, Tsai Fang-yin, Tsien H. S., Adam Pen-Tung Sah, Lo C. C.
Biology group	Zoology (10)	Wang Chia-Chi, Zhu Xi, Wu Hsien-Wen, Pai Sitsan, Ping Chi, Hu Jingfu, Chen Shixiang, Chen Shisan C., Tung Ti-Cho, Liu C. C.
	Botany (10)	Hu Hsen-hsu, Yin Hung-chang, Ching Ren Chang, Chang Chin-yueh, Pei Chien, Liou Tchen-Ngo, Chien Shung-shu, Tai Fon-Land, Loo Tsung-Le, Jao Chin-chih
	Medical science (10)	Wu Hsien, Li Zongen, Hu C. H., Hung S. L., Yuan I-Chin, Ma Wen-chao, Chang Hsiao-chien, Tang F. F. , Feng Lan-chou, Liu S. H.

(continued)

Table 4.2 (continued)

Group	Discipline and number	Member candidates examined and approved by the group
	Pharmacology (1)	Huang Minlon
	Anthropology (1)	Woo Ting-liang
	Psychology (3)	Wang Ging-Hsi, Lu Zhiwei, Zang Yugan
	Physiology (6)	Robert K. S. Lim, Hou Zonglian, Xu Fengyan, Tang P. S., Feng Tepei, Tsai Chiao
	Agronomy (7)	Li Hsien-wen, Yu T. F. , Chen Rong, Zhao Lianfang, Teng S. C., Teng Chih-I, Liu Chung-lo
Humanities group	Philosophy (5)	Woo Tsin-Hang, Chin Yuch-Lin, Tang Yung-tung, Feng You-lan, Chen Kang
	Literary history (6)	Yu Cha-Shih, Hu Shih, Zhang Yuanji, Yang Shu-Tah, Liu W. T., Tang Lan
	History (10)	Chen Yuan, Chen Yin-k'o, Fu Ssu-Nien, Ku Chieh-kan, Xu Bingchang, Xu Zhongshu, Chen Shouyi, Li Jiannong, Liu Yei-Tsen, Chiang Ting-fu
	Linguistics (4)	Chao Yuen Ren, Li Fang Kuei, Luo Changpei, Wang Li
	Archaeology and art history (6)	Kuo Mo-jo, Li Chi, Tung Tso-pin, Liang Ssu-yung, Liang Ssu-cheng, Xu Hongbao
	Jurisprudence (5)	Wang Chung Hwei, Wang Shih-Chieh, Yen Shu-Tang, Kuo Yun-kuan, Li Haopei
	Politics (7)	Chou Keng-Shang, Hsiao Kung-chuan, Tsien T. S. , Zhang Xiruo, Hsu Shuhsi, Chang C. F., Liu Naicheng
	Economics (8)	Ma Yin-chu, Lieu D. K., Ho L., Yang Duanliu, Chen Zong, Fong H. D., Ou Pao-san, Yang Hsi-meng
	Sociology (5)	Chen Ta, Tao L. K. , Pan Kuang-tan, Ling Chunsheng, Wu Ching-chao

Sources Zhongyang yanjiuyuan diyijie yuanshi houxuanren mingdan中央研究院第一届院士候选人名单(List of Member Candidates for the First Batch of Members of Academia Sinica). Nanjing: Zhongguo dier lishi danganguan中国第二历史档案馆(The Second Historical Archives of China), Total No. 393, File No. 2596

Rong Geng was deleted. In fact, during the PCE's plenary meeting on October 13, there had already been discussions about whether to remove people who had been forced to participate in pseudo-institutions and were therefore deemed suspected traitors of their educational institutes. In the end, due to there being a "difficult demarcation of boundaries", only those who had been executed or had previous convictions were eliminated. The examination committee's removal of Rong Geng showed that the group was determined to eliminate anyone who had participated in a pseudo-institution. However, the mathematics and physics group's examination committee did not investigate suspected participants or suspected traitors in depth. This was partly because Adam Pen-Tung Sah's brother, the chemist Peter P. T. Sah, had also been a member of the pseudo-National University of Peking. Along with Chou T. Q., a suspected traitor who was the director of Institute of Materia Medica of the National Academy of Peiping, he passed his examination successfully. Tsien H. S., however, was still retained. From this we can surmise that the mathematics and physics group's examination committee were lenient towards scholars who had never returned to work, but had made high academic achievements overseas in their early years. The biologist Li Zhuohao was removed from the medical science discipline of biology group, however. Ping Chi, the convener of the group's examination committee, still advocated that candidates should be removed from the nomination list if they "have not returned home or have continued not to make any contributions within a year or two of returning home."[56]

The Second General Meeting of the annual meeting began at 3 pm on October 16. Li Shu-Hua, Ping Chi and Hu Shih successively reported the results of candidates' examinations in the mathematics and physics, biology, and humanities groups. Subsequently, the council discussed the results of each group.[57] The specific results were as follows: 1: Yu C. S. and Chang Yu-Che were chosen for the astronomy list; Peter P. T. Sah and Chou T. Q. were chosen for chemistry and Tsien H. S. was removed from the engineering list; Kwei C. T. was added to the physics list, and Zhu Ruhua, Huang T. C., Chi Yuoh-Fong and Sun Xuewu were added to the chemistry list. 2: Hou Zonglian was selected for physiology; Chen Rong and Teng Chih-I were removed from the agronomy list; Chen Ko Kuei was selected for the pharmacology list; Feng Zefang was chosen for the agronomy list, and Wu Hsien was added to the chemistry list.[58] Peter P. T. Sah and Chou T. Q. were both removed because they

---

[56]Hu Shih Loo Tsung-Le dengren suo ti de yuanshi houxuanren mingdan ji youguan hanjian胡适、罗宗洛等人所提的院士候选人名单及有关函件(The Nomination Lists of Member Candidates Proposed by Hu Shih, Loo Tsung-Le, etc. and Related Correspondence). Nanjing: Zhongguo dier lishi danganguan中国第二历史档案馆(The Second Historical Archives of China), Total No. 393, File No. 1615.

[57]Zhongyang yanjiuyuan pingyihui dierjie disici nianhui jilu中央研究院评议会第二届第四次年会纪录(Minutes of the Fourth Annual Meeting of the Second Council of Academia Sinica). Nanjing: Zhongguo dier lishi danganguan中国第二历史档案馆(The Second Historical Archives of China), Total No. 393, File No. 1558.

[58]Zhongyang yanjiuyuan diyijie yuanshi houxuanren mingdan中央研究院第一届院士候选人名单(List of Member Candidates for the First Batch of Members of Academia Sinica). Nanjing:

had once worked at pseudo-institutes.[59] Tsien H. S. was removed because he had never returned to China for work and refused to do so. According to Shiah Nae's diary, there was the intention to add Li Zhuohao to the medical sciences list during the Second General Meeting on October 16. However, in the end, he failed to pass "because he did not return to work".[60] At the conference, the members present did not raise any objection to including scholars who had previously working overseas but had returned to the country for work. One example was Chen Ko Kuei, who worked for an extended period of time at the Lilly Pharmaceutical Company in the United States. He passed his examination smoothly.

At the Second General Meeting, the council approved and revised the results of candidates in the mathematics and physics, biology groups. They decided to discuss the examination report on the humanities group at the next general meeting. The council then went on to discuss the proposal for recommending people to write testimonials on the list of member candidates, i.e. the basis that each candidate is qualified a certain qualification. It was decided that 22 people wrote. The testimonial of each member candidate had a limit of between 31 and 50 characters.[61]

The Third General Meeting of the annual meeting was held at 9 am on October 17. Hu Shih, Chou Keng-Shang, Ou Pao-san and Li Chi successively drew up candidate lists for members in the philosophy, literary history, history, linguistics, jurisprudence, politics, economics, sociology, archaeology and art history disciplines of the humanities group. After that, the conference discussed "the examination report of the list of member candidates within the humanities group".[62] Subsequently, Hsu Shuhsi and Liu Naicheng were removed from the politics list and John C. H. Wu was added to the jurisprudence list.

Because there had been a civil war between the Kuomintang and the Communist Party, there was fierce dispute over whether to remove Kuo Mo-jo. As this was a controversial situation, Shiah Nae made detailed records in his diary on that day:

> In the morning, the council continued to examine the list. Hu Shizhi asked the chairman for his unofficial opinion on Kuo Mo-jo's nomination. Chu Chia-Hua said that Guo's participation in civil strife and crime made him a traitor and not suitable for the nomination list. Director-General Sah feared that stimulating the government would have a negative impact on future

---

Zhongguo dier lishi danganguan中国第二历史档案馆(The Second Historical Archives of China), Total No. 393, File No. 2596.

[59]Luo Feng罗丰(2004). Shiah Nae yu zhongyang yanjiuyuan diyije yuanshi xuanju夏鼐与中央研究院第一届院士选举(Shiah Nae and The Election of the First Batch ofMembers of Academia Sinica). *Kaogu yu wenwu*考古与文物(Archaeology and Cultural Relics)(4): 86.

[60]Shiah Nae夏鼐(2011). *Shiah Nae riji*夏鼐日记(Shiah Nae's Diary), Vol. 4. Shanghai: Huadong shifan daxue chubanshe华东师范大学出版社.150.

[61]Zhongyang yanjiuyuan pingyihui dierjie disici nianhui jilu中央研究院评议会第二届第四次年会纪录(Minutes of the Fourth Annual Meeting of the Second Council of Academia Sinica). Nanjing: Zhongguo dier lishi danganguan中国第二历史档案馆(The Second Historical Archives of China), Total No. 393, File No. 1558.

[62]Zhongyang yanjiuyuan pingyihui dierjie disici nianhui jilu中央研究院评议会第二届第四次年会纪录(Minutes of the Fourth Annual Meeting of the Second Council of Academia Sinica). Nanjing: Zhongguo dier lishi danganguan中国第二历史档案馆(The Second Historical Archives of China), Total No. 393, File No. 1558.

funding. Wu Zhengzhi[63] said that he was afraid Kuo might make negative remarks in the future as a member of Academia Sinica. Ou Pao-san stood up and voiced his opposition. One should not overlook his academic contributions because of his political relations. Tao L. K. said that if the government's will was taken as the standard, it was best to ask the government to make a judgement. Hu Shih said that an academic standpoint should be the focal point. Both sides expressed their opinions. In the end, they decided to vote by secret ballot. I was a non voting participant, and did not have the right to vote. I stood up and said that some people in the meeting equated heterodox parties with traitors. However, as Academia Sinica was the academy of sciences of China, besides academic contribution, the only condition was Chinese. If traitors were not Chinese at all, those who opposed the government were different from traitors. They should not be mentioned in the same breath. A China had existed before the national government (After the collapse of the national government, there was still China). I thought of this sentence, but I couldn't say it. Therefore, we could be strict towards traitors but lenient towards those with different political views. The result of the vote was 14 to 7, and Kuo Mo-jo stayed on the list.[64]

Although those in favor of removing Kuo Mo-jo's name from the list included Chu Chia-Hua and Adam Pen-Tung Sah, people who occupied high-ranking positions in Academia Sinica, Kuo Mo-jo finally passed his examination because Hu Shih and 14 other members insisted on academic independence being kept separate from politics. Ou Pao-san and Shiah Nae also expressed similar views. Kuo Mo-jo became an elected member of Academia Sinica in March 1948. On 30 May, 1949, Chu Coching wrote in his diary about Chen Yi: "He had read *the Records of Members of Academia Sinica*, and had seen Kuo Mo-jo's name. He knew that Academia Sinica could come to embrace anything."[65] Obviously, Chen Yi admired Academia Sinica for not ruling out "heterodox parties".

On the morning of October 17, the 23 council members who attended the Third General Meeting finally voted on the list of member candidates amended by the three groups by the way of standing up. The results were unanimously approved. As a result, the first official list of member candidates was produced. The council also decided that "when a list of member candidates is confirmed, each group of people will be arranged by disciplines. However, the names of their disciplines will not be listed, and each discipline will be ranked according to stroke order."[66]

The Fourth General Meeting of the annual meeting was held at 3 pm on October 17. The council passed "the list of member candidates to be published immediately before the announcement". Ping Chi, Li Shu-Hua, Mao T. E., Hsieh Chia-yung and Tao L. K. were invited to decide on the text, with Ping Chi taking charge. The council also discussed "how to write comments on the member candidates on the list, that is, how to draft neat and succinct text about why someone is qualified. The

[63] Wu Zhengzhi is Woo Y. H.

[64] Shiah Nae夏鼐(2011). *Shiah Nae riji*夏鼐日记(Shiah Nae's Diary), Vol. 4. Shanghai: Huadong shifan daxue chubanshe华东师范大学出版社. 150–151.

[65] Chu Coching竺可桢(2006). *Chu Coching quanji*竺可桢全集(The Complete Works of Coching Chu), Vol. 11. Shanghai: Shanghai keji jiaoyu chubanshe上海科技教育出版社.450.

[66] Zhongyang yanjiuyuan pingyihui dierjie disici nianhui jilu中央研究院评议会第二届第四次年会纪录(Minutes of the Fourth Annual Meeting of the Second Council of Academia Sinica). Nanjing: Zhongguo dier lishi danganguan中国第二历史档案馆(The Second Historical Archives of China), Total No. 393, File No. 1558.

council decided "to draft the text according to Articles 1 and 2 of *The Regulations for the Election of Members of Academia Sinica*". They authorized the chairperson, the honorary secretary and the director-general to be responsible for the final publication of the list. Finally, the council reviewed the language related to Academia Sinica staff members. They decided to translate "Yuan Shi" into "Member of Academia Sinica" and "Ming Yu Yuan Shi (名誉院士)" was translated into "Honorary Member of Academia Sinica."[67]

The annual meeting was fruitful, the official list of member candidates was examined and approved, and careful and sufficient preparations were made in the follow-up work related to announcing the list. On November 15, 1947, Academia Sinica announced its first official list of 150 member candidates in the government gazette and major Nanjing and Shanghai newspapers. There were 49 people in the mathematics and physics group, 46 people in the biology group and 55 people in the humanities group. The list of member candidates of various disciplines in each group was as follows:

Mathematics and Physics Group: (mathematics 8) Kiang Tsai-han, Chiang Li-Fu, Hsu Pao-Lu, Shiing-Shen Chern, Chen Kien Kwong, Hua Loo-Keng, Hiong King-lai, Su Buchin; (physics 9) Wu Ta-You, Woo Y. H. , Li Shu-Hua, Chou Pei-yuan, Kwei C. T., Yeh Chi-Sun, Chao Chung-Yoa, Ny Tsai-ze, Yao Y. T. ; (chemistry 8) Zhu Ruhua,Wu Hsien, Woo Sho-Chow, Chi Yuoh-Fong, Sun Xuewu, Chuang Chang-Kong, Tseng Chao-lun, Huang T. C.; (geology 12) Yin T. H., Wang C. C., Chu Chia-Hua, Lee J. S. , Lee S. P., Meng H. M., Yu C. C., Sun Y. C., Wong Wen-hao, Huang T. K., Young Chung-Chien, Hsieh Chia-yung; (meteorology 1) Chu Coching; (engineering 11) Wang C. Y., Wang H. C., Chow Jen, Shi Jiayang, Hou T. P., Mao T. E., Ling Hung-hsun, Cheng Xiaogang, Tsai Fang-yin, Adam Pen-Tung Sah, Lo C. C.

Biology group: (zoology 10) Wang Chia-Chi, Wu Hsien-Wen, Zhu Xi, Pai Sitsan, Ping Chi, Hu Jingfu, Chen Shixiang, Chen Shisan C., Tung Ti-Cho, Liu C. C.; (botany 10) Hu Hsen-hsu, Yin Hung-chang, Ching Ren Chang, Zhang Jingyu, Pei Chien, Liou Tchen-Ngo, Chien Shung-shu, Tai Fon-Land, Loo Tsung-Le, Jao Chin-chih; (medical science 9) Li Zongen, Hu C. H., Hung S. L., Yuan I-Chin, Ma Wen-chao, Chang Hsiao-chien, Tang F. F. , Feng Lan-chou, Liu S. H.; (pharmacology 2) Chen Ko Kuei and Huang Minlon; (physical anthropology 1) Woo Ting-liang; (psychology 3) Wang Ging-Hsi, Lu Zhiwei and Zang Yugan; (physiology 5) Robert K. S. Lim, Xu Fengyan, Tang P. S., Feng Teh-Pei, Tsai Chiao; (agronomy 6) Li Hsien-wen, Yu T. F. , Feng Zefang, Zhao Lianfang, Teng S. C., Liu Chung-lo.

Humanities group: (philosophy 5) Woo Tsin-Hang, Chin Yuch-Lin, Chen Kang, Tang Yung-tung, Feng You-lan; (literary History) Yu Cha-Shih, Hu Shih, Tang Lan, Zhang Yuanji, Yang Shu-Tah and Liu W. T.; (history 10) Li Jiannong, Liu Yei-Tsen, Xu Zhongshu, Xu Bingchang, Chen Yuan, Chen Yin-k'o, Chen Shouyi, Fu Ssu-Nien,

---

[67]Zhongyang yanjiuyuan pingyihui dierjie disici nianhui jilu中央研究院评议会第二届第四次年会纪录(Minutes of the Fourth Annual Meeting of the Second Council of Academia Sinica). Nanjing: Zhongguo dier lishi danganguan中国第二历史档案馆(The Second Historical Archives of China), Total No. 393, File No. 1558.

Chiang Ting-fu, Ku Chieh-kan; (linguisitics 4) Wang Li, Li Fang Kuei, Chao Yuen Ren, Luo Changpei; (archaeology 4) Li Chi, Liang Ssu-yung, Kuo Mo-jo and Tung Tso-pin; (art history 2) Liang Ssu-cheng and Xu Hongbao; (jurisprudence 6) Wang Shih-Chieh, Wang Chung Hwei, John C. H. Wu, Li Haopei, Kuo Yun-kuan, Yen Shu-Tang; (politics 5) Chou Keng-Shang, Chang C. F., Zhang Xiruo, Tsien T. S. and Hsiao Kung-chuan; (economics 8) Fong H. D., Ho L., Ou Pao-san, Ma Yin-chu, Chen Zong, Yang Hsi-meng, Yang Duan Liu, Lieu D. K.; (sociology 5) Wu Ching-chao, Ling Chunsheng, Chen Ta, Tao L. K. , Pan Kuang-tan.[68]

Out of the 150 official member candidates, 79 (52.67%) had primary academic qualifications. 53 people (35.33%) had primary and second academic qualifications. 17 people had secondary qualifications (11.33%). One person, Li Jiannong, his qualifications were unknown. Out of the 17 candidates that had secondary qualifications, two candidates, Li Shu-Hua and Yeh Chi-Sun, published two papers in international academic journals respectively. They both studied abroad in their early years, and had not obtained significant research results since returning back. Chiang Li-Fu had only published one academic paper in 1945. Although he nominally had primary and secondary qualifications, he actually only had secondary qualifications. In spite of this, the majority of the 150 chosen member candidates either had primary member qualifications or both primary and secondary qualifications.

4.5 Summary

The PCE played an important role in the nomination of member candidates. It not only promoted peoples' nomination, but it also drew on lessons from its earlier experience of electing members of the Second Council of Academia Sinica. The council drew up a reference list of candidates, and established a better framework for later selecting members of Academia Sinica. They also conducted preliminary examinations into the nominees submitted by academic institutions. Among these institutions, Academia Sinica, National University of Peking and National Tsing Hua University all had significant candidates, and these later served as important points of reference to the council. The National University of Peking played a particularly positive role to the PCE in promoting widespread nominees. At the same time, institutions mostly focused on nominating those who had made outstanding academic contributions. Their nominations were relatively concentrated, and they laid an important foundation for the future election of members of Academia Sinica, setting strict academic standards.

National Tsing Hua University put forward a list of worthy member candidates using a democratic but a strict selection process. They also, however, utilized

[68]Zhongyang yanjiuyuan yuanshi houxuanren mingdan ji xuanjupiao中央研究院院士候选人名单及选举票(The List of Candidates of Members of Academia Sinica and the Ballots). Nanjing: Zhongguo dier lishi danganguan中国第二历史档案馆(The Second Historical Archives of China), Total No. 393, File No. 1620.

their close relationship with Academia Sinica and National University of Peking. These social resources to some extent influenced the member candidate evaluation process. For example, Mei Yi-chi's view on the nomination of "overseas researchers" resonated with Adam Pen-Tung Sah, most members of the PCE, and the council. These social resources were an important reason for why some member candidates with high academic achievements were disqualified by the council—either because they had lived abroad for a long period of time and had not returned to work, or because they refused to return to the country.

Candidates needed to meet certain academic criteria, but the council also looked at whether they had taught or held positions in a so-called pseudo-institution in an occupied area. It was related with the special backgrounds that China had a cultural tradition of turning its back on traitors, and China had just experienced the Japanese invasion and the Anti-Japanese War. This reflects the nationalistic tendencies during the election of the first batch of members of Academia Sinica. Kuo Mo-jo, a member of a "heterodox party", passed his examinations and he was later appointed a member of Academia Sinica. This showed that the council prioritized academic rationality and an inclusive attitude over different political positions.

Chapter 5
From the Response to the Official List of Member Candidates to the Official Election

According to *The Regulations for the Election of Members of Academia Sinica*, the official list of member candidates needed to be publicized for 4 months in order to listen to the criticisms from all sides. During the period of publicity, Bo Shang, Wang Hefu, Yuan Hanqing, Fu Ssu-Nien and others put forward sharp criticisms. The PCE attached great importance to the criticisms of all parties and made the authentic-named criticisms extremely confidential and submitted them to the council for discussion. The Fifth Annual Meeting of the council was held in March 1948, and the formal election of the first batch of members of Academia Sinica was conducted according to the scheduled procedures. In the formal election, some criticisms had an impact. After the election, although some scholars raised objections to individual candidates, on the whole, the member election was very successful. Almost all those elected were outstanding scholars in the academic circle, and their academic achievements represented the highest academic level in China at that time.

5.1 Reactions from Bo Shang, Wang Hefu, Yuan Hanqing and Others

The announcement of the official list of member candidates of Academia Sinica could not be said to have aroused strong repercussions in the academic circle and society, but it did attract the attention of many parties. After the announcement was published in the government gazette and the major newspapers in Nanjing and Shanghai on November 15, 1947, 35 people and 3 institutions asked Academia Sinica for an official list of member candidates. The 35 people included not only personnel from universities, independent colleges and research institutions, but also personnel from education departments, secondary schools, hospitals, telecommunication bureaus,

J. Guo, *The Establishment and Reconstruction of the Academician System in China*,
https://doi.org/10.1007/978-981-15-7208-1_5

banks, money houses, factories, county councils, etc.[1] On the day the announcement was published, Ye Wenpei from the Weather Station of Nanjing China Aviation Company immediately asked Academia Sinica for the official list and said, "Your academia has created an academician system today and will hold the election of the first batch of members, which is really a rare event since the great changes in learning took place in our country. Compared with the constitution-making, constitution-establishing, and the national congress representative election that we Chinese people have devoted all our efforts to do for years, it is even more glorious and fortunate. It was reported in the newspaper that the list of candidates for the first batch of members was selected at the Fourth Annual Meeting of the Second Council and publicized according to law. Among them, many were my former teachers, and I am especially pleased with myself."[2] Ye Wenpei's remarks might have represented some people's positive impressions of the official list of member candidates.

However, there were also people who criticized the official list of member candidates. Wang Hefu, a doctor and biologist was among them. On March 6, 1948, he wrote to the PCE to point out the problems with the list, saying that it was "the opinion of mine as well as some doctors' in Peiping. Whether members are valued or not quite depends on the initial selection."[3] There were four opinions:

(1) "There are too many administrators among the candidates. The selection of members can only be based on academic contributions. Even if the head of a cultural and educational organization has made achievements, it is his job to do so and is not worthy of a prize. If a person ignorant and incompetent is selected, it will really cause shame in the academic circle. According to this principle, the following persons should not be selected: Chang Hsiao-chien, Tang F. F., Hung S. L., Li Zongen and Yu T. F..
(2) The biological group seems to be roughly divided into categories, and the number of people is hardly filled. In agronomy, no one has reached the academic level required, so there can be nobody selected in this field. And in the botanical discipline, there are few achievers.
(3) All the members of the Council in Academia Sinica are listed, which is the point mostly criticized by the people. It is better for them to retreat.

---

[1] Gefang hansuo diyici yuanshi houxuanren mingdan各方函索第一次院士候选人名单(People of All Sides Asked for the List of Candidates for the Election of the First Batch of Members of Academia Sinica). Nanjing: Zhongguo dier lishi danganguan中国第二历史档案馆(The Second Historical Archives of China), Total No. 393, File No. 1623.

[2] Gefang hansuo diyici yuanshi houxuanren mingdan各方函索第一次院士候选人名单(People of All Sides Asked for the List of Candidates for the Election of the First Batch of Members of Academia Sinica). Nanjing: Zhongguo dier lishi danganguan中国第二历史档案馆(The Second Historical Archives of China), Total No. 393, File No. 1623.

[3] Gefang dui diyici yuanshi xuanju dangxuan yuanshi houxuanren zhi piping ji yijian各方对第一次院士选举当选院士候选人之批评及意见(Criticisms and Opinions of All Sides on the Candidates for the Election of the First Batch of Members of Academia Sinica). Nanjing: Zhongguo dier lishi danganguan中国第二历史档案馆(The Second Historical Archives of China), Total No. 393, File No. 646.

(4) One member of the medical team should be specially 'corrected', that is, Dr. Ma Wen-chao. This man is old and dotty. He has no research accomplishment, enthusiasm or ability. All his several studies, which were neither fish nor fowl, were written by his technicians, assistants or friends instead of himself. His study on smoking avoidance is filled with fraud and offers no treatment. The lecithin he uses to avoid smoking is mixed with morphine and cannot let people give up smoking. Council members of Academia Sinica, Mr. Robert K. S. Lim and Mr. Wu Hsien, are familiar with this person and they can be asked to testify."[4]

In the first opinion, Wang Hefu believed that "the selection of members can only be based on academic contributions" and should not be based on academic leadership achievements. This view was based on the qualification of members in common use in the world and was worth affirming. He claimed that candidates for members of medicine and agronomy, such as Chang Hsiao-chien, Tang F. F., Hung S. L., Li Zongen and Yu T. F., should not be selected as "heads of cultural and educational organizations", mainly based on this view. At the same time, he also alleged that the five were "ignorant and incompetent" but gave no evidence. In fact, although these five people all had administrative positions, most of them had not completely dropped out of academic studies, and some had made significant achievements. In the formal election of members, Li Zongen and Chang Hsiao-chien were elected in the general election, and Yu T. F. was finally elected, too. They were not much affected by Wang Hefu's criticism. Different from them, Tang F. F. and Hongshilv both failed in the formal election.

Wang Hefu's second opinion was a little biased. For example, Li Hsien-wen, an member candidate of agronomy, had made important achievements in genetics, breeding and cultivation of crops such as wheat and millet. Hu Hsen-hsu, Chien Shung-shu, Loo Tsung-Le and Tai Fon-Land, member candidates of botany, had made important contributions in their respective research fields. Wang Hefu's third opinion was not entirely objective, because Chang Yu-Che and Jaw Jeou-Jang (ex officio members of the Council) and John Lee, Zhang Yun and Tang Yueh (elected members of the Council) were not formal candidates for members. In the fourth opinion, Wang Hefu especially "corrected" Ma Wen-chao, who was the dean of the Department of Anatomy of Medical College, National University of Peking. Among his studies, "avoiding smoking" referred to abstaining from smoking opium. Biographer Li Zhaote said that Ma Wen-chao's research on the diagnosis and treatment of abstaining from opium addiction had achieved results in experimental animals. The treatment was tried out in the hospital attached to Nanjing Smoking Cessation Committee and achieved the expected results.[5] It was not known if Wang Hefu's "correction" was true. However, in the formal election later, Ma Wen-chao won only

[4]Gefang dui diyici yuanshi xuanju dangxuan yuanshi houxuanren zhi piping ji yijian各方对第一次院士选举当选院士候选人之批评及意见(Criticisms and Opinions of All Sides on the Candidates for the Election of the First Batch of Members of Academia Sinica). Nanjing: Zhongguo dier lishi danganguan中国第二历史档案馆(The Second Historical Archives of China), Total No. 393, File No. 646.

[5]Li Zhaote李肇特(2005). Ma Wen-chao马文昭. In China Association for Science and Technology中国科学技术协会(Eds.). *Zhongguo kexue jishu zhuanjia zhuanlue*中国科学技术专家传

one vote in the general election of members and lost the election with 0 votes in all four by-elections.[6]

On March 9, 1948, Yuan Hanqing, professor of the Department of Chemistry and dean of the Department of Chemical Engineering at the National University of Peking, wrote to the Council of Academia Sinica with his name, criticizing the official list of member candidates. The spearhead referred to Chi Yuoh-Fong of chemistry discipline:

> I feel really gratified to hear that your association will soon gather in Nanjing to elect members of Academia Sinica, laying a foundation for the Chinese academic community. Reading the newspaper which published a list of candidates for members, including Mr. Chi Yuoh-Fong in the chemistry discipline, I have some negative ideas: Chemistry is an experimental science which has the least work in China and the least foundation. Therefore, it seems that the candidates for members in experimental science can adopt the attitude of "rather go without than have something shoddy", so as not to be laughed at in the international scientific community. Although Mr. Ji is a senior in chemistry and has published papers occasionally, none of the papers he has done are original. His experimental work is limited to derivatives of 'pyrimidine', which is the scope of work of American chemist Mr. Johnson (i.e. Mr. Ji's teacher). Mr. Johnson's studies have been considered as of second- and third-class papers in the chemical field. For 20 years, Mr. Ji has only limited himself to the continuation of his doctoral thesis. It is tantamount to admitting that he does not have the ability to seek his own path and conduct independent research. There are many organic chemists like Mr. Ji in China. The sole election of Mr. Ji as member today may not convince the domestic chemical circle but surprise the international academic circle.[7]

Chi Yuoh-Fong was a research fellow at the Institute of Chemistry, the National Academy of Peiping. He received a doctor's degree from Yale University in 1928. He was an earlier scholar engaged in organic chemistry research in China. According to Yuan Hanqing, Chi Yuoh-Fong's teacher Johnson was Yale University professor and organic chemist T. B. Johnson. Johnson was an international authority on pyrimidine research. From 1926 to 1928, Chi Yuoh-Fong was engaged in pyrimidine research under Johnson's guidance. After returning to China in 1928, he mainly did the research work in this field and had published more than 50 papers by 1947.[8] That Yuan Hanqing said about Ji "has published papers occasionally" should be

略(Biographies of Chinese Science and Technology Experts), Medicine Part, Volume 1 of Basic Medicine Book. Beijing: Renmin weisheng chubanshe人民卫生出版社.1–14.

[6]Zhongyang yanjiuyuan yuanshi houxuanren mingdan ji xuanjupiao中央研究院院士候选人名单及选举票(The List of Candidates of Members of Academia Sinica and the Ballots). Nanjing: Zhongguo dier lishi danganguan中国第二历史档案馆(The Second Historical Archives of China), Total No. 393, File No. 1620.

[7]Gefang dui diyici yuanshi xuanju dangxuan yuanshi houxuanren zhi piping ji yijian各方对第一次院士选举当选院士候选人之批评及意见(Criticisms and Opinions of All Sides on the Candidates for the Election of the First Batch of Members of Academia Sinica). Nanjing: Zhongguo dier lishi danganguan中国第二历史档案馆(The Second Historical Archives of China), Total No. 393, File No. 646.

[8]Guoli zhongyang yanjiuyuan diyici yuanshi xuanju houxuanren timingce国立中央研究院第一次院士选举候选人提名册(Nomination Lists of the Member Candidates for the Election of the First Batch of Members of Academia Sinica). Nanjing: Zhongguo dier lishi danganguan中国第二历史档案馆(The Second Historical Archives of China), Total No. 393, File No. 1597; Wu Yuanliu吴元鎏(1992). Chi Yuoh-Fong纪育沣. In the Editorial Team of the Complete Dictionary of Scientific

inconsistent with the actual situation. Judging from the voting results of the official election of members, Chi Yuoh-Fong's votes were few. He won only one vote in the general election. In the next four by-elections, he obtained one vote in one election and 0 votes in the other three.[9]

Regarding Yuan Hanqing's and Wang Hefu's criticisms, Adam Pen-Tung Sah instructed on March 15 that the criticisms be mimeographed as a reference record and submitted to the council for discussion and distributed to the participating council members. The mimeographed copies of the two critical opinions were also compiled as the No. 1 and No. 2 extremely confidential documents respectively.[10]

5.2 Response from Fu Ssu-Nien

After Fu Ssu-Nien went to the United States for medical treatment due to illness in June 1947, the PCE had little contact with him. What he received were only two documents, namely, the correspondence voting method for member election and the "Deliberation on the Way to Commemorate Mr. Tsai Yuan-Pei". However, neither of them mentioned the meeting date of the council for the election of the first batch of members of Academia Sinica, but the latter mentioned "the day for the meeting is coming near".[11] On March 9, 1948, after receiving a letter from Ho L., Fu Ssu-Nien knew that the election would be held in March. On that day, he quickly wrote a long letter to Chu Chia-Hua, Wong Wen-hao, Hu Shih, Adam Pen-Tung Sah and Li Chi, the main members of the PCE.[12]

In his letter, Fu Ssu-Nien mainly put forward suggestions and opinions on the upcoming member election and the published official list of member candidates. The

Biography《科学家传记大辞典》编辑组(Eds.). *Zhongguo Xiandai kexuejai zhuanji*中国现代科学家传记(Biographies of Modern Chinese Scientists), Vol. 2. Beijing: 科学出版社.224–232.

[9]Zhongyang yanjiuyuan yuanshi houxuanren mingdan ji xuanjupiao中央研究院院士候选人名单及选举票(The List of Candidates of Members of Academia Sinica and the Ballots). Nanjing: Zhongguo dier lishi danganguan中国第二历史档案馆(The Second Historical Archives of China), Total No. 393, File No. 1620.

[10]Gefang dui diyici yuanshi xuanju dangxuan yuanshi houxuanren zhi piping ji yijian各方对第一次院士选举当选院士候选人之批评及意见(Criticisms and Opinions of All Sides on the Candidates for the Election of the First Batch of Members of Academia Sinica). Nanjing: Zhongguo dier lishi danganguan中国第二历史档案馆(The Second Historical Archives of China), Total No. 393, File No. 646.

[11]Gefang dui diyici yuanshi xuanju dangxuan yuanshi houxuanren zhi piping ji yijian各方对第一次院士选举当选院士候选人之批评及意见(Criticisms and Opinions of All Sides on the Candidates for the Election of the First Batch of Members of Academia Sinica). Nanjing: Zhongguo dier lishi danganguan中国第二历史档案馆(The Second Historical Archives of China), Total No. 393, File No. 646.

[12]Gefang dui diyici yuanshi xuanju dangxuan yuanshi houxuanren zhi piping ji yijian各方对第一次院士选举当选院士候选人之批评及意见(Criticisms and Opinions of All Sides on the Candidates for the Election of the First Batch of Members of Academia Sinica). Nanjing: Zhongguo dier lishi danganguan中国第二历史档案馆(The Second Historical Archives of China), Total No. 393, File No. 646.

PCE mimeographed the letter and provided it to the council as extremely confidential document No. 3 together with extremely confidential documents No. 1 and No. 2. Fu Ssu-Nien's letter had the most weight among the three "extremely confidential documents" and it was divided into several parts: "correspondence voting method", "opinions on the list of candidates", "on principles" and "appendices". The contents of the letter touched on some important issues and details of the member election.

(1) **"Correspondence Voting Method"**

In the letter, Fu Ssu-Nien objected to the voting method of correspondence: "It seems quite questionable whether this method is appropriate." The reasonswere:

(a) Interlaced voting is really difficult to be correct. Therefore, we must first discuss and then vote. Those who did not come to the meeting did not take part in the discussion. They would feel very difficult to vote (Ssu-Nien myself didn't cast all the votes). It is what about recommending people you know.
(b) Although the list has been publicized, what are the criticisms from the academic circle? This review should be carefully considered, then a final list should be drawn up and passed before voting. This is not for correspondence voters to take part in, either.
(c) The letter of January 19th (i.e. the letter of voting sent) contains a proposal to subdivide the number of seats in each group, but it is only for reference. Since it has not been approved by the Council, it can only be referred to. However, if the matter is not approved in the next meeting, how will the votes be counted?
(d) The measure seems to have to be approved by the meeting before it has legal effect.[13]

The minutes of the Fourth Annual Meeting of the Second Council of Academia Sinica do not contain any information on correspondence voting method.[14] It shows that the correspondence voting method was decided by the PCE itself and had not been approved by the Council.Thus, it had no legal effect. The author believes that the reasons stated by Fu Ssu-Nien deserved full affirmation. First of all, the election of members involved more than 20 disciplines. Most Council Members were familiar with the situation of member candidates in the disciplines in which they themselves specialized, but may not know the situation of member candidates in other disciplines. As a result, it was difficult to ensure the fairness of the interweaving voting of Council Members by correspondence voting. So, Council Members would discuss before voting, which was a good way to avoid blind voting by the lay Members or voting

[13]Gefang dui diyici yuanshi xuanju dangxuan yuanshi houxuanren zhi piping ji yijian各方对第一次院士选举当选院士候选人之批评及意见(Criticisms and Opinions of All Sides on the Candidates for the Election of the First Batch of Members of Academia Sinica). Nanjing: Zhongguo dier lishi danganguan中国第二历史档案馆(The Second Historical Archives of China), Total No. 393, File No. 646.

[14]Zhongyang yanjiuyuan pingyihui dierjie disici nianhui jilu中央研究院评议会第二届第四次年会纪录(Minutes of the Fourth Annual Meeting of the Second Council of Academia Sinica). Nanjing: Zhongguo dier lishi danganguan中国第二历史档案馆(The Second Historical Archives of China), Total No. 393, File No. 1558.

without a target. Secondly, the Members conducted the correspondence voting before the end of the publicity of member candidates, they did not have time to refer to the critical opinions of the academic community. This would certainly affect the fairness of the election. Thirdly, there was no legal basis for the criteria for subdividing the number of members in the three groups, mathematical, biology and humanities, because they were not approved by the Council.

According to the official election of members, the Council mayhave considered Fu Ssu-Nien's opinions. Before the formal election, 11 council members took part in the correspondence voting. In the formal election, the Council only used the results of correspondence voting as a reference. Some elected members did not receive half of the votes from correspondence,such as Hsu Pao-Lu, 5 votes, in the mathematical group; Pai Sitsan, 5 votes, in the biology group; Woo Tsin-Hang, 5 votesand Yu Cha-Shih, 4 votes, in the humanities group. Some member candidates with higher correspondence votes lost the election instead, such as Sun Xuewu, who received 10 votes in the mathematical group, Hu C. H., who received 9 votes in the biology group, and Luo Changpei, who received 10 votes in the humanities group.[15]

(2) "**Opinions on the List of Candidates**"

In the letter, Fu Ssu-Nien expressed his attitude towards the official list of member candidates: "Since Ssu-Nien went abroad for medical treatment, I have received several documents concerning the member election. Although Ssu-Nien still have some opinions on the list of candidates, it is generally careful, fair and admirable. Your hard work has not been easy till this point."[16] Obviously, Fu Ssu-Nien basically approved of the official list, but it was not without opinions. First, he advocated the deletion of Liu W. T. Fu Ssu-Nien said:

> There are indeed candidates who should be deleted, such as Mr. Liu W. T. Mr. Liu's previous *Notes on Three Yus* is a masterpiece, but his contribution cannot be equal to that of Yu, Hu, Tang, Zhang and Yang.[17] Talking about the contribution of a scholar, his final work is the most important. Mr. Liu collated *Zhuang-zi* and was very conceited. Just by chance Mr. Wang Shu-min, a research assistant at the Institute of History and Philology, once reviewed it (Mr. Wang has also studied the subject) and found countless mistakes. Mr. Liu, with such an attitude towards collation, really cannot be taken as a model. If Mr. Liu is included in the list, young students should think differently. Even worse, Mr. Liu claimed to be "Er-Yun Jushi (meaning, 2-Yun Hermit)" in Kunming. Er Yun referred to Yunnan ham and Yunnan opium. He was once a guest of the chieftain who gave him a large amount of crude opium.

[15]Zhongyang yanjiuyuan diyici yuanshi xuanju houxuanren fenzu taolun toupiao qingkuangbiao中央研究院第一次院士选举候选人分组讨论投票情况表(Voting Tables of Groups Discussion on Candidates for the Election of the First Batch of Members of Academia Sinica). Nanjing: Zhongguo dier lishi danganguan中国第二历史档案馆(The Second Historical Archives of China), Total No. 393, File No. 1619.

[16]Gefang dui diyici yuanshi xuanju dangxuan yuanshi houxuanren zhi piping ji yijian各方对第一次院士选举当选院士候选人之批评及意见(Criticisms and Opinions of All Sides on the Candidates for the Election of the First Batch of Members of Academia Sinica). Nanjing: Zhongguo dier lishi danganguan中国第二历史档案馆(The Second Historical Archives of China), Total No. 393, File No. 646.

[17]That is, Yu Cha-Shih, Hu Shih, Tang Lan, Zhang Yuanji and Yang Shu-Tah.

> He smoked and sold the opium when he returned, and was dismissed by Tsinghua University and Southwest Associated University. This was known to everyone in Kunming at that time. Ssu-Nien, having written here in this letter, of course is legally responsible for what I have said. Won't it be a joke that Mr. Liu is included in the list of candidates and if he is elected?[18]

One of the main reasons that Fu Ssu-Nien pointed out for Liu W. T.'s insufficient member level was Wang Shu-min's examination of *Zhuang-zi*, which was collated by Liu W. T. When the PCE printed extremely confidential documents, it deleted the sentence in the original letter that "Just by chance Mr. Wang Shu-min, a research assistant at the Institute of History and Philology, once reviewed it (Mr. Wang has also studied the subject)". Regarding Wang Shu-min's criticism of Liu W. T., Shiah Nae mentioned it in his letter to Fu Ssu-Nien on October 20, 1947: Wang Shu-min once said that "Mr. Liu W. T.'s *Huai-nan-zi* and *Zhuang-zi*had very poor collation and textual research, and said if Mr. Fu attended,[19] he would not be recommended as a candidate".[20] Chien Mu, a colleague of Liu W. T.'s at Southwest Associated University, had detailed memories of Liu W. T.'s dismissal from Tsing Hua University and Southwest Associated University for smoking crude opium and taking the opium as his reward for writing tombstone's epitaph for the chieftain.[21] When Fu Ssu-Nien pointed out Liu W. T.'s misdeeds, he used the serious words"even worse". It showed that he thought because of this misdeed, Liu W. T. could not be selected as a member at all.

Secondly, Fu Ssu-Nien pointed out that there were still people on the official list of member candidates who had served in puppet universities and other institutions, and suggested that the PCE should carefully examine them. Fu Ssu-Nien said: "I heard that the last meeting decided that those who had taught in puppet schools or held puppet posts were not included (according to Mr. Shiah's letter). There are still some people on the list now (e.g. in medicine, more than one person). Ssu-Nien will not contribute any opinion on how to decide this matter. However, there must be equality. No one should be given preference over the other. I dare ask you to examine the matter carefully."[22] Mr. Shiah, namely Shiah Nae. During Fu Ssu-Nien's medical

---

[18]Gefang dui diyici yuanshi xuanju dangxuan yuanshi houxuanren zhi piping ji yijian各方对第一次院士选举当选院士候选人之批评及意见(Criticisms and Opinions of All Sides on the Candidates for the Election of the First Batch of Members of Academia Sinica). Nanjing: Zhongguo dier lishi danganguan中国第二历史档案馆(The Second Historical Archives of China), Total No. 393, File No. 646.

[19]It refers to the Fourth Annual Meeting of the Second Council held from October 15th to 17th, 1947.

[20]Pan Kuang-che潘光哲(2005). Zhongyanyuan de "yiqian ling yiye" 中研院的"一千零一夜"(Academia Sinica's "Arabian Nights"). *Shuwu*书屋(Book House) (2): 51.

[21]Chien Mu钱穆(2003). Yi Liu Shuya (Liu W. T.) 忆刘叔雅(Recall Liu Shuya). In Liu Pingzhang刘平章(Eds.). *Liu W. T. chuanwen yishi*刘文典传闻轶事(Anecdotes of Liu W. T.). Kunming: Yunnan meishu chubanshe云南美术出版社.21–23.

[22]Gefang dui diyici yuanshi xuanju dangxuan yuanshi houxuanren zhi piping ji yijian各方对第一次院士选举当选院士候选人之批评及意见(Criticisms and Opinions of All Sides on the Candidates for the Election of the First Batch of Members of Academia Sinica). Nanjing: Zhongguo dier lishi danganguan中国第二历史档案馆(The Second Historical Archives of China), Total No. 393, File No. 646.

treatment in the United States, Shiah Nae kept him informed of the preparations for the member election.[23] Fu Ssu-Nien hated the cultural traitors who worked in the puppet institutions or organizations in the occupied areas during the Anti-Japanese War. After acting as president of the National University of Peking in September 1945, he made successive statements, firmly refusing to employ any person from the puppet-school or puppet-organization to teach in the school, and reported on the crimes of traitors such as the Presidents of puppet- University of Peking Bao Jianqing, Wang Yintai and Qian Daosun,[24] He put forward the proposal to the PCE, which not only showed his idea of treating member candidates equally, but also showed his strong nationalism.

Thirdly, Fu Ssu-Nien believed that the candidates of medicine discipline were "the most problematic". He pointed out: "Some of them wrote only a few essays, which were not recognized by international scholars. Those who are responsible for medical education in China are well-known internationally for their contributions to the people and academic circles. It is unfair for those to be elected on the basis of articles that have not yet been decided and for those well-known leaders of the medical profession to be turned away from inclusion."[25] He suggested that when selecting members in applied disciplines such as medicine and agronomy, more consideration should be given to those who met the second qualification. Medical members "only three or four people will be elected this year, at least half of them should be those who are responsible for medicine." If the method was difficult, he suggested that Qi Shounan and James K. Shen be added to the candidates for medical members, and Xie Jiasheng be added to the candidates for agricultural members. In his view, mathematical science, biology and humanities "have different functions in society and academia, and are also different in theirdevelopment stagesin current China. Take humanities as an example, literature, history, philosophy, language, archaeology, etc., the second qualification may not be applicable, because it is not necessary at this stage of study. But this viewpoint is not appropriate for applied science. Both medicine and agronomyare applied sciences, and the second qualification should be considered."[26]

---

[23]Luo Feng罗丰(2004). Shiah Nae yu zhongyang yanjiuyuan diyije yuanshi xuanju夏鼐与中央研究院第一届院士选举(Shiah Nae and The Election of the First Batch ofMembers of Academia Sinica). *Kaogu yu wenwu*考古与文物(Archaeology and Cultural Relics)(4): 84–89.

[24]1946 nian 7 yue 17 ri Fu Ssu-Nien zhi Xie Guansheng xin 1946年7月17日傅斯年致谢冠生信(Fu Ssu-Nien's letter to Xie Guansheng on July 17th, 1946). In Ouyang Zhesheng欧阳哲生(Eds.) (2003). Fu Ssu-Nien quanji傅斯年全集(Complete Works of Fu Ssu-Nien), Vol. 7. Changsha: Hunan jiaoyu chubanshe湖南教育出版社.308.

[25]Gefang dui diyici yuanshi xuanju dangxuan yuanshi houxuanren zhi piping ji yijian各方对第一次院士选举当选院士候选人之批评及意见(Criticisms and Opinions of All Sides on the Candidates for the Election of the First Batch of Members of Academia Sinica). Nanjing: Zhongguo dier lishi danganguan中国第二历史档案馆(The Second Historical Archives of China), Total No. 393, File No. 646.

[26]Gefang dui diyici yuanshi xuanju dangxuan yuanshi houxuanren zhi piping ji yijian各方对第一次院士选举当选院士候选人之批评及意见(Criticisms and Opinions of All Sides on the Candidates for the Election of the First Batch of Members of Academia Sinica). Nanjing: Zhongguo dier lishi danganguan中国第二历史档案馆(The Second Historical Archives of China), Total No. 393, File No. 646.

At the same time, Fu Ssu-Nien pointed out that Robert K. S. Lim was the only council member of medical discipline in Academia Sinica, and his subject was not in clinical medicine. "If eight people(this number is mentioned in the letter) are elected from the one-member organ, it is inevitable to arouse criticism." Regarding the number of members to be selected for the medical discipline of the biology group, it was revised from 4 to 8 and 7.[27] Fu Ssu-Nien said "eight people" as mentioned in the letter, which referred to the number of members to be elected for medical science as revised by the PCE. Besides that, Fu Ssu-Nien stressed: "It is not a good idea to have more persons elected based on one person. Being partial to one school (Peiping Union Medical College) is not the way, either. This is a very straight remark, but I respectfully ask for consideration. All this is just for a great event and consideration for the future of Academia Sinica."[28] This showed that Fu Ssu-Nien was not only worried that the election of members of medicine discipline would provoke criticism, but also worried that Robert K. S. Lim, who had been teaching in Peiping Union Medical College for a long time, would prefer the people in the college to others in the election.

In this part, Fu Ssu-Nien also briefly introduced the situation of Qi Shounan, James K. Shen and Xie Jiasheng: "Qi Shounan has handled the Medical College of Central University for many years, and its contribution is comparable to that of Mr. Yeh Chi-Sun of the physics team (Ye has handled the Department of Physics in Tsing Hua University)"; "James K. Shen has presided over the technical matters of the Department of Health and the Surgery Department of Shanghai Medical College for many years, with remarkable achievements"; Xie jiasheng "has been responsible for administration for many years and has achieved fruitful results for a long time".[29]

Fu Ssu-Nien's main opinions on the official list of member candidates did indeed have an impact on the council. Although Fu Ssu-Nien's original letter about smoking and selling crude opium in Liu W. T. was deleted when the PCE printed the extremely confidential documents, in the formal election of members, the number of votes in the general election and the four by-elections for Liu W. T. were all 0.[30] The medical members were elected 3 from 9 candidates: Li Zongen, Yuan I-Chin and Chang

---

[27]Zhongyang yanjiuyuan diyici yuanshi xuanju timing biaoge中央研究院第一次院士选举提名表格(Nomination Forms for the Election of the First Batch of Members of Academia Sinica). Nanjing: Zhongguo dier lishi danganguan中国第二历史档案馆(The Second Historical Archives of China), Total No. 393, File No. 2600 (4).

[28]Gefang dui diyici yuanshi xuanju dangxuan yuanshi houxuanren zhi piping ji yijian各方对第一次院士选举当选院士候选人之批评及意见(Criticisms and Opinions of All Sides on the Candidates for the Election of the First Batch of Members of Academia Sinica). Nanjing: Zhongguo dier lishi danganguan中国第二历史档案馆(The Second Historical Archives of China), Total No. 393, File No. 646.

[29]Gefang dui diyici yuanshi xuanju dangxuan yuanshi houxuanren zhi piping ji yijian各方对第一次院士选举当选院士候选人之批评及意见(Criticisms and Opinions of All Sides on the Candidates for the Election of the First Batch of Members of Academia Sinica). Nanjing: Zhongguo dier lishi danganguan中国第二历史档案馆(The Second Historical Archives of China), Total No. 393, File No. 646.

[30]Zhongyang yanjiuyuan yuanshi houxuanren mingdan ji xuanjupiao中央研究院院士候选人名单及选举票(The List of Candidates of Members of Academia Sinica and the Ballots). Nanjing:

Hsiao-chien. Those three elected members were all directors of medical institutions: Li Zongen was the president of Peiping Union Medical College and had founded and presided over Guiyang Medical College for 7 years; Yuan I-Chin served as the president of Nanjing Tuberculosis Prevention and Control Institute of the Ministry of Health, and had been the director of the Laboratory of Epidemic Prevention of the Central Health Experimental Institute of the Ministry of Health since 1942; Chang Hsiao-chien was the president of Xiangya Medical College and had presided over it for more than 5 years. The unsuccessful candidates for medical members included Hu C. H., Hung S. L., Ma Wen-chao, Tang F. F., Feng Lan-chou and Liu S. H. Among them, Feng Lan-chou, a former professor at the puppet-Peking University medical school, won only one vote[31] in the general election and four by-elections.

However, Fu Ssu-Nien's proposal to add Qi Shounan, James K. Shen and Xie Jiasheng to the list of member candidates for the election was not accepted. It was because, according to *The Regulations for the Election of Members of Academia Sinica*, the addition of new members to the member candidate list required that the candidate had been nominated in advance, and could only be added to the member candidate list with the written proposal by 10 council members and the consent of more than half of the council members present.[32] However, only Qi alone in the three had been nominated.[33] When the official list of member candidates was approved at the Fourth Annual Meeting of the Second Council, it was not decided that the list could be amended by announcement.[34] What Shiah Nae later said in his report to

---

Zhongguo dier lishi danganguan中国第二历史档案馆(The Second Historical Archives of China), Total No. 393, File No. 1620.

[31]Pingjinqu riwei zuzhi suoban zhuanke yishang jiaozhuiyuan yilanbiao平津区日伪组织所办专科以上教职员一览表(A List of Staff at Colleges and Universities Run by Japanese and Puppet Organizations in Peiping and Tianjin). Nanjing: Zhongguo dier lishi danganguan中国第二历史档案馆(The Second Historical Archives of China), Total No. 5, File No. 2921; Zhongyang yanjiuyuan yuanshi houxuanren mingdan ji xuanjupiao中央研究院院士候选人名单及选举票(The List of Candidates of Members of Academia Sinica and the Ballots). Nanjing: Zhongguo dier lishi danganguan中国第二历史档案馆(The Second Historical Archives of China), Total No. 393, File No. 1620.

[32]Zhongyang yanjiuyuan juban diyici yuanshi xuanju tuijian houxanren mingdan ji youguan guizhang banfa he laiwang wenshu中央研究院举办第一次院士选举推荐候选人名单及有关规章办法和来往文书(List of Recommended Candidates for the Election of the First Batch of Members of Academia Sinica and Relevant Regulations, Methods and Correspondence). Beijing: Qinghua daxue danganguan清华大学档案馆(Archives of Tsinghua University), Total No. 1, Cat. No. 4-2, File No. 191.

[33]Dongbei daxue deng yuanxiao hansong yuanshi houxuanren timingdan ji youguan hanjian东北大学等院校函送院士候选人提名单及有关函件(The Nomination Lists of Member Candidates Sent by Northeastern University and Other Universities and Colleges and Related Correspondence). Nanjing: Zhongguo dier lishi danganguan中国第二历史档案馆(The Second Historical Archives of China), Total No. 393, File No. 1613.

[34]Zhongyang yanjiuyuan pingyihui dierjie disici nianhui jilu中央研究院评议会第二届第四次年会纪录(Minutes of the Fourth Annual Meeting of the Second Council of Academia Sinica). Nanjing: Zhongguo dier lishi danganguan中国第二历史档案馆(The Second Historical Archives of China), Total No. 393, File No. 1558.

Fu Ssu-Nien "just according to the regulations, no one can be added to the list of candidates, so Mr. Qi Shounan and others cannot be added"[35] referred to this matter.

(3) **"On Principles"**

In his letter, Fu Ssu-Nien also put forward suggestions and opinions on the principles to be adhered to in the member election. First of all, he suggested that the election "would rather be short of than have someone incompetent". He said: "This is pioneering work of Academia Sinica and also pioneering work of China. A quota of 100 can be filled with a few less, even a quota of 80 is not necessarily to be filled full. From the legal point of view, excess is illegal, while deficiency is not illegal. If there is any suspicion, just let it go without or leave it reserved for the General Assembly of Members or till the next meeting. To make up the number is to be blamed, but it won't hurt if the quotais not fully filled."[36] "A quota of 80" and "A quota of 100" referred to 80–100 members for the first term as stipulated in *The Organization Act of Academia Sinica*. The author believes that Fu Ssu-Nien's statement is an important principle that must be strictly abided by both the member election at that time and the academician election today. Because only by adhering to a strict and high-standard election system can the elected and the title ofmembers be guaranteed to match, thus maintaining the dignity of member, a noble academic honorary title. Indiscriminate selection will not only affect the leadership of Academia Sinica in China, but also affect China's reputation in international academia, and lead to endless harm to China's academic development.

Secondly, Fu Ssu-Nien once again made comments on the subdivision of the number of members in the three groups of mathematics and physics, biology, and humanities. He said: "The number of team members (i.e. the subdivision of the group number) seems to be stipulated, but it is very difficult to stipulate. Eight candidates for medicine seem to be too many from the list of candidates. It is really impossible to choose eight from nine (in the list). The third group should have more candidates in the department of literature and history than in the department of social sciences." At the same time, he also put forward suggestions for the election of young members: "Younger ones should be elected. However, if he/she has only one or two articles published in foreign magazines, or even goes abroad equivalent to being a student, and becomes a member suddenly, it will also surprise the international academic circle!"[37]

---

[35]Luo Feng罗丰(2004). Shiah Nae yu zhongyang yanjiuyuan diyije yuanshi xuanju夏鼐与中央研究院第一届院士选举(Shiah Nae and The Election of the First Batch ofMembers of Academia Sinica). *Kaogu yu wenwu*考古与文物(Archaeology and Cultural Relics)(4): 87.

[36]Gefang dui diyici yuanshi xuanju dangxuan yuanshi houxuanren zhi piping ji yijian各方对第一次院士选举当选院士候选人之批评及意见(Criticisms and Opinions of All Sides on the Candidates for the Election of the First Batch of Members of Academia Sinica). Nanjing: Zhongguo dier lishi danganguan中国第二历史档案馆(The Second Historical Archives of China), Total No. 393, File No. 646.

[37]Gefang dui diyici yuanshi xuanju dangxuan yuanshi houxuanren zhi piping ji yijian各方对第一次院士选举当选院士候选人之批评及意见(Criticisms and Opinions of All Sides on the Candidates for the Election of the First Batch of Members of Academia Sinica). Nanjing: Zhongguo dier lishi

From the official election results ofmembers later, only 3 medical scholars were elected; in the humanities group, 20 members were elected from "the literature and history department"and 8 from social sciences, the former is more than twice as many as the latter. Of the 81 people elected, 78 were over 40 years old, accounting for the vast majority; only 3 were under 40, namely Hua Loo-Keng (38), Hsu Pao-Lu (38) and Shiing-Shen Chern (37). At that time, they had all made world-class research results. Therefore, Fu Ssu-Nien's worries about the election of medical members, the number of members between the literature and history department and the social science department in the humanities group, and the election of young members did not appear in the election.

Thirdly, Fu Ssu-Nien suggested: "It seems that the election should not be concentrated on one party or one school (or one or two schools)". According to the distribution of the work of the elected members, the phenomenon did not occur. Of the 81 people elected,40 were distributed in Academia Sinica (20), the National University of Peking (10) and National Tsing Hua University (10);8 were distributed in National Central University (4) and the University of Chekiang (4); the other 33 were distributed in the National Academy of Peiping, the Fan Memorial Institute of Biology, the National Geological Survey of China, the Catholic University of Peking, National Wu-han University, Sichuan University, Hunan University, Shandong University, Fudan University, Taiwan University, National Social Education College, Peiping Union Medical College, National Health Experimental Institute, Institute for the Prevention and Cure of Tuberculosis in Nanjing of Ministry of Health, Xiangya Medical College, Resources Committee, Yongli Chemical Industry Company, Ministry of Communications, Ministry of Foreign Affairs, Commercial Press, and foreign universities and research institutions. However, as many as 20 people elected were working in Academia Sinica, which was somewhat concentrated. After the election of members, there was also some discussion in the academic circle. Xiang Da, a professor at Peking University, called this "arouse a kind of feeling that scholars come from royal family or officials". Shiah Nae believed that there was inevitably some"proximity" or "affection" effect.[38]

(4) "**Appendices**"

At the end of the letter, Fu Ssu-Nien put "various items attached". First of all, he suggested that Sun Pen-wen be included in the list of candidates for member of sociology. He said: Sociology "is not developed in China. If you want the next best thing, you should include Mr. Sun Pen-wen. His booksare well understood, and his efforts made cannot be compared by thosewho have only published one or two essays. Therefore, I dare to suggest that he be included in the candidate list."[39] Sun Pen-wen

danganguan中国第二历史档案馆(The Second Historical Archives of China), Total No. 393, File No. 646.

[38]Shiah Nae夏鼐(1948). Zhongyang yanjiuyuan diyijie yuanshi de fenxi中央研究院第一届院士的分析(The Analysis of the First Batch of Members of Academia Sinica). *Guancha*观察(Observation) 5 (14): 5.

[39]Gefang dui diyici yuanshi xuanju dangxuan yuanshi houxuanren zhi piping ji yijian各方对第一次院士选举当选院士候选人之批评及意见(Criticisms and Opinions of All Sides on the Candidates

was a professor at the National Central University and had made achievements in social psychology, modern sociology and the history of social thought. Both National Tsing Hua University and National Central University nominated him as a candidate for member of Academia Sinica. However, he was not approved when the council examined the qualification of member candidates. Despite Fu Ssu-Nien's suggestion, the council did not add Sun Pen-wen to the official list of member candidates. The reason was also that Fu Ssu-Nien's proposal did not conform to the provisions of the The *Regulations for the Election of Members of Academia Sinica*.

As a matter of fact, Fu Ssu-Nien's suggestion was made against Pan Kuang-tan, a candidate for member of social sciences. However, when the PCE printed the extremely confidential documents, it deleted the words about Pan Kuang-tan in Fu Ssu-Nien's original letter. Fu Ssu-Nien wrote in the original letter: "In sociology, there is Mr. Pan Kuang-tan. Mr. Pan is a wise man. However, the conclusion of his genealogy cannot be justified. He took the imperial examination as an example to prove that Suzhou people have superior talents, which shows that he really doesn't know what the imperial examination was. Suzhou people can be superior in talent, but Pan's statement cannot be true. There was a common practice in the imperial examination, which could be studied, thatthe examiners and the selected were in tune.Mr. Pan's statement has just ignored this historical fact. Therefore, Mr. Pan's work seemed unmatched with Chen Ta's."[40] Visibly, the main reason for Fu Ssu-Nien to disagree with Pan Kuang-tan was that Pan's conclusion of his genealogical study did not hold water. At the same time, Fu Ssu-Nien also disapproved of Pan Kuang-tan's articles published in newspapers, saying that "one's academic pursuits can't be judged by newspaper writings".[41] According to the results of the official election of members later, Pan Kuang-tan won 0 votes in the general election as well as four by-elections.[42] Because the passage about Pan Kuang-tan in Fu Ssu-Nien's original letter was deleted before the election, the result of the election was not directly related to Fu Ssu-Nien, but it showed that apart from Fu Ssu-Nien, other Council Members did not recognize Pan Kuang-tan's academic standards.

---

for the Election of the First Batch of Members of Academia Sinica). Nanjing: Zhongguo dier lishi danganguan中国第二历史档案馆(The Second Historical Archives of China), Total No. 393, File No. 646.

[40]Gefang dui diyici yuanshi xuanju dangxuan yuanshi houxuanren zhi piping ji yijian各方对第一次院士选举当选院士候选人之批评及意见(Criticisms and Opinions of All Sides on the Candidates for the Election of the First Batch of Members of Academia Sinica). Nanjing: Zhongguo dier lishi danganguan中国第二历史档案馆(The Second Historical Archives of China), Total No. 393, File No. 646.

[41]Gefang dui diyici yuanshi xuanju dangxuan yuanshi houxuanren zhi piping ji yijian各方对第一次院士选举当选院士候选人之批评及意见(Criticisms and Opinions of All Sides on the Candidates for the Election of the First Batch of Members of Academia Sinica). Nanjing: Zhongguo dier lishi danganguan中国第二历史档案馆(The Second Historical Archives of China), Total No. 393, File No. 646.

[42]Zhongyang yanjiuyuan yuanshi houxuanren mingdan ji xuanjupiao中央研究院院士候选人名单及选举票(The List of Candidates of Members of Academia Sinica and the Ballots). Nanjing: Zhongguo dier lishi danganguan中国第二历史档案馆(The Second Historical Archives of China), Total No. 393, File No. 1620.

Secondly, Fu Ssu-Nien suggested that the Council should select more members from geological disciplines of the mathematical group and literary and historical disciplines of humanities group. He said: "It seems that the subdivision of the number in each group should be based on the stage of development of the science in China,but not abstractly on the size of its scope. Therefore, the number of geological people in the first group should be large as this science has been developed in China for a long time. In the third group, there should also be a large number of people in literature and history, thestudy of which has a long tradition in China."[43] The development of humanities and social sciences in modern China is different. The former was developed by inheriting the system of down-to-earth learning in the Qing Dynasty, while the latter is a new discipline transplanted from the West and its foundation was weaker before the 1940s. Among many disciplines of natural science, geology developed the earliest in China. By the 1930s, the subject had developed quite well in China. Generally speaking, a relatively developed discipline has relatively more scholars with outstanding achievements. Therefore, it was natural for Fu Ssu-Nien to advocate that the Council elec more members in geology and humanities in the election of the first batch of members of Academia Sinica.

The results of the election show that there were indeed many members in geology and humanities. Among those elected, there were 28 members in the mathematical group, distributed in 6 disciplines: mathematics (5), physics (7), chemistry (4), geology (6), meteorology (1) and engineering (5). Among them, the number of members of geology is ranked second, only one less than that of physics. There were 28 people in the humanities group, distributed in 10 disciplines: philosophy (4), literature and history (4), historical science (5), linguistics (2), archaeology (4), art history (1), law (2), politics (3), economy (1) and society (2). Among them, there were only 8 members in social sciences (law, politics, economy and society).

At last, Fu Ssu-Nien explained his opinions on the official list of member candidates: "All the above-mentioned issues concerning the list of candidates assume that the list can be amended and approved by the announcement. The above suggestions are made on this assumption. These suggestions are Ssu-Nien's criticism of the list. If it is a foregone conclusion that the coming meeting cannot amend it (I don't know what the decision was last year), it will be abandoned."[44] As the council had not previously decided that the list of candidates for members could be amended by the announcement, Fu Ssu-Nien's criticism had to be dropped. Of course, this was mainly aimed at Fu Ssu-Nien's proposal to add new members to the official list of member

[43]Gefang dui diyici yuanshi xuanju dangxuan yuanshi houxuanren zhi piping ji yijian各方对第一次院士选举当选院士候选人之批评及意见(Criticisms and Opinions of All Sides on the Candidates for the Election of the First Batch of Members of Academia Sinica). Nanjing: Zhongguo dier lishi danganguan中国第二历史档案馆(The Second Historical Archives of China), Total No. 393, File No. 646.

[44]Gefang dui diyici yuanshi xuanju dangxuan yuanshi houxuanren zhi piping ji yijian各方对第一次院士选举当选院士候选人之批评及意见(Criticisms and Opinions of All Sides on the Candidates for the Election of the First Batch of Members of Academia Sinica). Nanjing: Zhongguo dier lishi danganguan中国第二历史档案馆(The Second Historical Archives of China), Total No. 393, File No. 646.

candidates, while his proposal to delete Liu W. T. still had significant repercussions in the Council.

5.3 Official Election of the First Batch of Members

The official list of member candidates was publicized till March 15, 1948, which was the expiration day of 4 months. From March 25 to 27, the Fifth Annual Meeting of the Second Council was held in the auditorium of Academia Sinica. The main task of the meeting was to formally elect the first batch of members of Academia Sinica. Chu Chia-Hua, Wong Wen-hao, Adam Pen-Tung Sah, Wang Shih-Chieh, Wang Chia-Chi, John Lee, Woo Ting-liang, Woo Sho-Chow, Li Chi, Li Shu-Hua, Ping Chi, Chou Keng-Shang, Chow Jen, Robert K. S. Lim, Hu Shih, Hu Hsen-hsu, Mao T. E., Chen Yuan, Chen Shisan C., Chang Yu-Che, Chuang Chang-Kong, Ling Hung-hsun, Chien Shung-shu, Jaw Jeou-Jang and Loo Tsung-Le, 25 council members attended. Tang Yueh, Tao L. K., Chen Yin-k'o, Hsieh Chia-yung and Tai Fon-Land asked for leave. Wang C. Y., Wang Ging-Hsi, Ho L., Woo Y. H., Chiang Li-Fu, Hou T. P., Zhang Yun, Tseng Chao-lun, Fu Ssu-Nien and Chao Yuen Ren, who were abroad and Lee J. S., who was on his way abroad, did not attend the meeting.[45]

The meeting opened at 10 am on March 25. Attending the opening ceremony were President Dai Jitao of the National Government's Examination Yuan, President Zhang Qun of the Executive Yuan, President Yu Youren and Vice President Liu Zhe of the Supervision Yuan, and Secretary General of the Executive Yuan, Gan Naiguang. Other participants included those research associate or above in Academia Sinica, such as Guo Baojun, Shiah Nae, Lao Gan and Yang Shifeng from the Institute of History and Philology, and Chen Zhiqiang and Chen Zongqifrom the Institute of Meteorology.[46] At the opening ceremony, Chairman Chu Chia-Hua first introduced the main tasks of the meeting, the purpose of having members in Academia Sinica and the significance of the meeting:

> The term of office of the current council members will expire on July 3 this year. The members of the next term will be elected by the General Assembly of Members of Academia Sinica. The list of candidates for members of Academia Sinica was sent to the government gazette and local newspapers on November 15th last year after the last meeting was drawn up. Four months have passed and the council should immediately discuss and vote on the publicized list according to law. Therefore, the main task of this meeting is to elect members of Academia Sinica. To have members is to complete the system of Academia Sinica itself and to set up our academic organization. It is the 20th anniversary of our establishment of

[45]Zhongyang yanjiuyuan dierjie diyi san si wuci nianhui jilu中央研究院评议会第二届第一、三、四、五次年会纪录(Minutes of the First, Third, Fourth and Fifth Annual Meetings of the Second Council of Academia Sinica). Nanjing: Zhongguo dier lishi danganguan中国第二历史档案馆(The Second Historical Archives of China), Total No. 393, File No. 2551.

[46]Zhongyang yanjiuyuan dierjie diyi san si wuci nianhui jilu中央研究院评议会第二届第一、三、四、五次年会纪录(Minutes of the First, Third, Fourth and Fifth Annual Meetings of the Second Council of Academia Sinica). Nanjing: Zhongguo dier lishi danganguan中国第二历史档案馆(The Second Historical Archives of China), Total No. 393, File No. 2551.

> Academia Sinica now. In the past 20 years, our academic research has made great progress, not only has the number of researchers increased day by day, but also many achievements and contributions have been made in our work. This election of members of Academia Sinica is still a pioneering work in our country. In fact, it was necessary a long time ago. However, it is only now that it is possible. In other words, it is only with the progress made in the past 20 years can our Academia Sinica hold the member election now. Therefore, this meeting is of epoch-making significance in our academy and even in the academic history of China.[47]

After that, Chu Chia-Hua pointed out the nature of member of Academia Sinica, it is "a lifetime honorary post, which is the highest academic honor". He introduced the two qualifications of members and the steps of the election. Finally, he said with deep emotion: "Since its establishment, our Academia Sinica has a history of 20 years. During these 20 years, there have been many national disasters and social unrest. The eight-year of the Anti-Japanese War was even more arduous. My colleagues here in the academy who are engaged in research have worked continually and made achievements even though the working conditions are impossible and life is uneasy. At the same time, various academic and research institutions in the country have also made great achievements in the past 20 years amid hardships. It is really gratifying. After the meeting produced members of Academia Sinica, the system of Academia Sinica will be completed. Then we'll promote international cooperation and make more efforts in Chinese academic research. Our research work will enter a new stage and its progress should be more rapid. We are looking forward to contributing more to the country and mankind."[48] Obviously, Chu Chia-Hua was full of expectation for the member election (Fig. 5.1).

At 3: 00 pm on March 25, the annual meeting held its First General Meeting. Among the council members who attended the opening ceremony in the morning, only Wang Shih-Chieh was absent. When reporting on the affairs of Academia Sinica at the meeting, Adam Pen-Tung Sah introduced the process of the publicity of member candidates, compilation of catalogs of member candidates' works, preparation and discussion of member election methods, the purpose of correspondence voting, and the drafting of *The Draft Regulations for the General Assembly of Members of Academia Sinica*.[49] The meeting amended and approved the *The Draft Regulations for the General Assembly of Members of Academia Sinica*, which was named *The Regulations for the General Assembly of Members of Academia Sinica*. The next day, the council began the formal election of members, which consisted

---

[47]Zhongyang yanjiuyuan dierjie pingyihui tiaoli jilu ti'an ji baogao中央研究院第二届评议会条例、记录、提案及报告(Regulations, Records, Proposals and Reports of the Second Council of Academia Sinica). Nanjing: Zhongguo dier lishi danganguan中国第二历史档案馆(The Second Historical Archives of China), Total No. 393, File No. 2926.

[48]Zhongyang yanjiuyuan dierjie pingyihui tiaoli jilu ti'an ji baogao中央研究院第二届评议会条例、记录、提案及报告(Regulations, Records, Proposals and Reports of the Second Council of Academia Sinica). Nanjing: Zhongguo dier lishi danganguan中国第二历史档案馆(The Second Historical Archives of China), Total No. 393, File No. 2926.

[49]Zhongyang yanjiuyuan dierjie diyi san si wuci nianhui jilu中央研究院评议会第二届第一、三、四、五次年会纪录(Minutes of the First, Third, Fourth and Fifth Annual Meetings of the Second Council of Academia Sinica). Nanjing: Zhongguo dier lishi danganguan中国第二历史档案馆(The Second Historical Archives of China), Total No. 393, File No. 2551.

Fig. 5.1 Photo taken at the Fifth Annual Meeting of the Second Council of Academia Sinica

of two parts: group qualification examination and collective voting of the council members. Prior to that, the PCE had submitted the official list of member candidates to the council for discussion along with the full volume of the extremely confidential documents of Yuan Hanqing, Wang Hefu and Fu Ssu-Nien's critical opinions.[50]

(1) Group Qualification Examination

Group qualification examination began at 10 am on March 26. Before that, 11 council members who were unable to attend had conducted correspondence voting on the official member candidates. Table 5.1 below is the summary results of votes obtained by the three groups, mathematics and physics, biology, humanities.

---

[50]Zhongyang yanjiuyuan diyijie yuanshi houxuanren mingdan中央研究院第一届院士候选人名单(List of Member Candidates for the First Batch of Members of Academia Sinica). Nanjing: Zhongguo dier lishi danganguan中国第二历史档案馆(The Second Historical Archives of China), Total No. 393, File No. 2596; Gefang dui diyici yuanshi xuanju dangxuan yuanshi houxuanren zhi piping ji yijian各方对第一次院士选举当选院士候选人之批评及意见(Criticisms and Opinions of All Sides on the Candidates for the Election of the First Batch of Members of Academia Sinica). Nanjing: Zhongguo dier lishi danganguan中国第二历史档案馆(The Second Historical Archives of China), Total No. 393, File No. 646.

Table 5.1 Summary results of votes obtained by three groups from correspondence voting: mathematics and physics, biology, humanities

Number of votes	Number of people in each group			Total
	Mathematics and physics	Biology	Humanities	
11	6	0	2	8
10 and above	14	4	13	31
9 and above	19	11	21	51
8 and above	26	19	26	71
7 and above	30	28	29	87
6 and above	35	30	31	96

Sources Zhongyang yanjiuyuan diyici yuanshi xuanju houxuanren fenzu taolun toupiao qingkuangbiao中央研究院第一次院士选举候选人分组讨论投票情况表(Voting Tables of Groups Discussion on Candidates for the Election of the First Batch of Members of Academia Sinica). Nanjing: Zhongguo dier lishi danganguan中国第二历史档案馆(The Second Historical Archives of China), Total No. 393, File No. 1619

There were 8 people who received 11 votes in the three groups, including 6 in the mathematics and physics group (Woo Y. H., Wu Hsien, Lee J. S., Chu Coching, Hou T. P. and Mao T. E.) and 2 in the humanities group (Hu Shih and Tao L. K.).[51] The number of people who obtained four fifths of the total votes, i.e.9 votes or more, was 51, including 19 in the mathematics and physics group, 11 in the biology group and 21 in the humanities group. That is to say, most of the 150 official member candidates all won less than 9 votes.

In the group qualification examination, like the group examination meeting in the Fourth Annual Meeting of the council in October 1947, the conveners of the three groups, mathematics and physics, biology, and humanities, were still Li Shu-Hua, Ping Chi and Hu Shih respectively. There were 11 members of the mathematics and physics group present: Jaw Jeou-Jang, Adam Pen-Tung Sah, Mao T. E., Woo Sho-Chow, Chuang Chang-Kong, Chu Chia-Hua, Ling Hung-hsun, Chang Yu-Che, Li Shu-Hua, Chow Jen and John Lee. The group voted on the basis of correspondence, and the 11 participating council members voted off the cuff. The votes of the 49 official member candidates are shown in Table 5.2 (the first number in brackets for each person was the number of correspondence votes obtained, and the second number was the number of group votes obtained on the morning of March 26).

Three group members, Woo Y. H. , Wu Hsien and Lee J. S. received a unanimous vote in this group review. Comparing the results of the official election of members

[51]Zhongyang yanjiuyuan diyici yuanshi xuanju houxuanren fenzu taolun toupiao qingkuangbiao中央研究院第一次院士选举候选人分组讨论投票情况表(Voting Tables of Groups Discussion on Candidates for the Election of the First Batch of Members of Academia Sinica). Nanjing: Zhongguo dier lishi danganguan中国第二历史档案馆(The Second Historical Archives of China), Total No. 393, File No. 1619; Zhongyang yanjiuyuan yuanshi houxuanren mingdan ji xuanjupiao中央研究院院士候选人名单及选举票(The List of Candidates of Members of Academia Sinica and the Ballots). Nanjing: Zhongguo dier lishi danganguan中国第二历史档案馆(The Second Historical Archives of China), Total No. 393, File No. 1620.

Table 5.2 Results of correspondence voting and group voting of mathematics and physics group

Number of votes	Name	Total (person)
22	Woo Y. H. (11 + 11), Wu Hsien (11 + 11), Lee J. S. (11 + 11)	3
21	Chu Coching (11 + 10)	1
20	Chiang Li-Fu (10 + 10), Shiing-Shen Chern (10 + 10), Wu Ta-You (10 + 10), Yeh Chi-Sun (10 + 10), Wong Wen-hao (9 + 11), Hsieh Chia-yung (9 + 11), Hou T. P. (11 + 9), Mao T. E. (11 + 9), Adam Pen-Tung Sah (10 + 10)	9
19	Hua Loo-Keng (9 + 10), Ny Tsai-ze (10 + 9), Chu Chia-Hua (9 + 10), Ling Hung-hsun (10 + 9)	4
18	Li Shu-Hua (8 + 10), Yao Y. T. (9 + 9), Woo Sho-Chow (8 + 10), Chuang Chang-Kong (8 + 10)	4
17	Su Buchin (8 + 9), Tseng Chao-lun (7 + 10), Young Chung-Chien (8 + 9)	3
16	Huang T. K. (7 + 9), Chow Jen (6 + 10)	2
10–15	Chao Chung-Yoa (8 + 7), Sun Xuewu (10 + 5), Kiang Tsai-han (8 + 6), Chou Pei-yuan (7 + 7), Chen Kien Kwong (7 + 6), Wang C. Y. (6 + 6), Hsu Pao-Lu (5 + 5)	7
Under 10 votes	Hiong King-lai (6 + 3), Yin T. H. (6 + 3), Kwei C. T. (5 + 3), Zhu Ruhua (6 + 2), Sun Y. C. (4 + 4), Huang T. C. (4 + 3), Meng H. M. (4 + 3), Wang C. C. (5 + 0), Shi Jiayang (4 + 1), Lo C. C. (4 + 1), Cheng Xiaogang (1 + 1), Chi Yuoh-Fong (1 + 2), Yu C. C. (3 + 0), Wang H. C. (0 + 2), Tsai Fang-yin (0 + 2), Lee S. P. (1 + 1)	16

Sources Zhongyang yanjiuyuan diyici yuanshi xuanju houxuanren fenzu taolun toupiao qingkuangbiao中央研究院第一次院士选举候选人分组讨论投票情况表(Voting Tables of Groups Discussion on Candidates for the Election of the First Batch of Members of Academia Sinica). Nanjing: Zhongguo dier lishi danganguan中国第二历史档案馆(The Second Historical Archives of China), Total No. 393, File No. 1619; Zhongyang yanjiuyuan yuanshi houxuanren mingdan ji xuanjupiao中央研究院院士候选人名单及选举票(The List of Candidates of Members of Academia Sinica and the Ballots). Nanjing: Zhongguo dier lishi danganguan中国第二历史档案馆(The Second Historical Archives of China), Total No. 393, File No. 1620

of Academia Sinica, it can be seen that those who received 15 votes or more in correspondence and group votes together were finally elected, while those who received less than 10 votes were all defeated. Of the seven member candidates who received between 10 and 15 votes, except Hsu Pao-Lu and Chao Chung-Yoa, the other five were not elected. Some of them received higher votes from correspondence, such as Sun Xuewu, 10 votes and Kiang Tsai-han, 8 votes.[52]

[52]Zhongyang yanjiuyuan diyici yuanshi xuanju houxuanren fenzu taolun toupiao qingkuangbiao中央研究院第一次院士选举候选人分组讨论投票情况表(Voting Tables of Groups Discussion on Candidates for the Election of the First Batch of Members of Academia Sinica). Nanjing: Zhongguo dier lishi danganguan中国第二历史档案馆(The Second Historical Archives of China), Total No. 393, File No. 1619.

Table 5.3 Correspondence voting results of biology group

Number of votes	Name	Total (person)
10	Wang Chia-Chi, Chang Chin-yueh, Robert K. S. Lim, Yu T. F.	4
9	Ping Chi, Tai Fon-Land, Loo Tsung-Le, Hu C. H., Wang Ging-Hsi, Lu Zhiwei, Feng Teh-Pei	7
8	Hu Jingfu, Chen Shisan C., Li Zongen, Chang Hsiao-chien, Tang F. F. , Woo Ting-liang, Tang P. S., Feng Zefang	8
7	Hu Hsen-hsu, Yin Hung-chang, Chien Shung-shu, Yuan I-Chin, Ma Wen-chao, Chen Ko Kuei, Tsai Chiao, Teng S. C., Liu Chung-lo	9
6	Wu Hsien-Wen, Li Hsien-wen	2
5	Pai Sitsan, Tung Ti-Cho, Hung S. L., Feng Lan-chou, Liu S. H., Xu Fengyan, Zhao Lianfang	7
4	Zhu Xi, Chen Shixiang, Jao Chin-chih, Zang Yugan	4
1–3	Ching Ren Chang (3 votes), Liou Tchen-Ngo (2 votes), Huang Minlon (2 votes), Liu C. C. (1 vote), Pei Chien (1 vote)	5

Sources Zhongyang yanjiuyuan diyici yuanshi xuanju houxuanren fenzu taolun toupiao qingkuangbiao中央研究院第一次院士选举候选人分组讨论投票情况表(Voting Tables of Groups Discussion on Candidates for the Election of the First Batch of Members of Academia Sinica). Nanjing: Zhongguo dier lishi danganguan中国第二历史档案馆(The Second Historical Archives of China), Total No. 393, File No. 1619

The biology group discussed on the basis of correspondence voting without impromptu voting. The correspondence voting results of 46 official member candidates are shown in Table 5.3.

After discussion, the group put forward a 33-member recommendation list of members to the council: zoology 7, Wang Chia-Chi, Wu Hsien-Wen, Zhu Xi, Pai Sitsan, Ping Chi, Chen Shisan C. and Tung Ti-Cho; botany 7, Hu Hsen-hsu, Yin Hung-chang, Chang Chin-yueh, Liou Tchen-Ngo, Chien Shung-shu, Tai Fon-Land and Loo Tsung-Le; medicine 6, Li Zongen, Hu C. H., Yuan I-Chin, Ma Wen-chao, Chang Hsiao-chien and Tang F. F. ; pharmacology 1, Chen Ko Kuei; Anthropology 1, Woo Ting-liang; psychology 2, Wang Ging-Hsi and Lu Zhiwei; physiology 4, Robert K. S. Lim, Tang P. S., Feng Teh-Pei and Tsai Chiao; agronomy 5, Li Hsien-wen, Yu T. F. , Feng Zefang, Zhao Lianfang and Teng S. C. This list included the members of the biology group who were elected later. The number of correspondence votes for those on the list was generally 8 or more, but there were some people who had lower votes from correspondence voting, such as Zhu Xi 4 and Liou Tchen-Ngo 2. However, the number of correspondence votes of some individual candidates not included in the list was not low, such as Hu Jingfu who obtained 8 votes.

The humanities group also discussed on the basis of correspondence voting and did not conduct impromptu voting. The correspondence voting results of 55 official member candidates are shown in Table 5.4.

After discussion, the humanities group also put forward a 33-member recommendation list of members: philosophy 4, Woo Tsin-Hang, Chin Yuch-Lin, Tang

Table 5.4 Correspondence voting results of humanities group

Number of votes	Name	Total (person)
11	Hu Shih, Tao L. K.	2
10	Chin Yuch-Lin, Feng You-lan, Zhang Yuanji, Chen Yuan, Chen Yin-k'o, Chao Yuen Ren, Luo Changpei, Li Chi, Wang Shih-Chieh, Wang Chung Hwei, Chou Keng-Shang	11
9	Tang Yung-tung, Xu Bingchang, Fu Ssu-Nien, Li Fang Kuei, Liang Ssu-yung, Kuo Mo-jo, Tung Tso-pin, Ma Yin-chu	8
8	Yang Shu-Tah, Ku Chieh-kan, Liang Ssu-cheng, Tsien T. S. , Chen Ta	5
7	Tang Lan, Zhang Xiruo, Ho L.	3
6	John C. H. Wu, Hsiao Kung-chuan	2
5	Woo Tsin-Hang, Liu Yei-Tsen, Xu Zhongshu, Chiang Ting-fu, Yen Shu-Tang, Fong H. D., Pan Kuang-tan	7
4	Yu Cha-Shih, Liu W. T., Wang Li, Kuo Yun-kuan, Ou Pao-san, Chen Zong, Yang Hsi-meng, Lieu D. K.	8
0–3	Chen Kang (3 votes), Chen Shouyi (3 votes), Xu Hongbao (3 votes), Yang Duanliu (3 votes), Li Jiannong (2 votes), Li Haopei (2 votes), Chang C. F. (2 votes), Wu Ching-chao (2 votes), Ling Chunsheng (0 votes)	9

Sources Zhongyang yanjiuyuan diyici yuanshi xuanju houxuanren fenzu taolun toupiao qingkuangbiao中央研究院第一次院士选举候选人分组讨论投票情况表(Voting Tables of Groups Discussion on Candidates for the Election of the First Batch of Members of Academia Sinica). Nanjing: Zhongguo dier lishi danganguan中国第二历史档案馆(The Second Historical Archives of China), Total No. 393, File No. 1619; Zhongyang yanjiuyuan yuanshi houxuanren mingdan ji xuanjupiao中央研究院院士候选人名单及选举票(The List of Candidates of Members of Academia Sinica and the Ballots). Nanjing: Zhongguo dier lishi danganguan中国第二历史档案馆(The Second Historical Archives of China), Total No. 393, File No. 1620

Yung-tung and Feng You-lan; Chinese literature 4, Yu Cha-Shih, Hu Shih, Zhang Yuanji and Yang Shu-Tah;historical science 7, Liu Yei-Tsen, Xu Bingchang, Chen Yuan, Chen Yin-k'o, Fu Ssu-Nien, Chiang Ting-fu and Ku Chieh-kan; linguistics 3, Li Fang Kuei, Chao Yuen Ren and Luo Changpei; archaeology 4, Li Chi, Liang Ssu-yung, Kuo Mo-jo and Tung Tso-pin; art history 1, Liang Ssu-cheng; law 2, Wang Shih-Chieh and Wang Chung Hwei; political science 3, Chou Keng-Shang, Tsien T. S. and Hsiao Kung-chuan; economy 3, Ho L., Ma Yin-chu and Chen Zong; sociology 2, Chen Ta and Tao L. K. This list included all the members of the humanities group elected later. All 26 member candidates who received more than 7 votes in correspondence were included in the list, while those who received 0 to 3 votes in correspondence were not. However, some included in the list also had less than half (6 votes) of the votes from correspondence, such as Yu Cha-Shih and Chen Zong, both received 4 votes, and Woo Tsin-Hang, Liu Yei-Tsen and Chiang Ting-fu, each received 5 votes (Fig. 5.2).

The group voting method adopted by mathematics and physics group should be related to the fact that the disciplines covered by the group were all precision

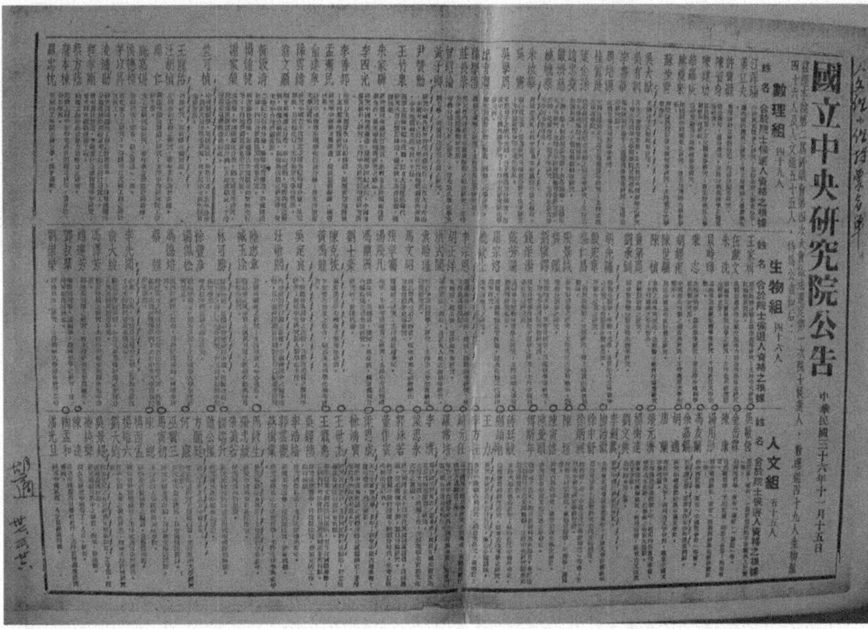
國立中央研究院公告

數理組

生物組

人文組

Fig. 5.2 Recommendation List of Members of the Humanities Group circled by Hu Shih, convener of the group

sciences and the academic contributions of member candidates could be easily evaluated through voting. The discussion method adopted by biology and humanities groups is mainly due to the fact that some disciplines of biology group and all the disciplines of humanities group belong to social sciences or humanities, and voting is unrealistic. Based on the results of the member election and the group examination, it can be judged that the correspondence voting was only a reference for the group examination, because the group qualification examination results were taken as the main basis in the following collective impromptu voting of the participating council members. It reflects that most council members were academically rational and basically respected the opinions of expert council members when voting for people of different disciplines.

(2) **Collective Voting**

At 3:00 pm on March 26, the Second General Meeting of the annual meeting was held. All the 25 council members present at the opening ceremony attended. Chu Chia-Hua served as the chairman and Wong Wen-hao as the secretary. Li Shu-Hua, Ping Chi and Hu Shih, conveners of the three groups of mathematics and physics, biology and humanities, successively reported the results of qualification examination of member candidates of each group. Adam Pen-Tung Sah explained the counting method and statistical list of correspondence votes. After that, the meeting discussed "The qualification of member candidates of the academy has been reviewed by groups according to law. Please decide on the discussion method and voting method". The result of the resolution was: "The election is stipulated to use the announcement list of member candidates, the quota for each group is determined according to law, and with secret ballot, those who obtain four-fifths of the votes cast by the attendees are elected and those elected rank behind are invalid if the quota is exceeded".[53] The so-called "announcement list of member candidates" referred to the *Announcement of Academia Sinica* on November 15, 1947. The legal quota for each group referred to the number of members in each group for the election of the first batch of members stipulated in Article 2 of *The Regulations for the Election of Members of Academia Sinica*, i.e. the number of members in mathematics and physics, biology groups was at least 27 respectively, up to 33; the number of members in humanities group was at least 27, up to 34. The meeting also presumed the drawer and scrutineer for each group: mathematics and physics group were Li Shu-Hua and Chou Keng-Shang; biology group were Hu Hsen-hsu and Mao T. E.; humanities group were Chien Shung-shu and Chuang Chang-Kong.[54] Each group had at least one council member from other group, which was naturally to avoid suspicion (Fig. 5.3).

In the collective voting election, council members cast a total of 5 rounds of votes, i.e. 1 general election and 4 by-elections. The general election was held at the Second General Meeting. According to the provisions of Article 14 of the *Regulations for the Election of Members of Academia Sinica*, candidates for members could only be elected if they had four-fifths of the votes of all the council members present. So the winner must get at least 20 votes. The ballot paper was the announcement of the official list of member candidates of Academia Sinica on November 15, 1947, with instructions: "Please circle 33 people from the mathematics and physics group, 33 from the biology group and 34 from the humanities group in the list of candidates. 1. The number of votes are considered by those with circles at the top of their names on the candidate list; 2. If the number of circles in each group is less than the number of the quota, the circles are valid and shall be counted according to the actual number. 3. If the number of people encircled in one group exceeds the quota, the number of people encircled in that group shall be deemed invalid".[55] The results of the general election are shown in Table 5.5 (the numbers in brackets are the votes for each member candidate).

---

[53]Zhongyang yanjiuyuan dierjie diyi san si wuci nianhui jilu中央研究院评议会第二届第一、三、四、五次年会纪录(Minutes of the First, Third, Fourth and Fifth Annual Meetings of the Second Council of Academia Sinica). Nanjing: Zhongguo dier lishi danganguan中国第二历史档案馆(The Second Historical Archives of China), Total No. 393, File No. 2551.

[54]Zhongyang yanjiuyuan dierjie diyi san si wuci nianhui jilu中央研究院评议会第二届第一、三、四、五次年会纪录(Minutes of the First, Third, Fourth and Fifth Annual Meetings of the Second Council of Academia Sinica). Nanjing: Zhongguo dier lishi danganguan中国第二历史档案馆(The Second Historical Archives of China), Total No. 393, File No. 2551.

[55]Zhongyang yanjiuyuan gonggao yuanshi houxuanren ji xuanding yuanshi mingdan中央研究院公告院士候选人及选定院士名单(The Announcement of Member Candidates and the List of Members of Academia Sinica). Nanjing: Zhongguo dier lishi danganguan中国第二历史档案馆(The Second Historical Archives of China), Total No. 393, File No. 2558.

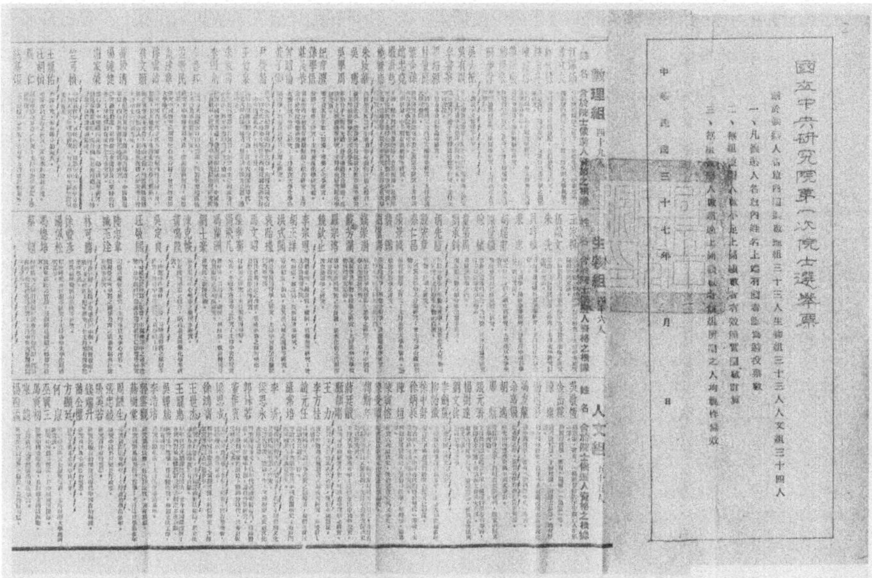
國立中央研究院第一次院士選舉票
數理組
生物組
人文組

Fig. 5.3 Election ticket for the first batch of members of Academia Sinica

A total of 67 people were elected by general election, of whom 24 were in the mathematics and physics group, leaving a gap of 9 to fill. There were 21 people in the biology group and 22 people in the humanities group, both of which were 12 short of full capacity. From the specific results, Chiang Li-Fu, Woo Y. H. , Lee J. S. , Chen Shisan C., Li Zongen, Robert K. S. Lim, Hu Shih, Chen Yin-k'o and Chao Yuen Ren received a total of 25 votes each. To a large extent, it reflects that their academic contributions or academic leadership achievements had been unanimously recognized in the council. Woo Sho-Chow, Huang T. K., Chow Jen, Wu Hsien-Wen, Yin Hung-chang, Chang Chin-yueh, Yuan I-Chin, Chang Hsiao-chien and Yang Shu-Tah just reached 20 votes and nearly lost the election. Tung Ti-Cho, Tang F. F. , Li Hsien-wen, Yu T. F. and Feng Zefang of the biology group all got 19 votes each, only one vote away from being elected (Fig. 5.4).

After the general election, each of the three groups did not reach the minimum number of 27 people stipulated in Article 2 of *The Regulations for the Election of Members of Academia Sinica*. At 9 am on March 27, the annual meeting held its Third General Meeting. There were still 25 council members present. Director-General Adam Pen-Tung Sah reported the results of the vote-counting for the Second General Meeting. Later, the meeting discussed ""What election method to choose for the proposed by-election of nine members in the mathematics and physics group, 12 members in the biology group and 12 members in the humanities group as well so as to fill the gap. Please make a decision." The result was: "Except for those who have obtained more than 20 votes, a by-election will be held on the announcement list of member candidates, and those who exceed the quota of each group will be invalid."

Table 5.5 General election results of the election of the first batch of members of Academia Sinica

Number of votes	Name	Total
20–25	**Mathematics and Physics group**: Chiang Li-Fu (25), Shiing-Shen Chern (24), Hua Loo-Keng (22), Wu Ta-You (23), Woo Y. H. (25), Li Shu-Hua (24), Yeh Chi-Sun (22), Ny Tsai-ze (23), Yao Y. T. (23), Wu Hsien (24), Woo Sho-Chow (20), Chuang Chang-Kong (23), Chu Chia-Hua (24), Lee J. S. (25), Wong Wen-hao (24), Huang T. K. (20), Young Chung-Chien (21), Hsieh Chia-yung (22), Chu Coching (23), Chow Jen (20), Hou T. P. (24), Mao T. E. (24), Ling Hung-hsun (24), Adam Pen-Tung Sah (23) **Biology group**: Wang Chia-Chi (24), Wu Hsien-Wen (20), Ping Chi (23), Chen Shisan C. (25), Hu Hsen-hsu (23), Yin Hung-chang (20), Chang Chin-yueh (20), Chien Shung-shu (22), Tai Fon-Land (23), Loo Tsung-Le (22), Li Zongen (25), Yuan I-Chin (20), Chang Hsiao-chien (20), Chen Ko Kuei (22), Woo Ting-liang (21), Wang Ging-Hsi (24), Robert K. S. Lim (25), Tang P. S. (23), FengDepei (22), Tsai Chiao (23), Teng S. C. (21) **Humanities group:** Chin Yuch-Lin (22), Tang Yung-tung (23), Feng You-lan s (22), Hu Shih (25), Zhang Yuanji (23), Yang Shu-Tah (20), Chen Yuan (24), Chen Yin-k'o (25), Fu Ssu-Nien (24), Li Fang Kuei (23), Chao Yuen Ren (25), Li Chi (23), Liang Ssu-yung (22), Tung Tso-pin (21), Liang Ssu-cheng (24), Wang Shih-Chieh (22), Wang Chung Hwei (22), Chou Keng-Shang (23), Tsien T. S. (21), Ma Yin-chu (21), Chen Ta (21), Tao L. K. (24)	67
Under 20 votes	**Mathematics and Physics group**: Kiang Tsai-han (13), Hsu Pao-Lu (18), Chen Kien Kwong (14), Hiong King-lai (3), Su Buchin (18), Chou Pei-yuan (15), Kwei C. T. (3), Chao Chung-Yoa (17), Zhu Ruhua (5), Chi Yuoh-Fong (1), Sun Xuewu (11), Tseng Chao-lun (18), Huang T. C. (2), Yin T. H. (13), Wang C. C. (1), Lee S. P. (0), Meng H. M. (1), Yu C. C. (2), Sun Y. C. (6), Wang C. Y. (15), Wang H. C. (0), Shi Jiayang (1), Cheng Xiaogang (3), Tsai Fang-yin (2), Lo C. C. (2) **Biology group**: Zhu Xi (14), Pai Sitsan (18), Hu Jingfu (3), Chen Shixiang (1), Tung Ti-Cho (19), Liu C. C. (0), Ching Ren Chang (7), Pei Chien (0), Liou Tchen-Ngo (9), Jao Chin-chih (0), Hu C. H. (1), Hung S. L. (11), Ma Wen-chao (1), Tang F. F. (19), Feng Lan-chou (0), Liu S. H. (0), Huang Minlon (5), Lu Zhiwei (17), Zang Yugan (0), Xu Fengyan (1), Li Hsien-wen (19), Yu T. F. (19), Feng Zefang (19), Liu Lianfang (16), Liu Chung-lo (5) **Humanities group:** Woo Tsin-Hang (17), Chen Kang (5), Yu Cha-Shih (16), Tang Lan (2), Liu W. T. (0), Li Jiannong (2), Liu Yei-Tsen (17), Xu Zhongshu (2), Xu Bingchang (16), Chen Shouyi (0), Chiang Ting-fu (12), Ku Chieh-kan (18), Wang Li (0), Luo Changpei (17), Kuo Mo-jo (18), Xu Hongbao (3), John C. H. Wu (2), Li Haopei (6), Kuo Yun-kuan (0), Yen Shu-Tang (4), Chang C. F. (1), Zhang Xiruo (1), Hsiao Kung-chuan (18), Fong H. D. (4), Ho L. (17), Ou Pao-san (0), Chen Zong (16), Yang Hsi-meng (1), Yang Duanliu (5), Lieu D. K. (2), Wu Ching-chao (3), Ling Chunsheng (2), Pan Kuang-tan (0)	83

Sources Zhongyang yanjiuyuan yuanshi houxuanren mingdan ji xuanjupiao中央研究院院士候选人名单及选举票(The List of Candidates of Members of Academia Sinica and the Ballots). Nanjing: Zhongguo dier lishi danganguan中国第二历史档案馆(The Second Historical Archives of China), Total No. 393, File No. 1620

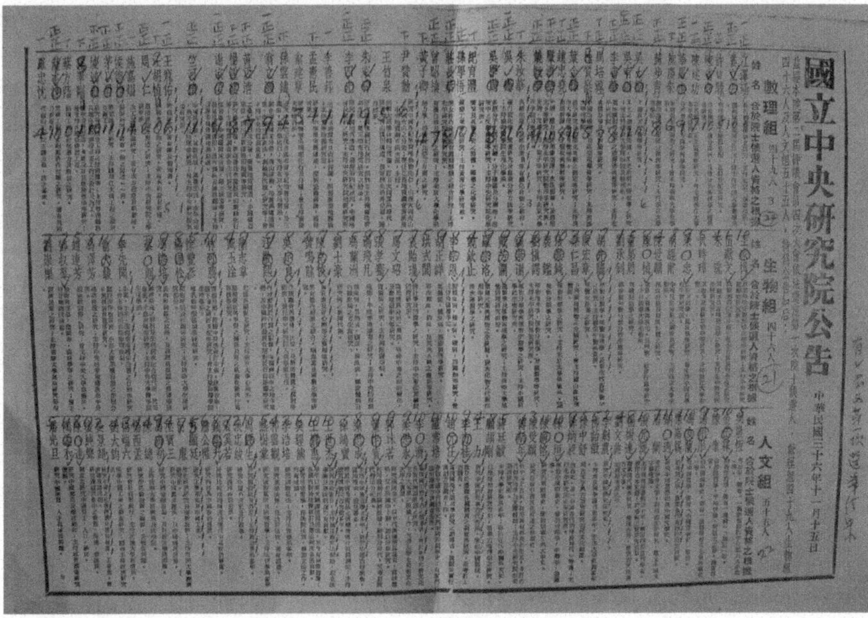
國立中央研究院公告

中華民國三十六年十一月十五日

數理組

生物組

人文組

Fig. 5.4 The elected by general election of the council in *the Announcement of Academia Sinica*

Li Shu-Hua and Chou Keng-Shang were chosen to be the drawer and scrutineer, and still those who obtained more than 20 votes would be the elected.[56]

The first by-election was held by the council members present. Eleven people, including Hsu Pao-Lu, Su Buchin, Chao Chung-Yoa, Tseng Chao-lun, Pai Sitsan, Tung Ti-Cho, Li Hsien-wen, Yu T. F. , Liu Yei-Tsen, Kuo Mo-jo and Hsiao Kung-chuan, got 20 votes. Among them, Tung Ti-Cho got 24, the highest number of votes; Su Buchin, Li Hsien-wen, Liu Yei-Tsen and Kuo Mo-jo all got 20 votes which was the exact number of votes needed to be selected. As a result, 78 people were elected by general election and the first by-election. As "it is still short of the predetermined number", the meeting discussed "whether to hold another by-election and what election method to apply for it". As a result, it was decided that "the way of the first by-election will be conducted again. The maximum number of people in the mathematics and physics group will be five, the biology group will be eight, and the humanities group will be nine. Those who exceed the number will be invalid." Mao T. E. and Chuang Chang-Kong were chosen to be the drawer and scrutineer. Since there were 23 council members present at the time, "19 votes are required by law to

[56]Zhongyang yanjiuyuan dierjie pingyihui tiaoli jilu ti'an ji baogao中央研究院第二届评议会条例、记录、提案及报告(Regulations, Records, Proposals and Reports of the Second Council of Academia Sinica). Nanjing: Zhongguo dier lishi danganguan中国第二历史档案馆(The Second Historical Archives of China), Total No. 393, File No. 2926.

be elected",[57] because 4/5 of 23 is 19. Only Ku Chieh-kan, a member candidate of the humanities group, had 20 votes in the second by-election. As a result, 79 people were elected, 28 in the mathematics and physics group, 25 in the biology group and 26 in the humanities group.

At 3:00 pm on March 27, the Fourth General Meeting of the annual meeting was held. 24 people including Chu Chia-Hua, Wong Wen-hao, Adam Pen-Tung Sah, Wang Chia-Chi, John Lee, Woo Ting-liang, Woo Sho-Chow, Li Chi, Li Shu-Hua, Ping Chi, Chow Jen, Chou Keng-Shang, Robert K. S. Lim, Hu Shih, Hu Hsen-hsu, Mao T. E., Chen Yuan, Chen Shisan C., Chang Yu-Che, Chuang Chang-Kong, Ling Hung-hsun, Jaw Jeou-Jang, Chien Shung-shu and Loo Tsung-Le attended the meeting. Adam Pen-Tung Sah reported the results of the second by-election. Wong Wen-hao reported that Woo Y. H. decided to give up the right to vote and to be voted in the member election by a letter from the United States. The meeting decided in its later agenda to "persuade him to stay and accept being the elected."[58] After Wong Wen-hao, the chairman of the meeting, Chu Chia-Hua, reported on the results of the general election and the two by-elections,"According to the law, there are still two vacancies in the biology group and one vacancy in the humanities group, so three more should be elected through by-election. In terms of the total number, at least 81 people should be elected. Therefore, two more should be added through by-election to meet the requirements."[59] The number of people still missing in both biology group and humanities group was calculated according to the stipulation in Article 2 of *The Regulations for the Election of Members of Academia Sinica* that each group should elect at least 27 people. There was no problem. However, that the total number ofmembers must reach at least 81 was slightly different from the lower limit of 80, which was set in Article 6 of *The Organization Act of Academia Sinica*, that the number of first batch of members of Academia Sinica was 80 to 100.

In the following agenda, the meeting discussed and passed "A whole ticket with groups divided, if there is a group of votes that exceed the quota, it is proposed to be invalid only in its own group, it will not affect the votes of other groups.Please make a decision." After the resolution was passed, discussions were held on "Should there be another by-election, and what method should be adopted to vote with the results of the second by-election, only Ku Chieh-kan from the humanities group was elected, still leaving some vacancies since a total of 79 people were elected?"

---

[57]Zhongyang yanjiuyuan dierjie pingyihui tiaoli jilu ti'an ji baogao中央研究院第二届评议会条例、记录、提案及报告(Regulations, Records, Proposals and Reports of the Second Council of Academia Sinica). Nanjing: Zhongguo dier lishi danganguan中国第二历史档案馆(The Second Historical Archives of China), Total No. 393, File No. 2926.

[58]Zhongyang yanjiuyuan dierjie pingyihui tiaoli jilu ti'an ji baogao中央研究院第二届评议会条例、记录、提案及报告(Regulations, Records, Proposals and Reports of the Second Council of Academia Sinica). Nanjing: Zhongguo dier lishi danganguan中国第二历史档案馆(The Second Historical Archives of China), Total No. 393, File No. 2926.

[59]Zhongyang yanjiuyuan dierjie pingyihui tiaoli jilu ti'an ji baogao中央研究院第二届评议会条例、记录、提案及报告(Regulations, Records, Proposals and Reports of the Second Council of Academia Sinica). Nanjing: Zhongguo dier lishi danganguan中国第二历史档案馆(The Second Historical Archives of China), Total No. 393, File No. 2926.

The decision was made after discussion: "Conduct the voting again according to the previous method, with a maximum of five members in the mathematics and physics group, eight members in the biology group and eight members in the humanities group." Hu Shih and Chien Shung-shu were chosen to be the drawer and scrutineer. Since there were 23 council members present, candidates should still "be elected with 19 votes."[60]

Apparently, the maximum number of people elected by voting in the three groups is 33 in both the mathematics and physics group and biology group and 34 in the humanities group as stipulated in Article 2 of the *Regulations for the Election of Members of Academia Sinica*. After a resolution was made on this issue, the present council members voted in the third by-election. The only person who got 19 votes this time was Yu Cha-Shih from the humanities group. Chou Pei-yuan, Yin T. H. and Feng Zefang each got 18 votes, only one vote short of being elected.[61]

After the third by-election, the number of elected members had reached the minimum of 80 members for the first term of Academia Sinica as stipulated in Article 6 of *The Organization Act of Academia Sinica*. After discussion, it was decided: "Make another by-election with the previous method, the number of votes to be cast shall be 5 for the mathematics and physics group, 8 for the biological group and 7 for the humanities group." Chien Shung-shu, Jaw Jeou-Jang, Ling Hung-hsun and Chow Jen were chosen to issue and supervise the votes. There were still 23 council members present, so the candidates should still be elected with 19 votes.[62] The only one who got 19 votes this time was again in the humanities group, whom was called by Hu Shih "the old timer in the existing ideological circle", Woo Tsin-Hang. This time Yin T. H. won 18 votes again and was finally defeated. In the general election, Tang F. F. and Feng Zefang, who were only 1 vote away from being elected, did not reach 19 votes in the four by-elections and were also not elected.

After the fourth by-election, the meeting discussed "Please make a decision on the proposal of stopping election since the number of members is already in line with the total of 80 to 100 members required by law, although the biology group is still short of two according to the allocation of seats in each group stipulated by the *Regulations for the Election of Members of Academia Sinica*." The case was

[60]Zhongyang yanjiuyuan dierjie pingyihui tiaoli jilu ti'an ji baogao中央研究院第二届评议会条例、记录、提案及报告(Regulations, Records, Proposals and Reports of the Second Council of Academia Sinica). Nanjing: Zhongguo dier lishi danganguan中国第二历史档案馆(The Second Historical Archives of China), Total No. 393, File No. 2926.

[61]Zhongyang yanjiuyuan yuanshi houxuanren mingdan ji xuanjupiao中央研究院院士候选人名单及选举票(The List of Candidates of Members of Academia Sinica and the Ballots). Nanjing: Zhongguo dier lishi danganguan中国第二历史档案馆(The Second Historical Archives of China), Total No. 393, File No. 1620.

[62]Zhongyang yanjiuyuan dierjie pingyihui tiaoli jilu ti'an ji baogao中央研究院第二届评议会条例、记录、提案及报告(Regulations, Records, Proposals and Reports of the Second Council of Academia Sinica). Nanjing: Zhongguo dier lishi danganguan中国第二历史档案馆(The Second Historical Archives of China), Total No. 393, File No. 2926.

unanimously passed. Later, Adam Pen-Tung Sah"read out the list of 81 members elected and asked for a referendum"which passed unanimously.[63]

Shiah Nae recorded the election of members on March 27 in his diary: "The council meeting was closed at night and the election of members was completed. I heard that there were 5 rounds of voting in total, and the number of the two groups, mathematics and physics, biology, did not increase after the second round. However, the humanities group added Liu Yei-Tsen for the second time, Ku Chieh-kan for the third time, Yu Cha-Shih for the fourth time, and Woo Tsin-Hang for the fifth time. The number reached 81 (the fourth time was 80, but at that time the number of people was 28 in the mathematics and physics group, 27 in the humanities group, and only 25 in the biology group. It was hoped that the number of the biology group could increase, but it failed in the end. Woo Tsin-Hang's accession began with President Chu's repeated entreaties before his last election)."[64] Shiah Nae did not attend the annual meeting of the council that day. The so-called second voting referred to the first by-election, which was only untrue in Liu Yei-Tsen's election. However, that he said Woo Tsin-Hang's election in the fifth ballot, i.e. the fourth by-election, was related to Chu Chia-Hua's "repeated entreaties" should not be groundless.

In short, 81 members were elected after five rounds of secret ballot on the spot by all the council members present in the annual meeting. The formal election process "can be said quite solemn and prudent."[65]

On April 1, 1948, Academia Sinica announced the list of the first batch of members (Fig. 5.5). The announcement read: "Members are hereby selected according to law by the Fifth Annual Meeting of the Second Council of Academia Sinica, with 28 members in the mathematics and physics group, 25 in the biology group and 28 in the humanities group."[66] The list of members in each discipline is as follows:

Mathematis and Physics Group: (mathematics 5) Chiang Li-Fu, Hsu Pao-Lu, Shiing-Shen Chern, Hua Loo-Keng and Su Buchin; (physics 7) Wu Ta-You, Woo Y. H. , Li Shu-Hua, Yeh Chi-Sun, Chao Chung-Yoa, Ny Tsai-ze and Yao Y. T. ; (chemistry 4) Wu Hsien, Woo Sho-Chow, Chuang Chang-Kong and Tseng Chao-lun; (geology 6) Chu Chia-Hua, Lee J. S. , Wong Wen-hao, Huang T. K., Young Chung-Chien and Hsieh Chia-yung; (meteorology 1): Chu Coching; (engineering 5) Chow Jen, Hou T. P., Mao T. E., Ling Hung-hsun and Adam Pen-Tung Sah.

---

[63]Zhongyang yanjiuyuan dierjie pingyihui tiaoli jilu ti'an ji baogao中央研究院第二届评议会条例、记录、提案及报告(Regulations, Records, Proposals and Reports of the Second Council of Academia Sinica). Nanjing: Zhongguo dier lishi danganguan中国第二历史档案馆(The Second Historical Archives of China), Total No. 393, File No. 2926.

[64]Shiah Nae夏鼐(2011). *Shiah Nae riji*夏鼐日记(Shiah Nae's Diary), Vol. 4. Shanghai: Huadong shifan daxue chubanshe华东师范大学出版社.179.

[65]Shiah Nae夏鼐(1948). Zhongyang yanjiuyuan diyijie yuanshi de fenxi中央研究院第一届院士的分析(The Analysis of the First Batch of Members of Academia Sinica). *Guancha*观察(Observation) 5 (14): 3.

[66]Yuanshi Tanhuahui院士谈话会(Symposium of Members of Academia Sinica). Nanjing: Zhongguo dier lishi danganguan中国第二历史档案馆(The Second Historical Archives of China), Total No. 393, File No. 569.

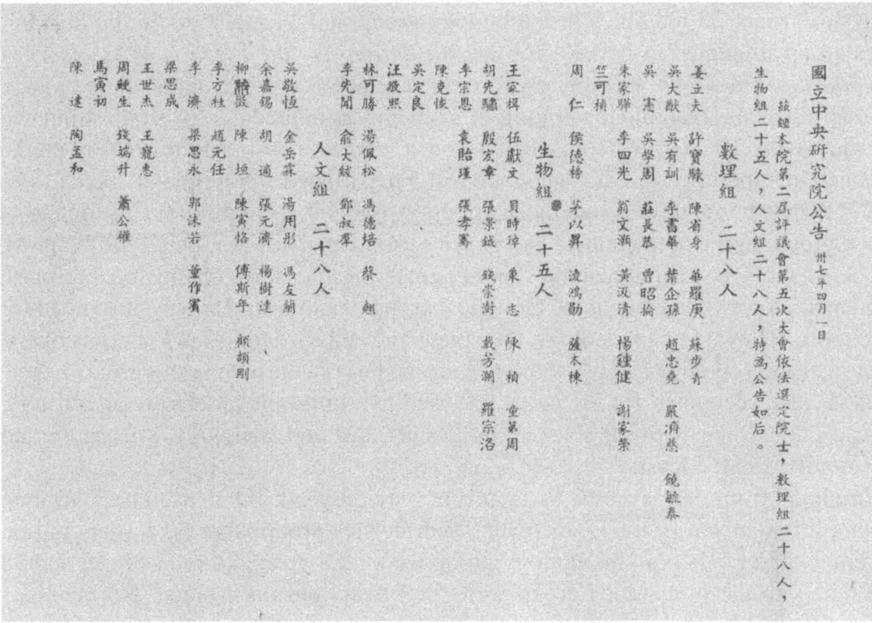

國立中央研究院公告 卅七年四月一日
茲經本院第二屆評議會第五次大會依法選定院士，數理組二十八人，
生物組二十五人，人文組二十八人，特為公告如后。
數理組　二十八人
姜立夫 許寶騄 陳省身 華羅庚 蘇步青
吳大猷 吳有訓 李書華 葉企孫 趙忠堯 嚴濟慈 饒毓泰
吳　憲 吳學周 莊長恭 曾昭掄
朱家驊 李四光 翁文灝 黃汲清 楊鍾健 謝家榮
竺可楨
周　仁 侯德榜 茅以昇 凌鴻勛 薩本棟
生物組　二十五人
王家楫 伍獻文 貝時璋 秉　志 陳　楨 童第周
胡先驌 殷宏章 張景鉞 錢崇澍 戴芳瀾 羅宗洛
李宗恩 袁貽瑾 張孝騫
陳克恢
吳定良
汪敬熙
林可勝 湯佩松 馮德培 蔡　翹
李先聞 俞大紱 鄧叔羣
人文組　二十八人
吳敬恆 金岳霖 湯用彤 馮友蘭
余嘉錫 胡　適 張元濟 楊樹達
柳詒徵 陳　垣 陳寅恪 傅斯年 顧頡剛
李方桂 趙元任
李　濟 梁思永 郭沫若 董作賓
梁思成
王世杰 王寵惠
周鯁生 錢端升 蕭公權
馬寅初
陳　達 陶孟和

Fig. 5.5 *The Announcement of Academia Sinica* on April 1, 1948

Biology Group: (zoology 6) Wang Chia-Chi, Wu Hsien-Wen, Pai Sitsan, Ping Chi, Chen Shisan C. and Tung Ti-Cho; (botany 6) Hu Hsen-hsu, Yin Hung-chang, Chang Chin-yueh, Chien Shung-shu, Tai Fon-Land and Loo Tsung-Le; (medicine 3) Li Zongen, Yuan I-Chin and Chang Hsiao-chien; (pharmacology 1) Chen Ko Kuei; (physical anthropology 1) Woo Ting-liang; (psychology 1) Wang Ging-Hsi; (physiology 4) Robert K. S. Lim, Tang P. S., Feng Teh-Pei and Tsai Chiao; (agronomy 3): Li Hsien-wen, Yu T. F. and Teng S. C.

Humanities Group: (philosophy 4) Woo Tsin-Hang, Chin Yuch-Lin, Tang Yung-tung and Feng You-lan; (literature and history 4) Yu Cha-Shih, Hu Shih, Zhang Yuanji and Yang Shu-Tah; (historical science 5) Liu Yei-Tsen, Chen Yuan, Chen Yin-k'o, Fu Ssu-Nien and Ku Chieh-kan; (linguistics 2) Li Fang Kuei and Chao Yuen Ren; (archaeology 4) Li Chi, Liang Ssu-yung, Kuo Mo-jo and Tung Tso-pin; (art history 1) Liang Ssu-cheng; (law 2) Wang Shih-Chieh and Wang Chung Hwei; (politics 3) Chou Keng-Shang, Tsien T. S. and Hsiao Kung-chuan; (economy 1) Ma Yin-chu; (sociology 2) Chen Ta and Tao L. K.[67]

The number of members in the three groups of mathematics and physics, biology and humanities differed from the upper limit of 33 members in the mathematics and

---

[67] Yuanshi Tanhuahui院士谈话会(Symposium of Members of Academia Sinica). Nanjing: Zhongguo dier lishi danganguan中国第二历史档案馆(The Second Historical Archives of China), Total No. 393, File No. 569.

physics group, 33 members in the biology group and 34 members in the humanities group stipulated in Article 2 of the *Regulations for the Election of Members of Academia Sinica* by 5, 8 and 6 respectively. The number of members in the biology group was 2 less than the lower limit of 27 members in the biology group stipulated in the article. The 81 members elected won at least four-fifths of the votes of all the council members present. It was in line with Fu Ssu-Nien's principle of "better to be short than to be shoddy". From the distribution of members of different disciplines in each group, in the mathematics and physics group, physics accounted for the most, geology the second, mathematics and engineering the third, chemistry the fourth and meteorology the least;in the biological group, zoology and botany accounted for the most, followed by physiology, then medicine and agronomy, and pharmacology, physical anthropology and psychology were the least;in the humanities group, historical science accounted for the most, followed by philosophy, literature and history, archaeology, then political science, linguistics, law and sociology, art history and economy were the least.

Judging from their age, the 81 members were 53 years old on average.[68] Among them, 28 members in the mathematics and physics group were 50.5 years old on average, 25 members in the biology group were 49.8 years old on average, and 28 members in the humanities group were 58.3 years old on average. No members were 60 or older in the mathematics and physics group, but 1 in the biology group (Chien Shung-shu, 65 years old) and 8 in the humanities group (Woo Tsin-Hang, 85 years old; Yu Cha-Shih, 64 years old; Zhang Yuanji,81 years old; Yang Shu-Tah, 63 years old; Liu Yei-Tsen,70 years old; Chen Yuan, 68 years old; Wang Chung Hwei, 71 years old and Tao L. K. , 60 years old). The oldest was 85-year-old Woo Tsin-Hang. There were three members under 40 years old, all of whom were in the mathematics of the mathematics and physics group, namely Hua Loo-Keng, Hsu Pao-Lu and Shiing-Shen Chern. If the average age of members in the mathematics and physics group, biological group was lower than that in the humanities group, it should be related to different optimal periods for innovative research activities in the fields of natural science, humanities and social sciences. Generally speaking, in the field of natural science, the best time for researchers to carry out innovative research activities is in the young and middle-aged period, especially in mathematics. In the field of humanities and social sciences, the accumulation of knowledge is required more, and the best time for researchers to carry out innovative research activities is often past middle age.

Judging from their birthplace, the 81 members were distributed in 13 provinces. Among them, Zhejiang and Jiangsu provinces (including Shanghai) had the largest number of 17 people each. The remaining 47 members were distributed in Guangdong Province 9, Jiangxi Province 7, Fujian Province 6, Hunan Province 6, Hubei Province

[68]For information on the age, birthplace, education, employment and published works of the first batch of members of Academia Sinica, please refer to Academia Sinica中央研究院(Eds.) (1948). *Guoli zhongyang yanjiuyuan yuanshilu*国立中央研究院院士录(The Records of Members of Academia Sinica), Vol. 1. Nanjing: Zhongyang yanjiuyuan中央研究院.

6, Sichuan Province 3, Shandong Province 3, Henan Province 3, Hebei Province (including Tianjin) 2, Gansu Province 1, Shaanxi Province 1.

Judging from the educational background, 75 of the 81 members had studied abroad, accounting for 92.6%. The countries they studied in included the United States (51 persons), Britain (11 persons), Germany (6 persons), France (5 persons), Japan (5 persons), Belgium (2 persons) and Switzerland (1 person), with the majority studying in the United States. Only six people in the humanities group, namely Yu Cha-Shih, Zhang Yuanji, Liu Yei-Tsen, Chen Yuan, Ku Chieh-kan and Tung Tso-pin, had no experience of studying abroad. That was related to their studies of literature, history or archaeology, which China has its own tradition and foundation and they could achieve important academic achievements without learning from the West. Among the 81 members, 58 had doctorates, accounting for 71.6%. Except for one person, Chang Hsiao-chien, who got his doctor's degree home, the other 57 obtained theirs abroad. Among the other 23 people, the highest 7 had master's degrees and all obtained abroad (2 in mathematics and physics group, 2 in biology group, and 3 in humanities group). Of the remaining 16 people, 4 definitely known to have obtained bachelor's degrees (2 obtained abroad, namely, Chien Shung-shu and Tai Fon-Land; 2 obtained in China, namely Ling Hung-hsun and Ku Chieh-kan); 2 were "Juren(举人)"[69] in the late Qing Dynasty, namely Woo Tsin-Hang and Yu Cha-Shih; one was a "Lin Sheng(廪生)"[70] in the late Qing Dynasty, namely Chen Yuan; one has only a junior high school diploma, namely Hua Loo-Keng. Overall, more than 80% of the members had doctor's or master's degrees, and more had studied abroad. The United States was the country where the largest number of members had studied.

Judging from their work units, the 81 members were widely distributed. They worked in Academia Sinica, the National Academy of Peiping, the National Geological Survey of China, the Fan Memorial Institute of Biology, the National University of Peking, National Tsing Hua University, National Central University, the University of Chekiang, the Catholic University of Peking, National Wu-han University, Sichuan University, Hunan University, Shandong University, Fudan University, Taiwan University, National Social Education College, Peiping Union Medical College, as well as the National Health Experimental Institute, Institute for the Prevention and Cure of Tuberculosis in Nanjing of Ministry of Health, Xiangya Medical College, Resources Committee, Yongli Chemical Industry Company, Ministry of Communications, Ministry of Foreign Affairs, Commercial Press and other institutions. In addition, Li Fang Kuei and Chao Yuen Ren both worked in foreign universities, while Chen Ko Kuei worked in a foreign research institution. Of the 81 members, the number of people working in Academia Sinica was 20, the largest, accounting for 24.7% of all members; the number of people working in the National University of Peking and National Tsing Hua University took the second place, 10 for each, accounting for 12.34%; 4 people worked in National Central University and the University of Chekiang respectively, accounting for 4.9% each. 2 people worked in the National Academy of Peiping, National Geological Survey of China and the

[69]Translator's note: successful candidate in the imperial exams at the provincial level.

[70]Translator's note: scholar living on government grands.

Catholic University of Peking respectively, accounting for 2.5% each, and one person worked in each of the other organizations, each accounting for 1.2%.

Judging from the published works, some of the 81 members had published a considerable number of papers, but some very few papers. Mathematics and physics group and biology group were especially noticeable. For example, in mathematics of the mathematics and physics group, Su Buchin published 95 papers, Hua Loo-Keng published 68 papers and a book, while Chiang Li-Fu published only one paper. In physics, Ny Tsai-ze published 53 papers, Wu Ta-You published 41 papers and a book, while Li Shu-Hua and Yeh Chi-Sun only published 2 papers each. In geology, Wong Wen-hao had published 65 papers, while Chu Chia-Hua had only 2 papers published. Another example was that Chen Ko Kuei in pharmacology of the biology group had published 225 papers, while Yuan I-Chin in medicine, which was close to Chen's discipline, only published 5 papers. In other words, there were obvious differences in the number of research results published by members of the same or similar disciplines.

Among the 81 members, 6 were elected with the second member qualification, namely, Chu Chia-Hua, Wong Wen-hao, Chow Jen, Ling Hung-hsun, Zhang Yuanji and Chou Keng-Shang,[71] accounting for only 7.4%. It shows that although the member qualifications provided a space for leaders of academic institutions to be elected members, the council did not open up a convenient door for them. Instead, it elected members towards the world-wide qualifications of academicians and rarely considered member candidates who only led and presided over academic institutions. Among the 81 members, 34 (41.98%) were elected with the first member qualification. At the same time, 42 people (51.85%) were elected as members with both the first and second qualifications. Moreover, of the 150 official member candidates, 79 who met the first qualification of a member were eliminated 45, while 53 who met both the first and second qualifications of a member were eliminated only 11. It also indicates that between member candidates who had only academic achievements and who had both academic achievements and academic leadership achievements, the council would consider the latter more. In other words, academic leadership was also an important bargaining chip for being elected as the first batch of members of Academia Sinica. Chiang Li-Fu, Yeh Chi-Sun and Li Shu-Hua were elected members for meeting both the first and second qualifications, their academic achievements were not significant, but they did make obvious academic leadership achievements, which can also illustrate the point.

After the list of the first batch of members of Academia Sinica was published, the academic circle received a good response. For example, in November 1948, Shiah Nae wrote: "If we say 'This list is quite enough to represent the situation of

[71]Guoli zhongyang yanjiuyuan diyici yuanshi xuanju houxuanren timingce国立中央研究院第一次院士选举候选人提名册(Nomination Lists of the Member Candidates for the Election of the First Batch of Members of Academia Sinica). Nanjing: Zhongguo dier lishi danganguan中国第二历史档案馆(The Second Historical Archives of China), Total No. 393, File No. 1597.

today's Chinese academia',it is probably not much of a mistake."[72] Young Chung-Chien recalled the election of members in his later years and said: "Being elected as a member was considered a special honor in China at that time, at least in terms of the mode of production, very cautious, lack of drawbacks.It wasthe reason that it attracted the attention of ordinary people."[73] However, there was some criticism from the academic community against the member election. Xiang Da and Shiah Nae had some opinions on the concentration ofmembers serving in Academia Sinica. Xiang Da said that the session ofmember election "gives people a feeling that scholars come from royal family or officials" and added: "The directors of our academy as well as most of the full-time researchers are almost all ex officio members."[74] Although Shiah Nae believed that there was inevitably some"proximity" or "affection"effect, he did not fully agree with Xiang Da's statements:

> Having checked the Academia Sinica, there are 13 institutes, of which the directors of 11 were elected members, the proportion is really big. However, if we assume that when Academia Sinica set up each institute, it considered the discipline had its research object and it had indeed produced authoritative figures with expertise in the field in China, otherwise, it would rather go without than have something shoddy; if the directors hired by Academia Sinica are indeed such authoritative figures, then it is rather a matter of course that almost all the directors are elected members. It is another matter whether the fact is as satisfactory as it is supposed to be, though. As for the full-time researchers (not counting part-time and associate researchers), there are currently 68 in Academia Sinica, of which the total number of members elected this time is 9, accounting for only 13%, which seems impossible to be called 'most'.[75]

Anyhow, the member election was a success. Because almost all the elected were outstanding scholars in their disciplines, their academic achievements essentially represented the highest academic level in China during that time.

5.4 Summary

The announcement of the official list of member candidates of Academia Sinica attracted the attention of various parties and drew sharp criticisms from Wang Hefu, Yuan Hanqing, Fu Ssu-Nien and others. Some of the criticisms were pertinent, but

[72]Shiah Nae夏鼐(1948). Zhongyang yanjiuyuan diyijie yuanshi de fenxi中央研究院第一届院士的分析(The Analysis of the First Batch of Members of Academia Sinica). *Guancha*观察(Observation) 5 (14): 3.

[73]Young Chung-Chien杨钟健(1983). *Young Chung-Chien huiyilu*杨钟健回忆录(Memoirs of Young Chung-Chien). Beijing: Dizhi chubanshe地质出版社.167.

[74]Shiah Nae夏鼐(1948). Zhongyang yanjiuyuan diyijie yuanshi de fenxi中央研究院第一届院士的分析(The Analysis of the First Batch of Members of Academia Sinica). *Guancha*观察(Observation) 5 (14): 5.

[75]Shiah Nae夏鼐(1948). Zhongyang yanjiuyuan diyijie yuanshi de fenxi中央研究院第一届院士的分析(The Analysis of the First Batch of Members of Academia Sinica). *Guancha*观察(Observation) 5 (14): 5.

others were not completely objective or even biased. The PCE attached great importance to the criticisms of all parties and made the real-named criticisms into extremely confidential documents and submitted to the council for discussion according to the *Regulations for the Election of Members of Academia Sinica*. Wang Hefu's criticism did not have a significant impact on Li Zongen, Chang Hsiao-chien or Yu T. F. . In the first official election of members of Academia Sinica, Li Zongen and Chang Hsiao-chien were elected in the general election, and Yu T. F. also passed in the first by-election. However, Tang F. F. and Hung S. L., both member candidates criticized by Wang Hefu, were ultimately defeated. Chi Yuoh-Fong, who was criticized by Yuan Hanqing, was not elected, either. Most of the criticisms and suggestions put forward by Fu Ssu-Nien had a positive impact on the member election. The election principles he put forward, such as "better to be short than to be shoddy" are still of referential significance to the current academician election. However, his proposal to add Qi Shounan, James K. Shen, Xie Jiasheng and Sun Pen-wen to the list of candidates for members was not adopted by the PCE and the council because it was inconsistent with the provisions of the *Regulations for the Election of Members of Academia Sinica.*

The election of the first batch members of Academia Sinica was successful although it was not impeccable.It can be regarded as a model of independent election of academic elites in modern Chinese academia. This is not only because those elected were almost all the excellent talents of the time, but also because the council had made effective institutional arrangements for the election in advance. In the election, the council stood in the position of objectivity, impartiality and academic independence from politics, upheld a solemn and prudent attitude, abided by academic norms, attached importance to academic standards, and adhered to democratic principles. Those are also the keys to the success of that member election.

The election of 81 members for the first term of Academia Sinica marked the birth of the first group of academicians in China, and the academic elite group enjoying the highest academic honorary title in their lives began to enter the Chinese historical stage. After the election of the first batch of members of Academia Sinica, the establishment of the academician system in China was just around the corner.

Chapter 6
The First General Assembly of Members of Academia Sinica

The System of General Assembly of Members of an academy or a society is the core of the academician system. Before the election of the first batch of members of Academia Sinica was officially held, the council had already made legislative preparations for the General Asssembly of Members by formulating *The Regulations of the General Assembly of Members of the National Acaemia Sinica*. After the election, the council turned its focus to the preparing and holding of the First General Assembly of Members. In September of 1948, the Kuomintang had already suffered defeat in the civil war and the political situation was grim; however, despite this, Academia Sinica held the First General Assembly of Members as scheduled. President Chiang Kai-shek of the Republic of China attended the opening ceremony and delivered a speech. According to democratic procedures, the assembly elected the Third Council members of Academia Sinica and discussed and approved a number of important proposals concerning the development of Academia Sinica and China's higher education and academic undertakings. However, due to drastic political changes after the assembly, most of the resolutions of the assembly could not be implemented. The first batch of members of Academia Sinica also had to make their own choices between leaving or staying.

J. Guo, *The Establishment and Reconstruction of the Academician System in China*,
https://doi.org/10.1007/978-981-15-7208-1_6

6.1 From the Members' Symposium to the Subcommittee's Meeting

(1) The Members' Symposium before the Assembly

On April 23, 1948, shortly after Academia Sinica officially announced the list of the first batch of Members, Chu Chia-Hua and Wong Wen-hao seized the chance of some members of Academia Sinica attending the national congress in Nanjing and invited the members from Nanjing, Shanghai and Hangzhou to attend the symposium at Academia Sinica.[1] At 4:00 pm on April 27, the symposium was held in the auditorium of Academia Sinica. One of the central issues discussed was how to hold the First General Assembly of Members of Academia Sinica. Chu Chia-Hua, Wong Wen-hao, Adam Pen-Tung Sah, Wang Chung Hwei, Wang Chia-Chi, Wu Hsien-Wen, Li Shu-Hua, Li Chi, Li Hsien-wen, Woo Ting-liang, Woo Sho-Chow, Chou Keng-Shang, Chow Jen, Hu Shih, Ma Yin-chu, Yuan I-Chin, Ling Hung-hsun, Chuang Chang-Kong, Shiing-Shen Chern, Huang T. K., Feng Teh-Pei, Young Chung-Chien, Tsai Chiao, Teng S. C., Chien Shung-shu, Hsiao Kung-chuan, and Loo Tsung-Le attended the symposium, reaching the quorum of "one-third of all members" stipulated in *The Regulations for the General Assembly of Members of Academia Sinica*. As the chairman of the symposium, Chu Chia-Hua delivered the first speech and put forth four issues to be discussed and decided: (1) The date of the First General Assembly of Members of Academia Sinica; (2) Whether the 20th anniversary of Academia Sinica would be held on June 9, and whether that time was too pressing or not; (3) How to prepare for the election of elected members of the Third Council when the Second Council was about to expire; (4) According to *The Organization Act of Academia Sinica*, "Thereafter fifteen Members will be elected by Members each year," and due to the time needed for nomination and the procedures of promulgation, whether the preparations should be made immediately or not. In his speech, Chu Chia-Hua reported, regarding the nomination process of the member candidates and the situation of unavoidable omission, "I hope to handle it more completely thereafter."[2]

In view of the fact that some members of Academia Sinica present at the meeting were not council members, "they are not very familiar with the situation of Academia Sinica," so Wong Wen-hao briefly reported on Article 9 of *The Organization Act of Academia Sinica*, regarding the functions and powers of members of Academia Sinica. And he reported the provisions of *The Regulations for the General Assembly of Members of Academia Sinica*. He stressed: "*The Regulations for the General Assembly of Members of Academia Sinica*, the current regulations for Academia Sinica, are very important to the progress of the *General Assembly of Members*." After

---

[1]Yuanshi Tanhuahui院士谈话会 (Symposium of Members of Academia Sinica). Nanjing: Zhongguo dier lishi danganguan中国第二历史档案馆 (The Second Historical Archives of China), Total No. 393, File No. 569.

[2]Yuanshi Tanhuahui院士谈话会 (Symposium of Members of Academia Sinica). Nanjing: Zhongguo dier lishi danganguan中国第二历史档案馆 (The Second Historical Archives of China), Total No. 393, File No. 569.

that, Adam Pen-Tung Sah made a report and raised the issue that "*The Regulations*[3] of Academia Sinica stipulate that members of Academia Sinica should be elected once a year, but the candidates for member should be elected by the *General Assembly of Members* after a four-month announcement period, which likely means that the *General Assembly of Members* will be held twice a year." Because of this, he invited participating members to exchange views on the announcement time and grouping situation. In the report, he also pointed out: "The Council Member may not be an Member of Academia Sinica; that is, the council member is not determined solely by academic contributions, and the academic researchers who advocate for them may also be elected." In addition, he talked about the election of honorary members of Academia Sinica and proposed that "according to laws and regulations, it is limited to Chinese nationality"; the Second Council would expire, and the *General Assembly of Members of Academia Sinica* must be held to elect the members of next term of council. "If it is not held at that time, the council will be interrupted; it will cost a lot and the weather will still be hot."[4]

During the discussion with the members of Academia Sinica attending the symposium, Hu Shih suggested that *The Regualations for the Election of the Council of Academia Sinica* seemed to be no longer applicable. He proposed that the members of the Third Council should be elected by correspondence in accordance with the provisions of Articles 2 & 6 of *The Regulations of the Council of Academia Sinica*, and the convocation be scheduled for autumn. At the same time, he also suggested that the honorary secretary of the council, the director-general, and members of Academia Sinica in Nanjing, Shanghai, Hangzhou, and Wuhan should set up a preparatory committee to prepare for the election of members of Academia Sinica and the council members. As a remedy, the preparatory committee will be granted greater powers to amend *The Regulations for the Election of the Council of Academia Sinica* with the corresponding method firstly.[5] Wong Wen-hao believed that there was no problem in convening the First General Assembly of Members of Academia Sinica and drafting and amending *The Regualations for the Election of the Council* by correspondence. However, due to the stipulation of group discussions and four-fifths of the participants voting for approval, the way of electing members of Academia Sinica by correspondence "fails to conform to the law, according to the previous discussion by

[3]"The Regulations" here refer to *The Organization Act of Academia Sinica* revised and promulgated on March 13, 1947. Article 6 of the Act stipulates that Members of Academia Sinica "shall be elected by members every year after the election of the first batch of members, the number of members to be elected each time shall not exceed 15."

[4]Yuanshi Tanhuahui院士谈话会 (Symposium of Members of Academia Sinica). Nanjing: Zhongguo dier lishi danganguan中国第二历史档案馆 (The Second Historical Archives of China), Total No. 393, File No. 569.

[5]Yuanshi Tanhuahui院士谈话会(Symposium of Members of Academia Sinica). Nanjing: Zhongguo dier lishi danganguan中国第二历史档案馆(The Second Historical Archives of China), Total No. 393, File No. 569.

the Council. It is easier to attend the election."[6] Since then, Hu Shih had proposed that: (1) Firstly, a personnel should be presumed to set up a drafting committee in order to amend *The Regualations for the Election of the Council*; (2) If the Council was changed into a standing review organ of the General Assembly of Members of Academia Sinica, the council members may be limited to selecting members of Academia Sinica from among themselves; (3) In order to increase efficiency, it was better to choose Councils Members among the people of Nanjing, Shanghai, Hangzhou, and Wuhan. As the participating members of Academia Sinica had different opinions on the nomination method of council members, Li Chi suggested: "There are different nomination methods for the election of Council Members. According to the election method of members of Academia Sinica, each group of members of Academia Sinica can propose their list first." In addition, he proposed that the form of badge for members of Academia Sinica should be decided.[7]

Finally, the symposium made the following decisions: (1) The First General Assembly of Members of Academia Sinica would be held in mid-September; (2) The election of the Third Council Members would be held at the First General Assembly of Members of Academia Sinica; (3) A subcommittee would be set up first with Wong Wen-hao, the Honorary Secretary of the Council, Adam Pen-Tung Sah, the Director-General, Li Shu-Hua, and Chou Keng-Shang as members. They also decided to amend *The Regualations for the Election of the Council* by correspondence, which would be drafted by Adam Pen-Tung Sah; (4) The election of the Third Council Members would focus on people from Nanjing, Shanghai, Hangzhou, and Wuhan; (5) The nomination for the election of the second batch of members of Academia Sinica in 1949 would begin after the First General Assembly of Members; (6) When electing council members, a Nomination Committee should be set up in order to decide on the principles and submit them to the drafting committee for reference.[8] After the symposium, the revision of *The Regualations for the Election of the Council* began immediately.

(2) **The Design of Certificate and the Badge for Members of Academia Sinica**

Before the First General Assembly of Members of Academia Sinica, Academia Sinica also carried out the work concerning design of certificate and the badge for the members of Academia Sinica. As for the design of Certificate, when the third session of the Fifth Annual Meeting was held by the Second Council of Academia Sinica on March 27, 1948, Wong Wen-hao, Adam Pen-Tung Sah, Li Chi, Chow Jen and Liang Ssu-cheng had already been recommended and determined to be in charge

[6] Yuanshi Tanhuahui院士谈话会 (Symposium of Members of Academia Sinica). Nanjing: Zhongguo dier lishi danganguan中国第二历史档案馆 (The Second Historical Archives of China), Total No. 393, File No. 569.

[7] Yuanshi Tanhuahui院士谈话会 (Symposium of Members of Academia Sinica). Nanjing: Zhongguo dier lishi danganguan中国第二历史档案馆 (The Second Historical Archives of China), Total No. 393, File No. 569.

[8] Yuanshi Tanhuahui院士谈话会 (Symposium of Members of Academia Sinica). Nanjing: Zhongguo dier lishi danganguan中国第二历史档案馆 (The Second Historical Archives of China), Total No. 393, File No. 569.

of the design and they were authorized to do the work as well.[9] As a member of the five-member team, Li Chi gave suggestions on the form of the elected certificate and badge that were to be decided at the aforementioned members'symposium. Adam Pen-Tung Sah was in charge of this group. On August 2, he sent the wording of his draft certificate to the other four members for comments, and asked Liang Ssu-cheng to design and compose it on his behalf. Liang was also instructed to design the badge for.

Liang Ssu-cheng was the head of the department of architecture at the National Tsinghua University then. He was a well-known scholar who had studied ancient Chinese architecture and was a suitable candidate for designing the member certificate and badge. By September 1, Liang Ssu-cheng had already designed the member certificate and badge. He said, in a letter to Adam Pen-Tung Sah on the same day: "I hereby present two types of member certificates: one is in a horizontal inscription style and the other is in the style of an album. They are only different in the mounting method. I think the album style is better." At the same time, he said: "The key to the certificate is to write well, so we must ask a calligrapher to write it. …The certificate paper can be white Xuan paper (Chinese art paper), but if you use old paper with a smoked topaz color and have it gilded with gold, it can also be very elegant. The surface of the album can be padded, and a docket shall be attached. Please also ask the same calligrapher to write for it." Later in the letter, he asked, "How about the design of the badge? Has it been made?"[10] The "badge" referred to the member badge designed by Liang Ssu-cheng, which was square, with nine squares inscribed, each with one of the nine convex characters of "国立中央研究院院士" written in official script. There was a pin on the back that was welded on the top of the badge.

Adam Pen-Tung Sah adopted Liang Ssu-cheng's suggestion, chose the album style for the member certificate, and specially invited calligrapher Shen Yinmo to write the certificate.[11] However, due to a lack of funds, the member badge was not cast with pure gold. As he said in his reply to Liang Ssu-cheng's letter on September 7, "Thank you very much for your letter from September 1 and for the two types of member certificates. It will be printed in accordance with the album style as a memorial. As for the member badge, it has been molded according to the model. Due to a limited budget, silver gilding was applied instead of gold gilding, and the result was not as beautiful as expected. In view of the time constraint, we have proposed to temporarily

[9]Zhongyang yanjiuyuan guanyu sheji yuanshi dangxuan zhengzhang shiyang de wenshu中央研究院关于设计院士当选证章式样的文书(Academia Sinica's Documents on Designing Certificates for Elected Members). Nanjing: Zhongguo dier lishi danganguan中国第二历史档案馆 (The Second Historical Archives of China), Total No. 393, File No. 1629.

[10]Zhongyang yanjiuyuan guanyu sheji yuanshi dangxuan zhengzhang shiyang de wenshu中央研究院关于设计院士当选证章式样的文书(Academia Sinica's Documents on Designing Certificates for Elected Members). Nanjing: Zhongguo dier lishi danganguan中国第二历史档案馆 (The Second Historical Archives of China), Total No. 393, File No. 1629.

[11]Zhongyanyuan shoujie yuanshihui jinchen shishi xing jiemuli中研院首届院士会今晨十时行揭幕礼 (The Opening Ceremony of the First General Assembly of Members of Academia Sinica Held at 10 o'clock This Morning). *Zhongyang ribao*中央日报 (Central Daily), 1948–09–23 (4).

Fig. 6.1 The badge for members of Academia Sinica in 1948

cope with the situation. Once the funds are slightly more sufficient, gold gilding will be used instead (Fig. 6.1)."[12]

(3) **The Opening Ceremony and Preliminary Meeting of the Assembly**

The First General Assembly of Members of Academia Sinica and the 20th Anniversary Commemoration Meeting of Academia Sinica were held from September 23 to 25, 1948. On the morning of September 23, the meeting opened in the Academia Sinica auditorium. In order to save money, the whole venue was arranged simply without flowers. Although it was raining, the lights were not turned on. Only bright red banners were hung in the auditorium, saying "The 20th Anniversary Commemoration Meeting of Academia Sinica and the First General Assembly of Members." The Director-General Adam Pen-Tung Sah stood at the door to welcome members and guests. After the assembly was started, the attendees signed either the left or right of the gate.[13] Participating members wore member badges.

Members attending the meeting included Chu Chia-Hua, Wong Wen-hao, Adam Pen-Tung Sah, Wang Chung Hwei, Wu Hsien-Wen, Yu Cha-Shih, Pai Sitsan, Li Chi, Li Shu-Hua, Li Zongen, Li Hsien-wen, Woo Ting-liang, Woo Sho-Chow, Robert K.

[12]Zhongyang yanjiuyuan guanyu sheji yuanshi dangxuan zhengzhang shiyang de wenshu中央研究院关于设计院士当选证章式样的文书(Academia Sinica's Documents on Designing Certificates for Elected Members). Nanjing: Zhongguo dier lishi danganguan中国第二历史档案馆(The Second Historical Archives of China), Total No. 393, File No. 1629.

[13]Zhongyanyuan yuanshihui jiemu中研院院士会揭幕(The Opening Ceremony of the First General Assembly of Members of Academia Sinica). *Zhongyang ribao*中央日报(Central Daily), 1948–09–24 (4).

S. Lim, Chow Jen, Chou Keng-Shang, Ping Chi, Chu Coching, Mao T. E., Liu Yei-Tsen, Hu Shih, Hu Hsen-hsu, Yu T. F., Chiang Li-Fu, Yuan I-Chin, Yin Hung-chang, Chuang Chang-Kong, Chen Yuan, Chen Ta, Shiing-Shen Chern, Ling Hung-hsun, Tao L. K., Zhang Yuanji, Liang Ssu-cheng, Fu Ssu-Nien, Feng You-lan, Feng Teh-Pei, Tang Yung-tung, Tang P. S., Young Chung-Chien, Yang Shu-Tah, Yeh Chi-Sun, Teng S. C., Tsai Chiao, Chien Shung-shu, Hsiao Kung-chuan, Hsieh Chia-yung, Tai Fon-Land, Su Buchin, Ny Tsai-ze, Yao Y. T., a total of 51 people. Among the 30 members who were not present, Woo Tsin-Hang, Chin Yuch-Lin, Ma Yin-chu, Chen Shisan C., Chen Yin-k'o, Liang Ssu-yung, Hsu Pao-Lu, Chang Hsiao-chien, Chang Chin-yueh, Loo Tsung-Le and Ku Chieh-kan took leave of absence. There are 19 people abroad, including Wang Shih-Chieh, Wang Chia-Chi, Wu Hsien, Woo Y. H., Wu Ta-You, Wang Ging-Hsi, Lee J. S., Li Fang Kuei, Hou T. P., Chen Ko Kuei, Kuo Mo-jo, Tseng Chao-lun, Hua Loo-Keng, Huang T. K., Tung Ti-Cho, Tung Tso-pin, Chao Yuen Ren, Chao Chung-Yoa and Tsien T. S.[14]

Guests attending the opening ceremony included Chiang Kai-shek, President of the Republic of China, President Chang Poling and Vice-President Chia Ching-teh of the Examination Yuan, Secretary-General of the Presidential Office Wu Ting-chang, Vice-President Liu Zhe of the Supervisory Institute, Minister of Defense Ho Ying-chin, National Government Member and Senior Minister Zeng Qi of the Presidential Office, President Ku Yu-hsiu of National Chengchi University, Professor Shi Zhiquan of Chaoyang University and the Standing Committee of the Central Party of the People's Society, President Wu Yi-fang of Ginling College, and President Chow Hung-ching of National Central University.[15] Due to Chiang Kai-shek's presence at the venue, the road from Jiming Mountain to the venue was "full of armed police, heavily guarded, and had strict security. The atmosphere was unusual."[16]

At 10:00 am, the bell rang and the chairman of the assembly, Chu Chia-Hua, announced the beginning of the assembly and said: "It is a great honor have the President present at today's assembly. In the past twenty years, China's academic progress has been remarkable, which can be attributed to the completion of the Northern Expedition and the promotion of scientific achievements after the government established Nanjing as its capital."[17] Then, Chiang Kai-shek delivered a speech. First of all, he

---

[14]Zhongyang yanjiuyuan zuzhi banfa yuanshi huiyi jilu jiqi xuanju guicheng deng youguan wenshu 中央研究院组织办法、院士会议纪录及其选举规程等有关文书(Relevant Documents such as the Organization Methods of Academia Sinica, Minutes of the General Assembly of Members and the Regulations for Election). Nanjing: Zhongguo dier lishi danganguan中国第二历史档案馆(The Second Historical Archives of China), Total No. 393, File No. 2824.

[15]Zhongyang yanjiuyuan zuzhi banfa yuanshi huiyi jilu jiqi xuanju guicheng deng youguan wenshu 中央研究院组织办法、院士会议纪录及其选举规程等有关文书(Relevant Documents such as the Organization Methods of Academia Sinica, Minutes of the General Assembly of Members and the Regulations for Election). Nanjing: Zhongguo dier lishi danganguan中国第二历史档案馆(The Second Historical Archives of China), Total No. 393, File No. 2824.

[16]Zhang Shunian张树年(2006). *Wode fuqin Zhang Yuanji*我的父亲张元济(My Father Zhang Yuanji).Tianjin: Baihua wenyi chubanshe百花文艺出版社. 213–214.

[17]Zhongyanyuan yuanshihui jiemu中研院院士会揭幕(The Opening Ceremony of the First General Assembly of Members of Academia Sinica). *Zhongyang ribao*中央日报(Central Daily), 1948–09–24 (4).

pointed out: "Today is the day when members of Academia Sinica meet for the first time. This is indeed a blessing for Chinese academia and the country."[18] Secondly, he analyzed the reason why the the election of the first batch of members and Genaral Assembly of Members of Academia Sinica were held only twenty years after the establishment of Academia Sinica:

> Academia Sinica has been set up for twenty years, and the election and assembly of members did not take place until this year. It was set up at the beginning of Nanjing's becoming the capital city of China, and it is one of the most important policies of founding the country after the Northern Expedition. However, the election of members has not been held in the past twenty years due to the invasion of the Japanese aggressors; it was not easy to take into account the long-term strategy of cultural, educational, and intellectual education during the military period. Moreover, every step of academic progress should be taken when the time is right. If the time is not right, it tends to have disadvantages. Academia Sinica has accumulated twenty years of experience, and only now can it fully demonstrate its gradual progress and seek truth from facts. According to the precedents of national institutes in various countries, academicians or members are chosen as the main body first, and then the development of research work is studied. Academia Sinica set up its research institute twenty years ago, and set up a national council thirteen years ago. Today, a national assembly of members has been set up, which seems to be contrary to the general order. This is precisely because, when Academia Sinica was first established, the development of various academic disciplines in China was not as rapid as ten years later, so the conditions for electing members were not prepared at that time. In the past ten years, although academics has been greatly influenced by the Japanese aggressors, some subjects are still on their way to steady progress. Therefore, today's event is a natural result (Fig. 6.2).[19]

Thus, Chiang Kai-shek believed that the invasion of Japan and China's low academic level were two main factors that restricted the establishment of the academician system in Academia Sinica. Jiang then praised Tsai Yuan-Pei's contribution to Academia Sinica, believing that Academia Sinica "started from the very beginning with this plan of steady progress, which was truly the vision of former president Tsai Jiemin."[20] Finally, he hoped that members would raise their academic standards, saying: "All of you have been admitted by the Chinese people and will be more specialized in your fields of study in the future. It is not difficult to catch up with the advanced scientific countries in ten years, as the father of our country has said." At the same time, he also hoped that members would "spread academic achievements",

[18]Zhongyang yanjiuyuan zuzhi banfa yuanshi huiyi jilu jiqi xuanju guicheng deng youguan wenshu 中央研究院组织办法、院士会议纪录及其选举规程等有关文书(Relevant Documents such as the Organization Methods of Academia Sinica, Minutes of the General Assembly of Members and the Regulations for Election). Nanjing: Zhongguo dier lishi danganguan中国第二历史档案馆(The Second Historical Archives of China), Total No. 393, File No. 2824.

[19]Zhongyang yanjiuyuan zuzhi banfa yuanshi huiyi jilu jiqi xuanju guicheng deng youguan wenshu 中央研究院组织办法、院士会议纪录及其选举规程等有关文书(Relevant Documents such as the Organization Methods of Academia Sinica, Minutes of the General Assembly of Members and the Regulations for Election). Nanjing: Zhongguo dier lishi danganguan中国第二历史档案馆(The Second Historical Archives of China), Total No. 393, File No. 2824.

[20]Jiemin(孑民) is Tsai Yuan-Pei's style name.

Fig. 6.2 Chiang Kai-shek delivered a speech at the First General Assembly of Members of Academia Sinica on September 23, 1948

"advocate for scientific cooperation", "communicate academic achievements with other countries' colleges", and "contribute to national construction."[21]

After Chiang Kai-shek delivered his speech, Chu Chia-Hua delivered the opening speech. He said first of all with deep emotion: "It has been six months since the members of Academia Sinica were elected by the council in March of this year. We have chosen today, the twentieth anniversary of Academia Sinica, to hold the meeting. We are both recalling and looking ahead; recalling that, while we have experienced difficulties for many years in the past, we have finally made some academic achievements and contributions; looking ahead as we consider how we will persevere in continuing to work on the distant research route." Then he pointed out: "Modern countries have many national colleges, and regardless of their organizational form, their history is very long and their content is very complete. Although our Academia Sinica is a national institute with insufficient equipment and human resources, and the

[21] Zhongyang yanjiuyuan zuzhi banfa yuanshi huiyi jilu jiqi xuanju guicheng deng youguan wenshu 中央研究院组织办法、院士会议纪录及其选举规程等有关文书(Relevant Documents such as the Organization Methods of Academia Sinica, Minutes of the General Assembly of Members and the Regulations for Election). Nanjing: Zhongguo dier lishi danganguan中国第二历史档案馆(The Second Historical Archives of China), Total No. 393, File No. 2824.

research work is hindered from time to time, it has been able to support the establishment of the General Assembly of Members of Academia Sinica and the completion of the academician system due to the dedication of the staff to their jobs and the love of the academic community." He also stressed: "In the past, Academia Sinica was focused on the council; in the future, it will focus on the General Assembly of Members." Later, he reviewed the process of the election of the first batch of members of Academia Sinica by the council and introduced the tasks of the General Assembly of Members. In the end, he raised two hopes: one was that "the government should, in line with the Constitution's basic national policy, reward scientific inventions and creations, and provide and perfect research facilities"; the second is that "all parties in the society should encourage more academic research, heighten the atmosphere of academic research, and expand the field of our knowledge."[22]

After Chu Chia-Hua's speech, Wong Wen-hao, who was the executive director of the national government and the Honorary Secretary of the Council of Academia Sinica, delivered a speech through his latter identity. In his speech, he talked about the current status and nature of members: "Members have a very high position. Our members hold honorary posts. Not only do they not receive any money, but they are very cautiously elected. They are nominated from various universities, independent colleges, special societies, and research institutions across the country. After many steps and discussions, they are elected after a careful voting by the council, which is indeed a great honor. …A member is a tenure. This is the square badge that we are wearing now. We can wear it for a lifetime." He also pointed out: "Due to some procedural regulations, there have been some important scholars across the country who were omitted. We haven't done enough work, so those of us who have been elected should, on the one hand, have the feeling of leaving pearls behind and, on the other hand, should double our self-encouragement to avoid becoming members who are little more than nameplates." In the end, he focused on the members' task and work, saying: "In addition to doing scientific research on his own, he should also make academic contacts and encouragement at home, and make academic cooperation and progress with the world." In his view, Academia Sinica "is not really sufficient for domestic contacts. I hope members will review and supplement it so that its name can match its reality and achieve successful academic research in our country." At the same time, he thought that Academia Sinica had "done a little in the past, but it is still better to increase its participation in international conferences," and hoped that China could invite "many foreign scholars to come to our country to have meetings and learn from each other, which can further enhance our enthusiasm for scientific research and thus make all kinds of work more advanced."[23]

[22]Zhongyang yanjiuyuan zuzhi banfa yuanshi huiyi jilu jiqi xuanju guicheng deng youguan wenshu 中央研究院组织办法、院士会议纪录及其选举规程等有关文书(Relevant Documents such as the Organization Methods of Academia Sinica, Minutes of the General Assembly of Members and the Regulations for Election). Nanjing: Zhongguo dier lishi danganguan中国第二历史档案馆(The Second Historical Archives of China), Total No. 393, File No. 2824.

[23]Zhongyang yanjiuyuan zuzhi banfa yuanshi huiyi jilu jiqi xuanju guicheng deng youguan wenshu 中央研究院组织办法、院士会议纪录及其选举规程等有关文书(Relevant Documents such as the Organization Methods of Academia Sinica, Minutes of the General Assembly of Members and

Next, members delivered speeches on behalf of Zhang Yuanji and Hu Shih. Zhang Yuanji was 82 years old and a Jinshi (a successful candidate of the highest imperial examinations) of the Guangxu Ren-Chen (the year of 1892) Imperial Examination. He served as Shu Ji Shi (equal to a doctoral student of the Chinese Academy of Science today) of the Imperial Academy and Zhang Jing (a kind of civil official) of the Prime Minister's Office. He supported the Reform Movement of 1898. He was dismissed after the revolt failed and became chairman of the Commercial Press. He talked about the Sino-Japanese War of 1894, the Reform Movement of 1898, the Revolution of 1911, Yuan Shikai proclaiming himself emperor, the eight-year war of resistance, and the civil war between the Kuomintang and the Communist Party. As for the civil war, he strongly opposed it, pointing out that "this is not an act of external aggression, but a struggle between brothers who turn against each other" and that "this war is really unnecessary." Moreover, he was deeply saddened by the turmoil caused by the civil war, the lack of livelihood from the people, and the country's state of trepidation, and affirmed that "if the war goes on, I am afraid, needless to say, Academia Sinica will inevitably close down." In the end, he also called for peace: "We need peace to preserve our country; we need peace to revive our nation; we need more peace to implement all kinds of academic research for the sake of our country and our nation!"[24] Hu Shih felt that Zhang Yuanji's speech was "a killjoy."[25] Shiah Nae, on the other hand, said that "the wording is painful but truly touching."[26]

In his speech, Hu Shih said that the opening day of the assembly "was a memorable day for Chinese academia." At the same time, he highly praised Wong Wen-hao's contribution to "member work" and believed that the event, which was presided over by Chu Chia-Hua "can be said to be a great success." In the end, he raised hope for the members with this encouragement:

> We can internally encourage from our position as well as encourage the next generation. It is not that we are members wearing a square badge, just sitting back and enjoying the benefits; nor, in the next half-century, will we eat on our own achievements instead of continuing to work. Academia Sinica is not an academic nursing home; rather, it should, firstly, encourage the next generation. We can serve as a model and continue our work so the academician system will not fail. Secondly, we should take more disciples. Today, among our members, the youngest are two mathematicians, who are already 40 years old. I think our past experiences and methods are mature enough to encourage the next generation. I hope that in the next twenty or 200 years, this spirit of our Academia Sinica will be carried forward. I hope our members will encourage each other.[27]

---

the Regulations for Election). Nanjing: Zhongguo dier lishi danganguan中国第二历史档案馆(The Second Historical Archives of China), Total No. 393, File No. 2824.

[24]Zhang Yuanji张元济(1948). Churao zhi yan: Guoli zhongyang yanjiuyuan diyici yuanshihui zhici 刍荛之言——国立中央研院第一次院士会致辞(My Superficial Remarks—Speech at the First General Assembly of Members of Academia Sinica). *Zhanwang*展望(Outlook)2(21): 5.

[25]Zhang Shunian张树年(2006). *Wode fuqin Zhang Yuanji*我的父亲张元济(My Father Zhang Yuanji).Tianjin: Baihua wenyi chubanshe百花文艺出版社. 214.

[26]Shiah Nae夏鼐(2011). *Shiah Nae riji*夏鼐日记(Shiah Nae's Diary), Vol. 4. Shanghai: Huadong shifan daxue chubanshe华东师范大学出版社. 205.

[27]1948 nian 9 yue 23 ri zai zhongyanyuan diyijie yuanshi huiyi shang de jianghua一九四八年九月廿三日在中研院第一届院士会议上的讲话(Speech at the First General Assembly of Members

After Hu Shih delivered his speech, the opening ceremony ended and the members took photos to mark the occasion. The preliminary meeting for the assembly was held at 3:00 pm on September 23, the remaining 47 members, with the exception of Wang Chung Hwei, Yu Cha-Shih, Zhang Yuanji, and Tsai Chiao, attended the opening ceremony, with Chu Chia-Hua as chairman. The preliminary meeting first discussed "how to arrange the agenda of the assembly proposed by Chu Chia-Hua and whether to discuss and review in groups or not." The members were asked to exchange views. At the meeting, it was decided that each of the three groups should have two members to organize the deliberation subcommittee. And the elected members were mathematics and physics group's Li Shu-Hua, Chu Coching, and Adam Pen-Tung Sah, biology group's Li Zongen and Ping Chi, and humanities group's Hu Shih and Li Chi. Among them, Hu Shih was the convener. The number of members of mathematics and physics group was one more than the originally planned number. Then, they discussed "how to decide the duration of the assembly of members proposed by Chu Chia-Hua." The meeting decided that the duration of the assembly would include the opening ceremony of the first day and the preliminary meeting, and it would last three days in all. The preliminary meeting also discussed five different proposals on the election of the Third Council members, the nomination and election of members in 1949, the election of honorary members, the presenting of important papers, and the drafting of national academic guidelines (Fig. 6.3).[28]

(4) **The Meeting of the Deliberation Subcommittee**

At 8:00 pm on September 23, the members of the Deliberation Subcommittee met in the conference room of Academia Sinica. Attendees included Li Shu-Hua, Chu Coching, Adam Pen-Tung Sah, Wu Hsien-Wen, Li Hsien-wen, Li Chi, Hu Shih, and Chen Ta, with Hu Shih as the chairman. Li Zongen and Ping Chi were absent. Wu Hsien-Wen, Li Hsien-wen, and Chen Ta, who were not members of the committee, were invited to attend the meeting. At first, the meeting drew up the points that were to be paid attention to in the election of the Third Council members, including that the council was the standing review organ of the General Assembly of Members of Academia Sinica, there was no need to have council members for each discipline, and no more than three persons should be elected as council members for each discipline. Secondly, the agenda of the assembly was drawn up, which mainly included the following items: (1) report on the election of honorary members: the election of honorary members would not be held at this session; please invite the Selection Committee of the assembly to draft the detailed rules for the election of honorary members and solicit the consent of the members through correspondence; (2) request

of Academia Sinica on September 23, 1948). In Geng Yunzhi耿云志(Eds.) (1994). *Hu Shih yigao ji micang shuxin*胡适遗稿及秘藏书信(Hu Shih's Posthumous Manuscripts and Secret Collection of Letters), Vol. 12. Hefei: Huangshan shushe黄山书社. 177–179.

[28]Zhongyang yanjiuyuan zuzhi banfa yuanshi huiyi jilu jiqi xuanju guicheng deng youguan wenshu 中央研究院组织办法、院士会议纪录及其选举规程等有关文书(Relevant Documents such as the Organization Methods of Academia Sinica, Minutes of the General Assembly of Members and the Regulations for Election). Nanjing: Zhongguo dier lishi danganguan中国第二历史档案馆(The Second Historical Archives of China), Total No. 393, File No. 2824.

Fig. 6.3 Photo of the First General Assembly of Members of Academia Sinica on September 23, 1948

the Second Council to promptly presume the preparatory committee for the election of the second batch of members of Academia Sinica; (3) request the assembly to set up various standing committees, one of which would be a thesis committee (three members in each group) and the other an academic speech committee (one member in each group); (4) propose that each group of members appoint a secretary to preside over academic contacts; (5) suggest that the assembly organize a preparatory committee for the next General Assembly of Members of Academia Sinica with the honorary secretary, the director-general, the secretary director, the general affairs director, and every group's secretary of the General Assembly of Members of Academia Sinica; (6) discuss the proposals.[29] These agreed proposals and agendas made further preparations for the official holding of the First General Assembly of Members of Academia Sinica.

---

[29]Zhongyang yanjiuyuan zuzhi banfa yuanshi huiyi jilu jiqi xuanju guicheng deng youguan wenshu 中央研究院组织办法、院士会议纪录及其选举规程等有关文书(Relevant Documents such as the Organization Methods of Academia Sinica, Minutes of the General Assembly of Members and the Regulations for Election). Nanjing: Zhongguo dier lishi danganguan中国第二历史档案馆(The Second Historical Archives of China), Total No. 393, File No. 2824.

6.2 The Implementation of Member Authority: The Election of the Third Council Members

At 9 am on September 24, the assembly held its First General Meeting. 48 members attended the meeting: Chu Chia-Hua, Adam Pen-Tung Sah, Wang Chung Hwei, Wu Hsien-Wen, Pai Sitsan, Li Chi, Li Shu-Hua, Li Zongen, Li Hsien-wen, Woo Ting-liang, Woo Sho-Chow, Robert K. S. Lim, Chow Jen, Chou Keng-Shang, Ping Chi, Chu Coching, Mao T. E., Liu Yei-Tsen, Hu Shih, Hu Xianzhi, Yu T. F., Chiang Li-Fu, Yuan I-Chin, Yin Hung-chang, Chuang Chang-Kong, Chen Yuan, Chen Ta, Shiing-Shen Chern, Zhang Yuanji, Ling Hung-hsun, Tao L. K., Liang Ssu-cheng, Feng You-lan, Feng Teh-Pei, Tang Yung-tung and Tang Wong Wen-hao, Yu Cha-Shih and Fu Ssu-Nien did not attend. Chu Chia-Hua was the chairman.[30] One of the main agendas of this meeting was to hold the election for the Third Council members. According to the regulations of *The Organization Act of Academia Sinica*, amended and promulgated by the National Government on March 13, 1947, the election of council members was one of the functions and authorities of the members of Academia Sinica.

Before this assembly, the council members voted by correspondence on May 31, 1948 to amend *The Regualations for the Election of the Council of Academia Sinica*, revised by the Subcommittee. The amended regulations were divided into four chapters, which are "*General Provisions*", "Election of the Elected Council Members", "Election of the Candidates for President of Academia Sinica", and "Supplementary Provisions". Chapters 1, 3, and 4 introduced the basis for formulating the regulations, the election methods of the candidates for president, and the revision methods, respectively. Chapter 2 was the main body of the new regulations; it stipulated that, after the expiration of the Second Council members, the members of Academia Sinica should elect the next term. Just like members of Academia Sinica, the council members were also divided into three groups—namely, mathematics and physics, biology, and humanities—with a total of up to 50 members, each with a quota of at least ten members, and their allocation determined by the previous council. As for the election of council members, there were two ways to produce candidates: one was to be nominated by the Nomination Committee of Council Members, and the other was to be jointly nominated by five members of Academia Sinica from each group. The number of candidates shall be at least a multiple of the number of candidates elected. At the assembly, the council members would be elected by secret ballot, and those with the votes of more than half of the members of Academia Sinica presenting would be elected. This chapter also provided regulations on how to deal with the situation in which a certain group had not yet filled the quota after two ballots: in the second ballot, the number of candidates who had obtained more votes

[30]Zhongyang yanjiuyuan zuzhi banfa yuanshi huiyi jilu jiqi xuanju guicheng deng youguan wenshu 中央研究院组织办法、院士会议纪录及其选举规程等有关文书(Relevant Documents such as the Organization Methods of Academia Sinica, Minutes of the General Assembly of Members and the Regulations for Election). Nanjing: Zhongguo dier lishi danganguan中国第二历史档案馆(The Second Historical Archives of China), Total No. 393, File No. 2824.

in the group was twice as many as the number of vacancies. In the run-off election, the candidate who got more votes should be elected and fill the vacancy.[31]

In 1935, *The Regualations for the Election of the Council of Academia Sinica*, amended and approved by the First Council, stipulated that candidates for the council should be elected by full-time professors (including full-time associate professors) and the Selection Committee from various departments of national universities and independent colleges. Candidates for the Second Council members were selected according to the regulations. The regulations had been amended by this correspondence vote, and the method of selecting candidates for council members changed. This was the result of the council's belief that the original mode of production already had disadvantages. As stated in Article 11 of *The Organization Act of National Academia Sinica*, which was amended and approved by the Fourth General Meeting of the Third Annual Meeting of the Second Council on October 23, 1946, the election method for the candidates was originally nomination by professors and associate professors of various universities, which had some drawbacks. First, higher education and academic research had both similarities and differences. Second, the standard for university professors had fallen over the years, and it would not be easy to improve in a couple of years. Third, the election procedure was very difficult and could not be discussed together."[32]

After the amendment of *The Regualations for the Election of the Council* of Academia Sinica was approved, the council set up the Nomination Committee for the Third Council members. Members included Chiang Li-Fu, Li Shu-Hua, Chu Coching, Mao T. E., Adam Pen-Tung Sah, Loo Tsung-Le, Ping Chi, Feng Teh-Pei, Li Chi, Fu Ssu-Nien, Chou Keng-Shang, Hu Shih, Wang Shih-Chieh, Wang Ging-Hsi, and Hsieh Chia-yung.[33] At the same time, the council informed all members of Academia Sinica to "submit their candidate elections to the Nomination Committee with the joint signature of five members of Academia Sinica" before August 10, 1948. The council also petitioned that the National Government extend the Second Council members until the end of 1948.[34]

---

[31]Guoli zhongyang yanjiuyuan gaikuang: minguo 17 nian 6 yue zhi 37 nian 6 yue国立中央研究院概况:民国十七年六月至三十七年六月(An Overview of National Academia Sinica: From June, 1928 to June,1948). In Liu Guiyun刘桂云 and Sun Chengrui孙承蕊(Eds.) (2008).*Guojia tushuguan cang guoli zhongyang yanjiuyuan shiliao congbian*国家图书馆藏国立中央研究院史料丛编(Collection of Historical Materials ofAcademia Sinica Collected by the National Library), Vol. 1. Beijing: Guojia tuchuguan chubanshe国家图书馆出版社. 76.

[32]Zhongyang yanjiuyuan zuzhifa pingyihui tiaoli pingyihui guicheng xianxing fagui zhangcheng mulu deng中央研究院组织法、评议会条例、评议会规程、现行法规章程目录等(The Organization Act, the Regulations of the Council, the Aritcles for the Council, the Catalogue of Current Regulations and Articles of Academia Sinica, etc.). Nanjing: Zhongguo dier lishi danganguan中国第二历史档案馆(The Second Historical Archives of China), Total No. 393, File No. 2539.

[33]Li Lairong李来容(2010). Yuanshi zhidu yu minguo xueshu: 1948 nian yuanshi zhidu de queli yu yunzuo院士制度与民国学术—1948年院士制度的确立与运作(Academician System in Republic of China: Establishment and Operation of the Academician System in 1948). Tianjin: Doctoral Dissertation of Nankai University. 267.

[34]Zhongyang yanjiuyuan dierjie pingyihui linshi huiyi jilu中央研究院第二届评议会临时会议纪录(Minutes of Provisional Meetings of the Second Council of Academia Sinica). Nanjing: Zhongguo

After the nomination deadline, the Nomination Committee met at 4:00 pm on August 27 to decide on the list of candidates for the council. Chiang Li-Fu, Mao T. E., Adam Pen-Tung Sah, Loo Tsung-Le, Feng Teh-Pei, Li Chi, Fu Ssu-Nien and Wang Shih-Chieh attended the meeting, with Adam Pen-Tung Sah as chairman. Ping Chi, Chou Keng-Shang, Hu Shih, and Li Shu-Hua did not attend, and were represented by Loo Tsung-Le, Adam Pen-Tung Sah, and Fu Ssu-Nien respectively.[35] The meeting decided: (1) ex officio members still had the right to be elected as council members; (2) if the nominated candidate was not a member of Academia Sinica and the reason had not been indicated by the nominator, the nominator should supplement the reason and take care not to exceed the limited number of persons to be nominated; (3) before the assembly was held, a council meeting was scheduled to be held on the afternoon of September 22. Each member of the Nomination Committee will request that the president convene a temporary meeting of the council by correspondence to discuss in detail the provisions of Articles 4 to 6 of *The Regualations for the Election of the Council of Academia Sinica*, which have been amended and approved by the members through correspondence voting; (4) the following principles should be followed when proposing the list of candidates for council membership: candidates who are not members of Academia Sinica should state the reasons, and the candidates jointly selected by five people should list out the five people originally nominated to show their responsibilities; since the number of members of Academia Sinica is small and most of them have been nominated by this time, the Nomination Committee will nominate all members of Academia Sinica as candidates for council membership, clarifying that "the situation will be different in the future, and this should not be taken as an example." The Nomination Committee will put forth suggestions to the assembly: there should be no more than three council members in each discipline. Members of Academia Sinica will independently propose council member candidates who are not members of Academia Sinica, and those who have not been approved by the Nomination Committee to obtain the number of joint signatures will not be proposed. At the meeting, the Nomination Committee endorsed the candidates that were proposed by the members and approved Ting S. L., Tang Yueh, Yen Shu-Tang, and Ho L.'s candidacy. At the same time, some members of the Nomination Committee jointly nominated Cheng Xiaogang, Luo Changpei, Kiang Tsai-han, and Chou Pei-yuan as candidates, who were all approved.[36] In this way, together with the candidates whose qualifications have been recognized before the meeting, a total of 99 people were nominated. Among them, there were 35 people from the mathematics and

dier lishi danganguan中国第二历史档案馆(The Second Historical Archives of China), Total No. 393, File No. 1559.

[35]Zhongyang yanjiuyuan pingyiyuan xuanju mingdan huiyi jilu deng wenjian中央研究院评议员选举名单、会议纪录等文件(Documents such as the Selection Lists and Minutes of Council Members of Academia Sinica, etc.). Nanjing: Zhongguo dier lishi danganguan中国第二历史档案馆(The Second Historical Archives of China), Total No. 393, File No. 2591.

[36]Zhongyang yanjiuyuan pingyiyuan xuanju mingdan huiyi jilu deng wenjian中央研究院评议员选举名单、会议纪录等文件(Documents such as the Selection Lists and Minutes of Council Members of Academia Sinica, etc.). Nanjing: Zhongguo dier lishi danganguan中国第二历史档案馆(The Second Historical Archives of China), Total No. 393, File No. 2591.

physics group, 26 people from the biology group, and 38 people from the humanities group. In addition, the Nomination Committee exchanged views on the election of honorary members of Academia Sinica, saying that honorary members "must have a high academic status in foreign countries and have ties with China or play a role in China's scientific work," but they did not make a decision.[37]

After the meeting on August 27, in order to make the election perfect, the Nomination Committee requested that Chu Chia-Hua convene an interim meeting of the council to discuss the provisions of Articles 4 to 6 of *The Regualations for the Election of the Council of Academia Sinica*, in accordance with the decision of the meeting. The meeting was held on September 22, the day before the opening of the assembly. This was the first temporary meeting of the Council since it was established in 1935. At 4 pm on the same day, 26 people—Chu Chia-Hua, Wong Wen-hao, Adam Pen-Tung Sah, John Lee, Woo Sho-Chow, Woo Ting-liang, Li Chi, Li Shu-Hua, Ping Chi, Chou Keng-Shang, Chow Jen, Robert K. S. Lim, Hu Shih,Mao T. E., Chiang Li-Fu, Chen Yuan, Tao L. K., Chuang Chang-Kong, Zhang Yun, Chang Yu-Che, Ling Hung-hsun, Fu Ssu-Nien, Jaw Jeou-Jang, Chien Shung-shu, Hsieh Chia-yung, andTai Fon-Land—attended the meeting. Chairman Chu Chia-Hua gave a brief report on the previous preparations for the election of the Third Council members. Adam Pen-Tung Sah said that the meeting would discuss the revised *The Regualations for the Election of the Council of Academia Sinica* in order to make it "more perfect and more legally based". After the discussion, the regulations were amended and passed. The amendment was to have the words "the number of candidates in each group shall be at least a multiple of the number of candidates elected" in Article 4 replaced with the words "each group shall have no less than twenty candidates", after the words "the Nomination Committee shall present each group of candidates together with the the proposals proposed by the members of Academia Sinica."[38]

Subsequently, according to the provisions of Article 3 of *The Regualations for the Election of the Council of Academia Sinica*, it was decided that the total number of the Third Council members was to be 32, including ten members each from the mathematics and physics group and biology group, and twelve members from the humanities group. Moreover, the meeting approved the Nomination Committee's suggestion that there should be no more than three council members in each discipline, and explained it as follows: the number of council members in each discipline should be three, including ex officio members; the disciplines to which non-members of Academia Sinica belonged to would be based on the reasons given when they are nominated. There was also a temporary proposal by Fu Ssu-Nien that "council members of this term should all be candidates for the next term," which was

[37]Zhongyang yanjiuyuan dierjie pingyihui linshi huiyi jilu中央研究院第二届评议会临时会议纪录(Minutes of Provisional Meetings of the Second Council of Academia Sinica). Nanjing: Zhongguo dier lishi danganguan中国第二历史档案馆(The Second Historical Archives of China), Total No.393, File No. 1559.

[38]Zhongyang yanjiuyuan dierjie pingyihui linshi huiyi jilu中央研究院第二届评议会临时会议纪录(Minutes of Provisional Meetings of the Second Council of Academia Sinica). Nanjing: Zhongguo dier lishi danganguan中国第二历史档案馆(The Second Historical Archives of China), Total No.393, File No. 1559.

approved.[39] As a result, the number of candidates for council members increased from 99 to a total of 102 and included Zhang Yun, John Lee, and Wang Chong You.[40]

Of the 102 people, Kiang Tsai-han, Chou Pei-yuan, Ting S. L., Cheng Xiaogang, Tang Yueh, Yen Shu-Tang, Ho L., and Luo Changpei were not nominated jointly by members of the same group. They were all nominated by the Nomination Committee on the spot at the August 27 meeting and jointly signed by five members. They were different from other candidates in the nomination method. Kiang Tsai-han, Zhang Yun, Ting S. L., Chou Pei-yuan, Sun Xuewu, John Lee, Yu Ta-wei, Mei Yi-chi, Cheng Xiaogang, Wang C. Y., Tang Yueh, Chien Mu, Zong Bai-hua, Ch'in Chung Mien, Luo Changpei, Deng Yi-zhe, Yen Shu-Tang, Zhang Xiruo, Chen Zong, Ho L., Pan Kuang-tan, and other 21 candidates were not members of Academia Sinica. As for them, the Nomination Committee asked nominators to indicate the reasons for nomination and their names. Judging from the list of candidates, the Nomination Committee did not deliberately focus on Peking, Shanghai, Hangzhou and Wuhan. Of the 21 candidates who were not members, only five were from Nanjing and Shanghai, while none were from Hangzhou and Wuhan.

After the opening of the First General Meeting of the assembly on the morning of September 24, Hu Shih, the convener of the Deliberation Subcommittee, reported on the process of drawing up the agenda of the assembly and the corresponding suggestions. The meeting adopted the report of the subcommittee and the agenda of its decisions. Then, the election of the elected Third Council members was held. Before the election, the meeting presumed that two members of Academia Sinica from each group would serve as the drawer and scrutineer. The mathematics and physics group was composed of Shiing-Shen Chern andMao T. E., the biology group was composed of Yin Hung-chang and Yuan I-Chin, and the humanities group was composed of Hsiao Kung-chuan and Feng You-lan. Both Chu Chia-Hua and Adam Pen-Tung Sah had waived the right to be elected as council members. In response to the suggestion that there should be no more than three council members for each discipline, decided at the temporary meeting of the council, the meeting affirmed that "if the number of council members elected in a discipline exceeds the limit of three, please discuss the case." After discussion, it was decided that "since it is not necessary to have council members in each discipline, the number of people elected in each group will take effect if it matches the total number of council members

[39]Zhongyang yanjiuyuan dierjie pingyihui linshi huiyi jilu中央研究院第二届评议会临时会议纪录(Minutes of Provisional Meetings of the Second Council of Academia Sinica). Nanjing: Zhongguo dier lishi danganguan中国第二历史档案馆(The Second Historical Archives of China), Total No. 393, File No. 1559.

[40]Zhongyang yanjiuyuan diyici yuanshi huiyi yubeihui cankao ziliao ji huiyi ti'an中央研究院第一次院士会议预备会参考资料及会议提案(Reference Materials for the Preliminary Meeting and Proposals of the First General Assembly of Members of Academia Sinica). Nanjing: Zhongguo dier lishi danganguan中国第二历史档案馆(The Second Historical Archives of China), Total No. 393, File No. 1625.

allocated in each group."[41] That is to say, the number of council members in each discipline was not necessarily limited to three, as long as the sum of the number of council members in each discipline reaches the total number of candidates in that group.

After the election began, the members of Academia Sinica voted by secret ballot for the candidates from the mathematics and physics group, the biology group, and the humanities group respectively. There were 22 candidates who obtained more than half of the votes of the 48 members of Academia Sinica present; that is, the 22 candidates obtained at least 25 votes. Among the 22 elected, there were eight from the mathematics and physics group, six from the biology group, and eight from the humanities group. The top three were Hu Shih with 46 votes, Chu Coching with 42 votes, and Ping Chi with 41 votes. The voting results are shown in Table 6.1.

After the vote, however, the groups had not met their quota: there were two vacancies in mathematics and physics group, four in the biology group, and four in the humanities group. The acting president, the director-general, and the directors of all institutes of Academia Sinica had been ex officio members and had the right to be elected as council members, but they did not obtain more than half of the votes. After the vote, Shiing-Shen Chern, Yin Hung-chang, and Feng You-lan reported the ballot results of the three groups respectively. The secretary re-reported the list and number of council members elected by each group. Subsequently, Chu Chia-Hua announced the results of the first election. Then the members of Academia Sinica voted for the other candidates in each group for the second time. Before the vote, Tao L. K. waived the right to be elected as council member. Chu Chia-Hua asked whether the drawer of each group should be changed. The resolution of the meeting was to "still let the original members serve, since they are more experienced."[42] After the result was announced, one vote of the humanities group elected more than five people in that round; the meeting decided that the vote was invalid. None of the candidates in each group won more than half of the votes, so no one was elected. The results are shown in Table 6.2.

On the afternoon of September 24, the assembly held its Second General Meeting. There were 49 council members present. Tsai Chiao was absent. Fu Ssu-Nien and Yu Cha-Shih, who were absent from the First General Meeting, attended the meeting. Feng You-lan reported the results of the second round of balloting for the election of council members, and Chu Chia-Hua announced it. The candidates in each group did not get more than half of the votes—a third ballot was to be held according to

[41]Zhongyang yanjiuyuan zuzhi banfa yuanshi huiyi jilu jiqi xuanju guicheng deng youguan wenshu 中央研究院组织办法、院士会议纪录及其选举规程等有关文书(Relevant Documents such as the Organization Methods of Academia Sinica, Minutes of the General Assembly of Members and the Regulations for Election). Nanjing: Zhongguo dier lishi danganguan中国第二历史档案馆(The Second Historical Archives of China), Total No. 393, File No. 2824.

[42]Zhongyang yanjiuyuan zuzhi banfa yuanshi huiyi jilu jiqi xuanju guicheng deng youguan wenshu 中央研究院组织办法、院士会议纪录及其选举规程等有关文书(Relevant Documents such as the Organization Methods of Academia Sinica, Minutes of the General Assembly of Members and the Regulations for Election). Nanjing: Zhongguo dier lishi danganguan中国第二历史档案馆(The Second Historical Archives of China), Total No. 393, File No. 2824.

Table 6.1 The results of the first vote for the third council members of Academia Sinica

Number of votes	Group	Names of elected	Number of elected	Total
Above 24 votes	Mathematics and physics group	Shiing-Shen Chern (28), Woo Y. H. (25), Li Shu-Hua (28), Yeh Chi-Sun (31), Chuang Chang-Kong (28), Wong Wen-hao (37), Chu Coching (42), Mao T. E. (32)	8	22
	Biology group	Ping Chi (41), Hu Hsen-hsu (35), Li Zongen (40), Robert K. S. Lim (34), Feng Teh-Pei (27), Yu T. F. (26)	6	
	Humanities group	Tang Yung-tung (26), Feng You-lan (34), Hu Shih (46), Chen Yuan (33), Chao Yuen Ren (27), Li Chi (37), Wang Chung Hwei (31), Chou Keng-Shang (33)	8	
Under 24 votes	Mathematics and physics group	Chiang Li-Fu (16), Hsu Pao-Lu (2), Hua Loo-Keng (7), Su Buchin (19), Kiang Tsai-han (1), Zhang Yun (4), Wu Ta-You (8), Chao Chung-Yoa (1), Ny Tsai-ze (14), Yao Y. T. (13), Ting S. L. (0), Chou Pei-yuan (0), Wu Hsien (13), Woo Sho-Chow (10), Tseng Chao-lun (10), Sun Xuewu (2), Chu Chia-Hua (3), Lee J. S. (13), Huang T. K. (5), Young Chung-Chien (12), Hsieh Chia-yung (10), John Lee (0), Chow Jen (4), Hou T. P. (18), Ling Hung-hsun (12), Adam Pen-Tung Sah (16), Yu Ta-wei (3), Mei Yi-chi (3), Cheng Xiaogang (0), Wang Chong You (0)	30	80

(continued)

Table 6.1 (continued)

Number of votes	Group	Names of elected	Number of elected	Total
	Biology group	Wang Chia-Chi (12), Wu Hsien-Wen (23), Pai Sitsan (20), Chen Shisan C. (15), Tung Ti-Cho (2), Yin Hung-chang (10), Zhang Jingyang (18), Chien Shung-shu (22), Tai Fon-Land (16), Loo Tsung-Le (10), Yuan I-Chin (10), Chang Hsiao-chien (4), Chen Ko Kuei (11), Woo Ting-liang (16), Wang Ging-Hsi (14), Tang Yueh (6), Tang P. S. (22), Tsai Chiao (7), Li Hsien-wen (8), Teng S. C. (11)	20	
	Humanities group	Woo Tsin-Hang (4), Chin Yuch-Lin (8), Chien Mu (3), Zong Bai-hua (1), Yu Cha-Shih (5), Zhang Yuanji (7), Yang Shu-Tah (8), Liu Yei-Tsen (14), Chen Yin-k'o (14), Fu Ssu-Nien (22), Ku Chieh-kan (11), Ch'in Chung Mien (0), Fang Kuei Li (13), Luo Changpei (1), Liang Ssu-yung (5), Kuo Mo-jo (4), Tung Tso-pin (8), Liang Ssu-cheng (23), Deng Yi-zhe (0), Wang Shih-Chieh (22), Yen Shu-Tang (2), Tsien T. S. (14), Hsiao Kung-chuan (15), Zhang Xiruo (0), Ma Yin-chu (23), Chen Zong (2), Ho L. (2), Pan Kuang-tan (1), Chen Ta (24), Tao L. K. (16)	30	

Sources Diyici yuanshi xuanju第一次院士选举 (The Election of the First Batch of Members). Nanjing: The Second Historical Archives of China, Total No. 393, File No. 494

Table 6.2 The results of the second vote for the third council members of Academia Sinica

Group	Name	Total
Mathematics and physics group	Chiang Li-Fu (1), Hsu Pao-Lu (1), Hua Loo-Keng (3), Su Buchin (15), Kiang Tsai-han (0), Zhang Yun (2), Wu Ta-You (0), Chao Chung-Yoa (0), Ny Tsai-ze (0), Yao Y. T. (0), Ting S. L. (0), Chou Pei-yuan (0), Wu Hsien (5), Woo Sho-Chow (3), Tseng Chao-lun (5), Sun Xuewu (0), Chu Chia-Hua (0), Lee J. S. (5), Huang T. K. (2), Young Chung-Chien (6), Hsieh Chia-yung (6), John Lee (1), Chow Jen (2), Hou T. P. (6), Ling Hung-hsun (8), Adam Pen-Tung Sah (22), Yu Ta-wei (1), Mei Yi-chi (0), Cheng Xiaogang (0), Wang Chong You (0)	30
Biology group	Wang Chia-Chi (6), Wu Hsien-Wen (13), Pai Sitsan (9), Chen Shisan C. (11), Tung Ti-Cho (8), Yin Hung-chang (10), Chang Chin-yueh (10), Chien Shung-shu (14), Tai Fon-Land (8), Loo Tsung-Le (7), Yuan I-Chin (12), Chang Hsiao-chien (3), Chen Ko Kuei (7), Woo Ting-liang (14), Wang Ging-Hsi (5), Tang P. S. (4), Tsai Chiao (3), Li Hsien-wen (5), Teng S. C. (6)	20
Humanities group	Woo Tsin-Hang (0), Chin Yuch-Lin (4), Chien Mu (2), Zong Bai-hua (0), Yu Cha-Shih (1), Zhang Yuanji (6), Yang Shu-Tah (3), Liu Yei-Tsen (9), Chen Yin-k'o (8), Fu Ssu-Nien (23), Ku Chieh-kan (8), Ch'in Chung Mien (1), Fang Kuei Li (3), Luo Changpei (0), Liang Ssu-yung (2), Kuo Mo-jo (2), Tung Tso-pin (4), Liang Ssu-cheng (15), Deng Yi-zhe (0), Wang Shih-Chieh (9), Yen Shu-Tang (1), Tsien T. S. (10), Hsiao Kung-chuan (7), Zhang Xiruo (1), Ma Yin-chu (24), Chen Zong (0), Ho L. (2), Pan Kuang-tan (0), Chen Ta (21), Tao L. K. (23)	33

Sources Diyici yuanshi xuanju第一次院士选举 (The Election of the First Batch of Members). Nanjing: The Second Historical Archives of China, Total No. 393, File No. 494

the regulations. Fu Ssu-Nien waived the right to be elected as a council member. According to *The Regualations for the Election of the Council of Academia Sinica*, this ballot would be a run-off, with twice as many candidates as the number of vacancies from the candidates who obtained more votes in each group in the second ballot. The elected people did not require a majority in votes. Judging from the results of the second ballot, Adam Pen-Tung Sah received the highest number of votes in mathematics and physics group; however, he had already waived his right. Instead of proposing him, the five people after him were proposed: Su Buchin, Ling Hung-hsun, Young Chung-Chien, Hsieh Chia-yung, and Hou T. P. There were two vacancies in this group, so four more should be elected. The reason for proposing five people was that Young Chung-Chien, Hsieh Chia-yung, and Hou T. P. all received six votes.

Table 6.3 The results of the third vote for the third council members of Academia Sinica

Group	Name	Total
Mathematics and physics group	Su Buchin (23), Ling Hung-hsun (20), Young Chung-Chien (13), Hsieh Chia-yung (14), Hou T. P. (20)	5
Biology group	Chien Shung-shu (31), Woo Ting-liang (20), Tang P. S. (26), Wu Hsien-Wen (28), Yuan I-Chin (16), Chen Shisan C. (20), Yin Hung-chang (19), Chang Chin-yueh (17)	8
Humanities group	Chen Ta (36), Liang Ssu-cheng (26), Tsien T. S. (20), Liu Yei-Tsen (14), Wang Shih-Chieh (41), Chen Yin-k'o (11), Ku Chieh-kan (14), Hsiao Kung-chuan (16)	8

Sources Diyici yuanshi xuanju第一次院士选举 (The Election of the First Batch of Members). Nanjing: The Second Historical Archives of China, Total No. 393, File No. 494

The biology group was currently short four people. Chien Shung-shu, Woo Ting-liang,Tang P. S., Wu Hsien-Wen, Yuan I-Chin, Chen Shisan C., Yin Hung-chang, and Chang Chin-yueh, who were in the top eight in the group, were put forward. The highest number of votes in humanities group were for Ma Yin-chu, but he was not put forward because he had "waived his right by correspondence."[43] Fu Ssu-Nien and Tao L. K. received the same number of votes, second to Ma Yin-chu, but because they had both given up their right, they were not put forward. In the absence of four members in this group, the meeting put forth the next eight members after the first three people: Chen Ta, Liang Ssu-cheng, Tsien T. S., Liu Yei-Tsen, Wang Shih-Chieh, Chen Yin-k'o,Ku Chieh-kan, and Hsiao Kung-chuan. As a result, a total of 21 people was proposed. Then, a third vote was held, at which 45 members of Academia Sinica were present. The results are shown in Table 6.3.

Su Buchin, who was from mathematics and physics group, was elected with the highest number of votes. After him, Ling Hung-hsun and Hou T. P., both with twenty votes each, had the second highest number of votes in the same group. As there was only one vacancy left in this group, Ling Hung-hsun was elected by the general meeting. Chien Shung-shu, Wu Hsien-Wen, andTang P. S., who came in top three in the biology group, were elected. After them, Woo Ting-liang and Chen Shisan C. won the second highest number of votes in the same group, both with twenty votes each. As there was only one vacancy in this group, Chen Shisan C. was elected after being decided on by the general meeting. Wang Shih-Chieh, Chen Ta, Liang Ssu-cheng, and Tsien T. S., the top four in humanities group, were elected. This group did not have to hold another meeting because none of the candidates had the same

[43]Chu Coching竺可桢(2006). *Chu Coching quanji*竺可桢全集(The Complete Works of Coching Chu), Vol.11. Shanghai: Shanghai keji jiaoyu chubanshe上海科技教育出版社.214.

number of votes. After the voting, Shiing-Shen Chern, Yin Hung-chang, and Feng You-lan reported the ballot results of the three groups respectively.[44]

After three rounds of voting, the meeting elected a total of 32 council members. Mathematics and physical group included (mathematics) Shiing-Shen Chern, Su Buchin, (physics) Woo Y. H., Li Shu-Hua and Yeh Chi-Sun, (chemistry) Chuang Chang-Kong, (geology) Wong Wen-hao, (meteorology) Chu Coching, (engineering)Mao T. E. and Ling Hung-hsun. Biology group included (zoology) Ping Chi, Wu Hsien-Wen, Chen Shisan C., (botany) Hu Hsen-hsu, Chien Shung-shu, (medical science) Li Zongen, (physiology) Linkesheng, Feng Teh-Pei,Tang P. S., and (agronomy) Yu T. F.. Humanities group included (philosophy) Tang Yung-tung, Feng You-lan, (literature and history) Hu Shih, (history) Chen Yuan, (linguistics) Chao Yuen Ren, (archaeology) Li Chi, (art history) Liang Ssu-cheng, (jurisprudency) Wang Chung Hwei, Wang Shih-Chieh, (politics) Chou Keng-Shang, Tsien T. S., and (economics) Chen Ta. They, together with 13 ex officio members of Chu Chia-Hua, Adam Pen-Tung Sah, Chiang Li-Fu, Chang Yu-Che, Woo Sho-Chow, Lee J. S., Wang Chia-Chi, Loo Tsung-Le, Jaw Jeou-Jang, Fu Ssu-Nien, Tao L. K., Chow Jen and Wang Ging-Hsi, formed the third term of the Council of Academia Sinica.[45]

In the election of council members, no candidate was elected in the five disciplines of astronomy, pharmacology, physical anthropology, psychology and economics. There were no more than three candidates in each discipline. Although Chen Yin-k'o, Hou T. P. and other first-class scholars won higher votes in the election of the first batch of members of Academia Sinica, they were defeated this time. The elected members were all members of Academia Sinica, and only two were not elected by general election in the election of the first batch of members of Academia Sinica. This showed that the scholars highly recognized by the Second Council were also basically recognized by the General Assembly of Members of Academia Sinica. Among those elected, Woo Y. H., Li Shu-Hua, Chuang Chang-Kong, Wong Wen-hao,Mao T. E., Ling Hung-hsun, Ping Chi, Chen Shisan C., Hu Hsen-hsu, Chien Shung-shu, Robert K. S. Lim, Hu Shih, Chen Yuan, Chao Yuen Ren, Li Chi, Wang Shih-Chieh and Chou Keng-Shang were all 17 members elected in the previous term. They were re-elected in this term. Of the remaining 15 people, except Yeh Chi-Sun and Chu Coching, who had served as council members, the remaining 13 people were Shiing-Shen Chern, Su Buchin, Wu Hsien-Wen, Li Zongen, Feng Teh-Pei,Tang P. S., Yu T. F., Tang Yung-tung, Feng You-lan, Liang Ssu-cheng, Wang Chung Hwei, Tsien T. S. and Chen Ta, who were all entering the threshold of the Council for the first time.

---

[44]Zhongyang yanjiuyuan zuzhi banfa yuanshi huiyi jilu jiqi xuanju guicheng deng youguan wenshu 中央研究院组织办法、院士会议纪录及其选举规程等有关文书(Relevant Documents such as the Organization Methods of Academia Sinica, Minutes of the General Assembly of Members and the Regulations for Election). Nanjing: Zhongguo dier lishi danganguan中国第二历史档案馆(The Second Historical Archives of China), Total No. 393, File No. 2824.

[45]Disanjie pingyiyuan chansheng mingyu yuanshi zan bu xuanju第三届评议员产生名誉院士暂不选举(The Third Council Members Came into Being and the Honorary Members of Academia Sinica will not be Elected). *Zhongyang ribao*中央日报(Central Daily), 1948–09–25 (4).

Based on the academic backgrounds of these council members, there was no major expansion in the disciplines of this term, and the distribution of disciplines in the previous two terms of councils remained comprehensive. However, its academic authority was slightly higher than that of the second term. The fifteen newly appointed council members were all members of Academia Sinica; however, five of the eleven members of last term—Hou T. P., Tseng Chao-lun, Wang Chong You,Tai Fon-Land, Hsieh Chia-yung, Zhang Yun, John Lee, Tang Yueh, Ho L., Woo Ting-liang, and Chen Yin-k'o—who failed to be elected this time, were notmembers of Academia Sinica. Of the 32 people elected in the assembly, seven people —Shiing-Shen Chern, Wu Hsien-Wen, Robert K. S. Lim, Feng Teh-Pei, Yu T. F., Wang Shih-Chieh, and Chen Ta—were elected in the election of the first batch of members of Academia Sinica according to the first qualification, and three people—Wong Wen-hao, Ling Hung-hsun, and Chou Keng-Shang—were elected in the election of the first batch of members of Academia Sinica according to the second qualification. The remaining 22 people were elected in the election of the first batch of members of Academia Sinica according to both the first and second qualifications, accounting for 68.75% of those elected.[46] It can be seen that, like the election of the second term, the scholars who had outstanding academic and leadership achievements still accounted for the majority in this term election.

Compared to the previous two terms of councils, the comprehensiveness of representation in this council had shrunk. The first and second terms of council members of Academia Sinica were generally evenly distributed amongst important academic institutions. Each organization usually had one council member, and a few organizations had more than three; this showed that the Council had the characteristic of extensive representativeness. By the third term, this situation had changed. The 15 newly appointed council members including three from Academia Sinica, five from National Tsing Hua University, three from the National University of Peking, two from the University of Chekiang, one from Peiping Union Medical College and one from National Government. In addition to the above-mentioned institutions, the institutions of the Members of the Second Council who failed to be elected included Yongli Chemical Industry Company, Daye Iron Mine, National Sun Yatsen University, Central Weather Bureau, Nankai University, and Mineral Survey Office of the Resources Committee. In addition, among the 17 re-elected council members, there were two from Academia Sinica, two from the National Academy of Peiping, two from National Central University, one from the National University of Peking, one from National Tsing Hua University, one from National Wu-han University, one from the Catholic University of Peking, one from the Science Society of China, one from Fan Memorial Institute of Biology, one from Resources Committee, one from Qiantang River Engineering Office, two from National Ministries and Commissions, and one from a foreign university. Therefore, although this term of the Council still

[46]Guoli zhongyang yanjiuyuan diyici yuanshi xuanju houxuanren timingce国立中央研究院第一次院士选举候选人提名册(Nomination Lists of the Member Candidates for the Election of the First Batch of Members of Academia Sinica). Nanjing: Zhongguo dier lishi danganguan中国第二历史档案馆(The Second Historical Archives of China), Total No. 393, File No. 1597.

had the characteristic of extensive representativeness, it was not as comprehensive as the previous two terms. At the same time, National Tsing Hua University (six people) and the National University of Peking (four people) had already occupied strong positions in the council.

The Members of the Third Council were mainly distributed in Nanjing, Shanghai, and Peking. There were fourteen elected Members of the Council from Peking, nine from Nanjing, five from Shanghai, two from Zhejiang, one from Hubei and one from the United States. Almost all of the thirteen ex officio members were in Nanjing and Shanghai. This was somewhat at odds with the previous decision of the Council to focus on Peking, Shanghai, Hangzhou, and Wuhan; however, according to strict democratic procedures, this result was inevitable because Peking, Nanjing, and Shanghai were, after all, the main gathering places for academic elites in the north and south at that time.

In general, the election for the members of the Third Council of Academia Sinica had experienced detailed planning with earnest preparation. In the General Assembly, the members of Academia Sinica attending the meeting adhered to the strict democratic electoral procedure and had their rights exercised seriously on the election of the member of the Council seriously in accordance with the quota defined by the council. With regard to the candidates being selected, emphasis had been put on those scholars who had both outstanding academic achievements and great academic leadership during the election. This followed the tradition achieved from the election of the elected members of the Second Council of Academia Sinica. As the new and old members of the council alternated, the Third Council of Academia Sinica appeared to keep comprehensive academic distribution with slightly elevated academic authority and contracted the broadness of representatives.

6.3 Other Agendas and Resolutions of the Assembly

In addition to the election of the Third Council members, the assembly also agreed on and approved of a number of bills and proposals. At the First General Meeting held on the morning of September 24, the discussion approved the Deliberation Subcommittee's proposal that "the election of honorary members of Academia Sinica will not be held at this assembly, please discuss." The general meeting also decided to set up a committee to draw up *The Draft Regulations for the Election of Honorary Members*. Each group elected one person—Shiing-Shen Chern from the mathematics and physics group, Chien Shung-shu from the biology group, and Fu Ssu-Nien from the humanities group; this was convened by Chien Shung-shu. The assembly also approved the proposal: "Request the Second Council to Presume the Preparatory Committee for the Election of Members of Academia Sinica in 1949 as Soon as Possible." Moreover, it approved the establishment of a thesis committee and academic lecture committee. With regards to the task of the thesis committee, it was decided that "the papers of the members of Academia Sinica papers will not be reviewed. If they are sent to Academia Sinica for publication, they will be reviewed and revised by the people who have the right to, according to the general publication procedures."[47]

At the Second General Meeting held on the afternoon of September 24, it was presumed that the members of the thesis committee would be Chiang Li-Fu, Woo Sho-Chow, and Hsieh Chia-yung of the mathematics and physics group,Tang P. S., Pai Sitsan, andTeng S. C. of the biology group, and Fu Ssu-Nien, Chen Ta, and Tao L. K. of the humanities group. The convener of each group and the general convener of the three groups would be selected on their own. The meeting also selected the members of the academic lecture committee —Shiing-Shen Chern of the mathematics and physics group, Wu Hsien-Wen of the biology group, and Li Chi of the humanities group, all convened by Li Chi—with a term of one year and another possible term if re-elected.[48] Regarding the organization of the General Assembly of Members of Academia Sinica, the meeting discussed the case of "inviting each group of members of Academia Sinica to appoint a secretary that would preside over the academic contact," and elected Chu Coching of the mathematics and physics group, Li Zongen of the biology group, and Tang Yung-tung of humanities group. After that, the meeting approved "the plan to appoint the honorary secretary of thecouncil, the director-general, the director of the secretaries, the director of general affairs, and the secretaries of all groups of the assembly to organize the preparatory committee for the next General Assembly of Members of Academia Sinica."[49]

These decisions were all made in preparation for Academia Sinica's future general assemblies of members and academic exchange activities. Among them, academic lecture was not a topic of the meeting stipulated in *The Regulations of the General Assembly of Members of Academia Sinica*, but upon the recommendation of the deliberation subcommittee, the general meeting decided to establish an academic lecture committee, which is related to the fact that the importance of academic lecture in academic research had been recognized by many members of Academia Sinica. For example, on August 30, 1947, at the joint annual meeting of seven scientific

[47]Zhongyang yanjiuyuan zuzhi banfa yuanshi huiyi jilu jiqi xuanju guicheng deng youguan wenshu 中央研究院组织办法、院士会议纪录及其选举规程等有关文书(Relevant Documents such as the Organization Methods of Academia Sinica, Minutes of the General Assembly of Members and the Regulations for Election). Nanjing: Zhongguo dier lishi danganguan中国第二历史档案馆(The Second Historical Archives of China), Total No. 393, File No. 2824.

[48]Zhongyang yanjiuyuan zuzhi banfa yuanshi huiyi jilu jiqi xuanju guicheng deng youguan wenshu 中央研究院组织办法、院士会议纪录及其选举规程等有关文书(Relevant Documents such as the Organization Methods of Academia Sinica, Minutes of the General Assembly of Members and the Regulations for Election). Nanjing: Zhongguo dier lishi danganguan中国第二历史档案馆(The Second Historical Archives of China), Total No. 393, File No. 2824; Zhongyang yanjiuyuan yuanshi huiyi tuiding lunwen weiyuanhui deng weiyuan mingdan中央研究院院士会议推定论文委员会等委员名单(Lists of Members of the Thesis Committee and Other Committees Selected at the First General Assembly of Members of Academia Sinica). Nanjing: Zhongguo dier lishi danganguan中国第二历史档案馆(The Second Historical Archives of China), Total No. 393, File No. 1630.

[49]Zhongyang yanjiuyuan zuzhi banfa yuanshi huiyi jilu jiqi xuanju guicheng deng youguan wenshu 中央研究院组织办法、院士会议纪录及其选举规程等有关文书(Relevant Documents such as the Organization Methods of Academia Sinica, Minutes of the General Assembly of Members and the Regulations for Election). Nanjing: Zhongguo dier lishi danganguan中国第二历史档案馆(The Second Historical Archives of China), Total No. 393, File No. 2824.

organizations, which included the Science Society of China, the Natural Science Society of China, and Chinese Astronomical Society, etc.,Wong Wen-hao pointed out that, based on his personal experience of participating in a grand gathering of scientific organizations from various countries in Europe in 1937, "speeches made by famous experts to the public are extremely important, and many famous scientists have been inspired by this and successfully furthered their research work."[50]

At 9 am on September 25, the Third General Meeting of the assembly was held. Fu Ssu-Nien suggested that when the papers presented by Academia Sinica were to be given as public academic lectures in the future, Peking should also be expected, besides Nanjing. As a result, he requested that houses in Peking be prepared as places for socializing, gathering, and reading papers. Chu Chia-Hua said: "Let the head office handle it and invite Fu Ssu-Nien,Tang P. S., and Li Shu-Hua to contact any related people and make it a social gathering place."[51] Chien Shung-shu reported on the progress of the subcommittee in drafting the *Draft Rules for the Election of Honorary Members of Academia Sinica*. Additionally, the highlight of the meeting was to discuss members' proposals; in addition to the proposals mentioned above, the meeting discussed the following eight proposals, as shown in Table 6.4.

The fourth proposal had the largest number of proposers. Fourteen members, including Yao Y. T. and Ny Tsai-ze, proposed it because of "the lack of funds and the lack of progress in academic research over the past years". They "appealed" in the proposal: "imitate the precedent set by the U.S. for National Research Fellows and make appropriate decisions according to the situation in our country. Grant subsidies and research schedules to researchers with experience so that their plans could be realized. In addition, a 'primary subsidy' would be set up for university graduates with excellent results to train junior talents."[52] The proposal concerned not only scientists but also young talents. From the "resolution", it can be seen that theassembly attached great importance to this case, and selected some members to draw up the measures on the spot.

The eighth proposal had the second largest number of proposers. This proposal addressed the problem of naming large numbers. Li Shu-Hua, Yao Y. T., and twelve other members proposed the case because hyperinflation had occurred after the outbreak of the civil war, and people were using large numbers more and more frequently. However, the naming standard of large numbers that was used in our country was not consistent with the internationally accepted method, which was not

[50]Guonei xiaoxi国内消息(Domestic News). *Kexue*科学(Science)1947, 29 (10): 310.

[51]Zhongyang yanjiuyuan zuzhi banfa yuanshi huiyi jilu jiqi xuanju guicheng deng youguan wenshu 中央研究院组织办法、院士会议纪录及其选举规程等有关文书(Relevant Documents such as the Organization Methodsof Academia Sinica, Minutes of the General Assembly of Members and the Regulations for Election). Nanjing: Zhongguo dier lishi danganguan中国第二历史档案馆(The Second Historical Archives of China), Total No.393, File No. 2824.

[52]Zhongyang yanjiuyuan diyici yuanshi huiyi yubeihui cankao ziliao ji huiyi ti'an中央研究院第一次院士会议预备会参考资料及会议提案(Reference Materials for the Preliminary Meeting and Proposals of the First General Assembly of Members of Academia Sinica). Nanjing: Zhongguo dier lishi danganguan中国第二历史档案馆(The Second Historical Archives of China), Total No. 393, File No. 1625.

Table 6.4 The proposals of members and resolutions of the First General Assembly of Members of Academia Sinica

Serial Number	Proposals	Proposers	Resolution
1	Doctoral degrees should be awarded through university examination	Feng You-lan, Hu Shih, Yin Hung-chang, Chou Keng-Shang, Tang Yung-tung, and Li Zongen	Approved
2	Please set up a case concerning the Institute of Humanities as soon as possible in accordance with *The Organization Act* of Academia Sinica	Feng You-lan, Yang Shu-Tah, Tang Yung-tung, Liang Ssu-cheng, Chen Ta	Transfer to the Council
3	The seventh Pacific Science Conference sent a letter inviting participants to attend the conference that was to be held in New Zealand in February 1949. Please discuss the proposal	Wong Wen-hao, Adam Pen-Tung Sah	In principle, the Council is in favor of sending a small number of scientists to participate, and authorizes the Honorary Secretary of the Council and the Director-General of Academia Sinica, together with Mr. Hu Hsen-hsu, Mr. Chu Coching, and Mr. Young Chung-Chien, who have participated in the Pacific Science Conference before, to make preparations and decide on the candidates, who will be convened by Director-General Sa (after the decision, four people, Wu Hsien-Wen, Hsieh Chia-yung, Zheng Wanjun and Shen Zonghan, were invited to participate.)

(continued)

Table 6.4 (continued)

Serial Number	Proposals	Proposers	Resolution
4	A request for the government to allocate special funds in order to set up "national academic grants" and reward research projects	Yao Y. T., Ny Tsai-ze, Su Buchin, Woo Sho-Chow, Chu Coching, Chow Jen, Shiing-Shen Chern, Chuang Chang-Kong, Yeh Chi-Sun, Li Shu-Hua, Young Chung-Chien, Mao T. E., Hsieh Chia-yung, Chiang Li-Fu	The proposal was generally approved, and the method will be drawn up by another subcommittee. It is presumed that all the secretaries of the assembly along with Hsiao Kung-chuan, Feng Teh-Pei, Shiing-Shen Chern and Adam Pen-Tung Sah would be members
5	In order to maintain academic independence, please let this assembly issue a declaration to express its views	Tang P. S., Ny Tsai-ze, Feng You-lan, Liang Ssu-cheng, Li Chi, Yin Hung-chang, Tao L. K., Yao Y. T., Chen Ta	The amendment was passed. It read: in order to maintain academic independence, the assembly will send a letter to the government for full attention
6	Should the selection of honorary members of Academia Sinica next year be carried out by a group meeting organized by the General Assembly of Members of Academia Sinica?		The subcommittee that originally drafted the detailed rules will push the secretaries of each group and the director-general of Academia Sinica to prepare for next year's election of honorary members. It will be convened by Director-General Sah
7	Discuss and formulate the national academic policy, and submit it to the academic subsidy group meeting to study it together and put forward the motion in the next year's General Assembly of Members of Academia Sinica	Chu Chia-Hua	Approved

(continued)

Table 6.4 (continued)

Serial Number	Proposals	Proposers	Resolution
8	It was proposed to ask the government to determine "million" as "zhao(兆)" in order to simplify the calculation of large numbers	Li Shu-Hua, Yao Y. T., Yin Hung-chang, Chen Ta, Chow Jen, Chiang Li-Fu, Woo Sho-Chow, Chien Shung-shu, Tao L. K., Yeh Chi-Sun, Shiing-Shen Chern, Adam Pen-Tung Sah	Determine the three-digit entry method, mark the three-digit entry sharp points, use dots for decimal points, and reiterate that "million" is "zhao (兆)," and the number giga shall be renamed separately. Submit to the council (solicit opinions from all parties) for drafting and then submit to the government for promulgation

Sources Zhongyang yanjiuyuan diyici yuanshi huiyi yubeihui cankao ziliao ji huiyi tian中央研究院第一次院士会议预备会参考资料及会议提案(Reference Materials for the Preliminary Meeting and Proposals of the First General Assembly of Members of Academia Sinica). Nanjing: Zhongguo dier lishi danganguan中国第二历史档案馆(The Second Historical Archives of China), Total No.393, File No. 1625; Zhongyang yanjiuyuan zuzhi banfa yuanshi huiyi jilu jiqi xuanju guicheng deng youguan wenshu 中央研究院组织办法、院士会议纪录及其选举规程等有关文书(Relevant Documents such as the Organization Methods of Academia Sinica, Minutes of the General Assembly of Members and the Regulations for Election). Nanjing: Zhongguo dier lishi danganguan中国第二历史档案馆(The Second Historical Archives of China), Total No. 393, File No. 2824

convenient for large number conversion and had a negative impact on disciplines who required large numbers.[53]

The fifth proposal had the third largest number of proposers. The case was: "Since the victory of the Anti-Japanese War, national academic research and higher education institutions have not been fully equipped, the lives of research staff and teachers have not been guaranteed, and academic and educational institutions have been influenced by politics. In order to maintain academic independence and research freedom, a declaration should be issued to express it at the assembly."[54] Nine members of

[53]Zhongyang yanjiuyuan diyici yuanshi huiyi yubeihui cankao ziliao ji huiyi ti'an中央研究院第一次院士会议预备会参考资料及会议提案(Reference Materials for the Preliminary Meeting and Proposals of the First General Assembly of Members of Academia Sinica). Nanjing: Zhongguo dier lishi danganguan中国第二历史档案馆(The Second Historical Archives of China), Total No. 393, File No. 1625.

[54]Zhongyang yanjiuyuan diyici yuanshi huiyi yubeihui cankao ziliao ji huiyi ti'an中央研究院第一次院士会议预备会参考资料及会议提案(Reference Materials for the Preliminary Meeting and Proposals of the First General Assembly of Members of Academia Sinica). Nanjing: Zhongguo dier lishi danganguan中国第二历史档案馆(The Second Historical Archives of China), Total No. 393, File No. 1625.

Academia Sinica, includingTang P. S. and Ny Tsai-ze, strongly hoped that the government would "maintain academic independence and freedom of research" while also enriching research and teaching institutions and improving the treatment of scholars. The case aroused wide sympathy among members of Academia Sinica; "it was discussed for the longest time, and it involved military and police encircling schools and research funding as well".[55]

The number of proposals ranked fourth was the first one. The "doctoral degree awarding" issue was not raised for the first time by Members Feng You-lan and Hu Shih. On April 20, 1933, the National Government promulgated the *Degree Awarding Law* with corresponding provisions.[56] However, after its implementation on July 1, 1935, the issue of awarding bachelor's and master's degrees was formally implemented in the same year and in 1937 respectively. After the outbreak of the Anti-Japanese War in 1937, there was an acute shortage of senior scientific talents. The National Government, considering that "the cultivation of talents can never be fully supported by furthering study abroad", was determined to solve the problem of doctorate degree awarding.[57] The Academic Review Committee of the Ministry of Education put this issue on its agenda in the autumn of 1940, and considered and approved two drafts of *The Organization Act of the Doctoral Degree Evaluation Committee* and *The Detailed Rules for Doctoral Degree Examination.* After the Examination Yuan proposed a revision principle, the Academic Review Committee made amendments. In May of 1942, the Ministry of Education submitted the amendments to the Executive Yuan and asked them to look over them for approval before enacting a legislation.[58] However, on May 12, 1943, the Executive Yuan instructed them saying: "in recent years, the difficulties of schools have increased, especially in cultivating talents. The awarding of the doctoral degree should be postponed."[59] In this way, the issue of doctoral degree awarding was stranded again.

---

[55]Chu Coching竺可桢(2006). *Chu Coching quanji*竺可桢全集(The Complete Works of Coching Chu), Vol. 11. Shanghai: Shanghai keji jiaoyu chubanshe上海科技教育出版社. 215.

[56]Zhongyang yanjiuyuan diyici yuanshi huiyi yubeihui cankao ziliao ji huiyi ti'an中央研究院第一次院士会议预备会参考资料及会议提案(Reference Materials for the Preliminary Meeting and Proposals of the First General Assembly of Members of Academia Sinica). Nanjing: Zhongguo dier lishi danganguan中国第二历史档案馆(The Second Historical Archives of China), Total No. 393, File No. 1625.

[57]Jiaoyubu xueshu shenyi weiyuanhui gongzuo gaikuang教育部学术审议委员会工作概况(Overview of the Work of the Academic Review Committee of the Ministry of Education). Nanjing: Zhongguo dier lishi danganguan中国第二历史档案馆(The Second Historical Archives of China), Total No. 5, File No. 1347.

[58]Zhang Jin张瑾(1998). Kangzhan shiqi jiaoyubu xueshu shenyi weiyunahui shulun抗战时期教育部学术审议委员会述论(A Study on the Academic Review Committee of the Ministry of Education during the Anti-Japanese War). *Jindaishi yanjiu*近代史研究(Modern Chinese History Studies) (2): 185–186.

[59]Qing shenyi boshi xuewei pingdinghui zuzhifa ji kaoshi xize liangxiang xiuzheng caoan请审议博士学位评定会组织法及考试细则两项修正草案(Please Review the Two Draft Amendments to the Organization Act and Examination Rules of the Doctoral Degree Review Committee). Nanjing: Zhongguo dier lishi danganguan中国第二历史档案馆(The Second Historical-Archives of China), Total No. 5, File No. 1429.

After the Anti-Japanese War ended, the Executive Yuan paid attention to this matter again. On August 16, 1946, the 754th meeting of the Executive Yuan approved the aforementioned two drafts and decided to send them to the council of Academia Sinica for deliberation. In particular, the Executive Yuan discussed this: "Whether a doctoral degree should be awarded by a unified government examination or whether it should be awarded through separate examinations by government-approved universities is a major issue that needs to be studied."[60] On October 23, 1946, the Second Council of Academia Sinica held discussions at the Fourth General Meeting of the Third Annual Meeting. Hu Shih, Chou Keng-Shang, Ping Chi, Loo Tsung-Le, Ho L., Chen Yuan, Wang Chia-Chi, and Chien Shung-shu were selected to be members of the Doctoral Degree Review Committee, which was convened by Hu Shih. In the afternoon of the following day, the council decided at its Fifth General Meeting: "Normal research work and the doctoral thesis of doctoral candidates should be examined and tested by universities or independent colleges, on their own, that are approved by the government to set up a research institute for more than five years and authorized to accept doctoral candidates. Those who pass the examination will be awarded a doctoral degree by the university after being approved by the Ministry of Education."[61] On November 10, the council delivered the resolution to the Executive Yuan, but there was no response. Members Feng You-lan, Hu Shih, and others proposed that this was closely related to the fact that many universities and research institutions in China already had the ability of training doctoral students. With the establishment of the academician system in Academia Sinica, China's academic system was becoming mature, and it was imperative to solve the problem of doctoral degree awarding. They proposed in this case:

> We request the president of Academia Sinica and the honorary secretary of the council to consult with the government on the basis of the resolution of the Third Annual Meeting of the Second Council, and request that the government convene with the relevant authorities to discuss the issue of doctoral degree awarding again as early as possible. We admit the principle that universities should examine and approve of the award of doctoral degree separately, and request the Court to amend the *Degree Awarding Law* (or Articles 6 and 7 thereof).[62]

---

[60]Xingzhengyuan jiaoyubu guanyu boshi xuewei kaoshi ji kexue yanjiu huodong deng wenti yu Zhongyang yanjiuyuan pingyihui laiwang wenshu行政院、教育部关于博士学位考试及科学研究活动等问题与中央研究院评议会来往文书(Correspondence between Executive Yuan and Ministry of Education on Doctoral Degree Examination and Scientific Research Activities with the Council of Academia Sinica). Nanjing: Zhongguo dier lishi danganguan中国第二历史档案馆(The Second Historical Archives of China), Total No. 393, File No. 1546.

[61]Zhongyang yanjiuyuan pingyihui dierjie disanci nianhui jilu中央研究院评议会第二届第三次年会纪录(Minutes of the Third Annual Meeting of the Second Council of Academia Sinica). Nanjing: Zhongguo dier lishi danganguan中国第二历史档案馆(The Second Historical Archives of China), Total No. 393, File No. 1557.

[62]Zhongyang yanjiuyuan diyici yuanshi huiyi yubeihui cankao ziliao ji huiyi ti'an中央研究院第一次院士会议预备会参考资料及会议提案(Reference Materials for the Preliminary Meeting and Proposals of the First General Assembly of Members of Academia Sinica). Nanjing: Zhongguo dier lishi danganguan中国第二历史档案馆(The Second Historical Archives of China), Total No. 393, File No. 1625.

The number of proposals ranked fifth was the second one. Members Feng You-lan, Yang Shu-Tah, and others said in the case: "Only the Institute of History and Philology and the Institute of Social Sciences have been established by Academia Sinica. Regarding the other aspects of humanities, there are no specialized research institutes serving as national leaders. It is proposed that all the research institutes on the humanities stipulated in the articles of *The Organization Act of Academia Sinica* be established first."[63] This referred to the provisions of *The Organization Act of Academia Sinica*, which was revised and promulgated on March 13, 1947. Academia Sinica's institutes of the humanities included not only the established Institute of History and Philology and the Institute of Social Sciences, but also the seven institutes of philosophy, pedagogy, Chinese literature, jurisprudence, economics, archaeology, and ethnology.

Of the remaining three proposals, one was about selecting the personnel to participate in the Seventh Pacific Science Conference. The meeting would be held in New Zealand from February 2 to 23, 1949. On March 27, 1948, the Second Council of Academia Sinica decided to send representatives to participate in the Fourth General Meeting of the Fifth Annual Meeting, authorizing Wong Wen-hao and Adam Pen-Tung Sah to decide on the candidates.[64] However, due to a shortage of foreign exchange, the Ministry of Foreign Affairs did not actively support this and the selection work was postponed. At the assembly, the case was put forth mainly because of the wide scope of the Pacific Science Conference and the need to participate. According to the case: "The Ministry of Foreign Affairs sent a letter to us, saying the principle is not to participate in international conferences and meetings that have no significant relationship with our country. However, the Pacific Science Conference places special emphasis on the specific observations and research of meteorology, geology, biology, ethnic groups, land use, and so on in various parts of the Pacific Ocean, covering a large aspect, so we need to have a discussion on it. In the previous conferences held in Japan, Java, and Canada, many experts and scholars from our country attended, presented papers, and joined various research organizations in the conference. According to the case: "It is advisable for Academia Sinica and other institutes to select a small number of experts to attend the conference, taking into account the two policies of external contact and saving foreign exchange."[65] The

---

[63]Zhongyang yanjiuyuan diyici yuanshi huiyi yubeihui cankao ziliao ji huiyi ti'an中央研究院第一次院士会议预备会参考资料及会议提案(Reference Materials for the Preliminary Meeting and Proposals of the First General Assembly of Members of Academia Sinica). Nanjing: Zhongguo dier lishi danganguan中国第二历史档案馆(The Second Historical Archives of China), Total No. 393, File No. 1625.

[64]Zhongyang yanjiuyuan dierjie pingyihui linshi huiyi jilu中央研究院第二届评议会临时会议纪录(Minutes of Provisional Meetings of the Second Council of Academia Sinica). Nanjing: Zhongguo dier lishi danganguan中国第二历史档案馆(The Second Historical Archives of China), Total No. 393, File No. 1559.

[65]Zhongyang yanjiuyuan diyici yuanshi huiyi yubeihui cankao ziliao ji huiyi ti'an中央研究院第一次院士会议预备会参考资料及会议提案(Reference Materials for the Preliminary Meeting and Proposals of the First General Assembly of Members of Academia Sinica). Nanjing: Zhongguo dier lishi danganguan中国第二历史档案馆(The Second Historical Archives of China), Total No. 393, File No. 1625.

assembly attached great importance to this case, and decided on the spot to authorize Wong Wen-hao and Adam Pen-Tung Sah to conduct this matter, together with members of Academia Sinica who had already participated in the Pacific Science Conference such as Hu Hsen-hsu, Chu Coching, and Young Chung-Chien. That night, Wu Hsien-Wen, Hsieh Chia-yung, Zheng Wanjun, and Shen Zonghan were selected to attend the conference as representatives.[66]

The other two proposals were directly related to the Second General Assembly of Members of Academia Sinica that was to be held in 1949. One was to propose the establishment of a Preparatory Committee for the election of honorary members of Academia Sinica, and the other was to propose the instruction of the academic subsidy and bonus team to study the national academic policy. If these proposals were implemented, the Second General Assembly of Members of Academia Sinica would become more formal and substantial.

As for the aforementioned proposal, the participating members of Academia Sinica had a heated discussion and ended it in the afternoon. In the end, Chu Chia-Hua delivered a brief closing speech, fully affirming the effectiveness of the assembly, and pointed out: "The assembly was held for three days and attended by 51 people, with more than 48 people attending each meeting.[67] Everyone was concentrated on participating in the meeting, discussing each proposal carefully and thoroughly. The agenda of the assembly was even drafted by a deliberation group in the preparatory meeting, so it was very fruitful."[68] He thought that the proposals approved by the assembly "are quite meaningful in awarding domestic academic research and promoting international scientific cooperation." The assembly itself set up a thesis committee, an academic lecture committee, and a secretary for each group of academicians who would preside over academic contacts frequently in order to "strengthen the organization and carry out work that establishes a system and creates a new atmosphere for the academic circles in our country."[69] Finally, he stressed that "this assembly is of epoch-making significance for the academic history of our country." Moreover, he had great hopes for the future scientific development of China. He said:

> China's import of western science began in the Ming Dynasty, but it had been repeatedly influenced by traditional ideas and the environment, and was without achievement. The late

[66] Chu Coching竺可桢(2006). *Chu Coching quanji*竺可桢全集(The Complete Works of Coching Chu), Vol. 11. Shanghai: Shanghai keji jiaoyu chubanshe上海科技教育出版社. 215.

[67] The number of people attending each meeting said by Chu Chia-Hua was slightly inconsistent with the facts. On the afternoon of September 23, 1948, 47 members of Academia Sinica attended the preliminary meeting for the assembly.

[68] Zhongyang yanjiuyuan zuzhi banfa yuanshi huiyi jilu jiqi xuanju guicheng deng youguan wenshu 中央研究院组织办法、院士会议纪录及其选举规程等有关文书(Relevant Documents such as the Organization Methods of Academia Sinica, Minutes of the General Assembly of Members and the Regulations for Election). Nanjing: Zhongguo dier lishi danganguan中国第二历史档案馆(The Second Historical Archives of China), Total No. 393, File No. 2824.

[69] Zhongyang yanjiuyuan zuzhi banfa yuanshi huiyi jilu jiqi xuanju guicheng deng youguan wenshu 中央研究院组织办法、院士会议纪录及其选举规程等有关文书(Relevant Documents such as the Organization Methods of Academia Sinica, Minutes of the General Assembly of Members and the Regulations for Election). Nanjing: Zhongguo dier lishi danganguan中国第二历史档案馆(The Second Historical Archives of China), Total No. 393, File No. 2824.

> Qing Dynasty reform movement advocated new means of education and introduced western culture to us, but science had still failed to take root in our country. It has also been twenty years since the National Government established Academia Sinica, after Nanjing became the capital city of China.. But as of today, members of Academia Sinica have been elected and our academician system has become truly complete. The foundation for scientific research has been laid; scientific research has entered a new stage. We hope that scientific research will be accelerated in the future under the leadership of all of you and better achievements and contributions will be made in 20 years.[70]

At this point, the First General Assembly of Members of Academia Sinica ended. On the night of September 25, Chiang Kai-shek hosted a banquet for all the members of Academia Sinica at the presidential palace. About 40 members were present, accompanied by Chiang Kai-shek's wife Song Meiling, He Jingzhi, Chia Ching-teh, Chang Poling, Wu Ting-chang, and other dignitaries. The dinner was "extremely simple and delicate, but everyone under the age of 50 felt that it was not enough to eat." The banquet did not last long and ended at 9:15.[71]

6.4 The Implementation of the Resolution After the Assembly and the Scattering of the Members

As for the resolutions of this assembly, Academia Sinica began to implement them one after another after the meeting. For example, concerning the resolution "ask the government to determine 'million' as 'zhao' in order to simplify the calculation of the large numbers", after the meeting "letters will be sent to all the ministries and departments to ask for their support." On October 5, 1948, the Executive Yuan agreed in a letter to confirm "million" as "zhao", saying that our country had established a rule regarding this; "the application of the three-digit decimal system has actually been approved." It was hoped that Academia Sinica would work out a special name for "10^9 (千兆)" as soon as possible, so that it would be approved and implemented by the Executive Yuan's government affairs conference. On October 7, Chu Chia-Hua sent a letter to Adam Pen-Tung Sah to urge the process, stating that "the opinions expressed by the members of Academia Sinica have been very comprehensive" during the assembly, and proposed to designate "10^9" as "jing (京)" and "10^{12}" as "gai (垓)".[72] Since then, the national government had taken on the recommendation of Academia Sinica and adopted a three-digit decimal system for the naming of

[70]Zhongyang yanjiuyuan zuzhi banfa yuanshi huiyi jilu jiqi xuanju guicheng deng youguan wenshu 中央研究院组织办法、院士会议纪录及其选举规程等有关文书(Relevant Documents such as the Organization Methods of Academia Sinica, Minutes of the General Assembly of Members and the Regulations for Election). Nanjing: Zhongguo dier lishi danganguan中国第二历史档案馆(The Second Historical Archives of China), Total No. 393, File No. 2824.

[71]Chu Coching竺可桢(2006). *Chu Coching quanji*竺可桢全集(The Complete Works of Coching Chu), Vol. 11. Shanghai: Shanghai keji jiaoyu chubanshe上海科技教育出版社. 215–216.

[72]Zhongyang yanjiuyuan diyici yuanshi huiyi yubeihui cankao ziliao ji huiyi ti'an中央研究院第一次院士会议预备会参考资料及会议提案(Reference Materials for the Preliminary Meeting and Proposals of the First General Assembly of Members of Academia Sinica). Nanjing: Zhongguo dier

large numbers. Another example was on October 5, when Academia Sinica sent a letter to convener Adam Pen-Tung Sah to formally notify the resolution on "whether or not it is necessary to organize a group meeting to prepare for the election of honorary members of Academia Sinica next year."[73] On October 5, a letter was also sent to Adam Pen-Tung Sah to formally inform the honorary secretary, the director-general, the secretary director, the general affairs director, and secretaries of all groups of General Assembly of Members to organize the Preparatory Committee for the Second General Assembly of Members of Academia Sinica.[74]

Not only that, but Academia Sinica had set up a Preparatory Committee for the election of members in October of 1948, in order to begin preparations for the election of the second batch of members. On October 27, the Preparatory Committee sent a letter to any relevant institutes, requesting that "the nomination of member candidates be favorably received", and that "the nomination period is scheduled to end on December 15, 1949". It also stated that "the number of members to be elected by Academia Sinica next year shall be up to fifteen people, according to the regulations, and the number of members in each of the three groups shall be up to five". A nomination form for member candidates, Articles 5 to 7 of *The Organization Act of National Academia Sinica* and Articles 3, 5–8 of *The Regulations for the Election of Members of Academia Sinica* was also attached to the letter.[75] Since then, some institutions have also carried out the work of nominating member candidates. For example, Nankai University that did not participate in the nomination of candidates for the first batch of members established a Nomination Committee. On December 8, the unibersity nominated Chang Poling and Ho L. as member candidates to the Preparatory Committee.[76] Li Yan, deputy chief engineer of Longhai Railway,

lishi danganguan中国第二历史档案馆(The Second Historical Archives of China), Total No. 393, File No. 1625.

[73]Zhongyang yanjiuyuan hanxun ge danwei zhengji diyici yuanshi houxuanren nianji zili deng xiang ziliao中央研究院函询各单位征集第一次院士候选人年籍资历等项资料(Academia Sinica Sent Letters to Ask Each Unit for Information on the Ages, Hometowns and Qualifications of Candidates for the First Batch of Members). Nanjing: Zhongguo dier lishi danganguan中国第二历史档案馆(The Second Historical Archives of China), Total No. 393, File No. 1611.

[74]Zhongyang yanjiuyuan diyici yuanshi huiyi yubeihui cankao ziliao ji huiyi ti'an中央研究院第一次院士会议预备会参考资料及会议提案(Reference Materials for the Preliminary Meeting and Proposals of the First General Assembly of Members of Academia Sinica). Nanjing: Zhongguo dier lishi danganguan中国第二历史档案馆(The Second Historical Archives of China), Total No. 393, File No. 1625.

[75]Guoli nankai daxue guanyu tuixuan houxuanren he yuanshi timing wenti yu lianheguo jiaoyu kexue wenhua zuzhi weiyuanhui mishu ji zhongyang yanjiuyuan de laiwang wenshu国立南开大学关于推选候选人和院士提名问题与联合国教育科学文化组织委员会秘书处及中央研究院的来往文书(Correspondence between National Nankai University and the Secretariat of the United Nations Educational, Scientific and Cultural Organization Committee and Academia Sinica on the Selection of Candidates and the Nomination of Members. Tianjin: Nankai daxue danganguan南开大学档案馆(Archives of Nankai University), File No. 1-DZ-01–529.

[76]Guoli nankai daxue guanyu tuixuan houxuanren he yuanshi timing wenti yu lianheguo jiaoyu kexue wenhua zuzhi weiyuanhui mishu ji zhongyang yanjiuyuan de laiwang wenshu国立南开大学关于推选候选人和院士提名问题与联合国教育科学文化组织委员会秘书处及中央研究院的来往文书(Correspondence between National Nankai University and the Secretariat of the United

heard that "Academia Sinica intends to invite universities to nominate members of Academia Sinica before the end of this year." He hoped that, based on his research achievements for the history of Mathematics in China, National Tsing Hua University could recommend him as a candidate for member.[77] National Tsing Hua University proposed to nominate him and communicated with Academia Sinica on the issue of which group the history of mathematics in China belonged.[78]

However, when Academia Sinica was implementing the resolutions of the First General Assembly of Members, China's political situation began to undergo drastic changes. The People's Liberation Army (PLA) went with irresistible force to win the Liao-Shen Campaign on November 2, 1948, liberating the entire northeast, and then launched the Huai-Hai Campaign and the Pingjin Campaign. The Kuomintang troops were defeated successively. In the face of this precarious situation, the National Government decided to relocate some academic institutes to Taiwan and some coastal provinces. Academia Sinica secretly decided to temporarily relocate to Guangzhou, Guilin, and Taiwan at its executive meeting.[79] As relocation became a top priority, the implementation of the resolutions of the First General Assembly of Members had to be interrupted. Academia Sinica, which had just established its academician system and entered a new stage of development, had fallen into the dilemma of relocation and reconstruction. Regarding the election of the second batch of members, after the deadline for the nomination of member candidates on December 15, 1948, Academia Sinica decided to "extend one term and hold it with the election of the next batch of members in 1951", in view of the fact that "it is difficult to smoothly continue the member election at this time."[80] However, the second election of members was not held until 1958. On New Year's Eve of the solar calendar—that is, on the night

Nations Educational, Scientific and Cultural Organization Committee and Academia Sinica on the Selection of Candidates and the Nomination of Members. Tianjin: Nankai daxue danganguan南开大学档案馆(Archives of Nankai University), File No. 1-DZ-01–529.

[77]Zhongyang yanjiuyuan juban diyici yuanshi xuanju tuijian houxanren mingdan ji youguan guizhang banfa he laiwang wenshu中央研究院举办第一次院士选举推荐候选人名单及有关规章办法和来往文书(List of Recommended Candidates for the Election of the First Batch of Members of Academia Sinica and Relevant Regulations, Methods and Correspondence). Beijing: Qinghua daxue danganguan清华大学档案馆(Archives of Tsinghua University), Total No.1, Cat. No. 4–2, File No. 191.

[78]Zhongyang yanjiuyuan juban diyici yuanshi xuanju tuijian houxanren mingdan ji youguan guizhang banfa he laiwang wenshu中央研究院举办第一次院士选举推荐候选人名单及有关规章办法和来往文书(List of Recommended Candidates for the Election of the First Batch of Members of Academia Sinica and Relevant Regulations, Methods and Correspondence). Beijing: Qinghua daxue danganguan清华大学档案馆(Archives of Tsinghua University), Total No. 1, Cat. No. 4–2, File No. 191.

[79]Chu Chia-Hua朱家骅. 30 nian lai de zhongyang yanjiuyuan (minguo 47 nian 6 yue 9 ri) 三十年来的中央研究院 (民国四十七年六月九日) (Academia Sinica for 30 Years, June 9, 47, Republic of China). In Wang Yujun王聿均 and Sun Bin孙斌(Eds.) (1977). *Chu Chia-Hua xiansheng yanlun ji*朱家骅先生言论集(Speech by Mr. Chu Chia-Hua). Taipei: Zhongyang yanjiuyuan jinaishi yanjiusuo 中央研究院近代史研究所(Institute of Modern History of Academia Sinica). 113.

[80]Zhongyang yanjiuyuan zuzhi banfa yuanshi huiyi jilu jiqi xuanju guicheng deng youguan wenshu 中央研究院组织办法、院士会议纪录及其选举规程等有关文书(Relevant Documents such as the Organization Methods of Academia Sinica, Minutes of the General Assembly of Members and

of December 31, 1948— Hu Shih had arrived Nanjing from Peking by a special plane sent by Chiang Kai-shek. He celebrated the spring festival with Fu Ssu-Nien together. They mournfully looked at each other, drinking while reciting the ninth poem of "Imitation of the Ancients," written by Tao Yuanming, a poet of the Jin Dynasty.

They planted mulberry trees by the Yangzi's bank,
In three years expecting to pluck their leaves.
But just when the twigs and foliage were about to turn lush,
Suddenly the hills and rivers happened to change.
The branches and leaves snapped by themselves,
The roots and trunks floated on the blue sea.
Where spring silkworms have no food,
Whom can one look to for winter clothes?
Since first they did not plant on the high plain,
What is there to regret today?[81]

At that time, both of them shed tears.[82] They were aware that the National Government was on the decline, and the reconstruction of Academia Sinica in Taiwan would be extremely difficult. On April 23, 1949, the People's Liberation Army crossed the Yangtze River and liberated Nanjing. Then Academia Sinica closed in mainland China. This historical process confirmed the assertion made by Zhang Yuanji in his speech at the First General Assembly of Members of Academia Sinica. On October 1, the People's Republic of China was founded, and the Kuomintang's rule in the mainland completely collapsed.

As China's political situation changed, the first batch of members from Academia Sinica made their own choice between leaving or not. They scattered between Taiwan, abroad (mainly the United States), and the mainland. Among the 81 members, 9 chose to go to Taiwan: Chu Chia-Hua, Ling Hung-hsun, Li Hsien-wen, Woo Tsin-Hang, Fu Ssu-Nien, Li Chi, Tung Tso-pin, Wang Shih-Chieh and Wang Chung Hwei. Thirteen people chose to go abroad or stay abroad: Shiing-Shen Chern, Wu Ta-You, Adam Pen-Tung Sah, Li Shu-Hua, Wu Hsien, Yuan I-Chin, Chen Ko Kuei, Wang Ging-Hsi, Linkesheng, Hu Shih, Fang Kuei Li, Chao Yuen Ren and Hsiao Kung-chuan. The remaining 59 people stayed in the mainland or returned from overseas shortly after the founding of new China, accounting for 72.8% of the members. In other words, only a few of the members left the mainland and even fewer went to Taiwan.

---

the Regulations for Election). Nanjing: Zhongguo dier lishi danganguan中国第二历史档案馆(The Second Historical Archives of China), Total No. 393, File No. 2824.

[81]The original text of the poem is as follows. 种桑长江边,三年望当采。枝条始欲茂,忽值山河改。柯叶自摧折,根株浮沧海。春蚕既无食,寒衣欲谁待。本不植高原,今日复何悔! It was translated by Charles Yim-tze Kwong. See Charles Yim-tze Kwong(1994). *Tao Qian and the Chinese Poetic Tradition: the Quest for Cultural Identity*. Ann Arbor, Michigan: Center for Chinese Studies, the University of Michigan. 99.

[82]Hu Songping胡颂平(1990). *Hu Shizhi xiansheng nianpu changbian chugao*胡适之先生年谱长编初稿(The First Draft of Mr. Hu Shizhi's Chronology), Vol. 6. Taipei: Lianjing chuban shiye gongsi联经出版事业公司. 2065–2066.

Among the members who left the mainland, Fu Ssu-Nien, Hu Shih, and Chu Chia-Hua were all loyal followers of Chiang Kai-shek. They knew that staying in the mainland would rarely have good results and naturally they choose not to stay. On November 15, 1948, the Institute of History and Philology of Academia Sinica held an executive meeting. Fu Ssu-Nien reported on the current situation and said: "In the event of the fall of Nanjing, only two and a half people in Academia Sinica are willing to do what the Communist Party wants; namely, President Zhu and myself. As for Director-General Sa, he can only be counted as a half." Moreover, he said to himself, "I cannot stay in Nanjing, or I will be in the hands of the Communist Party."[83]

Of course, besides political orientation, there were other factors that prompted some members to leave the mainland and choose to go to Taiwan. After Ling Hung-hsun withdrew from the political path in March 1949, he "delayed going to Taiwan" because he was "afraid of going back to the track of politics." Due to his detention in Hong Kong, he did not adapt himself to impersonal, commercial competition. He was also threatened by the spirit of his old friends who were close to the Communist Party of China and was disturbed by "some political forces supported by the United States." At the invitation of Fu Ssu-Nien, he finally went to Taiwan in 1950.[84] Li Hsien-wen chose to go to Taiwan mainly because he was worried that his research on crop genetics would not be carried out in the New China, which believed in "Lysenkoism". As he said in his old age self-report: "When leaving the country at the end of November 1948, I said goodbye to my colleagues. Almost every colleague unanimously advised me not to leave, saying that this was just a 'change of dynasties'. I answered them that each person had his own thoughts. Since Lysenko's style is dominant in the research of genetics here, I have to leave."[85]

The reason why most members chose to stay in mainland China was complicated. Wong Wen-hao was mainly disappointed with the Kuomintang regime. After the failure of issuing the "gold yuan" notes, Weng resigned as president of the Executive Yuan in November 1948. At this time, although Weng had realized the decadence and decline of the Kuomintang regime, he still had hope in it. On January 21, 1949, Chiang Kai-shek went to his hometown after the political downfall. Weng went to Fenghua, Zhejiang Province, determined to persuade Jiang to give up government affairs completely and retire. However, he was disappointed to see that Jiang continued to manipulate the Kuomintang's important military and political affairs as well as sabotage the peace talks, so he immediately took leave. After that, Weng turned his hope to acting president Li Zongren, who openly advocated for peace, and was invited to serve as acting secretary-general of the presidential palace in early February 1949. Until April 20, Li Zongren had also refused to sign the *Domestic*

---

[83]Shiah Nae夏鼐(2011). *Shiah Nae riji*夏鼐日记(Shiah Nae's Diary), Vol. 4. Shanghai: Huadong shifan daxue chubanshe华东师范大学出版社. 215.

[84]Dictated by Ling Hung-hsun凌鸿勋, Interviewed by Shen Yunlong沈云龙(2011). *Ling Hung-hsun koushu zizhuan*凌鸿勋口述自传(The Oral Autobiography of Ling Hung-hsun). Changsha: Hunan jiaoyu chubanshe湖南教育出版社. 223.

[85]Li Hsien-wen李先闻(2009). Li Hsien-wen zishu李先闻自述(The Autobiography of Li Hsien-wen). Changsha: Hunan jiaoyu chubanshe湖南教育出版社. 202.

Peace Agreement. Weng lost hope in the Kuomintang regime and refused to take office in Taiwan.[86] Chu Coching and Ping Chi also chose to stay in the mainland because they were disappointed with the Kuomintang regime. On May 26, 1949, Chu Coching and Ping Chi met in Shanghai. When the two talked about the political situation, they voiced the same opinions and both of them thought that "The failure of the Kuomintang is to be blamed on itself. In the year 1938, Chiang Kai-shek was widely popular among the Chinese people, but for last ten years he has been headstrong and selfish, taking charge of everything, and indulging in corruption, causing his reputation to fall apart. Is it not a pity?"[87] After the liberation of Shanghai, Mayor Chen Yi paid great attention to the work of scientists, which also had an important impact on Chu Coching's choice to stay in the mainland.[88]

There were multiple reasons for Hu Hsen-hsu's decision to stay on the mainland. According to his autobiography, "I have decided not to leave Peking because I care for the business of the Fan Memorial Institute of Biology."[89] He also he loved his botanical research career. His major was Chinese plant taxonomy, and Chinese plants were mainly distributed on the mainland—to study Chinese plants, he could only stay in the mainland. In addition, when the Fan Memorial Institute of Biology was demobilized, he repeatedly applied to the National Government for financial assistance, but he did not get any support from the Chinese Educational History Fund's board of directors; he became disappointed with the government. This was also one of the reasons why he stayed in the mainland.[90]

Among the eleven members who were the directors of Academia Sinica, eight of them—Chiang Li-Fu, Woo Sho-Chow, Lee J. S., Chow Jen, Loo Tsung-Le, Wang Chia-Chi, Woo Y. H., and Tao L. K.—stayed in the mainland. Among the eight, Tao L. K., director of the Institute of Social Sciences, unhesitatingly stayed in the mainland because he was disappointed with the Kuomintang regime, but also because of his long-term personal discord with Fu Ssu-Nien as well as the personal discord between

---

86 Yan Ruping严如平. Wong Wen-hao翁文灏. In Li Xin李新(Editor-in-Chief), Zhongguo shehui kexueyuan jindaishi yanjiusuo Zhoghua minguoshi yanjiushi中国社会科学院近代史研究所中华民国史研究室 (Research Office of History of the Republic of China of Institute of Modern History, Chinese Academy of Social Sciences)(Eds.) (2011). *Zhonghua minguoshi renwuzhuan*中华民国史·人物传(History of the Republic of China, Biographies of Characters), Vol. 6. Beijing: Zhonghua shuju中华书局.3903–3904.

87 Chu Coching竺可桢(2006). *Chu Coching quanji*竺可桢全集(The Complete Works of Coching Chu), Vol. 11. Shanghai: Shanghai keji jiaoyu chubanshe上海科技教育出版社. 447.

88 Fan Hong-ye樊洪业, Wang De-lu王德禄, and Wei Hong-ning尉红宁(2000). Huang Zongzhen fangtanlu黄宗甄访谈录(An Interview with Huang Zongzhen). *Zhongguo keji shiliao*中国科技史料(China Historical Materials of Science and Technology) 21 (4): 322.

89 Hu Zonggang胡宗刚(2005). *Jingsheng shengwu diaochasuo shigao*静生生物调查所史稿(Historical Manuscript of Fan Memorial Institute of Biology). Jinan: Shandong jioayu chubanshe山东教育出版社. 218.

90 Hu Zonggang胡宗刚(2005). *Jingsheng shengwu diaochasuo shigao*静生生物调查所史稿(Historical Manuscript of Fan Memorial Institute of Biology). Jinan: Shandong jioayu chubanshe山东教育出版社. 218.

the Institute of History and Philology and the Institute of Social Sciences.[91] Feng Teh-Pei, the acting director of the preparatory department of the Institute of Medical Science at Academia Sinica, resolved to stay as firmly as Tao L. K. did. At the end of 1948, Chu Chia-Hua mobilized to move to Taiwan at a meeting of Academia Sinica. Feng Teh-Pei said at the meeting: "The institute of Academia Sinica is a national institute. Do not follow anyone."[92]

The active mobilization and struggling of the Chinese Communist Party played a positive role in the return of Lee J. S., the director of the Institute of Geological Research, to the mainland. Li was lecturing in England when Nanjing was liberated. In early April 1949, Kuo Mo-jo led a Chinese delegation to Prague to attend the world peace-keeping conference. Before the meeting, Kuo Mo-jo brought a letter to Li according to Zhou En-lai's instructions, asking Li to return to China as soon as possible. Li was very excited after receiving the letter in England.[93] Earlier, when Academia Sinica was planning to move, hidden communists such as Xu Jie and Zhao Jinke from the Institute of Geological Research drew up a pledge to boycott the move in an organized way, and sent the pledge to Li in a special letter. Li supported it.[94] This reflected Li's favorable impression of the Communist Party.

The lack of effective organizers and difficulties in funding, transportation, and other aspects of the relocation, as well as the difficult living and working conditions in Taiwan, also led some directors and personnel of Academia Sinica to choose to stay in the mainland. This is seen from Chu Chia-Hua's memoirs of the relocation of Academia Sinica to Taiwan in 1958: "In September of 1949, when Academia Sinica moved to Taiwan, it asked the government for four million 'gold yuan' notes to prepare for the relocation, and half of it was sent to Taiwan successively. ...Until the end of January 1950, all other personnel and equipment had been concentrated in official residences in Shanghai, ready for the next action; at that time, director Fu wasappointed to take over as president of Taiwan University, and Yu Yousun, the general affairs director, was appointed as the general affairs director of Taiwan University. The government had asked Director Fu to serve as president of Taiwan University. Mr. Fu's resolute acceptance of the appointment was originally to facilitate the transfer of Academia Sinica. However, at this critical point, Director General Sa passed away. When Director Fu came to Taiwan, the position of the Director of General Affairs had been vacant, which affected the relocation work. However, the

[91]Chen Shiwei陈时伟(2011). Cong Nanjing dao Nangang: Zhongyang yanjiuyuan 1949 nian qianhou liangci yuanshi xuanju shulun从南京到南港: 中央研究院1949年前后两次院士选举述论(From Nanjing to Nangang: On the Two Academia SinicaElections of 1949). *Zhongguo xueshu* 中国学术(China Scholarship) 8 (2), Total 28: 53–54.

[92]Yao Shuping姚蜀平(1989). Zhongguo kexueyuan de choubei yu jianli中国科学院的筹备与建立(The Preparation and Establishment of the Chinese Academy of Sciences). *Zhongguo keji shiliao* 中国科技史料(China Historical Materials of Science and Technology)10 (3): 57.

[93]Chen Qun陈群et al. (Eds.) (1996). *Li Siguang (Lee J. S.) zhuan*李四光传(Biography of Lee J. S.). Beijing: Renmin chubanshe.174.

[94]Li Yang李扬(1999). Jiefang qianxi Nanjing kejijie fanbanqian douzheng解放前夕南京科技界反搬迁斗争(The Anti-relocation Struggle of Nanjing Science and Technology Circles on the Eve of Liberation). *Yanhuang chunqiu*炎黄春秋(The Spring and Autumn of Yan Huang) (3): 21.

most important reason was that, due to the drastic depreciation of the "gold yuan" notes, the originally-prepared relocation fee was no longer sufficient; the overall situation became more difficult, it was impossible to raise funds separately, and means of transportation were extremely scarce and could not be moved in batches. At that time, it was extremely difficult for individuals in Shanghai to buy steamer tickets to Taiwan, not to mention buying for an entire group. Psychological factors were also reasons that delayed the migration. Additionally, those who went to Taiwan first corresponded with their counterparts in Shanghai and told them about the various difficulties of their lives in Taiwan; research work could not be carried out. Moreover, those who postponed coming to Taiwan made the latecomers more frustrated. Therefore, they concentrated their personnel in Shanghai and did not make urgent preparations for moving to Taiwan."[95] Chu Chia-Hua once sent letters and telegrams "to persuade those who stayed in Shanghai to leave the mainland as soon as possible, but he did not make it."[96] After the liberation of Shanghai in May 1949, the personnel of Academia Sinica that was concentrated in Shanghai generally stayed in the mainland.

In addition to Lee J. S., Hua Loo-Keng was also a heavyweight among members of Academia Sinica returning from overseas. Before returning to China in 1950, Hua taught at the University of Illinois in the United States. He received a good salary of about $10,000 USD a year and had four teaching assistants.[97] The reason why Hua gave up his high salary and favorable working conditions in the United States to return to China was complicated. According to Wang Yuan's research, the main reason was that he was convinced of China's reunification and that China had the conditions for peaceful, democratic nation-building. He was willing to contribute to catching up China's mathematics to the world level. Racial discrimination in American society, loneliness brought about by different cultural backgrounds, and the influence of the Chinese Communist Party on his work all affected him.[98] According to Hsu L. C., who had close ties with Hua Loo-Keng, there were other reasons for Hua's return to China: first, he was determined to realize his lofty ambition of promoting independence in Chinese mathematics; second, Hua hoped to become the head of the mathematics department in Tsinghua University after returning to China and the

[95] Chu Chia-Hua朱家骅. 30 nian lai de zhongyang yanjiuyuan (minguo 47 nian 6 yue 9 ri) 三十年来的中央研究院 (民国四十七年六月九日) (Academia Sinica for 30 Years, June 9, 47, Republic of China). In Wang Yujun王聿均 and Sun Bin孙斌(Eds.) (1977). *Chu Chia-Hua xiansheng yanlun ji*朱家骅先生言论集(Speech by Mr. Chu Chia-Hua). Taipei: Zhongyang yanjiuyuan jinaishi yanjiusuo 中央研究院近代史研究所(Institute of Modern History of Academia Sinica). 113–114.

[96] Chu Chia-Hua朱家骅. 30 nian lai de zhongyang yanjiuyuan (minguo 47 nian 6 yue 9 ri) 三十年来的中央研究院 (民国四十七年六月九日) (Academia Sinica for 30 Years, June 9, 47, Republic of China). In Wang Yujun王聿均 and Sun Bin孙斌(Eds.) (1977). *Chu Chia-Hua xiansheng yanlun ji*朱家骅先生言论集(Speech by Mr. Chu Chia-Hua). Taipei: Zhongyang yanjiuyuan jinaishi yanjiusuo 中央研究院近代史研究所(Institute of Modern History of Academia Sinica) . 115.

[97] Dictated by Xu Lizhi(Hsu L. C.)徐利治, Interviewed by Yuan Xiangdong袁向东 and Guo Jinhai 郭金海(2017). *Xu Lizhi fangtanlu*徐利治访谈录(An Interview with Xu Lizhi). Changsha: Hunan jiaoyu chubanshe湖南教育出版社. 212.

[98] Wang Yuan王元(1999). *Hua Loo-Keng*华罗庚. Nanchang: Jiangxi jiaoyu chubanshe江西教育出版社. 169–170.

head of the Institute of Mathematics of the Chinese Academy of Sciences (CAS) after its establishment; third, Hua was worried that his children would have to join the United States army when they reached adulthood.[99]

However, regardless of their reasons for leaving or staying, the choice made by the first batch of members of Academia Sinica at this turning point in history meant that they placed their hopes on their futures and also affected the development of academic undertakings in Taiwan and the mainland.

6.5 Summary

In 1948, the civil war in China between the Kuomintang and the Communist Party was raging; society was in turmoil, and the economy was about to collapse. There were many difficulties in holding the General Assembly of Members of Academia Sinica, such as large funds, difficulty in preparation, and difficulty in setting a quorum. However, out of respect for the academician system, the council still decided to hold the First General Assembly of Members of Academia Sinica and made adequate preparations. This assembly was fruitful. The holding of the assembly marked the establishment of Academia Sinica's academician system, which meant that Academia Sinica had formed a sound academic system with the academy's members as the main, the council as permanent appraisal organ of the General Assembly of Members of Academia Sinica and the institutes conducting academic research. This made up for the defects of the previous academic system of Academia Sinica and led it to enter a new stage of development. The establishment of academician system of Academia Sinica marked the establishment of the academician system in China and was a milestone in the process of the systematization of science in China.

However, due to the drastic political changes, with the exception of the resolution "ask the government to determine 'million' as 'zhao' in order to simplify the calculation of the large numbers", almost none of the other resolutions of the assembly were implemented. Moreover, under the changeable political situation, the first batch of Members was forced to disintegrate and the academician system of Academia Sinica was severely damaged. Thanks to the efforts of Chu Chia-Hua, Hu Shih, and others, the academician system was restored in 1958 after Academia Sinica moved to Taiwan and the second batch of members was elected.[100] After the closing down of Academia Sinica in 1949, with the exception of the Institute of History and Philology and the Institute of Mathematics, the other institutes of Academia Sinica and most of its

[99]Dictated by Xu Lizhi(Hsu L. C.)徐利治, Interviewed by Yuan Xiangdong袁向东 and Guo Jinhai 郭金海(2017). *Xu Lizhi fangtanlu*徐利治访谈录(An Interview with Xu Lizhi). Changsha: Hunan jiaoyu chubanshe湖南教育出版社. 212–216.

[100]For the process of resuming the academician system after Academia Sinica moved to Taiwan, please refer to Chen Shiwei陈时伟(2011). Cong Nanjing dao Nangang: Zhongyang yanjiuyuan 1949 nian qianhou liangci yuanshi xuanju shulun从南京到南港: 中央研究院1949年前后两次院士选举述论(From Nanjing to Nangang: On the Two Academia Sinica Elections of 1949). *Zhongguo xueshu*中国学术(China Scholarship) 8 (2), Total 28: 57–71.

personnel stayed in the mainland and were accepted by the Chinese Academy of Sciences. The members of Academia Sinica who stayed in the mainland or returned from overseas after the founding of new China generally became leaders or the backbone of the CAS and some institutions of higher learning or scientific research. In the process of building up the foundation for new China's scientific cause, they played the historical role of inheriting and transforming.[101] However, under the influence of regime change, the academician system was not inherited by the new regime as the scientific heritage of the Republic of China.

[101] Fan Hongye樊洪业(2005), Yuanshi zhidu qibian院士制度七辩(Seven Debates on the Academician System). *Kexue shibao*科学时报(Science Times), 2005–06–06 (6).

Chapter 7
From the Abolition of Academician System to the Establishment of the Membership System of Academic Divisions

On November 1, 1949, just one month after the founding of the People's Republic of China, the Chinese Academy of Sciences was established. The institutes of Academia Sinica in Chinese mainland and the National Academy of Peiping had become an important foundation for the CAS to set up institutes. The CAS inherited Academia Sinica's basic systems like the two-level structure of the Academy, the organizational structure of the institutes and the professional titles and ranks,[1] but its academician system was abandoned. In the national "one-sided" craze to learn from the Soviet Union, the CAS, after more than a year's preparations, established the Academic Divisions and its membership system in 1955. This chapter begins with the founding of the CAS, and discusses the complete abandonment of academician system of Academia Sinica in the mainland China, the preparations of the CAS for establishing the Academic Divisions and its membership system, so as to reveal the origin and prelude of the reconstruction of the academician system in China.[2]

---

[1]Fan Hongye樊洪业(2005). Zhongguo kexueyuan zaoqi de gaizao zhilu中国科学院早期的"改造"之路(Recasting Academia Sinica: 1949–1955). *Kexue wenhua pinglun*科学文化评论(Science & Culture Review) 2 (6): 23–26.

[2]Wang Yangzong's *From Academicians to Academic Division Members: Difficulties in Establishing Academic System of the CAS* inspired the author to write this chapter and provided an important reference. See Wang Yangzong王扬宗. Cong yuanshi dao xuebu weiyuan: Zhongguo kexueyuan xueshu tizhi jianli de kunjing从院士到学部委员—中国科学院学术体制建立的困境(From Academicians to Academic Division Members: Difficulties in Establishing Academic System of the CAS). In Yu Miin-ling余敏玲(Ed.) (2012). *Liangan fenzhi: Xueshu jianzhi tuxiang xuanchuan yu zuqun zhengzhi(1945–2000)*两岸分治:学术建制、图像宣传与族群政治(1945–2000)(Divided Rule across the Taiwan Straits: Educational Reorganization, Visual Propaganda, and Ethnic Politics, 1945–2000). Taipei: Institute of Modern History, Academia Sinica. 65–135.

J. Guo, *The Establishment and Reconstruction of the Academician System in China*,
https://doi.org/10.1007/978-981-15-7208-1_7

7.1 Abandonment of Academician System of Academia Sinica and System Selection of the CAS

The Communist Party of China attached great importance to science and technology as early as Yan'an period. Yan'an Institute of Natural Sciences was established in 1939 and changed into Yan'an Academy of Natural Sciences in 1940 to train "technical cadres and specialized technical talents for the Anti-Japanese War and the founding of the People's Republic of China".[3] In 1940, Mao Tse-tung also stressed at the inaugural meeting of the Shaanxi-Gansu-Ningxia Border Region Natural Science Research Association: "Natural science is a very good thing. It can solve the living problems of clothing, food, housing and transportation. Therefore, everyone should approve of it and everyone should study natural science."[4] In the spring of 1949, on the eve of the founding of the People's Republic of China, the CPC Central Committee had "the intention of establishing a unified academy of sciences as the highest scientific institution in the country after the founding of New China".[5] In order to realize this intention, Lu Dingyi, minister of the Publicity Department of the CPC Central Committee, was appointed to take charge of the preparatory work for the Academy of Sciences. The first to take part in the work were Yun Ziqiang and Ding Zan, followed by Tsien San-Tsiang and then Huang Zongzhen.[6]

Before they formally started the preparatory work for the CAS in June 1949, the preparatory work for the National Congress of Natural Scientists of China (hereinafter referred to as the "Congress of Scientists") began. Yun Ziqiang, Ding Zan, Tsien San-Tsiang and Huang Zongzhen were all members of the Preparatory Committee of the Congress of Scientists. The committee had a standing committee with Yun as the publicity department minister, Ding as the organization department

---

[3]Ziran kexueyuan zhaosheng qishi (1940 nian 5 yue) 自然科学院招生启事(1940年5月)(Notice of Admission to the Academy of Natural Sciences, May 1940). In *Yan'an ziran kexueyuan shiliao* bianji weiyuanhui《延安自然科学院史料》编辑委员会(Eds.) (1986). *Yanan ziran kexueyuan shiliao*延安自然科学院史料(Historical Materials of Yan'an Academy of Natural Sciences). Beijing: Zhonggong dangshi ziliao chubanshe, Beijing gongye xueyuan chubanshe中共党史资料出版社、北京工业学院出版社.97.

[4]Mao Tse-tung Chenyun tongzhi zai ziran kexue yanjiuhui chengli dahui shang zuo zhongyao jianghua (1940 nian 2 yue 5 ri) 毛泽东、陈云同志在自然科学研究会成立大会上作重要讲话(1940年2月5日)(Comrades Mao Tse-tung and Chen Yun Made Important Speeches at the Inaugural Meeting of the Natural Sciences Research Association, February 5th, 1940). In *Yan'an ziran kexueyuan shiliao* bianji weiyuanhui《延安自然科学院史料》编辑委员会(Eds.) (1986). *Yanan ziran kexueyuan shiliao*延安自然科学院史料(Historical Materials of Yan 'an Academy of Natural Sciences). Beijing: Zhonggong dangshi ziliao chubanshe, Beijing gongye xueyuan chubanshe中共党史资料出版社、北京工业学院出版社.32.

[5]Tsien San-Tsiang钱三强(2000). Choujian kexueyuan qianhou wo canyu de yixie shiqing筹建科学院前后我参与的一些事情(Preparation I Have Done for the Founding of Academia Sinica). *Zhongguo keji shiliao*中国科技史料(China Historical Materials of Science and Technology) 21 (4): 339.

[6]Fan Hong-ye樊洪业, Wang De-lu王德禄, and Wei Hong-ning尉红宁(2000). Huang Zongzhen fangtanlu黄宗甄访谈录(An Interview with Huang Zongzhen). *Zhongguo keji shiliao*中国科技史料(China Historical Materials of Science and Technology) 21 (4): 317–319.

deputy minister and Tsien as the planning committee member. On July 13, 1949, the plenary meeting of the Preparatory Committee of the Congress of Scientists was held. Zhou En-lai, vice-chairman of the People's Revolutionary Military Commission, said at the meeting: "In the near future, we must set up an academy of sciences owned by the people, and we hope everyone will participate in the planning."[7] At that time, many scientists also hoped that the government could come up with a new plan to reorganize the national scientific research institutions. After many discussions, the Planning Committee was ready to put forward the following proposal to the People's Political Consultative Conference:

> The National Academy of Sciences shall be established to co-ordinate and lead the research in the national natural and social sciences so as to closely cooperate with production and scientific education. The National Academy of Sciences shall also be responsible for reviewing and rewarding scientific creations, works and inventions throughout the country. The National Academy of Sciences shall set up various research institutions to meet special needs. When such research institutions are developed to a certain stage, independent institutions shall be established to further coordinate with production.[8]

In early September, Huang Zongzhen drew up the *Draft for the Establishment of the People's Academy of Sciences* according to his discussions with Ding Zan and Tsien San-Tsiang as the central government requested to come up with a plan for the establishment of the Academy of Sciences as soon as possible. On the basis of the first draft, Ding Zan and Tsien San-Tsiang formally wrote out the *Draft for the Establishment of the People's Academy of Sciences* (hereinafter referred to as the "*Draft*"). The *Draft* explained the necessity of establishing the Academy of Sciences and its name, affiliation, tasks and organization. It was the reference basis for the Central Government to make decisions on the establishment of the Academy of Sciences before the National Political Consultative Conference.[9] According to the plan of the *Draft,* in the organizational system of the Academy of Sciences, the first level was the president and vice presidents, the second level was the general office and the secretariat, the third level was the Council, General Assembly and special research institutions, and the fourth level was the general affairs department (including affairs section and accounting section), editing and publishing department, research planning department, foreign culture department, special committees and heads of research institutions under General Assembly (Fig. 7.1).[10]

---

[7]Zhou fuzhuxi enlai zai kedai choubeihui shang jianghua zhaiyao周副主席恩来在科代筹备会上讲话摘要(The Summary of Speech Made by Vice Chairman Zhou En-lai at the Preparatory Meeting of the National Conference of Workers of Natural Sciences in China). *Kexue tongxun*科学通讯(Science News) 1949, (2): 3.

[8]Jianli zhongkeyuan youguan wenjian cailiao建立中科院有关文件材料(Relevant Documents on Establishing the Chinese Academy of Sciences). Beijing: Zhongguo kexueyuan danganguan中国科学院档案馆(Archives of Chinese Academy of Sciences), 1950-02-001.

[9]Fan Hongye樊洪业(2000). Jianli renmin kexueyuan caoan de lailong qumai《建立人民科学院草案》的来龙去脉(The Origin and Development of the "Draft for the Establishment of the People's Academy of Sciences"). *Zhongguo keji shiliao*中国科技史料(China Historical Materials of Science and Technology) 21 (4): 326–328.

[10]Jianli zhongkeyuan youguan wenjian cailiao建立中科院有关文件材料(Relevant Documents on Establishing the Chinese Academy of Sciences). Beijing: Zhongguo kexueyuan danganguan中国科学院档案馆(Archives of Chinese Academy of Sciences), 1950-02-001.

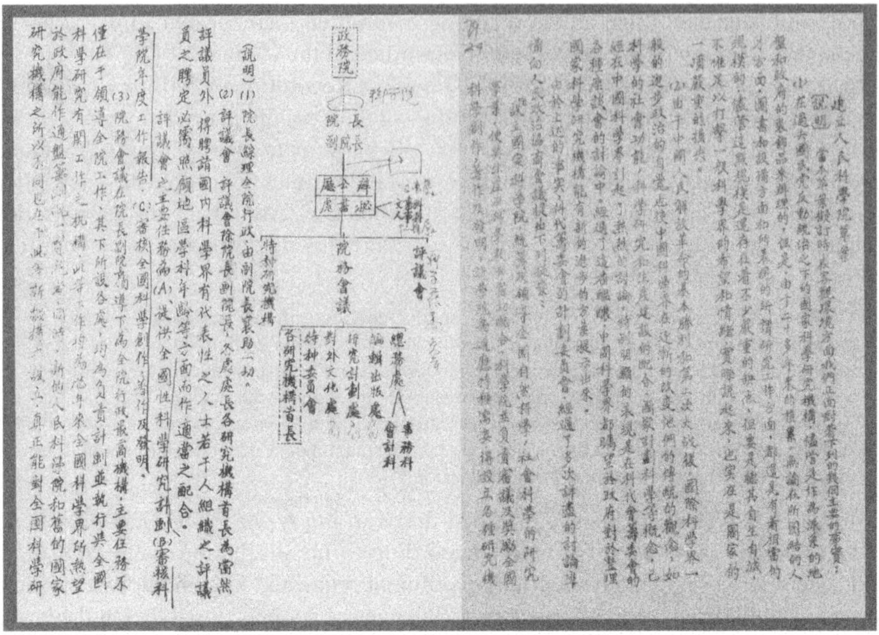

Fig. 7.1 *Draft for the Establishment of the People's Academy of Sciences* (part)

The *Draft* provided for the establishment of the Council, following the practice of Academia Sinica.[11] However, the *Draft* did not cover the General Assembly, indicating that Ding Zan, Tsien San-Tsiang and Huang Zongzhen did not follow the example of Academia Sinica in incorporating the academician system into the organizational system of the Academy of Sciences. Having finished the *Draft*, Ding Zan and Tsien San-Tsiang sent it to Lu Dingyi. On September 27, the First Plenary Session of the Chinese People's Political Consultative Conference passed the *Organic Law of the Central People's Government of the People's Republic of China*, which stipulated that the "Academy of Sciences" was one of the departments of the Administration Council, under the leadership of the Administration Council and under the guidance of the Culture and Education Commission.[12] Soon there was a change in setting up the Council by the Academy of Sciences. On October 13, in a memo to Lu Dingyi, Ding Zan proposed to change the Council into a scientific working

[11]Song Zhenneng宋振能(2011). Zhongguo kexueyuan jianli zhuanmen weiyuan zhidu de huigu中国科学院建立专门委员制度的回顾(Review of the Setting up of the System of Special Committee Members in the Chinese Academy of Sciences). In Song Zhenneng. *Zhongguo kexueyuan yuanshi shiling*中国科学院院史拾零(Sidelights of the History of the Chinese Academy of Sciences). Beijing: Kexue chubanshe科学出版社.3.

[12]Zhongyang renmin zhengfu fazhi weiyuanhui中央人民政府法制委员会(he Legal Committee of the Central People's Government) (Eds.) (1952). *Zhongyang renmin zhengfu faling huibian(1949–1950)* 中央人民政府法令汇编(1949–1950) (Compilation of Decrees of the Central People's Government, 1949–1950). Beijing: Renmin chubanshe人民出版社.4–6.

committee, believing "The Council used to be limited to the election of the president and should now participate in planning issues. It is better to change it to a scientific working committee".[13] On October 19, the Central People's Government Committee held its third meeting and appointed Kuo Mo-jo as president of the CAS and Chen Boda, Lee J. S., Tao L. K. and Chu Coching as vice presidents. Except the nominal Chen Boda, the other 4 were academicians of Academia Sinica.

After the establishment of the CAS on November 1, the CAS solicited opinions from the Central Committee on the establishment of a scientific working committee. To this, the leaders of the CPC Central Committee held a negative attitude, believing that the Academy of Sciences had just been established and the conditions were not yet ripe and that organizations of the nature of former academicians should not be set up and scientists should not be given too much power.[14] This opinion reflected that the new regime's bad feelings towards the academician system of Academia Sinica determined the fate of the system being abandoned after the founding of new China.

After the abortion of the establishment of the scientific working committee, the CAS decided on November 13 to set up special committees for various disciplines. The special committee member had the nature of an academic consultant and was an honorary post. The main responsibilities were to be responsible for discussing the work plans, the implementation process and the work reports of various research departments of the CAS, the appointment and promotion of senior researchers and technicians, the cooperation between the CAS and research work outside the Academy, the subsidies provided for various academic studies outside the Academy by the CAS, as well as the examination of scientific discoveries, inventions and works, and the promotion of international academic cooperation. After the establishment of the special committees, although many members of the special committees performed their respective duties, the collective role played by the special committees had basically not been brought into play.[15]

At the same time, the greater problem encountered by the CAS was the lack of regular business leadership and ideological leadership in its research units, which were actually in a state of self-flow. The Planning Bureau in charge of research business was mostly young cadres who had just graduated from university, except for a

---

[13]Song Zhenneng宋振能(2011). Zhongguo kexueyuan jianli zhuanmen weiyuan zhidu de huigu中国科学院建立专门委员制度的回顾(Review of the Setting up of the System of Special Committee Members in the Chinese Academy of Sciences). In Song Zhenneng. *Zhongguo kexueyuan yuanshi shiling*中国科学院院史拾零(Sidelights of the History of the Chinese Academy of Sciences). Beijing: Kexue chubanshe科学出版社.3.

[14]Song Zhenneng宋振能(2011). Zhongguo kexueyuan jianli zhuanmen weiyuan zhidu de huigu中国科学院建立专门委员制度的回顾(Review of the Setting up of the System of Special Committee Members in the Chinese Academy of Sciences). In Song Zhenneng. *Zhongguo kexueyuan yuanshi shiling*中国科学院院史拾零(Sidelights of the History of the Chinese Academy of Sciences). Beijing: Kexue chubanshe科学出版社.4.

[15]Song Zhenneng宋振能(2011). Zhongguo kexueyuan jianli zhuanmen weiyuan zhidu de huigu中国科学院建立专门委员制度的回顾(Review of the Setting up of the System of Special Committee Members in the Chinese Academy of Sciences). In Song Zhenneng. *Zhongguo kexueyuan yuanshi shiling*中国科学院院史拾零(Sidelights of the History of the Chinese Academy of Sciences). Beijing: Kexue chubanshe科学出版社. 16–18.

few directors who were familiar with the scientific situation and had considerable academic standards. In fact, it could not be a powerful assistant in academic leadership of the General Assembly, which led to a serious situation of disconnection between the top and bottom, hectic work, more administrative leadership and less academic leadership.[16] As a result, although the CAS was the national science center, its overall academic leadership was weak and lacked the authority of academic leadership, which made it difficult to be competent for the important task of organizing and leading the national academic work. Moreover, "many scientists believed that the real leadership of the scientific community is the Scientific Federation,[17] that the Academy of Sciences is run by the Communist Party, and people there do not understand science."[18] However, with the large-scaled construction of the country, the central government actually hoped that the CAS would shoulder the important task of organizing and leading the academic work of the country.

Under such circumstances, it was necessary for the CAS to properly adjust its organizational structure, strengthen its academic leadership and enhance its authority. At that time, New China implemented a one-sided foreign policy and was setting off an upsurge of learning from the Soviet Union. In that upsurge, the CAS sent a delegation to visit the Soviet Union in February 1953 to learn how the Soviet Union organized and led scientific research work, especially the Soviet Union's experience in how science developed and expanded from the old foundation after the October Revolution, as well as the current situation of scientific development and its development direction, and to exchange views on scientific cooperation between the two countries.[19] The delegation was composed of 26 people, with Tsien San-Tsiang as its head, Zhang Jiafu as the temporary branch secretary, who was a vice-president and Secretary of the Party group of the CAS, and Wu Heng as the secretary general, who was the Secretary-general of the Northeast Branch (Fig. 7.2).[20]

The Soviet Academy of Sciences was one of the main institutions visited by the delegation. In the academic system, the Soviet Academy of Sciences had implemented the academician system since its predecessor, the Petersburg Royal Academy of Sciences, was formally established in 1725. Its highest body was the Academician Assembly composed of all academicians and honorary academicians. The main

---

[16]Zhongkeyuan dangzu guanyu muqian benyuan gongzuo jiben qingkuang he jinhou gongzuo renwu de baogao ji zhongyang pishi中科院党组关于目前本院工作基本情况和今后工作任务的报告及中央批示(Report of the Party Group of the Chinese Academy of Sciences on the Basic Situation of the Work of the Academy at Present and Its Future Tasks and Instructions from the Central Committee). Beijing: Zhongguo kexueyuan danganguan中国科学院档案馆(Archives of Chinese Academy of Sciences), 1954-01-001.

[17]Scientific Federation(科联) refers to All-China Federation of Specialized Societies in Natural Sciences(中华全国自然科学专门学会联合会).

[18]Zhang Jiafu张稼夫(1984). *Gengshen yishi*庚申忆逝(Recalling the Passed in 1980). Taiyuan: Shanxi renmin chubanshe山西人民出版社.131.

[19]Zhongguo kexueyuan guanyu fangsu daibiaotuan gongzuo de baogao中国科学院关于访苏代表团工作的报告(Report of the Chinese Academy of Sciences on the Work of the Delegation to Soviet Union). Kexue tongbao科学通报(Chinese Science Bulletin)1954, (4): 12–14.

[20]Wu Heng武衡(1992). *Keji zhanxian 50 nian*科技战线五十年(50 Years of Science and Technology Front). Beijing: Kexue jishu wenxian chubanshe科学技术文献出版社.118–119.

Fig. 7.2 Some leaders of the CAS took photos with members of the delegation visiting the Soviet Union in 1953

functions and powers of the Academician Assembly included formulating the general guidelines for scientific work implemented within the system of the Soviet Academy of Sciences, guiding the research work of the institutes and electing academicians, honorary academicians and corresponding academicians. Academician assembly were held several times a year, at least once to hear detailed progress reports on current research and other activities. When the academician congress was not in session, the Presidium elected at the academician assembly exercised the decision-making power and led all daily business and organizational work in the Academy. The Presidium consisted of the president, vice president, academic secretary-general, academician secretary and members of the Presidium.[21] The Presidium had an executive body, the academic secretariat, which was an assistant to the leaders of the Academy and was responsible for the scientific organization of the entire Academy.[22]

The Soviet Academy of Sciences had eight divisions according to its main disciplines: physics and mathematics, geology and geography, chemistry, biology, technology, history, economics, philosophy and law, and literature and language. The Academic Divisions were academic leading bodies attached to the Presidium and were composed of Members and corresponding members of the Academy related to each Division. The Division Assembly and its Standing Committee were the leading organs of the Academic Division. The Standing Committee of the Academic Division consisted of the Secretary and Deputy Secretary of the Academic Division Members, the Academic Secretary and the members of the Standing Committee. The work of

[21] Alexander Vucinich(1956). *The Soviet Academy of Sciences*. Stanford: Stanford University Press. 21–22.

[22] Kovda guwen dui youguan xuebu ji mishuchu de gongzuo de baogao柯夫达顾问对有关学部及秘书处的工作的报告(Consultant Kovda's Report on the Work of the Academic Divisions and Secretariat). In Zhongguo kexueyuan xueshu mishuchu中国科学院学术秘书处(Academic Secretariat of the Chinese Academy of Sciences) (Eds.). *Zhongguo kexueyuan nianbao*中国科学院年报(The Annals of the Chinese Academy of Sciences) 1955: 98.

the division was led by the bureau of each division, and the bureau members need not be academicians or corresponding academicians.[23]

The organizational structure and the system of the Soviet Academy of Sciences impressed the members of the delegation. After the delegation returned to Beijing, Tsien San-Tsiang reported on the visit to the Soviet Union at the executive meeting of the CAS on June 20, and introduced the organizational structure of the Soviet Academy of Sciences, including the General Assembly, the Presidium, the Academic Secretariat, Academic Divisions, etc.[24] Later, the delegation summarized in the work report the experience of the Soviet Union and the Soviet Academy of Sciences in their great achievements in just 30 years. There were two of them: (1) "clear division of labor and coordination among scientific institutions were integrated into an organic whole"; (2) "to cultivate a healthy academic atmosphere". The academician system played a positive role in Soviet's experience. It was mainly manifested in the practice that university professors and scientists in the industrial sector could be elected as academicians and corresponding academicians of the Soviet Academy of Sciences, which made the cooperation between the Soviet Academy of Sciences and universities and the industrial sector closer. The academic atmosphere that Soviet government and people all respected the scientists who had won the highest academic titles of academicians or corresponding academicians promoted the continuous development of science.[25]

Through the visit to the Soviet Union, Zhang Jiafu "deeply felt that China's scientific foundation is far weaker than that of the Soviet Union".[26] On July 21, after deliberation with various parties, especially with President Kuo Mo-jo and several vice presidents, Zhang jiafu put forward *Opinions on Future Scientific Work* at the executive meeting of the Academy of Sciences. The *Opinions* suggested "the establishment of Academic Divisions to improve academic leadership and expand academic leadership institution". There would be four divisions, "basic science, technical science, biological science and social science". At the same time, "adjust the organization and leadership of the Central Office of Administration appropriately. Under the current circumstances, it is generally believed that it is still difficult to set up an academician system or a full-scaled academic committee. Therefore, there will be no major changes in the organization of Central Office of Administration. The

[23]Alexander Vucinich, *The Soviet Academy of Sciences*. Stanford: Stanford University Press, 1956: 25–26.

[24]Zhongkeyuan 1953 nian zhaokai dishiyi zhi sanshici yuanwu changwu huiyi jilu ji youguan wenjian中科院一九五三年召开第十一至三十次院务常务会议记录及有关文件(Minutes and Related Documents of the 11th to 30th Executive Meetings of the Chinese Academy of Sciences Held in 1953). Beijing: Zhongguo kexueyuan danganguan中国科学院档案馆(Archives of Chinese Academy of Sciences), 1953-02-003.

[25]Zhongguo kexueyuan guanyu fangsu daibiaotuan gongzuo de baogao中国科学院关于访苏代表团工作的报告(Report of the Chinese Academy of Sciences on the Work of the Delegation to Soviet Union). Kexue tongbao科学通报(Chinese Science Bulletin)1954, (4):12—14.

[26]Guanyu fangsu daibiaotuan xiang zhongyang de gongzuo baogao关于访苏代表团向中央的工作报告(Report to the Central Committee on the Work of the Delegation to the Soviet Union). Beijing: Zhongguo kexueyuan danganguan中国科学院档案馆(Archives of Chinese Academy of Sciences), 1953-02-026.

General Assembly of the Academy takes over everything and acts as the Presidium of the Soviet Academy of Sciences. Under it, there are one office, three bureaus and special committees."[27] It was the Party group of the CAS that played a substantial role in decision-making and operation. As the Party secretary of the Academy, Zhang jiafu's opinions were actually the basic decision. As a result, the CAS decided to establish the Academic Divisions and the division membership system, but not to establish an academician system for the time being.

As a matter of fact, some scientists advocated the establishment of academician system under the influence of Soviet experience after the Soviet delegation returned home. Such as Wu Heng recalled: "After the Soviet delegation returned to China in 1953, the idea of the Academy of Sciences as the country's highest academic institution was gradually accepted by the majority of scientists and leading cadres. Therefore, there were also discussions among scientists about the establishment of an academician system. However, shortly after the founding of New China, scientists' ideas had yet to be reformed. To regard the Academy of Sciences as the organization of academicians and the General Assembly of the CAS Members as the highest authority meant to break away from the leadership of the Party, and it was difficult to ensure progress in accordance with the socialist direction and meet the needs of socialist construction. Some even thought that scientists advocated the academician system was to "seize power" from the Party."[28] This also revealed the influence of political factors behind the decision of the CAS not to establish the academician system for the time being.

On November 19, 1953, the Party group of the CAS submitted to Chairman Mao Tse-tung and the Central Committee the *Report to the Central Committee on the Basic Situation of the Current Work of the Academy of Sciences and Its Future Tasks* (hereinafter referred to as the "*Report*"). In order to solve the problem of academic leadership and authority, the *Report* pointed out that the CAS would establish divisions based on Soviet experience to lead the work of various research institutes, and at the same time establish and enrich the academic leadership of the Academy, divisions and research institutes. Also, the *Report* proposed to set up a secretariat under the General Assembly, with one secretary-general, two deputy secretaries-general and several secretaries. It also proposed to establish four academic divisions: physics, mathematics and chemistry, biology and earth science, technological science, and social sciences. Each division consisted of the directors of its affiliated institutes and relevant experts inside and outside the Academy, forming a committee of 20 to 50 members. The Division Committee had a Standing Committee to handle daily work, while the Division only cared about academic leadership, not administrative affairs.[29]

---

[27]Guanyu fangsu daibiaotuan xiang zhongyang de gongzuo baogao关于访苏代表团向中央的工作报告(Report to the Central Committee on the Work of the Delegation to the Soviet Union). Beijing: Zhongguo kexueyuan danganguan中国科学院档案馆(Archives of Chinese Academy of Sciences), 1953-02-026.

[28]Wu Heng武衡(1992) . *Keji zhanxian 50 nian*科技战线五十年(50 Years of Science and Technology Front). Beijing: Kexue jishu wenxian chubanshe科学技术文献出版社.147.

[29]Zhongkeyuan dangzu guanyu muqian benyuan gongzuo jiben qingkuang he jinhou gongzuo renwu de baogao ji zhongyang pishi中科院党组关于目前本院工作基本情况和今后工作任务

The names of the Division of Basic Science and the Division of Biological Science proposed by Zhang Jiafu on July 21 were changed to the Division of Physics, Mathematics and Chemistry and the Division of Biology and Earth Sciences respectively in the report. On January 28, 1954, Kuo Mo-jo made a *Report on the Basic Situation and Future Tasks of the CAS* at the 204th Government Affairs Meeting of the State Council, which was approved at the same meeting. This report was a revised edition of the *Report*. The organization and leadership issues raised and the plans to establish the Academic Divisions and the secretariat were not fundamentally different from the *Report*. On March 8, 1954, the Central Government approved the *Report*.[30] This provided a legal basis for the establishment of the CAS Academic Divisions and its membership system.

On October 12, 1954, the Soviet general consultant and soil scientist Kovda (B. A. Ковда), who was hired by the CAS, came to work in China. Koffida expressed opposition to the decision of the CAS to establish the Academic Divisions and the member system instead of an academician system. In January 1955, he formally submitted to the Party group of the CAS a proposal *On Some Measures for Planning and Organizing the National Scientific Research Work of the People's Republic of China (draft)*. One of the proposals was to establish an academician system. He thought that without academicians, "it cannot be called an academy of sciences, but a joint administrative organization of various research institutes."[31]

In response, the Party group of the CAS put forward two plans to Premier Zhou Enlai, Vice Premier Chen Yi and the Central Committee on February 12 after studying. One was to prepare for the establishment of the academician system as soon as possible and to postpone the establishment of the Academic Divisions; the other was to still set up the Academic Divisions, and to use more time to prepare for

---

的报告及中央批示(Report of the Party Group of the Chinese Academy of Sciences on the Basic Situation of the Work of the Academy at Present and Its Future Tasks and Instructions from the Central Committee). Beijing: Zhongguo kexueyuan danganguan中国科学院档案馆(Archives of Chinese Academy of Sciences), 1954-01-001.

[30]Zhonggong zhongyang dui Zhongguo kexueyuan dangzu guanyu muqian kexueyuan gongzuo de jiben qingkuang he jinhou gongzuo renwu gei zhongyang de baogao de pishi中共中央对中国科学院党组《关于目前科学院工作的基本情况和今后工作任务给中央的报告》的批示(Instructions from the Central Committee of the CPC to the Party Group of the Chinese Academy of Sciences' Report to the Central Committee on the Basic Situation of the Work of the Academy at Present and Its Future Tasks). In Zhonggong zhongyang wenxian yanjiushi中共中央文献研究室(Literature Research Office of the Central Committee of the CPC) (Eds.) (2011). *Jianguo yilai zhongyao wenxian xuanbian*建国以来重要文献选编(Selected Important Documents Since the Founding of the People's Republic of China), Vol. 5. Beijing: Zhongyang wenxian chubanshe中央文献出版社.144–147.

[31]Guanyu sulian guwen Kovda dui zuzhi quanguoxing kexue yanjiu gongzuo de yijian ji yu Zhang fuyuanzhang tanhua jiyao关于苏联顾问柯夫达对组织全国性科学研究工作的意见及与张副院长谈话纪要(About Soviet Consultant Kovda's Opinions on Organizing National Scientific Research Work and Summary of His Conversation with Vice President Zhang). Beijing: Zhongguo kexueyuan danganguan中国科学院档案馆(Archives of Chinese Academy of Sciences), 1955-01-036.

maturity before establishing the academician system.[32] Kuo Mo-jo agreed with the establishment of an academician system. On March 19, when Chen Yi, Vice Premier of the State Council in charge of the work of the CAS, attended the first enlarged meeting of the Party group of the CAS, he also agreed, saying: "On the issue of the academician system, I fully support the opinions of Comrade Koffida and President Guo. This problem cannot be blamed on the Party group of the CAS. I also advocated postponing the implementation. At this meeting, you have persuaded me to implement it earlier, and now I can fully propose to the Central Committee on your side."[33] However, the CAS later decided to set up the Academic Divisions and the membership system first and postpone the establishment of the academician system.

7.2 Preparations and Establishment of Academic Divisions and the Membership System

Before the 204th government affairs meeting of the State Council on January 28, 1954, the CAS had already begun to build its academic secretariat and divisions. On January 30, the executive meeting of the CAS discussed and preliminarily approved the names of the secretary-general, deputy secretaries-general of the Academic Secretariat, the secretaries, the chief and deputy directors of the Academic Divisions. The secretary-general was Tsien San-Tsiang, the deputy secretaries-general were Qin Lisheng and Wu Heng, and the secretaries were Pai Sitsan, Tsao Pen-hsi, Ding Zan, Liu Danian, Liu Ta-kong, Wang Zhihua, Chien Wei-zang, Ye Du-zheng, Yap Chu-Phay and Zhang Wen-you. The director of the Division of Physics, Mathematics and Chemistry was Woo Y. H., and the deputy directors were Chen Kangbai and Hua Loo-Keng. The director of the Division of Biology and Earth Science was Chu Coching, and the deputy directors were Xu Jie and Chen Fengtong. The director of the Technological Division was Ny Tsai-ze and the deputy director was Wu Heng. Kuo Mo-jo was the director of the Social Science Division and Zhang Jiafu was the

[32] Guanyu sulian guwen Kovda dui zuzhi quanguoxing kexue yanjiu gongzuo de yijian ji yu Zhang fuyuanzhang tanhua jiyao关于苏联顾问柯夫达对组织全国性科学研究工作的意见及与张副院长谈话纪要(About Soviet Consultant Koffida's Opinions on Organizing National Scientific Research Work and Summary of His Conversation with Vice President Zhang). Beijing: Zhongguo kexueyuan danganguan中国科学院档案馆(Archives of Chinese Academy of Sciences), 1955-01-036.

[33] Chen Yi tongzhi zai kexueyuan dangzu diyici kuoda huiyi shang de zongjie (1953 nian 3 yue 19 ri) 陈毅同志在科学院党组第一次扩大会议上的总结(1955年3月19日) (Comrade Chen Yi's Summary at the First Enlarged Meeting of the Party Group of the Chinese Academy of Sciences, March 19th, 1955). In Wang Zhongjun王忠俊(Eds.) (1995). *Zhongguo kexueyuan shiliao huibian (1955)* 中国科学院史料汇编(1955年) (Compilation of Historical Materials of the Chinese Academy of Sciences, 1955). Beijing: Zhongguo kexueyuan yuanshi wenwu ziliao zhengji weiyuanhui bangongshi中国科学院院史文物资料征集委员会办公室. 58–59.

deputy director.[34] On April 8, the CAS held its first meeting of the Division directors and announced the establishment of the Academic Secretariat.[35] In that month, the CAS drafted a written opinion on the formation of the Academic Divisions, and immediately solicited the opinions of scientists in and out of Beijing. Scientists thought it was necessary to set up the Divisions.[36] Since then, the CAS began to prepare for the selection of the first batch of academic division members.

(1) **Selection Criteria, Selection Methods and Number of Members**

The selection criteria, selection methods and number of candidates were the main preparatory work for the selection of members of the Academic Division. The Academic Secretariat was responsible for the work. At 2:00 pm, May 7, 1954, Tsien San-Tsiang presided over a meeting of Academic secretaries to discuss relevant issues. Before the meeting, the CAS set up four preparatory groups according to the proposed divisions to prepare for the establishment of each division separately. Sun Keming, Deng Jiaxian, Yan Jichi and Liu Guiwu reported on the preparatory work for the selection of members of the divisions of biology and earth sciences, physics, mathematics and chemistry, technological science, and social sciences.[37]

Judging from the reports of each group, the criteria for selecting the members of the division proposed by the three groups in the field of natural science were roughly the same, and there were three criteria in both academic and political aspects: (1) having more important academic achievements or contributions; (2) having a driving effect on the discipline or on the academic level; (3) loyal to the people's cause. The criteria for selecting candidates drawn up by the social science group also included academic and political aspects, with a total of 3 articles: (1) "academic level—having certain achievements in the discipline"; (2) "Solving Problems with Marxism-Leninism"; (3) "Taking due care of representatives from all walks of life".[38] However, the social science group did not put forward the academic standard of promoting the subject or

---

[34]Zhongkeyuan zhaokai diyi zhi shisi ci yuanwu changwu huiyi de tongzhi jiyao ji youguan cailiao中科院召开第1—14次院务常务会议的通知、纪要及有关材料(Notices, Summaries and Related Materials of the 1st–14th Executive Meetings of the Chinese Academy of Sciences). Beijing: Zhongguo kexueyuan danganguan中国科学院档案馆(Archives of Chinese Academy of Sciences), 1954-02-003.

[35]Chu Coching竺可桢(2007). *Chu Coching quanji*竺可桢全集(The Complete Works of Coching Chu), Vol. 13. Shanghai: Shanghai keji jiaoyu chubanshe上海科技教育出版社.412.

[36]Zhongkeyuan guanyu tuixuan xuebu weiyuan jingguo qingkuang xiang zhongyang de baogao 中科院关于推选学部委员经过情况向中央的报告(Report of the Chinese Academy of Sciences to the Central Committee on the Selection of Academic Division Members). Beijing: Zhongguo kexueyuan danganguan中国科学院档案馆(Archives of Chinese Academy of Sciences), 1955-01-005.

[37]Zhongkeyuan xueshu mishuchu 1954 nian chuwu huiyi jiyao gongzuo jihua (yi zhi ershijiu ci) 中科院学术秘书处一九五四年处务会议纪要、工作计划(一至二十九次) (Office Meeting Summaries and Work Plans (1st–29th) of the Academic Secretariat of the Chinese Academy of Sciences in 1954). Beijing: Zhongguo kexueyuan danganguan中国科学院档案馆(Archives of Chinese Academy of Sciences), 1954-02-022.

[38]Zhongkeyuan xueshu mishuchu 1954 nian chuwu huiyi jiyao gongzuo jihua (yi zhi ershijiu ci) 中科院学术秘书处一九五四年处务会议纪要、工作计划(一至二十九次) (Office Meeting

academic ability, and emphasized "solving problems with Marxism-leninism" on the political standard of candidates. In the course of discussion, the participants "thought that there should be" the criterion of "loyalty to the cause of the people", "but there was no clear conclusion as to whether it should be open or not and whether there would be any reaction."[39]

Regarding the method of selection, no consensus was reached at the meeting, but there were roughly two opinions: (1) the initial list of academic division members was first decided in principle by directors and some others at the meeting, and then the minority was consulted, during which the number of people on the list increased by 50%; (2) the number of people on the list would not increase by 50%. Regarding the number of academic division members, the Division of Physics, Mathematics and Chemistry initially set at 30 to 40, with about 35 members. In the field of biology and earth sciences, 30 to 35 people in biology and 15 to 20 people in earth sciences were initially determined. The Division of Technological Science initially designated 40 to 50 people. In the social sciences, 40 to 45 people were initially selected. At this meeting, the Academic Secretariat only changed the initial quota of the preparatory committee of the biology and earth science division, increasing the number of biology and earth science to 35–40 and 20–25 respectively.[40]

On May 11, the CAS held an academic division symposium to further discuss the selection criteria, selection methods and quota of academic division members. Woo Y. H., Chu Coching, Ny Tsai-ze and Liu Danian respectively reported on the relevant preparations of the Physics, Mathematics and Chemistry Group, Biology and Earth Science Group, Technological Science Group and Social Science Group. Regarding the selection criteria, according to Chu Coching's diary, "Each group has put forward three items, namely, offering contributions to the academic discipline (specialized in), promoting the discipline (past or present) and being loyal to the people's cause."[41] It showed that the Social Science Group had changed the selection criteria to make them consistent with that raised by the groups in the natural science field. When

Summaries and Work Plans (1st–29th) of the Academic Secretariat of the Chinese Academy of Sciences in 1954). Beijing: Zhongguo kexueyuan danganguan中国科学院档案馆(Archives of Chinese Academy of Sciences), 1954-02-022.

[39]Zhongkeyuan xueshu mishuchu 1954 nian chuwu huiyi jiyao gongzuo jihua (yi zhi ershijiu ci) 中科院学术秘书处一九五四年处务会议纪要、工作计划(一至二十九次) (Office Meeting Summaries and Work Plans (1st–29th) of the Academic Secretariat of the Chinese Academy of Sciences in 1954). Beijing: Zhongguo kexueyuan danganguan中国科学院档案馆(Archives of Chinese Academy of Sciences), 1954-02-022.

[40]Zhongkeyuan xueshu mishuchu 1954 nian chuwu huiyi jiyao gongzuo jihua (yi zhi ershijiu ci) 中科院学术秘书处一九五四年处务会议纪要、工作计划(一至二十九次) (Office Meeting Summaries and Work Plans (1st–29th) of the Academic Secretariat of the Chinese Academy of Sciences in 1954). Beijing: Zhongguo kexueyuan danganguan中国科学院档案馆(Archives of Chinese Academy of Sciences), 1954-02-022.

[41]Chu Coching竺可桢(2007). *Chu Coching quanji*竺可桢全集(The Complete Works of Coching Chu), Vol. 13. Shanghai: Shanghai keji jiaoyu chubanshe上海科技教育出版社.429.

discussing the number of candidates, Kuo Mo-jo advocated "better to be short than to be shoddy", which was unanimously supported by the participants.[42]

After the meeting, physics, mathematics and chemistry group, technological science group, biology and earth sciences with the group divided held symposiums. At 9 am on May 28, Tsien San-Tsiang presided over a meeting of academic secretaries, asking each group to report on the preparations for the establishment of the division. In the report, each group set different criteria for the selection of academic division members, but both included academic and political aspects. The selection methods of academic division members were also different, but they had decided that the final list would be determined by the CAS. At the end of the meeting, Tsien San-Tsiang pointed out according to the reports at the meeting: "There can be inconsistencies on how to proceed with the next work because it is still a brewing work, and the way of doing it can be different. But the last step is to draw up a list to be determined by the Academy, and everyone is in agreement." At the same time, he proposed to the selection of academic division members: the social science division should consider it individually, the technology division should reconsider it, and the biology and earth science division and the physics, mathematics and chemistry division should discuss it jointly.[43]

The academic secretariat also held an executive meeting on June 19 and decided to invite experts from outside the Academy to recommend academic division members of natural science in the following 11 areas temporarily: earth science, zoology and botany, medicine and basic medicine, agriculture and forestry, mining and metallurgy, chemistry and chemical engineering, civil engineering (including geodesy) and water conservancy, machinery, electric machinery, mathematics and mechanics, and physics. At the same time, the Academic Secretariat required that when inviting experts outside the Academy who would recommend members of the Academic Division, attention should be paid to the balance of the number of experts in various disciplines and units.[44] This made further preparations for the selection of the first group of academic division members of the CAS.

(2) Initial Selection of Academic Division Members

After the above deliberation, the CAS decided to invite some representative experts to recommend academic division members in their specialized subjects of Natural

[42]Chu Coching竺可桢(2007). *Chu Coching quanji*竺可桢全集(The Complete Works of Coching Chu), Vol. 13. Shanghai: Shanghai keji jiaoyu chubanshe上海科技教育出版社.429.

[43]Zhongkeyuan xueshu mishuchu 1954 nian chuwu huiyi jiyao gongzuo jihua (yi zhi ershijiu ci) 中科院学术秘书处一九五四年处务会议纪要、工作计划(一至二十九次) (Office Meeting Summaries and Work Plans (1st–29th) of the Academic Secretariat of the Chinese Academy of Sciences in 1954). Beijing: Zhongguo kexueyuan danganguan中国科学院档案馆(Archives of Chinese Academy of Sciences), 1954-02-022.

[44]Zhongkeyuan xueshu mishuchu 1954 nian chuwu huiyi jiyao gongzuo jihua (yi zhi ershijiu ci) 中科院学术秘书处一九五四年处务会议纪要、工作计划(一至二十九次) (Office Meeting Summaries and Work Plans (1st–29th) of the Academic Secretariat of the Chinese Academy of Sciences in 1954). Beijing: Zhongguo kexueyuan danganguan中国科学院档案馆(Archives of Chinese Academy of Sciences), 1954-02-022.

Science. At the beginning of July 1954, 645 letters were sent to experts in the name of President Kuo Mo-jo, inviting them to recommend members in the field of Natural Science. These experts included special members of the CAS, standing members of All-China Federation of Specialized Societies in Natural Science, chairpersons of various specialized societies and directors engaged in research work, department heads and professors with research achievements in national institutions of higher learning, major scientific and technical personnel in industrial departments, and major researchers of the CAS.[45] The letter explained the criteria for selecting members of academic divisions:

> Having studied the opinions of many scientists at the symposium and in writing, we think: in order for the academic divisions to fully embody its academic leadership function, the criteria for selecting members of academic divisions should not be too wide. First, academic achievements should be taken as the main basis for consideration, and second, the role in promoting the scientific cause in our country should also be considered. Due to the weak scientific foundation and unbalanced development in our country, it is difficult to require various aspects (such as disciplines, regions, etc.) to be taken care of under the current situation. In the future, there will be various specialized academic organizations within the scope of academic divisions to invite experts from various fields to participate.[46]

The two criteria for selection were both academic and basically consistent with the first two proposed by each group at the symposium of the CAS on May 11. The letter did not specify the criterion of "loyalty to the cause of the people", but judging from the selection later, it did serve as a political criterion.

By November 22, the CAS had received 527 replies, with 665 people recommended. Among the recommended, the candidates in physics, mathematics, chemistry, geology, zoology and botany were relatively concentrated, while the candidates in agronomy, forestry, soil science, medicine and technological science were relatively scattered. In determining the initial candidates, the CAS formulated the selection method:

> In principle, the selection criterion for various disciplines with relatively concentrated recommendations is those with more than one-fourth of the total number of recommended votes. At the same time, the following points are also noted: (1) If the business of the discipline is relatively developing better and there are more talents, more members in this field will be selected. For those disciplines with low professional level, no suitable candidates, and which must be developed, the better ones in the discipline should be selected to participate in the

[45] Zhongkeyuan guanyu tuixuan xuebu weiyuan jingguo qingkuang xiang zhongyang de baogao 中科院关于推选学部委员经过情况向中央的报告(Report of the Chinese Academy of Sciences to the Central Committee on the Selection of Academic Division Members). Beijing: Zhongguo kexueyuan danganguan中国科学院档案馆(Archives of Chinese Academy of Sciences), 1955-01-005.

[46] Zhongkeyuan zhaokai dishiwu zhi ershijiu ci yuanwu changwu huiyi de tongzhi jiyao ji youguan cailiao中科院召开第15–29次院务常务会议的通知、纪要及有关材料(Notices, Summaries and Related Materials of the 15th–29th Executive Meetings of the Chinese Academy of Sciences). Beijing: Zhongguo kexueyuan danganguan中国科学院档案馆(Archives of Chinese Academy of Sciences), 1954-02-004.

> Academic Division. (2) To add a few Party members and cadres who do scientific organization work; (3) Those who are politically suspected and are generally dissatisfied with in the scientific community will not be included in the list despite their academic achievements.[47]

By the end of November, the CAS had identified 133 initial candidates in the field of natural science. Among them, 33 were from the Division of Physics, Mathematics and Chemistry, 65 from the Division of Biology and Earth Science, and 35 from the Division of Technology. In terms of mathematics, mechanics and astronomy, Hua Loo-Keng, Su Buchin, Chen Kien Kwong, Chou Pei-yuan, Hsu Pao-Lu, Tuan Hsio-fu and Kiang Tsai-han all met the selection criteria.[48] As Li Guoping, a professor of mathematics department at National Wu-han University, had made achievements in differential equations and complex functions, "the subject of differential equations is relatively weak in our country and needs to be developed, and his votes are close to the selection criteria, so it is still considered for inclusion". Chang Yu-Che, director of Zijinshan Observatory of the CAS, was "not high in academic level, but since astronomy is an important science and six people in astronomy have recommended him, he is considered to be put in the list."[49]

In physics, Tsien San-Tsiang, Chao Chung-Yoa, Peng Huan-wu, Wang Ganchang, Woo Y. H., Yeh Chi-Sun, Wang Jwu-Shi, Ny Tsai-ze, Yao Y. T., Yu Ruihuang, Ge Tingsui and Shih J. W. met the selection criteria. Lu S. S., acting director of the Institute of Applied Physics of the CAS, was still listed as he was "close to being elected" and "presiding over all achievements in applied physics, and having played a certain role in promoting the development of physics".[50]

In chemistry, Chuang Chang-Kong, Huang T. C., Woo Sho-Chow, Tseng Chao-lun, Fu Ying, Chang Tsinglien, Huang Minlon, Chou T. Q., Yang Shixian, Tang Au-chin and Liang Shuquan met the selection criteria. Woo Sho-Chow, director of

---

[47]Zhongkeyuan guanyu tuixuan xuebu weiyuan jingguo qingkuang xiang zhongyang de baogao 中科院关于推选学部委员经过情况向中央的报告(Report of the Chinese Academy of Sciences to the Central Committee on the Selection of Academic Division Members). Beijing: Zhongguo kexueyuan danganguan中国科学院档案馆(Archives of Chinese Academy of Sciences), 1955-01-005.

[48]Zhongkeyuan youguan dui xuebu weiyuan gongzuo de yijian ji xuebu weiyuan mingdan zhuanchang jianli he dui mingdan de yijian中科院有关对学部委员工作的意见及学部委员名单、专长、简历和对名单的意见(The Opinions of the Chinese Academy of Sciences on the Work of Academic Division Members and the Lists of Academic Division Members and Their Specialities and Curriculum Vitae, and Opinions on the Lists). Beijing: Zhongguo kexueyuan danganguan中国科学院档案馆(Archives of Chinese Academy of Sciences), 1955-02-036. The list of those who met the selection criteria in the following are all referred to this archive.

[49]Zhongkeyuan guanyu tuixuan xuebu weiyuan jingguo qingkuang xiang zhongyang de baogao 中科院关于推选学部委员经过情况向中央的报告(Report of the Chinese Academy of Sciences to the Central Committee on the Selection of Academic Division Members). Beijing: Zhongguo kexueyuan danganguan中国科学院档案馆(Archives of Chinese Academy of Sciences), 1955-01-005.

[50]Zhongkeyuan guanyu tuixuan xuebu weiyuan jingguo qingkuang xiang zhongyang de baogao 中科院关于推选学部委员经过情况向中央的报告(Report of the Chinese Academy of Sciences to the Central Committee on the Selection of Academic Division Members). Beijing: Zhongguo kexueyuan danganguan中国科学院档案馆(Archives of Chinese Academy of Sciences), 1955-01-005.

the Institute of Applied Chemistry of the CAS, got more votes, but should belong to the technological division, so he was included in that division. Huang Minlon was not included in the list because of "complicated social relations", although he "won a lot of votes and made high academic achievements". Lu Jia-xi, Dean of Xiamen University was listed for "getting close to the selection criteria, being more politically progressive and capable of working". Wang You, deputy director of the Institute of Organic Chemistry of the CAS, was also included in the list after being "specially recommended" by Chuang Chang-Kong, director of the Institute, although he "got fewer votes" but "made some achievements in antibiotic work, especially in the stereochemical structure of citrinin after liberation".[51]

In zoology of the Division of Biology and Earth Sciences, Ping Chi, Pai Sitsan, Zhu Xi, Chen Shisan C., Wu Hsien-Wen, Tung Ti-Cho, Hu Jingfu, Wang Chia-Chi, Chen Shixiang and Liu Chung-lo met the selection criteria. Liu C. C., President of Sichuan Medical College, "though got fewer votes", was included in the list because "he is the better of the only two experts in amphibian research in China and is a more insightful person in the field of zoology", and his being included "can appropriately increase the proportion of academic division members of zoology outside the Academy".[52]

In botany, fewer people were recommended concentratedly than in zoology. In order to strike a balance between the number of members in botany and zoology, five members were included, namely, Teng S. C., Yin Hung-chang, Wu Zheng-yi, Zhang Zhaoqian and Li Tsi-tung. And Hu Hsen-hsu, who was "politically very bad before liberation", but "senior in botany and of relatively higher academic level" was also listed.[53]

In geology and paleontology, Huang T. K., Hsieh Chia-yung, Lee J. S., Cheng Y. C., Zhang Wen-you, Yin T. H., Sun Y. C., Young Chung-Chien, Hou Teh-feng, Meng H. M., Si Xingjian, Xu Jie, Tien C. C. came in first.[54] Yu T. Y., Vice-president of

---

[51]Zhongkeyuan guanyu tuixuan xuebu weiyuan jingguo qingkuang xiang zhongyang de baogao 中科院关于推选学部委员经过情况向中央的报告(Report of the Chinese Academy of Sciences to the Central Committee on the Selection of Academic Division Members). Beijing: Zhongguo kexueyuan danganguan中国科学院档案馆(Archives of Chinese Academy of Sciences), 1955-01-005.

[52]Zhongkeyuan guanyu tuixuan xuebu weiyuan jingguo qingkuang xiang zhongyang de baogao 中科院关于推选学部委员经过情况向中央的报告(Report of the Chinese Academy of Sciences to the Central Committee on the Selection of Academic Division Members). Beijing: Zhongguo kexueyuan danganguan中国科学院档案馆(Archives of Chinese Academy of Sciences), 1955-01-005.

[53]Zhongkeyuan guanyu tuixuan xuebu weiyuan jingguo qingkuang xiang zhongyang de baogao 中科院关于推选学部委员经过情况向中央的报告(Report of the Chinese Academy of Sciences to the Central Committee on the Selection of Academic Division Members). Beijing: Zhongguo kexueyuan danganguan中国科学院档案馆(Archives of Chinese Academy of Sciences), 1955-01-005.

[54]Zhongkeyuan youguan dui xuebu weiyuan gongzuo de yijian ji xuebu weiyuan mingdan zhuanchang jianli he dui mingdan de yijian中科院有关对学部委员工作的意见及学部委员名单、专长、简历和对名单的意见(The Opinions of the Chinese Academy of Sciences on the Work of Academic Division Members and the Lists of Academic Division Members and Their Specialities and Curriculum Vitae, and Opinions on the Lists). Beijing: Zhongguo kexueyuan danganguan中国科学院档案馆(Archives of Chinese Academy of Sciences),1955-02-036.

Northeast Institute of Geology, was also included in the list as "although he received fewer votes, Lee J. S. and Xu Jie both specially recommended him for his achievements in non-ferrous metals". Ding Daoheng, discoverer of Baotou Iron Mine, was also included because "the Ministry of Higher Education specially recommended him".[55]

In geophysics, Jaw Jeou-Jang, Tu Chang-wang, Chu Coching, Fu C. Y. and Lee S. P. met the selection criteria. Gu Gong-xu won "a relatively small number of votes", but after liberation he "has done a lot of work in geophysical prospecting and played a certain role in the exploration of geological resources in our country" and was still included. Although Fu C. Y. and Lee S. P. obtained more votes than Gu Gong-xu, they were not included.

Only Huang Bingwei, acting director of the Institute of Geography of the CAS, was included in the initial selection for geography. The recommended results in agronomy and forestry were not very concentrated. After discussing with the Ministry of Agriculture and the Ministry of Agriculture and Forestry, the CAS decided to include Jin Shanbao, Liang Xi, Ding Ying and Dai Songen.[56]

In the field of soil science, "there are no outstanding talents and it is not easy to choose", only Hou Guangjiong, professor of Southwest Agricultural College, and Li Lianjie, professor of Beijing Agricultural University were included, because Hou was "a senior scientific worker in the soil field and has made some achievements in the investigation of soil in our country", and Li "participated in the work of the Tibet team and made some contributions."[57]

In terms of medicine and its basic branches, except for 3 physiologists, Feng Teh-Pei, Zhang Xijun and Tsai Chiao, all the candidates were agreed upon by the CAS and the Party group of the Ministry of Health.[58] Western medicine chose Chung H. L. In traditional Chinese medicine, the Ministry of Health recommended Xiao Longyou and Lu Yuanlei. In addition, Shen Qizhen, president of the Central Health

---

[55] Zhongkeyuan guanyu tuixuan xuebu weiyuan jingguo qingkuang xiang zhongyang de baogao 中科院关于推选学部委员经过情况向中央的报告(Report of the Chinese Academy of Sciences to the Central Committee on the Selection of Academic Division Members). Beijing: Zhongguo kexueyuan danganguan中国科学院档案馆(Archives of Chinese Academy of Sciences), 1955-01-005.

[56] Zhongkeyuan guanyu tuixuan xuebu weiyuan jingguo qingkuang xiang zhongyang de baogao 中科院关于推选学部委员经过情况向中央的报告(Report of the Chinese Academy of Sciences to the Central Committee on the Selection of Academic Division Members). Beijing: Zhongguo kexueyuan danganguan中国科学院档案馆(Archives of Chinese Academy of Sciences), 1955-01-005.

[57] Zhongkeyuan guanyu tuixuan xuebu weiyuan jingguo qingkuang xiang zhongyang de baogao 中科院关于推选学部委员经过情况向中央的报告(Report of the Chinese Academy of Sciences to the Central Committee on the Selection of Academic Division Members). Beijing: Zhongguo kexueyuan danganguan中国科学院档案馆(Archives of Chinese Academy of Sciences), 1955-01-005.

[58] Zhongkeyuan guanyu tuixuan xuebu weiyuan jingguo qingkuang xiang zhongyang de baogao 中科院关于推选学部委员经过情况向中央的报告(Report of the Chinese Academy of Sciences to the Central Committee on the Selection of Academic Division Members). Beijing: Zhongguo kexueyuan danganguan中国科学院档案馆(Archives of Chinese Academy of Sciences), 1955-01-005.

Research Institute, was listed, which was related to "strengthening the ties between the Central Health Research Institute and the Academy of Sciences".[59]

In terms of technological science, the recommended votes were scattered. Of the 35 people initially selected, only Tsai Fang-yin and Hou T. P. were recommended relatively concentrated. The other nominees were mainly selected by the CAS in consultation with relevant business departments.[60]

"Because the selection criteria are not easy to master", the initial selection of members of the Division of Social Sciences did not go through the process of recommendation. Instead, the Party group of the CAS first put forward a preliminary list of members in history and archaeology, economics, philosophy, literature and linguistics. After studying with the Science Section of the Publicity Department of the CPC Central Committee, they respectively solicited opinions from relevant experts in Beijing, and finally 44 people were selected.[61] Experts consulted included Yang Xianzhen, Ai Siqi, Zhang Ruxin, Tang Yung-tung, Wang Li, Luo Changpei, Ding Shengshu, Lü Shuxiang, Ji Xian-lin and Di Chaobai[62]. In determining the candidates, the Party group of the CAS was hesitant to include Chen Yin-k'o, because Chen was an authority in studying the history and history of the Sui, Tang and Five Dynasties, his non-election would cause controversy, but he had repeatedly stated that he did not believe in Marxism. After Chairman Mao Tse-tung instructed for him "to be elected", the Party group of the Academy decided to include him.[63] Ma Yin-chu was

[59]Zhongkeyuan guanyu tuixuan xuebu weiyuan jingguo qingkuang xiang zhongyang de baogao 中科院关于推选学部委员经过情况向中央的报告(Report of the Chinese Academy of Sciences to the Central Committee on the Selection of Academic Division Members). Beijing: Zhongguo kexueyuan danganguan中国科学院档案馆(Archives of Chinese Academy of Sciences), 1955-01-005.

[60]Zhongkeyuan guanyu tuixuan xuebu weiyuan jingguo qingkuang xiang zhongyang de baogao 中科院关于推选学部委员经过情况向中央的报告(Report of the Chinese Academy of Sciences to the Central Committee on the Selection of Academic Division Members). Beijing: Zhongguo kexueyuan danganguan中国科学院档案馆(Archives of Chinese Academy of Sciences), 1955-01-005.

[61]Zhongkeyuan guanyu tuixuan xuebu weiyuan jingguo qingkuang xiang zhongyang de baogao 中科院关于推选学部委员经过情况向中央的报告(Report of the Chinese Academy of Sciences to the Central Committee on the Selection of Academic Division Members). Beijing: Zhongguo kexueyuan danganguan中国科学院档案馆(Archives of Chinese Academy of Sciences), 1955-01-005.

[62]Liu Danian刘大年(1991). Xuebu de choujian jiqi lishi zuoyue学部的筹建及其历史作用(The Preparations and Establishment of Academic Divisions and Its Historical Role). *Yuanshi ziliao yu yanjiu*院史资料与研究(Historical Materials and Research of the Chinese Academy of Sciences) (6): 39–40; Liu Lu刘潞(1998). Liu Danian yi Kuo Mo-jo刘大年忆郭沫若(Liu Danian recalling Kuo Mo-jo). *Bainian chao*百年潮(Hundred Years Tide) (4): 61–63.

[63]Zhang Jiafu张稼夫(1984). *Gengshen yishi*庚申忆逝(Recalling the Passed in 1980). Taiyuan: Shanxi renmin chubanshe山西人民出版社.131.

chosen "for the sake of the United Front relationship, and also for the cooperation between the Chinese Academy of Sciences and Peking University".[64]

(3) **Addition and Deletion of the Initial List**

At the end of November 1954, the CAS submitted to the Central Committee the "Report on the Selection Process of Academic Division Members and the List of Members". Since then, some colleges and universities, business divisions, provincial and municipal Party committees and publicity departments were consulted on the initial list of 177 academic division members. As for institutions of higher learning, Beijing Agricultural University proposed to add Xiong Daren, Lou Chenghou, Tang P. S., Zhou Mingzang and Huang Ruilun, and recommended Lin Chuanguang as a reference. Beijing Medical College did not approve of Liu Sichi, believing that his political thoughts were backward and his attitude towards the Party was antagonistic. Peking University believed that the Party's strength was relatively weak among the academic division members in the field of natural science, and some divisions did not take good care of scientists from the South, and put forward some specific suggestions, such as adding Li Ju-chi, Tang P. S., Le Tianyu, Zhao Yibing, Wang Li, Wei Jiangong and You Guo-en, replacing Ding Zan with Pan Shu and not considering Feng You-lan. Beijing Normal University suggested that Zhang Ruxin should not be an academic division member. Renmin University of China believed that Chin Yuch-Lin, Feng You-lan and Tang Yung-tung were not suitable to be the members for they did not meet the requirements of "the standard of excellent scientists", and proposed to add Shang Yue, He Ganzhi and Song Tao.

In terms of business departments, the Party group of the First Ministry of Machinery Industry proposed to add Tao Hengxian, an engineer of the Ministry. The Party group of the Second Ministry of Machinery Industry proposed to add Liu ding, an engineer of the Ministry, and Qian Zhidao, a technical director of the Ministry. The Ministry of Construction and Engineering proposed the addition of Liu Dunzhen, a professor at Nanjing Institute of Technology, and Qian Zhengying, an expert in water conservancy engineering. The Ministry of Agriculture proposed adding 4 more people, Zhou Shilu, Tu Zhi, Wang Shou and Zhong Jinglin.

In terms of provincial and municipal Party committees or publicity departments, the Beijing Municipal Party Committee publicity department thought that Zhang Guangdou had some skills in water conservancy and suggested that he be included. It also suggested that Chu F. T., Lin Qiaozhi and Huang Kun be added. However, it thought that Xie Shaowen on the list "has complicated historical, ideological and social relations". The publicity department of Nanjing Municipal Party Committee proposed to add Liu Dunzhen, Fang Guangtao, Lee S. P., Zheng Baojun, Li F. H., Zhang Jiangshu and Geng Yili, famous figures in Nanjing scientific circle. The

[64]Zhongkeyuan guanyu tuixuan xuebu weiyuan jingguo qingkuang xiang zhongyang de baogao 中科院关于推选学部委员经过情况向中央的报告(Report of the Chinese Academy of Sciences to the Central Committee on the Selection of Academic Division Members). Beijing: Zhongguo kexueyuan danganguan中国科学院档案馆(Archives of Chinese Academy of Sciences), 1955-01-005.

publicity department of Shaanxi Provincial Party Committee suggested adding Yu Hongzheng. The Sichuan Provincial Party Committee publicity department proposed to take Yang Yunkui, Yang Kaiqu, Xu Zhongshu, Ko Chao, Liu Weitao and Huang Kewei as reference candidates. Liaoning Provincial Party Committee publicity department believed it should be cautious to select Shao Xianghua as a member of the Academic Division and pointed out that he had a history that was unclear and the cause of several accidents with him was not clear either.[65]

After repeated weighing, the CAS adopted some opinions, added and deleted the members in the initial list, and then made a second list. The number of candidates on this list increased to 220, and "political, legal and educational" candidates were added separately to the members of the Division of Social Sciences, namely Chou Keng-Shang, Hu Xikui, Huang Songling, Zhang Zhirang, Yang Xiufeng, Jiang Nanxiang and Tsien T. S.[66] Later, members of these subjects were not listed separately, and except Huang Songling who was added to the "economics", the other 6 were deleted.

On March 15, 1955, the CAS established the Preparatory Committees for the divisions of Physics, Mathematics and Chemistry, Biology and Earth Sciences, Technological Science and Social Sciences. Before the divisions were formally established, "each Preparatory Committee will temporarily take over the division's functions and powers and begin the work of establishing the division."[67] Director of the Preparatory Committee of the Division of Physics, Mathematics and Chemistry was Woo Y. H., deputy directors were Chuang Chang-Kong, Hua Loo-Keng and Yun Ziqiang, academic secretary was Yun Ziqiang (concurrently), deputy academic secretary was Deng Jiaxian, and the committee consisted of 16 members. Director of the Preparatory Committee of the Biology and Earth Science Division was Chu Coching, deputy directors were Xu Jie and Chen Fengtong, academic secretary was Guo Xingxian, and deputy academic secretary was Shi Yafeng, with a total of 19 members. The head of the Preparatory Committee of the Technological Science Division was Ny Tsai-ze, the deputy heads were Mao T. E. and Zhao Feike, the academic secretary was Zhao Feike (concurrently), the deputy academic secretary was Yang Liangui, and there were 17 members. Kuo Mo-jo was the director of the Preparatory Committee of the

---

[65]Zhongkeyuan youguan dui xuebu weiyuan gongzuo de yijian ji xuebu weiyuan mingdan zhuanchang jianli he dui mingdan de yijian中科院有关对学部委员工作的意见及学部委员名单、专长、简历和对名单的意见(The Opinions of the Chinese Academy of Sciences on the Work of Academic Division Members and the Lists of Academic Division Members and Their Specialities and Curriculum Vitae, and Opinions on the Lists). Beijing: Zhongguo kexueyuan danganguan中国科学院档案馆(Archives of Chinese Academy of Sciences),1955-02-036.

[66]Zhongkeyuan youguan dui xuebu weiyuan gongzuo de yijian ji xuebu weiyuan mingdan zhuanchang jianli he dui mingdan de yijian中科院有关对学部委员工作的意见及学部委员名单、专长、简历和对名单的意见(The Opinions of the Chinese Academy of Sciences on the Work of Academic Division Members and the Lists of Academic Division Members and Their Specialities and Curriculum Vitae, and Opinions on the Lists). Beijing: Zhongguo kexueyuan danganguan中国科学院档案馆(Archives of Chinese Academy of Sciences), 1955-02-036.

[67]Zhongkeyuan xuebu chengli dahui richeng ji canjia renyuan mingdan中科院学部成立大会日程及参加人员名单(The agenda of the Inaugural Assembly of the Academic Divisions of the Chinese Academy of Sciences and the List of Participants). Beijing: Zhongguo kexueyuan danganguan中国科学院档案馆(Archives of Chinese Academy of Sciences), 1955-02-017.

Division of Social Sciences, Pan Zinian was the deputy director, Liu Danian was the academic secretary, Liu Guiwu was the deputy academic secretary, and there were 19 members.[68] The preparatory committee of each division was not limited to the personnel of the CAS, but absorbed the personnel from outside the Academy.

In March 1955, according to the instructions of the Publicity Department of the CPC Central Committee and many discussions at the Preparatory Committee meetings and joint meetings of the various divisions, the CAS increased the number of academic division members to 249. The additional candidates included Chiang Li-Fu, Wang Xianghao, Wu Wen-Tsün, Huang Kun, Hu Ning, Sun Cheng-e, Yuan Hanqing, Cui Zhilan, Tsai Pang-hwa, Feng Zefang, Tang F. F., Chang Hsiao-chien, James K. Shen, Lin Qiaozhi, Yu C. C., Hsia Chien-pei, Sun Jingzhi, Yan Kai, Zhou Zhihong, Zhong Lin, Sun Dehe, Hou Hsiong-Lin, Maa Dah-you, Wu Xuelin, Jiang Zemin, Zhu Wuhua, Ma Heng, Deng Tuo, Yang Xiangkui, Zhang Xiruo, Qian Junrui, Wang Yanan, Ma Te, Ma Xulun, Mao Dun, etc.[69] After discussion by the Preparatory Committee of each division and the Party group of the Academy, the CAS passed a list of 238 academic division members at its executive meeting on April 7. The list deleted Chiang Li-Fu, Wu Wen-Tsün, Sun Cheng-e, Cui Zhilan, James K. Shen, Ma Xingyuan, Sun Jingzhi, Zhong Lin, Jiang Zemin, Ma Heng, Deng Tuo, Yang Xiangkui, Zhang Xiruo, Ma Te, etc., and added Ding Zan, Chi Yuoh-Fong, Yue Senxun, etc.[70]

(4) **Refinement of the Selection Criteria**

On April 27, 1955, the Politburo meeting of the Central Committee discussed the list of academic division members and the report of the Party group of the CAS on the establishment of Academic Divisions. Liu Shaoqi talked about the list of academic division members at the meeting, "It is necessary to be very careful, they must really be persons with academic status; the arrangement of Communist Party members must also be those with academic contributions, not those just relying on their qualifications and status. People sent by our Party to serve in scientific institutions cannot appear as scholars and must honestly serve science. Communist Party members cannot rely on the Party's qualifications to become academicians!"[71] This was the requirement that academic achievements should be taken as the criteria for selecting academic division members, and Party members were no exception.

---

[68]Zhongkeyuan xuebu chengli dahui richeng ji canjia renyuan mingdan中科院学部成立大会日程及参加人员名单(The agenda of the Inaugural Assembly of the Academic Divisions of the Chinese Academy of Sciences and the List of Participants). Beijing: Zhongguo kexueyuan danganguan中国科学院档案馆(Archives of Chinese Academy of Sciences), 1955-02-017.

[69]Chu Coching竺可桢(2008). *Chu Coching quanji*竺可桢全集(The Complete Works of Coching Chu), Vol. 14. Shanghai: Shanghai keji jiaoyu chubanshe上海科技教育出版社.45.

[70]Chu Coching竺可桢(2008). *Chu Coching quanji*竺可桢全集(The Complete Works of Coching Chu), Vol. 14. Shanghai: Shanghai keji jiaoyu chubanshe上海科技教育出版社.45, 51, 60.

[71]Yang Shangkun杨尚昆(2001). *Yang Shangkun riji* (I) 杨尚昆日记(上册) (Yang Shangkun's Diary), Vol. I. Beijing: Zhongyang wenxian chubanshe中央文献出版社.199.

After the meeting of the Political Bureau of the Central Committee, the CAS reviewed and revised the list of academic division members again, further clarifying the general principle of the selection criteria: "Only those who have a high academic level, have a high reputation in the discipline and are not politically suspected of existing counter-revolutionaries can serve as academic division members."[72] At the same time, taking into account "the specific circumstances, such as China's current scientific foundation is still very weak, the academic level is generally not high, the development of various disciplines is uneven, and the political situation among the old scientists is quite complicated", the CAS put forward 7 principles based on the spirit of the general principle and the status quo of the Chinese scientific community:

(1) Those with high academic level but serious political suspicion of the current counter-revolution should not be listed as academic division members. (2) Those who have a relatively high academic level and no current suspicious circumstances but are dissatisfied by the scientific community due to serious evil deeds in history are not suitable to be listed as academic division members. (3) Those with general academic level and low prestige in the discipline are not suitable to be listed as members of the Academic Division. (4) Those with high academic level and some suspicious political circumstances, but without proper reasons to explain to the scientific community publicly or because of the needs of national construction, who are currently holding important positions in state organs, enterprises, factories or mines, and cannot be dispensed with, should still be listed as academic division members. (5) Those who have a not high academic level and no suspicion in politics, but must be properly cared for due to the shortage of talents in their disciplines should still be included as members of the Academic Division. (6) In order to take care of the old scientific tradition, representative figures of the old scientific community should also be listed as members of the Academic Division, although some of them are not high in academic level, and some hold outdated or even reactionary academic views. (7) Due to the needs of work, Party members who come to various academic divisions to engage in academic organizational work should also be listed as members of the Academic Division, although their academic level is not high or they know little about the subject.[73]

According to the first three principles, 15 of the 238 people were deleted: Bao Guobao, Zhao Zongyu, Tsai Chin-tao, Huang Minlon, Hu Jingfu and Xie Shaowen (according to Principle 1); Chen Rong and Liu Sichi (according to Principle 2); Wei Jiangong, Ding Zan, Liu Hui-xian, Liu Ding, Ji Chaoding, Lu Zhijun and Chen

---

[72]Zhongkeyuan guanyu tuixuan xuebu weiyuan jingguo qingkuang xiang zhongyang de baogao 中科院关于推选学部委员经过情况向中央的报告(Report of the Chinese Academy of Sciences to the Central Committee on the Selection of Academic Division Members). Beijing: Zhongguo kexueyuan danganguan中国科学院档案馆(Archives of Chinese Academy of Sciences), 1955-01-005.

[73]Zhongkeyuan guanyu tuixuan xuebu weiyuan jingguo qingkuang xiang zhongyang de baogao 中科院关于推选学部委员经过情况向中央的报告(Report of the Chinese Academy of Sciences to the Central Committee on the Selection of Academic Division Members). Beijing: Zhongguo kexueyuan danganguan中国科学院档案馆(Archives of Chinese Academy of Sciences), 1955-01-005.

Kangbai [74] (according to Principle 3). According to the following principles 4 to 7, there were still 23 persons retained or included: Wang Li, Huang T. K., Cheng Y. C., Hsieh Chia-yung, Tang F. F., Chow Jen, Hu Hsen-hsu and Tang P. S. (according to Principle 4); Chang Yu-Che, Liang Shuquan, Chang Tsinglien, Huang Bingwei (according to Principle 5); Tao L. K., Ma Yin-chu, Chin Yuch-Lin, Feng You-lan (according to Principle 6); Zhang Jiafu, Yun Ziqiang, Wu Heng, Zhao Feike, Liu Danian, Li Yanong, Li Wencai (according to Principle 7).[75]

According to the statement of the Party group of the CAS, the review and revision of the membership list of the Academic Divisions was "in accordance with the spirit of the Central Committee meeting".[76] However, Principle 7 actually contradicted Liu Shaoqi's instructions at the Politburo meeting of the Central Committee on April 27.

In addition, at the joint meeting of the heads of all specialized societies of the Scientific Federation, the Medical Association proposed that Wu Y. K., a thoracic surgery expert from the Union Medical College, had been hired as an honorary member of the Soviet Surgical Society and should be listed as a member of the Academic Division. At the meeting of the Political Bureau of the Central Committee, some leaders suggested that Zhou Zezhao, vice-president of the Beijing Hospital, should also be listed as an academic division member because of his high medical level and good achievements. The CAS agreed with these opinions.[77] As a result of this review and revision, the membership of the Academic Division was reduced from 238 to 224.

---

[74]Liu Ding, Ji Chaoding, Lu Zhijun and Chen Kangbai are all members of the CPC. The deletion of Chen Kangbai this time is also related to its transfer from the Chinese Academy of Sciences. See Zhongkeyuan guanyu tuixuan xuebu weiyuan jingguo qingkuang xiang zhongyang de baogao 中科院关于推选学部委员经过情况向中央的报告(Report of the Chinese Academy of Sciences to the Central Committee on the Selection of Academic Division Members). Beijing: Zhongguo kexueyuan danganguan中国科学院档案馆(Archives of Chinese Academy of Sciences), 1955-01-005.

[75]Zhongkeyuan guanyu tuixuan xuebu weiyuan jingguo qingkuang xiang zhongyang de baogao 中科院关于推选学部委员经过情况向中央的报告(Report of the Chinese Academy of Sciences to the Central Committee on the Selection of Academic Division Members). Beijing: Zhongguo kexueyuan danganguan中国科学院档案馆(Archives of Chinese Academy of Sciences), 1955-01-005.

[76]Zhongkeyuan guanyu tuixuan xuebu weiyuan jingguo qingkuang xiang zhongyang de baogao 中科院关于推选学部委员经过情况向中央的报告(Report of the Chinese Academy of Sciences to the Central Committee on the Selection of Academic Division Members). Beijing: Zhongguo kexueyuan danganguan中国科学院档案馆(Archives of Chinese Academy of Sciences), 1955-01-005.

[77]Zhongkeyuan guanyu tuixuan xuebu weiyuan jingguo qingkuang xiang zhongyang de baogao 中科院关于推选学部委员经过情况向中央的报告(Report of the Chinese Academy of Sciences to the Central Committee on the Selection of Academic Division Members). Beijing: Zhongguo kexueyuan danganguan中国科学院档案馆(Archives of Chinese Academy of Sciences), 1955-01-005.

(5) **Determination and Approval of the Final Candidates**

On May 9, the Party group of the CAS sent a letter to the Publicity Department of the CPC Central Committee reporting the list of academic division members and their review and revision.[78] On May 12, the Publicity Department of the CPC Central Committee held a meeting attended by relevant departments to discuss and revise the list of academic division members reported by the Party group of the CAS. This time, the first of the above-mentioned seven principles was revised to read: "Those with high academic level but serious political suspicion of the current counter-revolution and proper reasons for public explanation to the scientific community should not be listed as academic division members."[79] This raised the conditions for the principle. Through this meeting, the deleted or retained candidates in the original 224-member list changed: according to the revised Principle 1, the deleted candidates became Bao Guobao, Zhao Zongyu, Tsai Chin-tao and Tang F. F., while according to Principle 4, the deleted Huang Minlon, Hu Jingfu and Xie Shaowen were retained; according to Principle 2, Chen Rong, Liu Sichi and Hu Hsen-hsu were deleted; according to Principle 3, Ting S. L., Ding Zan, Yu T. Y., Liu Hui-xian, Liu Ding, Ji Chaoding, Lu Zhijun, Lu Yuanlei and Chen Kangbai were deleted, while Wei Jiangong, who had been deleted, was re-listed.[80]

After this revision, Yu Fuchun, Ching Ren Chang, Yang We-I, Zhao Hongzhang, Chen Wengui, Huang C. S., Chu F. T., Liang Boqiang, Ye Juquan, Cheng Dan'an, Feng Zhi and Wei Xi were added to the original addition of Wu Y. K. and Zhou Zezhao.[81] Among them, Chen Wengui, Ye Juquan and Liang Boqiang were once recommended by the Party group of the Ministry of Health and included in the list of members of the Academic Division by the Party group of the CAS.[82] They were

[78]Zhongkeyuan guanyu tuixuan xuebu weiyuan jingguo qingkuang xiang zhongyang de baogao 中科院关于推选学部委员经过情况向中央的报告(Report of the Chinese Academy of Sciences to the Central Committee on the Selection of Academic Division Members). Beijing: Zhongguo kexueyuan danganguan中国科学院档案馆(Archives of Chinese Academy of Sciences), 1955-01-005.

[79]Zhongkeyuan guanyu tuixuan xuebu weiyuan jingguo qingkuang xiang zhongyang de baogao 中科院关于推选学部委员经过情况向中央的报告(Report of the Chinese Academy of Sciences to the Central Committee on the Selection of Academic Division Members). Beijing: Zhongguo kexueyuan danganguan中国科学院档案馆(Archives of Chinese Academy of Sciences), 1955-01-005.

[80]Zhongkeyuan guanyu tuixuan xuebu weiyuan jingguo qingkuang xiang zhongyang de baogao 中科院关于推选学部委员经过情况向中央的报告(Report of the Chinese Academy of Sciences to the Central Committee on the Selection of Academic Division Members). Beijing: Zhongguo kexueyuan danganguan中国科学院档案馆(Archives of Chinese Academy of Sciences), 1955-01-005.

[81]Zhongkeyuan guanyu tuixuan xuebu weiyuan jingguo qingkuang xiang zhongyang de baogao 中科院关于推选学部委员经过情况向中央的报告(Report of the Chinese Academy of Sciences to the Central Committee on the Selection of Academic Division Members). Beijing: Zhongguo kexueyuan danganguan中国科学院档案馆(Archives of Chinese Academy of Sciences), 1955-01-005.

[82]Zhongkeyuan guanyu tuixuan xuebu weiyuan jingguo qingkuang xiang zhongyang de baogao 中科院关于推选学部委员经过情况向中央的报告(Report of the Chinese Academy of Sciences

later deleted and this time included again. So academic division members increased from 224 to 235.

On the afternoon of May 31, the 10th Plenary Meeting of the State Council approved 233 of the 235-member list. Yu Fuchun, a professor at Peking University, and Xie Shaowen, a professor at the Union Medical College, were not approved. The meeting also approved the "Report of the CAS on the Process of Organizing the Academic Divisions s and Convening the Founding Conference of the Academic Divisions".[83] Premier Zhou En-lai issued a State Council order on June 3rd agreeing to publish the list of 233 approved academic division members.[84] Among them, 48 were from the Division of Physics, Mathematics and Chemistry, 84 from the Division of Biology and Earth Sciences, 40 from the Division of Technological Science and 61 from the Division of Philosophy and Social Sciences.[85] The academic division members were appointed by Kuo Mo-jo, president of the CAS.[86] On June 4, *People's Daily* published the list of first academic division members as shown in Table 7.1.

The number of academic division members was nearly three times that of the first academicians of Academia Sinica. That was related to the large quota, the lax academic standards in the selection process, and the failure to implement Kuo Mo-jo's principle "better to be short than to be shoddy". Judging from the candidates, most had high academic level and had made outstanding academic achievements, but there were a few experts and leading cadres with low academic level.

Experts with low academic level included some elders and prominent figures in the scientific community. The appointment of these old scientists was to unite all sectors of the scientific community and was an expression of respect for the elderly. This practice was criticized as "playing the United Front" by some members of the higher academic level and later generations.[87] There were 42 members of the Communist Party of China (CPC) in this group of academic division members, and 15 of them

to the Central Committee on the Selection of Academic Division Members). Beijing: Zhongguo kexueyuan danganguan中国科学院档案馆(Archives of Chinese Academy of Sciences), 1955-01-005.

[83]Guowuyuan juxing dishici quantihuiyi国务院举行第十次全体会议(The State Council Held Its 10th Plenary Meeting). *Renmin ribao*人民日报(People's Daily), 1955-06-03 (1).

[84]Zhonghua renmin gongheguo Guowuyuan mingling中华人民共和国国务院命令(Order of the State Council of the People's Republic of China). *Renmin ribao*人民日报(People's Daily), 1955-06-04 (1).

[85]The name of the Division of Philosophy and Social Sciences was determined after May 15th, 1955. It was previously called the Division of Social Sciences.

[86]Wang Yangzong王扬宗. Cong yuanshi dao xuebu weiyuan: Zhongguo kexueyuan xueshu tizhi jianli de kunjing从院士到学部委员—中国科学院学术体制建立的困境(From Academicians to Academic Division Members: Difficulties in Establishing Academic System of the CAS). In Yu Miin-ling余敏玲(Ed.) (2012). *Liangan fenzhi: Xueshu jianzhi tuxiang xuanchuan yu zuqun zhengzhi(1945–2000)*两岸分治:学术建制、图像宣传与族群政治(1945–2000)(Divided Rule across the Taiwan Straits: Educational Reorganization, Visual Propaganda, and Ethnic Politics, 1945–2000). Taipei: Institute of Modern History, Academia Sinica. 97.

[87]Wang Yangzong王扬宗. Cong yuanshi dao xuebu weiyuan: Zhongguo kexueyuan xueshu tizhi jianli de kunjing从院士到学部委员—中国科学院学术体制建立的困境(From Academicians to Academic Division Members: Difficulties in Establishing Academic System of the CAS). In Yu Miin-ling余敏玲(Ed.) (2012). *Liangan fenzhi: Xueshu jianzhi tuxiang*

Table 7.1 List of first academic division members of CAS (1955)

Division name	Name of academic division members
Division of physics, mathematics and chemistry	Wang Jwu-Shi, Wang Ganchang, Wang Xianghao, Kiang Tsai-han, Yu Ruihuang, Woo Y. H., Woo Sho-Chow, Li F. H., Li Guoping, Wang You, Chow Tung-ching, Chou Pei-yuan, Shih J. W., Ko Chao, Liu Ta-kong, Tuan Hsio-fu, Chi Yuoh-Fong, Hu Ning, Tang Au-chin, Yuan Hanqing, Chang Tsinglien, Chang Yu-Che, Liang Shuquan, Chuang Chang-Kong, Hsu Pao-Lu, Chen Kien Kwong, Lu S. S., Fu Ying, Peng Huan-wu, Yun Ziqiang, Tseng Chao-lun, Hua Loo-Keng, Huang T. C., Huang Kun, Huang Minlon, Yang Shixian, Yeh Chi-Sun, Ge Tingsui, Yu Hongzheng, Chao Chung-Yoa, Chou T. Q., Lu Jia-xi, Tsien San-Tsiang, Chien Wei-zang, Tsien L. C., Ny Tsai-ze, Su Buchin, Yao Y. T.
Division of biology and earth sciences	Ding Ying, Yin T. H., Wang Chia-Chi, Wang Ying-Lai, Tien C. C., Wu Hsien-Wen, Zhu Xi, Ho Tso-lin, Wu Y. K., Wu Zheng-yi, Lee J. S., Li Lianjie, Lee Ching-kwei, Li Tsi-tung, Shen Qizhen, Pai Sitsan, Zhou Zezhao, Meng H. M., Cheng Dan'an, Lin Qiaozhi, Lin Rong, Wu Heng, Ping Chi, Chu Coching, Jin Shanbao, Hou Guangjiong, Hou Teh-feng, Yu T. F., Yu C. C., Hu Jingfu, Hsia Chien-pei, Sun Y. C., Yin Hung-chang, Tu Zhi, Tu Chang-wang, Ching Ren Chang, Ma Wen-chao, Zhang Wen-you, Chang Hsiao-chien, Chang Chin-yueh, Zhang Zhaoqian, Zhang Xijun, Liang Xi, Liang Boqiang, Xu Jie, Chen Wengui, Chen Shixiang, Chen Shisan C., Chun W. Y., Chen Fengtong, Si Xingjian, Tang P. S., Sheng Tongsheng, Cheng Y. C., Tung Ti-Cho, FengDepei, Feng Zefang, Huang T. K., Huang Bingwei, Huang C. S., Yang We-I, Young Chung-Chien, Ye Juquan, Pei W. C., Jaw Jeou-Jang, Zhao Hongzhang, Liu C. C., Liu Chung-lo, Yue Senxun, Pan Shu, Tsai Pang-hwa, Tsai Chiao, Teng S. C., Zheng Wanjun, Xiao Longyou, Chu F. T., Chien Shung-shu, Hsieh Chia-yung, Chung H. L., Tai Fon-Land, Dai Song'en, Wei Xi, Loo Tsung-Le, Gu Gong-xu
Division of technological science	Wang Daheng, Wang Zhixi, Shi Zhiren, Zhu Wuhua, Wu Xuelin, Li Wencai, Li Qiang, Li Guohao, Li Xun, Wang H. C., Chow Jen, Zhou Zhihong, Meng Chao-ying, Shao Xianghua, Hou Hsiong-Lin, Hou T. P., Mao T. E., Sun Dehe, Maa Dah-you, Zhang Dayu, Zhang Guangdou, Zhang Wei, Zhang Deqing, Liang Ssu-cheng, Zhang Mingtao, Tao Hengxian, Cheng Xiaogang, Huang Wenxi, Yang Tingbao, Yap Chu-Phay, Lei Tianjue, Jin Shuliang, Zhao Feike, Liu Hsian-chou, Liu Dunzhen, Tsai Fang-yin, Chu Yinghuang, Qian Ling-xi, Qian Zhidao, Yan Kai

(continued)

Table 7.1 (continued)

Division name	Name of academic division members
Division of philosophy and social sciences	Ding Shengshu, Qian Jiaju, Yu Guang-yuan, Yin Da, Wang Li, Wang Yanan, Wang Xuewen, Xiang Da, Ai Siqi, He Qifang, Wu Yuzhang, Wu Han, Lü Shuxiang, Lü Zhenyu, Li Yanong, Li Da, Li Yan, Du Guoxiang, Shen Zhiyuan, Di Chaobai, Zhou Yang, Ji Xian-lin, Chin Yuch-Lin, Hou Wai-lu, Hu Qiaomu, Hu Sheng, Fan Wenlan, Mao Dun, Shiah Nae, Ma Yin-chu, Ma Xulun, Zhang Ruxin, Zhang Jiafu, Xu Dixin, Guo Dali, Kuo Mo-jo, Chen Boda, Chen Yuan, Chen Yin-k'o, Chen Wangdao, Chen Han-seng, Tao L. K., Tang Yung-tung, Feng You-lan, Feng Zhi, Feng Ding, Huang Songling, Yang Shu-Tah, Yang Xianzhen, Liu Danian, Pan Zinian, Jian Bozan, Deng Tuo, Zheng Zhenduo, Li Jinxi, Qian Junrui, Luo Gengmo, Bao Erhan, Xue Muqiao, Wei Jiangong, Luo Changpei

Sources Zhongguo kexueyuan xuebu weiyuan mingdan中国科学院学部委员名单(List of Academic Division Members of the Chinese Academy of Sciences). *Renmin ribao*人民日报(People's Daily), 1955-06-04 (1)

in the three divisions of natural science: Yun Ziqiang, Shen Qizhen, Zhou Zezhao, Chen Fengtong, Wu Heng, Li Qiang, Zhao Feike and Qian Zhidao were from the liberated area like Yan'an, Wu Zheng-yi and Li Wencai had been the underground CPC members, Tsien San-Tsiang, Huang C. S., Tu Zhi and Tao Hengxian were new CPC members, and Hou Hsiong-Lin joined the CPC secretly in 1938. With the exception of four experts who newly joined the Party, other Party members did not make any important academic achievements, and were mostly included in the academic division members by virtue of their positions.[88] Due to the needs of the work, Zhang Jiafu, Liu Danian and other Party members selected as members of the Academic Division of Philosophy and Social Sciences also had low academic standards. Chi Yuoh-Fong, Ma Wen-chao and other member candidates who were questioned before the election of the first batch of members of Academia Sinica and received few votes in that election were also among the members of this group.

46 of the first batch of members of Academia Sinica who stayed in the mainland were selected as members of this group. As Liang Ssu-yung and Yu Cha-Shih

*xuanchuan yu zuqun zhengzhi(1945–2000)*两岸分治:学术建制、图像宣传与族群政治(1945–2000)(Divided Rule across the Taiwan Straits: Educational Reorganization, Visual Propaganda, and Ethnic Politics, 1945–2000). Taipei: Institute of Modern History, Academia Sinica. 111.

[88]Wang Yangzong王扬宗. Cong yuanshi dao xuebu weiyuan: Zhongguo kexueyuan xueshu tizhi jianli de kunjing从院士到学部委员—中国科学院学术体制建立的困境(From Academicians to Academic Division Members: Difficulties in Establishing Academic System of the CAS). In Yu Miin-ling余敏玲(Ed.) (2012). *Liangan fenzhi: Xueshu jianzhi tuxiang xuanchuan yu zuqun zhengzhi(1945–2000)*两岸分治:学术建制、图像宣传与族群政治(1945–2000)(Divided Rule across the Taiwan Straits: Educational Reorganization, Visual Propaganda, and Ethnic Politics, 1945–2000). Taipei: Institute of Modern History, Academia Sinica. 101–102.

had died, there were 57 academicians in the mainland at that time. That means 11 academicians had been voted out of the academic division members, namely, Chiang Li-Fu, Wong Wen-hao, Hu Hsen-hsu, Li Zongen, Woo Ting-liang, Chou Keng-Shang, Tsien T. S., Chen Ta, Zhang Yuanji, Ku Chieh-kan and Liu Yei-Tsen. The first nine were elected as the first academicians of Academia Sinica by general election.

(6) The Inaugural Assembly of the Academic Divisions

From June 1 to 10, 1955, the Inaugural Assembly of the Academic Divisions of the CAS was held in Beijing. It marked the establishment of the Academic Divisions of the CAS and its membership system. A total of 199 members attended the assembly, together with the heads of the CAS and research institutes, as well as representatives of government ministries, commissions, universities and foreign guests. 739 people attended the assembly, including Premier Zhou En-lai, Vice Premier Chen Yi, central leaders such as Li Jishen and Dong Biwu, and representatives of the Soviet Academy of Sciences and Polish Academy of Sciences. It was quite grand.[89] In addition to the formal establishment of the Academic Divisions and the election of their standing committees, the assembly discussed Kuo Mo-jo's work report, the work report of each Academic Division, the draft outline of the five-year plan of the CAS and the main points of the 1955 work plan of each Academic Division, and read and discussed 39 academic papers.[90]

During the assembly, the divisions of physics, mathematics and chemistry, biology and earth sciences, technological science and philosophy and social sciences respectively elected members of their Standing Committees.[91] After the assembly, the candidates for the division directors and deputy directors were officially confirmed. Among them, Woo Y. H. was the director of the Division of Physics, Mathematics and Chemistry, Chuang Chang-Kong, Hua Loo-Keng and Yun Ziqiang were deputy directors; Chu Coching was the director of the Division of Biology and Earth Sciences,

---

[89] Xue bu chengli dahui gongzuo zongjie baogao学部成立大会工作总结报告(Summary Report on the Work of the Inaugural Assembly of the Academic Divisions). In Zhongguo kexueyuan xueshu mishuchu中国科学院学术秘书处(Academic Secretariat of the Chinese Academy of Sciences) (Eds.). *Zhongguo kexueyuan nianbao*中国科学院年报(The Annals of the Chinese Academy of Sciences) 1955: 50.

[90] Xue bu chengli dahui gongzuo zongjie baogao学部成立大会工作总结报告(Summary Report on the Work of the Inaugural Assembly of the Academic Divisions). In Zhongguo kexueyuan xueshu mishuchu中国科学院学术秘书处(Academic Secretariat of the Chinese Academy of Sciences) (Eds.). *Zhongguo kexueyuan nianbao*中国科学院年报(The Annals of the Chinese Academy of Sciences) 1955: 50.

[91] Xue bu chengli dahui gongzuo zongjie baogao学部成立大会工作总结报告(Summary Report on the Work of the Inaugural Assembly of the Academic Divisions). In Zhongguo kexueyuan xueshu mishuchu中国科学院学术秘书处(Academic Secretariat of the Chinese Academy of Sciences) (Eds.). *Zhongguo kexueyuan nianbao*中国科学院年报(The Annals of the Chinese Academy of Sciences) 1955: 50; Zhongguo kexueyuan xuebu chengli dahui zongjie bing xiang guowuyuan de baogao中国科学院学部成立大会总结并向国务院的报告(Summary of the Inaugural Assembly of the Academic Divisions of Chinese Academy of Sciences and Its Report to the State Council). Beijing: Zhongguo kexueyuan danganguan中国科学院档案馆(Archives of Chinese Academy of Sciences), 1955-02-037.

Huang T. K., Tung Ti-Cho, Xu Jie, Chen Fengtong and Yin T. H. were deputy directors; Ny Tsai-ze was the director of the Division of Technological Science, and Mao T. E. and Zhao Feike were deputy directors; Kuo Mo-jo was the director of the Division of Philosophy and Social Sciences, and Pan Zinian was the deputy director. Only the people for deputy directors of the Division of Biology and Earth Sciences were different from those for the deputy directors of the Preparatory Committee of the division determined by the CAS on March 15, 1955.

Before and during the assembly, many scientists expressed dissatisfaction or doubts about the selection and the members of the Academic Divisions. For example, among the scientists who were not elected to the Academic Divisions, Shou Zhenhuang knew long before that he was not on the list of the academic division members, and was "very dissatisfied" and said that since Chao Chung-Yoa was a member of the Academic Divisions, he should also be. When Lee S. P. and Fu C. Y. attended the assembly as nonvoting delegates, their "mood was not high"; Hu Hsen-hsu made inquiries as early as before the Inaugural Assembly of the Academic Divisions. He was very dissatisfied after receiving the attendance certificate. Xin Shuzhi had a lot of complaints about not being a member of the Academic Divisions, grumbling about the difference between the materials distributed to him and the members of the Academic Divisions. Si Xingjian said: "Zheng Wanjun is also a member of the Academic Divisions? There are too many such people."[92]

Among the members of the Academic Divisions, high-levelled scientists such as Hua Loo-Keng, Lu S. S., Young Chung-Chien and Wang You all reflected that the Academic Division members "are too much of a United Front" and "there are many people who don't do research work". Hua once said: "This is not for the sake of research work. If we do it this way, the scientific cause will be delayed for 50 years." He also said: "If those who don't do research work are removed, I promise to achieve results in a few years. I can assure the Chinese Academy of Sciences and the Party. If not, I won't do it, either." Young Chung-Chien said: "There are still those people, the so-called old scientists, in where things often go bad because of them. They don't do scientific work but do more harm than good. "Lu S. S. thought that biology and earth sciences were especially well cared for. Hua and Lu had many opinions about Woo Y. H. They thought that he did not work, was lazy and had bad style. Hua was also dissatisfied with Xu Jie, who did not know the business but was deputy director of the Division of Biology and Earth Sciences. Lu was also dissatisfied with Ny Tsai-ze's position as an academic division member, saying: "He has not done physics work since he came back from abroad. This person is only good at making money and does not understand science."[93]

---

[92]Zhongkeyuan xuebu chengli dahui qingkuang jianbao中科院学部成立大会情况简报(Bulletins on the Inaugural Assembly of the Academic Divisions of Chinese Academy of Sciences). Beijing: Zhongguo kexueyuan danganguan中国科学院档案馆(Archives of Chinese Academy of Sciences), 1955-02-038.

[93]Zhongkeyuan xuebu chengli dahui qingkuang jianbao中科院学部成立大会情况简报(Bulletins on the Inaugural Assembly of the Academic Divisions of Chinese Academy of Sciences). Beijing: Zhongguo kexueyuan danganguan中国科学院档案馆(Archives of Chinese Academy of Sciences), 1955-02-038.

In addition, Li Xun expressed his opinion on the large number of academic division members in Beijing and the small number of them in the Northeast, saying: "Beijing has 150, Nanjing has more than 20, and the Northeast has only 15. Are all the talents concentrated in Beijing?" Zhang Dayu thought: "The majority of the members are in chemistry, but the number for chemical engineering is too small, which is not appropriate." Zhu Xi and Loo Tsung-Le reflected that "there are too many members".[94]

In response to the dissatisfaction or doubts of the scientists, Chen Yi stated in the report on the third day of the Inaugural Assembly of the Academic Divisions on the afternoon of June 3rd: "I think the choice of candidates is basically correct, but it may not be 100% correct. The State Council and the Central Committee of the Communist Party of China are prepared to approve this list. According to the degree to which we are now familiar with scientists and the scientists with us, and according to our mutual understanding and friendship, we have put forward this list… But there may be omissions, real talents may be buried. It may be his/her reason or ours, so we have a reservation. The list of academic division members will be announced tomorrow or the day after tomorrow, so it cannot be filled this year, but those who/that should be filled can be proposed so that they can be filled with consideration at the second division conference next year. The Communist Party is always ready to be supported and always ready to admit its mistakes as well. I try my best to defend the list today but admit that there might be omissions. On the other hand, will a few members be drawn down next year? It is also possible… Those who are physically unfit, of inadequate standards, or unwilling to engage in scientific research work may not be put up next year, and those who are "genuine talents" can be added." While defending the list of academic division members, Chen Yi also admitted: "Our Academic Divisions also include the consideration of the United Front. Some people say this is the Political Consultative Conference of the scientific community. I think there is nothing wrong with this statement. Our country is in consultation, why can't the scientific community negotiate?"[95]

Chen Yi's explanation basically solved the ideological problems of some scientists who were not selected as academic division members. For example, after hearing Chen Yi's report, Xin shuzhi said: "Shortcomings are inevitable at the beginning.""Vice Premier Chen said so thoroughly that even if there are opinions, there will be no opinions." Lee S. P. and Fu C. Y. became more emotional after hearing

[94]Zhongkeyuan xuebu chengli dahui qingkuang jianbao中科院学部成立大会情况简报(Bulletins on the Inaugural Assembly of the Academic Divisions of Chinese Academy of Sciences). Beijing: Zhongguo kexueyuan danganguan中国科学院档案馆(Archives of Chinese Academy of Sciences), 1955-02-038.

[95]Vice Premier Chen Yi's Political Report at the Inaugural Assembly of the CASAD. From Wang Yangzong王扬宗. Cong yuanshi dao xuebu weiyuan: Zhongguo kexueyuan xueshu tizhi jianli de kunjing从院士到学部委员—中国科学院学术体制建立的困境(From Academicians to Academic Division Members: Difficulties in Establishing Academic System of the CAS). In Yu Miin-ling余敏玲(Ed.) (2012). *Liangan fenzhi: Xueshu jianzhi tuxiang xuanchuan yu zuqun zhengzhi(1945–2000)*两岸分治:学术建制、图像宣传与族群政治(1945–2000)(Divided Rule across the Taiwan Straits: Educational Reorganization, Visual Propaganda, and Ethnic Politics, 1945–2000). Taipei: Institute of Modern History, Academia Sinica. 113–114.

Chen Yi's report and attended the group meeting on June 4th.[96] At the same time, Chen Yi's explanation also served as the basic guidline for how the CAS dealt with the opinions of scientists on the list of academic division members. Therefore, the CAS, in its "*Summary Report on the Work of the Inaugural Assembly of the Academic Divisions*" to Chen Yi and the State Council wrote on the opinions of scientists:

> After a year of repeated deliberation and negotiation, the list of academic division members is basically appropriate. However, due to our incomplete understanding of scientists across the country, there are still some people who should have been selected but left out. As a result, some scientists, at the beginning of the assembly, made comments on the list of the members. We are prepared to propose another list of supplementary members at an appropriate time and announce them before the Second General Assembly of the CASAD Members next year after the approval from the State Council.[97]

In addition to the formal agenda, the assembly was more concerned with the implementation of the academician system. On May 31, Kuo Mo-jo said at the preparatory meeting of the assembly: The Chinese Academy of Sciences should be based on academicians and corresponding academicians. The Central Committee has entrusted us with the task of establishing an academician system and a degree system. We plan to draft regulations within the year and submit them to the State Council for approval.[98] In his opening speech at the assembly on June 1, Kuo Mo-jo also said: "The Chinese Academy of Sciences is carrying out a series of work in order to make the CAS a center that unites the scientists of the entire country and promotes the scientific cause of our country… In all these measures, the establishment of the Academic Divisions is a central link. This is due to the concentration of representative scientists to carry out their work separately, which makes it possible for the Chinese Academy of Sciences to strengthen academic leadership or guidance for scientific research both inside and outside the Academy in a more organized and planned way. It also prepares further the conditions for the CAS to establish the academician system (Fig. 7.3)."[99] On the morning of June 2, Kuo Mo-jo made a report on the

---

[96]Zhongkeyuan xuebu chengli dahui qingkuang jianbao中科院学部成立大会情况简报(Bulletins on the Inaugural Assembly of the Academic Divisions of Chinese Academy of Sciences). Beijing: Zhongguo kexueyuan danganguan中国科学院档案馆(Archives of Chinese Academy of Sciences), 1955-02-038.

[97]Xue bu chengli dahui gongzuo zongjie baogao学部成立大会工作总结报告(Summary Report on the Work of the Inaugural Assembly of the Academic Divisions). In Zhongguo kexueyuan xueshu mishuchu中国科学院学术秘书处(Academic Secretariat of the Chinese Academy of Sciences) (Eds.). *Zhongguo kexueyuan nianbao*中国科学院年报(The Annals of the Chinese Academy of Sciences) 1955:52.

[98]Zhongkeyuan xuebu chengli dahui choubei weiyuanhui ji dahui zhuxituan huiyi jilu中科院学部成立大会筹备委员会及大会主席团会议记录(Meeting minutes of the Preparatory Committee and the Presidium of the Inaugural Assembly of the Academic Divisions of Chinese Academy of Sciences). Beijing: Zhongguo kexueyuan danganguan中国科学院档案馆(Archives of Chinese Academy of Sciences), 1955-02-029.

[99]Kuo Mo-jo yuanzhang zai Zhongguo kexueyuan xue bu chengli dahui shang de kaimuci郭沫若院长在中国科学院学部成立大会上的开幕词(President Kuo Mo-jo's Opening Speech at the Inaugural Assembly of the Academic Divisions of Chinese Academy of Sciences). In Zhongguo kexueyuan xueshu mishuchu中国科学院学术秘书处(Academic Secretariat of the

Fig. 7.3 Kuo Mo-jo, president of the CAS delivering a speech at the Inaugural Assembly of the Academic Divisions

work of the CAS and put forward five suggestions on the realization of the tasks of the Academy. One of them was "to establish a degree system, an academician system and an academic award system". Kuo Mo-jo stressed on the establishment of academician system:

> The Chinese Academy of Sciences is an organization of scientists. The best form of promoting scientific democracy is the academician system, that is, the system in which the General Assembly of CAS Members is the highest organ of the Academy of Sciences. We will also draft a law on the academician system with relevant departments. After approval, we will select the best scientists in our country as the first academicians and petition the government for appointment. According to international advanced experience, the first batch of academicians are appointed by the State, and the future academicians will be elected by the General Assembly of CAS Members. After the establishment of the academician system, the CAS will soon hold its first academician assembly to establish the leading body of the Academy through election. The convening of the academician assembly will further democratize the Chinese Academy of Sciences on the basis of the existing ones, thus leading the development of science in our country better.[100]

Chinese Academy of Sciences) (Eds.). *Zhongguo kexueyuan nianbao*中国科学院年报(The Annals of the Chinese Academy of Sciences) 1955: 2.

[100]Kuo Mo-jo yuanzhang zai Zhongguo kexueyuan xue bu chengli dahui shang de baogao郭沫若院长在中国科学院学部成立大会上的报告(President Kuo Mo-jo's Report at the Inaugural

It showed that Kuo Mo-jo had a preliminary idea of establishing the academician system. On June 11, the day after the closing of the assembly, the CAS held a joint meeting of the Academic Divisions. At the meeting, Chen Yi also mentioned the "academician system" and said, "we are now engaged in ideological reform first, then in the Academic Divisions, and then in the academician system."[101]

On September 20, the CAS raised the issue of implementing the academician system in its above-mentioned "*Summary Report on the Work of the Inaugural Assembly of the Academic Divisions*" submitted to Chen Yi and the State Council. The issue is as follows:

> Scientists at this assembly attached great importance to the implementation of the academician system. According to the actual situation of the scientific community in our country and the instructions of the Premier and Vice Premier Chen Yi, we believe that the implementation of the academician system is still more appropriate to postpone for one to three years. In this way, we can have a deeper and more comprehensive understanding of the academic level of scientists, and the mutual understanding between scientists will improve. Meanwhile, we can have plenty of time to encourage everyone to work hard on scientific research and strive for being elected as an academician (Since many scientists have not carried out research well in recent years, they themselves have such opinions). After the implementation of the academician system, the division membership system can still run in parallel because academic division members have a wider range of unity than academicians, which is beneficial to uniting scientists across the country, making better communication between the Chinese Academy of Sciences and various divisions, and promoting the development of science.[102]

In other words, the CAS decided to establish an academician system within one to three years through the Inaugural Assembly of the Academic Divisions of the Chinese Academy of Sciences, which is compatible with the division membership system of the Academy. In order to establish the academician system, the CAS began to make relevant legislative preparations after the assembly. On October 29, 1955, the CAS made provisions on the academician system in its *First Draft of Draft Constitution of the Chinese Academy of Sciences* (the second revision) submitted to the State Council.[103] However, the legislative work to establish the academician system was interrupted after that.

Assembly of the Academic Divisions of Chinese Academy of Sciences). In Zhongguo kexueyuan xueshu mishuchu中国科学院学术秘书处(Academic Secretariat of the Chinese Academy of Sciences) (Eds.). *Zhongguo kexueyuan nianbao*中国科学院年报(The Annals of the Chinese Academy of Sciences) 1955: 9.

[101]Chen Yi fuzongli 1955 nian 6 yue zai kexueyuan xuebu lianxihui shang de jianghua陈毅副总理一九五五年六月在科学院学部联席会上的讲话(Vice Premier Chen Yi's Speech at the Joint Meeting of the Academic Divisions of Chinese Academy of Sciences in June 1955. Beijing: Zhongguo kexueyuan danganguan中国科学院档案馆(Archives of Chinese Academy of Sciences), 1955-01-010.

[102]Xue bu chengli dahui gongzuo zongjie baogao学部成立大会工作总结报告(Summary Report on the Work of the Inaugural Assembly of the Academic Divisions). In Zhongguo kexueyuan xueshu mishuchu中国科学院学术秘书处(Academic Secretariat of the Chinese Academy of Sciences) (Eds.). *Zhongguo kexueyuan nianbao*中国科学院年报(The Annals of the Chinese Academy of Sciences) 1955: 53–54.

[103]Youguan yuanshi zhidu cankao ziliao guanyu xuewei xuexian tiaoli deng wenti de baogao有关院士制度参考资料,关于学位、学衔条例等问题的报告(Reference Materials on Academician

7.3 Summary

Under the influence of political changes, the academician system of Academia Sinica was abandoned in the mainland in 1949 and failed to be incorporated into the academic system of New China. In the upsurge of "one-sided" study of the Soviet Union, the CAS chose the latter between the academician system and the system of academic division members through instructions from the Central Government. The establishment of the Academic Divisions of the CAS and the membership system was to give full play to the role of the Communist Party of China in leading science based on China's reality. Judging from the original orientation of the Party group of the CAS to the Academic Divisions and their members, the division was a working organization, and academic division members were working titles rather than academic honorary titles. Therefore, academic division members were not equivalent to academicians.[104] After the Inaugural Assembly of the Academic Divisions, the CAS tried to establish an academician system, but failed.

In 1955, the selection of academic division members of the CAS was led directly by the Publicity Department of the CPC Central Committee, and within the CAS by the Party group of the Academy, with the participation of the Academic Secretariat and the Preparatory Committees of the various divisions. Some universities and colleges, business departments, provincial and municipal Party committees or publicity departments were also involved. This batch of academic division members were not elected by strict and standardized democratic election procedures. Before the selection, no selection rules were formulated on legal basis. The selection of members of the physics, mathematics and chemistry division and the biology and earth science division was relatively standardized, and most of the candidates were presumed according to the results recommended by experts. The candidates for the technological division and the philosophy and social science division were basically determined through negotiation. The selection of this group of academic division members emphasized academic standards, and especially advocated academic achievements as the main basis for considering candidates. However, in addition to academic standards, candidates must be loyal to the people's cause and had no serious historical problems in politics. This determined that the selection of this group of academic division members could not avoid the influence of political and ideological factors. In the selection process, the needs of the United Front, the work and the development of disciplines, and taking care of the old scientific traditions were also the conditions for considering candidates.

The need for the United Front and the consideration of the old scientific traditions were the conditions for considering candidates, which were related to the CPC's

System, Reports on Regulations of Academic Degrees, Academic Titles, etc.). Beijing: Zhongguo kexueyuan danganguan中国科学院档案馆(Archives of Chinese Academy of Sciences), 1964-02-041.

[104]Li Zhenzhen李真真and Wei Hongning尉红宁 (1991). Zhang Jiafu tongzhi fangtanku张稼夫同志访谈录(An Interview with Comrade Zhang Jiafu). *Yuanshi ziliao yu yanjiu*院史资料与研究(Historical Materials and Research of the Chinese Academy of Sciences) (6): 22–27.

important policy of uniting scientists in scientific work. On March 8, 1954, the Central Government put special emphasis on this policy in its instructions to the *Report* submitted by the Party group of the CAS on November 19, 1953 and pointed out: "Scientists are the precious wealth of the country and society. They must be valued and respected, and all the scientists must be enlisted and united to serve the people. For a small number of scientists who carried out counter-revolutionary activities in history, we should also strive for and use them appropriately."[105] The academicians Tao L. K. , Ma Yin-chu, Chin Yuch-Lin and Feng You-lan,[106] regarded by the Party group of the CAS as "representatives of the old scientific community" with "low academic level" or "outdated or even reactionary academic views", were the typical beneficiaries of this policy. The need for work and discipline development as a condition for consideration of candidates was influenced by the orientation of the CAS to the academic division members and the membership system.

[105]Zhonggong zhongyang dui Zhongguo kexueyuan dangzu guanyu muqian kexueyuan gongzuo de jiben qingkuang he jinhou gongzuo renwu gei zhongyang de baogao de pishi中共中央对中国科学院党组《关于目前科学院工作的基本情况和今后工作任务给中央的报告》的批示(Instructions from the Central Committee of the CPC to the Party Group of the Chinese Academy of Sciences' Report to the Central Committee on the Basic Situation of the Work of the Academy at Present and Its Future Tasks). In Zhonggong zhongyang wenxian yanjiushi中共中央文献研究室(Literature Research Office of the Central Committee of the CPC) (Eds.) (2011). *Jianguo yilai zhongyao wenxian xuanbian*建国以来重要文献选编(Selected Important Documents Since the Founding of the People's Republic of China), Vol. 5. Beijing: Zhongyang wenxian chubanshe中央文献出版社.144.

[106]Zhongkeyuan guanyu tuixuan xuebu weiyuan jingguo qingkuang xiang zhongyang de baogao 中科院关于推选学部委员经过情况向中央的报告(Report of the Chinese Academy of Sciences to the Central Committee on the Selection of Academic Division Members). Beijing: Zhongguo kexueyuan danganguan中国科学院档案馆(Archives of Chinese Academy of Sciences), 1955-01-005.

Chapter 8 The Expansion of the Members of Academic Divisions of the Chinese Academy of Sciences Before and After the Cultural Revolution

Although the Chinese Academy of Sciences decided to establish an academician system within one to three years after the Academic Divisions and the membership system were set up, it failed to implement it. However, through supplementing more members of the Academic Divisions of the Chinese Academy of Sciences (CASAD) in 1957, the identity of those members began to change into CAS members. After this expansion, the CAS did not recruit any more members before the Cultural Revolution. Its Academic Divisions suffered ups and downs in the political movements and collapsed after the outbreak of the Cultural Revolution. When the Revolution was over and the National Science Congress and the Third Plenary Session of the 11th CPC Central Committee were held in 1978, China's scientific cause started to embrace vigorous development. In 1979, the CAS resumed its activities among Academic Divisions. With the CAS co-opting more academic division members in 1980, it took a step further towards the academician system. This chapter studies the 1957 supplement of members of academic divisions and the 1980 co-optation of members of the academic divisions of Chinese Academy of Sciences, how the selection system was improved before and after the Cultural Revolution, and how membership came to include academic honors of academicians.

8.1 The 1957 Supplement of Members of the Academic Divisions

(1) The Background of the Supplement

The establishment of the Academic Divisions and the membership system was an important achievement for the CAS in constructing the academic system. It helped strengthen the academic leadership. However, some scholars were dissatisfied that some candidates were left out during the selection of academic division members in 1955. To deal with the problem, Vice Premier Chen Yi of the State

J. Guo, *The Establishment and Reconstruction of the Academician System in China*,
https://doi.org/10.1007/978-981-15-7208-1_8

Council instructed at the Inaugural Assembly of the Academic Divisions that by-elections could be considered. The CAS decided to "prepare to propose another list of supplementary members at an appropriate time" and announce it before the Second General Assembly of Members of the Academic Divisions in 1956.[1] Moreover, "some people in the scientific community have been suggesting adding academic division members."[2]

In 1956, Zhou En-lai, on behalf of the Central Government, put forward the task of drawing up a long-term plan for the development of science and technology for 12 years at the "Meeting on Intellectual Issues" held in January. He issued a call for "marching into modern sciences" and stressed that "the CAS should be strengthened with great strength so that it can become the locomotive leading the country to raise its scientific level and cultivate new forces."[3] Zhou En-lai also clearly put forward:

> The most advanced achievements in world science must be introduced as quickly as possible to China's science departments, national defense departments, production departments and education departments according to the possibilities and needs, and the categories that China's scientific community is most short of and that China's national construction needs most urgently must be made up as quickly as possible, so that in 12 years' time, China's scientific and technological level in these categories can be close to that of the Soviet Union and other world powers.[4]

At this meeting, Mao Tse-tung called on the whole Party to study hard scientific knowledge, unite with non-Party intellectuals and strive to quickly catch up with the

---

[1]Xue bu chengli dahui gongzuo zongjie baogao学部成立大会工作总结报告(Summary Report on the Work of the Inaugural Assembly of the Academic Divisions). In Zhongguo kexueyuan xueshu mishuchu中国科学院学术秘书处(Academic Secretariat of the Chinese Academy of Sciences) (Eds.). *Zhongguo kexueyuan nianbao*中国科学院年报(The Annals of the Chinese Academy of Sciences) 1955: 52.

[2]Guanyu zhaokai dierci xuebu weiyuan dahui he zengbu xuebu weiyuan wenti xiang zhongyang de qingshi baogao关于召开第二次学部委员大会和增补学部委员问题向中央的请示报告(Report to the Central Committee for Instructions on Convening the Second General Assembly of Members of the Academic Divisions and the Problems of Supplementing Academic Division Members). Beijing: Zhongguo kexueyuan danganguan中国科学院档案馆(Archives of Chinese Academy of Sciences), 1957-02-020.

[3]Zhou En-lai周恩来(1984). Guanyu zhishi fenzi wenti de baogao关于知识分子问题的报告(Report on Intellectual Issues). In Zhonggong zhongyang wenxian bianji weiyuanhui中共中央文献编辑委员会(the Document Editing Committee of the CPC Central Committee) (Eds.). *Zhou En-lai xuanji*(II) 周恩来选集(下卷) (Selected Works of Zhou En-lai), Vol. II. Beijing: Renmin chubanhse人民出版社.181–187.

[4]Zhou En-lai周恩来(1984). Guanyu zhishi fenzi wenti de baogao关于知识分子问题的报告(Report on Intellectual Issues). In Zhonggong zhongyang wenxian bianji weiyuanhui中共中央文献编辑委员会(the Document Editing Committee of the CPC Central Committee) (Eds.). *Zhou En-lai xuanji*(II) 周恩来选集(下卷) (Selected Works of Zhou En-lai), Vol. II. Beijing: Renmin chubanhse人民出版社.184.

world's advanced scientific level.[5] Those reflected that the Party and the Government attached great importance to science and aroused the initiative and work enthusiasm of a vast number of intellectuals.[6] Under the influence of this conference, many Chinese scientists living in Britain, the United States, France and other countries returned to work.[7] The CAS, as the "locomotive" for "marching into modern sciences", needed to shoulder not only a large amount of research work but also the important task of academic leadership in the scientific cause of New China. Although the CAS had strengthened its overall academic leadership through the Academic Divisions and the membership system, its academic leadership in many scientific fields, especially important and urgent ones, was still insufficient.

After this meeting, the Central Committee issued the "Double Hundred" policy. On May 26, at Kuo Mo-jo's invitation, Lu Dingyi, Minister of the Publicity Department of the CPC Central Committee, delivered a report entitled *Let a Hundred Flowers Bloom and a Hundred Schools of Thought Contend* to the scientific and literary circles in the capital, clarifying that "let a hundred schools of thought contend" was the fundamental policy for developing the scientific cause of New China.[8] As a new policy aimed at vigorously promoting the development of scientific and cultural undertakings, the "Double Hundred" policy advocated academic freedom and clearly opposed interference in academic affairs by administrative orders and methods, bringing unprecedented vitality to the academic circles of New China. New China, which was still in a period of turbulence, saw a relatively relaxed and free political and academic environment, and intellectuals ushered in the "warm sunny spring".

(2) **Determination of Criteria and Scope of Candidates**

On January 21, 1956, the day after the "Meeting on Intellectual Issues", the Party group of the CAS decided to include supplementing academic division members into the work for that year.[9] Since all divisions needed to concentrate on formulating

[5] Zhongguo gongchandang zhongyang weiyuanhui zhaokai guanyu zhishi fenzi wenti huiyi中国共产党中央委员会召开关于知识分子问题会议(The CPC Central Committee Convened a Meeting on Intellectual Issues). In Zhongguo kexueyuan bangongting中国科学院办公厅(General Office of the Chinese Academy of Sciences) (Eds.). *Zhongguo kexueyuan nianbao*中国科学院年报(The Annals of the Chinese Academy of Sciences) 1956: 1.

[6] Zhang Jiafu Wu Heng tongzhi zai zhongyang zhaokai de zhishi fenzi wenti huiyi shang fayan ji kexueyuan xiaozu huibao张稼夫、武衡同志在中央召开的知识分子问题会议上发言及科学院小组汇报(Speeches made by Comrade Zhang Jiafu and Comrade Wu Heng at the Meeting on Intellectual Issues held by the Central Committee and the Report of the Group of Chinese Academy of Sciences). Beijing: Zhongguo kexueyuan danganguan中国科学院档案馆(Archives of Chinese Academy of Sciences), 1956-1-010.

[7] Zhang Jiafu张稼夫(1984). *Gengshen yishi*庚申忆逝(Recalling the Passed in 1980). Taiyuan: Shanxi renmin chubanshe山西人民出版社.134.

[8] Lu Dingyi陆定一(1956). Baihua qifang baijia zhengming百花齐放，百家争鸣(Let a Hundred Flowers Bloom and a Hundred Schools of Thought Contend). *Renmin ribao*人民日报(People's Daily), 1956-06-13 (2, 3).

[9] Dangzu huiyi jiyao dangzu bangong huiyi jilu党组会议纪要、党组办公会议记录(Summaries of the Party Group Meetings, Minutes of the Party Group Office Meetings). Beijing: Zhongguo

the 12-year long-term plan for the development of science and technology, and the awards for the 1956 science prize had been launched, the recruitment work did not start immediately. On May 10, the CAS issued a notice requesting the Standing Committee of each division to discuss the preliminary list of additional academic division members and submit the list to the Academy for review before May 20.[10]

Regarding the selection criteria, the Academy did not intend to follow the 1955 selection criteria. Before October 25, 1956, the Academy formulated two new standards for the recruitment work: (1) "academic conditions—those with high academic level, important achievements in scientific work and high prestige among scientists in the discipline"; (2) "political conditions—those with real sufficient academic level can be included in the consideration list as long as they are not current anti-revolutionaries and sentencing regulators in politics".[11] The new standards were quite different from that of 1955. First, in terms of academic conditions, the criterion of promoting the discipline or academics was deleted, which meant that scientists who only promoted scientific undertakings but had low academic achievements would not be included in the list of candidates. Second, political standards were diluted, and scientists with reactionary ideas or political and historical problems were no longer excluded.

However, after discussion at the General Assembly, the CAS decided that "some members of Academic Divisions will be added according to the selection conditions in 1955".[12] It could be seen that the academic organization ability of the candidates, like their academic achievements and contributions, was also valued by the General Assembly. At the same time, the old standards would still be adopted for political conditions, which showed that the General Assembly was very cautious about political standards.

In 1956, when the CAS decided to recruit more academic division members, it also considered the scope of the candidates: First, scientists who "already had the qualifications for academic division members, and were omitted or not selected for some reason in 1955"; second, "scientists who returned from abroad with the qualifications of academic division members after selecting in 1955"; and third, scientists who were required "for disciplines newly developing or in urgent need of development in which academic division members selected in 1955 are not sufficient

kexueyuan danganguan中国科学院档案馆(Archives of Chinese Academy of Sciences), 1956-01-008.

[10]Jishu kexuebu wei zhongkeyuan dierci xuebu dahui zhunbei de huiyi wenjian ji ti'an de chuli yijian技术科学部为中科院第二次学部大会准备的会议文件及提案的处理意见(The Documents and Processing Opinions on the Proposals Prepared by the Division of Technological Sciences for the Second General Assembly of Members of the Academic Divisions of the Chinese Academy of Sciences). Beijing: Zhongguo kexueyuan danganguan中国科学院档案馆(Archives of Chinese Academy of Sciences), 1957-17-001.

[11]Guanyu zengpin xuebu weiyuan de yijian关于增聘学部委员的意见(Opinions on the Supplement of Academic Division Members). Beijing: Zhongguo kexueyuan danganguan中国科学院档案馆(Archives of Chinese Academy of Sciences), 1957-15-002.

[12]Guanyu zengpin xuebu weiyuan de yijian关于增聘学部委员的意见(Opinions on the Supplement of Academic Division Members). Beijing: Zhongguo kexueyuan danganguan中国科学院档案馆(Archives of Chinese Academy of Sciences), 1957-15-002.

and need to be supplemented." On the other hand, for agronomy and medicine, "since the Academy of Agricultural Sciences and the Academy of Medical Sciences are preparing for the establishment of their academic leadership organizations, and the number of scientific research institutions related to them in our Academy is very small, so it is difficult to combine their work. Therefore, except basic medicine, the supplement of academic division members may be postponed." When discussing at the General Assembly, Article 1 was changed to "scientists who already had the qualifications of an academic division member and were omitted in the 1955 selection".[13] Article 1 of the original plan referred to "those who were not elected for some reason" mainly referred to those who were not elected due to unqualified political conditions in 1955. The removal of this point by the General Assembly also showed that it was very cautious about political standards.

The CAS was considering recruiting more scientists who had recently returned to China, not only to strengthen the leadership of the Academic Divisions in science, but also to strengthen the unity of the scientific community. On May 3, 1957, the Party group of the CAS said in its report for instructions to the Publicity Department of the CPC Central Committee, Deputy Premier Nie Rongzhen and the Central Committee on the recruitment of additional members of the Academic Divisions: "Some newly returned scientists from abroad should also be added to the list so as to strengthen the academic leadership of scientific cause and the unity of the scientific community."[14] Zhou En-lai's explicit requirements for the scientific community on behalf of the Central Committee at the "Meeting on intellectual Issues" was directly related to an emphasis on recruitment of scientists for disciplines that were newly developing or in urgent need of development.

(3) **Initial Preparation of Each Academic Division for Candidates**

Around October 29, 1956, the CAS sent "Opinions on the Supplement of Academic Division Members" (hereinafter referred to as *Opinions*) to the four divisions. In addition to the aforesaid scope of candidates, it also stipulated the steps and methods of recruitment: (1) "Each division shall hold a Standing Committee meeting before the end of October to discuss the methods of adding academic division members and the list of additions, propose a preliminary plan and issue it to all academic division members for deliberation and comment before November 10." (2) "Each division will gather the opinions prepared by its members at the end of November, then hold a meeting of the Standing Committee of the division to discuss and formally propose a list (draft) of additional members of the division." (3) "On December 15,

---

[13] Guanyu zengpin xuebu weiyuan de yijian关于增聘学部委员的意见(Opinions on the Supplement of Academic Division Members). Beijing: Zhongguo kexueyuan danganguan中国科学院档案馆(Archives of Chinese Academy of Sciences), 1957-15-002.

[14] Guanyu zhaokai dierci xuebu weiyuan dahui he zengbu xuebu weiyuan wenti xiang zhongyang de qingshi baogao关于召开第二次学部委员大会和增补学部委员问题向中央的请示报告(Report to the Central Committee for Instructions on Convening the Second General Assembly of Members of the Academic Divisions and the Problems of Supplementing Academic Division Members). Beijing: Zhongguo kexueyuan danganguan中国科学院档案馆(Archives of Chinese Academy of Sciences), 1957-02-020.

after the joint deliberation of all divisions, the list will be submitted to the General Assembly for discussion and approval, and finally submitted to the State Council for approval."[15] It could be seen that the Academy planned that the final selection of the additional members would be approved by the State Council as in 1955. In terms of recruitment time, the Academy had more stringent requirements on each division. However, each division did not carry out the recruitment work in strict accordance with the *Opinions*. It was because there was no unified and specific method for recruitment, and the methods for preparing initial list of candidates were different.

The Division of Physics,[16] Mathematics and Chemistry made a preliminary selection earlier. At 7:00 pm on May 28, 1956, the division held a Standing Committee meeting at Xijiao Hotel and decided that the Standing Committee members of the division would nominate the candidates after deliberation and discussion with the academic division members separately. Regarding the selection criteria, it was required that "the criteria of the existing division members should be referred to, and consideration should be made among the scientists in the recommended list who have been extensively consulted in the past and the scientists who have recently returned from abroad."[17] At the same time, the meeting provided a preliminary list of candidates for "reference in preparation": (mathematics, mechanics and astronomy) Wu Wen-Tsün, Shen Yu-cheng, Ko Chao,[18] Cheng Min-Teh, Min Szu-hoa, Tseng Y. Y. and Chiang Li-Fu; (physics) Zhao Guangzeng, Chang T. S. and Feng Ping-chuan; (chemistry) Sun Cheng-e, Chiang Ming-chien, Xing Qi-yi and Gao Ji-yu. According to the note attached to the 14-member list, all candidates were recommended in the 1955 selection, "ranking is based on the number of recommended votes."[19]

After the meeting, the Standing Committee members of the Division of Physics, Mathematics and Chemistry and members of the division in Beijing recommended 26[20] candidates preliminarily after deliberation and discussion: (Mathematics) Wu

[15]Guanyu zengpin xuebu weiyuan de yijian关于增聘学部委员的意见(Opinions on the Supplement of Academic Division Members). Beijing: Zhongguo kexueyuan danganguan中国科学院档案馆(Archives of Chinese Academy of Sciences), 1957-15-002.

[16]On July 28th, 1956, the Division of Physics, Mathematics and Chemistry was renamed the Division of Mathematics, Physics and Chemistry.

[17]Shulihua xuebu zhaokai diyi er wu ci xuebu changwei huiyi jilu ji youguan cailiao数理化学部召开第一、二、五次学部常委会议记录及有关材料(Minutes of the First, Second and Fifth Standing Committee Meetings held by the Division of Mathematics, Physics and Chemistry and Related Materials). Beijing: Zhongguo kexueyuan danganguan中国科学院档案馆(Archives of Chinese Academy of Sciences), 1956-15-005.

[18]Ko Chao was elected as an academic division member in 1955. This is a mistake in the original document.

[19]Shulihua xuebu zhaokai diyi er wu ci xuebu changwei huiyi jilu ji youguan cailiao数理化学部召开第一、二、五次学部常委会议记录及有关材料(Minutes of the First, Second and Fifth Standing Committee Meetings held by the Division of Mathematics, Physics and Chemistry and Related Materials). Beijing: Zhongguo kexueyuan danganguan中国科学院档案馆(Archives of Chinese Academy of Sciences), 1956-15-005.

[20]The numbers in parentheses after the recommended persons: before the plus sign is the number recommended by people in his/her subject, and the number after the plus sign is the number recommended by people in other subjects.

Wen-Tsün (4+1), Shen Yu-cheng (3+1), Tseng Y. Y. (3+1), Cheng Min-Teh (3+0) and Min Szu-hoa (2+0); (Mechanics)Tsien H. S. (1+3); (Physics) Chang T. S. (4+2), Chang Wen-yu (3+0), Zhao Guangzeng (2+1), Ho Zah-wei (2+0), Hoff Lu (2+0), Hu Ji-min (1+0), Kwei C. T. (1+0), Feng Ping-chuan (1+0), Yu Fuchun (1+0), Young C. T. (1+0); (Chemistry) Sun Cheng-e (4+1), Gao Ji-yu (4+0), Wang Pao-jen (3+0), Tai A. P. (3+0), Xing Qi-yi (2+0), Tsai Liu-sheng (2+0), Chiang Ming-chien (2+0), Yang Jeng-tsong (1+1), Chu Tze-tsin (1+0), Chien Jen-yuan (1+0).[21]

The number of recommended candidates nearly doubled compared with the number of candidates on the reference list. The candidates for supplement were concentrated in mechanics, physics and chemistry, especially the latter two, includingTsien H. S., who returned to China in 1955. With the exception of Chiang Li-Fu, all the candidates listed in the reference list became nominees. After losing the selection in 1955, Chiang Li-Fu was not recommended this time because of Chen Kien Kwong's opposition. Chen Kien Kwong said: "In 1949, all the mathematics books of the Pseudo-Academia Sinica were sent to Taiwan, but they have not been returned yet, causing serious losses in the mathematics field". Chen Kien Kwong held that Chiang Li-Fu should be responsible for that. At the same time, he blamed Chiang Li-Fu for the failure of some young researchers from the Institute of Mathematics of Academia Sinica to return to the mainland after transferring from Taiwan to the United States, saying that it "caused great losses in the field of mathematics".[22]

After the list of nominees was produced, the CAS issued the*Opinion*, and the Division of Mathematics, Physics and Chemistry then solicited opinions from all members of the division and the director of the Division of Technological Sciences, Ny Tsai-ze, on November 5th, 1956.[23] By the end of the month, 36 people, or 75% of the division members,[24] had given feedback and Ny Tsai-ze did not respond. The specific results are shown in Table 8.1.

A total of 33 people were recommended, an increase of 7 over the original 26, namely, Hiong King-lai, Li Hen, Dai Wensai, Kuo Y. H., Wang Pu, Chien Pao-kung and Wu C. K. Among them, Kuo Y. H. and Wang Pu were scientists who returned to China in 1956 studying advanced national defense technologies, aerodynamics and nuclear physics respectively; Hiong King-lai was a veteran of Chinese mathematics

---

[21]Guanyu zengpin xuebu weiyuan de yijian关于增聘学部委员的意见(Opinions on the Supplement of Academic Division Members). Beijing: Zhongguo kexueyuan danganguan中国科学院档案馆(Archives of Chinese Academy of Sciences), 1957-15-002.

[22]Guanyu zengpin xuebu weiyuan de yijian关于增聘学部委员的意见(Opinions on the Supplement of Academic Division Members). Beijing: Zhongguo kexueyuan danganguan中国科学院档案馆(Archives of Chinese Academy of Sciences), 1957-15-002.

[23]Guanyu zengpin xuebu weiyuan de yijian关于增聘学部委员的意见(Opinions on the Supplement of Academic Division Members). Beijing: Zhongguo kexueyuan danganguan中国科学院档案馆(Archives of Chinese Academy of Sciences), 1957-15-002.

[24]The number of members who give feedback from various disciplines is: 4 in mathematics, 1 in mechanics, 15 in physics and 16 in chemistry. Except for Ny Tsai-ze, 12 people did not give feedback: (mathematics) Li Guoping, Hsu Pao-Lu, Wang Xianghao and Su Buchin; (mechanics) Chien Wei-zang; (physics) Lu S. S., Wang Ganchang, Hu Ning, Tseng Chao-lun, Yu Ruihuang and Peng Huan-wu; (chemistry) Huang T. C. Different from other people who did not give feedback, Hsu Pao-Lu gave feedback on January 16th, 1957, and it was after the deadline and not counted.

Table 8.1 Results of the Division of Mathematics, Physics and Chemistry's solicitation of comments on initial referees at the end of November 1956

Discipline	Referee	Number of people recommended from the discipline	Number of people recommended from other parties	Total number of references
Mathematics	Wu Wen-Tsün	4	15	19
	Tseng Y. Y.	4	7	11
	Shen Yu-cheng	3	7	10
	Cheng Min-Teh	3	5	8
	Min Szu-hoa	2	4	6
	Hiong King-lai	1	0	1
Astronomy	Li Hen	1	0	1
	Dai Wensai	1	0	1
Mechanics	Tsien H. S.	1	16	17
	Kuo Y. H.	0	3	3
Physics	Chang T. S.	10	12	22
	Chang Wen-yu	9	9	18
	Zhao Guangzeng	6	3	9
	Hoff Lu	4	3	7
	Ho Zah-wei	4	2	6
	Hu Ji-min	2	2	4
	Feng Ping-chuan	2	2	4
	Kwei C. T.	1	2	3
	Young C. T.	2	1	3
	Yu Fuchun	1	1	2
	Wang Pu	1	0	1
Chemistry	Gao Ji-yu	12	5	17
	Sun Cheng-e	11	6	17
	Chu Tze-tsin	6	4	10
	Wang Pao-jen	7	2	9
	Tai A. P.	7	2	9
	Chien Jen-yuan	4	3	7
	Xing Qi-yi	3	1	4

(continued)

Table 8.1 (continued)

Discipline	Referee	Number of people recommended from the discipline	Number of people recommended from other parties	Total number of references
	Tsai Liu-sheng	3	1	4
	Yang Jeng-tsong	2	2	4
	Chiang Ming-chien	2	1	3
	Chien Pao-kung	1	0	1
	Wu C. K.	1	0	1

Sources 1957 nian xuebu changwei huiyi ji xueke xiaozu huiyi youguan wenjian一九五七年学部常委会议及学科小组会议有关文件(Relevant Documents of the 1957 Meetings of the Standing Committees of the Academic Divisions and the Meetings of the Discipline Teams). Beijing: Zhongguo kexueyuan danganguan中国科学院档案馆(Archives of Chinese Academy of Sciences), 1957-15-001

who was still in France but had decided to return home. According to the number of recommended persons, the first 26 recommended persons had little change in ranking after this consultation. Wu Wen-Tsün,Tsien H. S., Chang T. S., Chang Wen-yu, Sun Cheng-e and Gao Ji-yu were still in the top ranking, with only Chu Tze-tsin and Chien Jen-yuan in chemistry ranking ahead of schedule.

On January 17, 1957, the Division of Mathematics, Physics and Chemistry held an enlarged meeting of the Standing Committee to further discuss the recommended candidates in groups. After group discussions, the candidates focused on the following 24 people: (mathematics) Wu Wen-Tsün, Shen Yu-cheng, Tseng Y. Y., Cheng Min-Teh and Min Szu-hoa; (astronomy) Li Hen; (mechanics)Tsien H. S. and Kuo Y. H.; (physics) Chang T. S., Chang Wen-yu, Zhao Guangzeng, Hoff Lu, Ho Zah-wei, Wang Pu and Wang De-zhao; (chemistry) Sun Cheng-e, Gao Ji-yu, Chu Tze-tsin, Wang Pao-jen, Tai A. P., Chien Jen-yuan, Xing Qi-yi, Tsai Liu-sheng and Zhang Jin. With the exception of Wang De-zhao and Zhang Jin who recently returned to China and were added temporarily, other candidates were basically taken from Table 8.1 and ranked the former in each division. The only exception was Wang Pu, who ranked last in physics. For the sake of prudence, the meeting finally did not produce formal preliminary candidates for recruitment, but decided to further collect the candidates' "topics of personal works, and then distribute them to all the academic division members for comments."[25]

[25] 1957 nian xuebu changwei huiyi ji xueke xiaozu huiyi youguan wenjian一九五七年学部常委会议及学科小组会议有关文件(Relevant Documents of the 1957 Meetings of the Standing Committees of the Academic Divisions and the Meetings of the Discipline Teams). Beijing: Zhongguo kexueyuan danganguan中国科学院档案馆(Archives of Chinese Academy of Sciences), 1957-15-001.

On May 18, 1956, the Standing Committee of the Division of Biology and Earth Sciences held a meeting to discuss the issue of recruitment. It was believed that "the academic division members of biology and earth sciences may consider and put forward suggestions on the candidates for recruitment, and then make a summary for consideration."[26] Based on the opinions of the academic division members, on November 28, Director Chu Coching, Deputy Directors Tung Ti-Cho, Yin T. H. and Secretary Guo Xingxian initially agreed to recruit 20 more people: (biology 13) Tang F. F., Hu Hsen-hsu, Zhu Yuanding, Chang Hsiang-tung, Liu Sichi, Liou Tchen-Ngo, Ma Jung-chih, Tan Chia-chen, Li Ju-chi, Chen Hsin-tao, Wu Guang, Tseng C. K. and Jao Chin-chih; (earth sciences 7) Feng Jinglan, Fu C. Y., Ren Mei-e, Yu T. Y., Wang Hong-zhen, Wang Zhizhuo and Sun Jingzhi.[27] Among them, Tang F. F., Hu Hsen-hsu and Liu Sichi were all voted out of the academic division members in 1955 mainly because of political and historical problems. Only Chang Hsiang-tung returned to China after 1955. After agreeing on the initial candidates, the biology and earth science division sent letters to all the members for advice. Based on the feedback, Lee S. P. and Weng Wen-bo were added to the earth sciences, temporarily presumed 22.[28]

On December 15, Chu Coching hosted an enlarged meeting of the Standing Committee of the Division of Biology and Earth Sciences at the Institute of Geophysics of the CAS to discuss the issue of recruiting more members of the division. First, Chu Coching introduced and explained the issue of recruiting more academic division members, and then divided the people into two groups, biology and earth sciences, to discuss the tentative list of candidates. Results in biology, 5 people were proposed: (physiology) Chang Hsiang-tung, (biochemistry) Liu Sichi, (botany) Hu Hsen-hsu, (microbiology) Tang F. F., and (genetics) Li Ju-chi. There were 13 people proposed in earth sciences: (pedology) Xiong Yi and Ma Jung-chih; (geography) Ren Mei-e and Sun Jingzhi; (geodesy) Wang Zhizhuo and Fang Jun; (geology) Feng Jinglan, Wang C. C., Wang Hengsheng, Li Chunyu, Xu Keqin, Yu T. Y. and Wang Yuelun. The meeting decided to "send the list to all the division members for comments."[29] Compared with the candidates in the list recommended at this meeting, many additions and deletions were made to those in the provisional list produced before the meeting. 11 people, Zhu Yuanding, Liou Tchen-Ngo, Tan Chia-chen, Chen Hsin-tao, Wu Guang, Tseng C. K., Jao Chin-chih, Fu C. Y., Wang

[26]Shengwu dixue bu dishici zhi shisici changwu kuoda huiyi jiyao生物地学部第十次至十四次常务扩大会议纪要(Summaries of the 10th to 14th Expanded Meetings of the Standing Committee of the Division of Biology and Earth Sciences). Beijing: Zhongguo kexueyuan danganguan中国科学院档案馆(Archives of Chinese Academy of Sciences), 1956-16-003.

[27]Chu Coching竺可桢(2008). *Chu Coching quanji*竺可桢全集(The Complete Works of Coching Chu), Vol. 14. Shanghai: Shanghai keji jiaoyu chubanshe上海科技教育出版社.460.

[28]Chu Coching竺可桢(2008). *Chu Coching quanji*竺可桢全集(The Complete Works of Coching Chu), Vol. 14. Shanghai: Shanghai keji jiaoyu chubanshe上海科技教育出版社.467.

[29]Shengwu dixue bu dishici zhi shisici changwu kuoda huiyi jiyao生物地学部第十次至十四次常务扩大会议纪要(Summaries of the 10th to 14th Expanded Meetings of the Standing Committee of the Division of Biology and Earth Sciences). Beijing: Zhongguo kexueyuan danganguan中国科学院档案馆(Archives of Chinese Academy of Sciences), 1956-16-003.

Hong-zhen, Lee S. P. and Weng Wen-bo were deleted and 7 people, Xiong Yi, Wang C. C., Wang Hengsheng, Li Chunyu, Xu Keqin, Wang Yuelun and Fang Jun were added.

In addition, the zoological group proposed to hire one more member each for entomology, parasitology, invertebrate zoology and ichthyology at the enlarged meeting of the Standing Committee, and put forward a preliminary list: (entomology) Chu Hung-fu and Liu Chi-ying; (parasitology) Feng Lan-chou, Chen Hsin-tao and Wu Guang; (invertebrate zoology) Chen Fei, Tchang Si and Shen Chia-jui. However, the meeting did not determine the candidates, but instructed the group to further consider and put forward opinions.[30]

On January 14, 1957, it was decided at the 15th Meeting of the Standing Committee of the Biology and Earth Science Division to add parasitologist Feng Lan-chou, proposed by the Zoological group, to the preliminary list of candidates.[31] A preliminary list of 23 candidates was confirmed by January 16: (biology) Chang Hsiang-tung, Liu Sichi, Hu Hsen-hsu, Tang F. F., Li Ju-chi, Tseng C. K., Feng Lan-chou, Xiong Yi and Ma Jung-chih; (earth sciences) Ren Mei-e, Sun Jingzhi, Wang Zhizhuo, Fang Jun, Feng Jinglan, Wang C. C., Wang Hengsheng, Li Chunyu, Xu Keqin, Yu T. Y., Wang Yuelun, Fu C. Y., Weng Wen-bo and Lee S. P.[32] Among them, Tseng C. K., Fu C. Y., Weng Wen-bo and Lee S. P. should have been added after consulting opinions on December 15, 1956. On January 16, the Division of Biology and Earth Sciences sent a letter to all its members, asking for their opinions on the 23-person list, stressing that the recruitment of additional personnel should be "focused on addendum, there are still places in the preliminary list", and their opinions should be put forward "in the strict spirit of recommendation".[33]

From November 5 to 9, 1956, the Division of Technological Sciences at the 14th Enlarged Meeting of its Standing Committee began to discuss the recruitment issue.

---

[30]Shengwu dixue bu dishici zhi shisici changwu kuoda huiyi jiyao生物地学部第十次至十四次常务扩大会议纪要(Summaries of the 10th to 14th Expanded Meetings of the Standing Committee of the Division of Biology and Earth Sciences). Beijing: Zhongguo kexueyuan danganguan中国科学院档案馆(Archives of Chinese Academy of Sciences), 1956-16-003.

[31]Shengwu dixue bu changweihui changwei kuoda huiyi yiji xukezu huodong wenjian生物地学部常委会、常委扩大会议以及学科组活动文件(Documents of the Standing Committee Meetings and Enlarged Meetings of the Division of Biology and Earth Sciences, and of the Activities of the Discipline Teams). Beijing: Zhongguo kexueyuan danganguan中国科学院档案馆(Archives of Chinese Academy of Sciences), 1957-16-006.

[32]Shengwu xuebu changwei kuodahui tanlun zengbu dierci xuebu dahui weiyuan mingdan ji tuijian lunwen生物学部常委扩大会讨论增补第二次学部大会委员名单及推荐论文(Discussion at the Enlarged Meeting of the Standing Committee of the Biology Division for List of Supplementing Academic Division Members and Recommending Papers for the Second General Assembly of the Academic Divisions). Beijing: Zhongguo kexueyuan danganguan中国科学院档案馆(Archives of Chinese Academy of Sciences), 1957-16-004.

[33]Shengwu xuebu changwei kuodahui tanlun zengbu dierci xuebu dahui weiyuan mingdan ji tuijian lunwen生物学部常委扩大会讨论增补第二次学部大会委员名单及推荐论文(Discussion at the Enlarged Meeting of the Standing Committee of the Biology Division for List of Supplementing Academic Division Members and Recommending Papers for the Second General Assembly of the Academic Divisions). Beijing: Zhongguo kexueyuan danganguan中国科学院档案馆(Archives of Chinese Academy of Sciences), 1957-16-004.

The discussion was conducted in five subject groups: civil engineering and water, machinery, fuel and chemical engineering, metallurgy, and electrical engineering. At the meeting, people in each group exchanged views on the preliminary candidates and decided to consult with each other at another time before proposing a specific list.[34] After the meeting, except the group of electrical engineering, the other four groups again discussed about the candidates. In the end, 26 people were recommended from the five groups, namely 4 people from the civil engineering and water group: Shi Jiayang, Tong Jun, Wang Juqian and Liu Hui-xian; 2 from the mechanical group: Wu Chung-hua andTsien H. S.; 4 from the fuel and chemical engineering group: Zhang Hongyuan, Tsao Pen-hsi, Wang Dexi and Zhao Zongyu; 4 from the metallurgical group: T. Ko, Huang Peiyun, Hu Yitong and Wang Shifeng; 12 from the electrical engineering group: Feng Ping-chuan, Ming Nai-Ta, Bi Dexian, Luo Peilin, Lin Wei-gan, Chang Tung, Bao Guobao, Mao Henian, Chung S. M., Ding Shunnian, Ge Helin and Xu Shigao.[35]

Among the nominees, no one butTsien H. S. returned to China after 1955. However, there were indeed some scientists in emerging disciplines or those in urgent need of development, such as nuclear chemical engineering experts Tsao Pen-hsi and Wang Dexi, electronics experts Feng Ping-chuan and Luo Peilin, and telecommunications network expert Ming Nai-Ta. In addition, Zhao Zongyu and Bao Guobao who were recommended for "high academic level", but they were not selected as academic division members in 1955 because of "serious political suspicion of the current counter-revolution."[36]

After the meeting, the Division of Technological Sciences put forward six additional candidates according to the opinions of relevant parties, including one for civil engineering and water, one for metallurgy, one for electrical engineering and three for machinery. The Preparatory Committee of the Wuhan Branch of the CAS also recommended 2 candidates for civil engineering and water, and the Ministry of Coal Industry recommended 6 candidates.[37] Together with the candidates recommended by the five groups, a total of 40 were recommended. Later, the division studied the

---

[34]Jishu kexuebu changwei huiyi de youguan wenjian (1–15 ci) 技术科学部常委会议的有关文件(1–15次) (Relevant Documents of the Standing Committee Meetings of the Division of Technological Sciences (No. 1–15). Beijing: Zhongguo kexueyuan danganguan中国科学院档案馆(Archives of Chinese Academy of Sciences), 1956-17-001.

[35]Jishu kexuebu changwei huiyi de youguan wenjian (1–15 ci) 技术科学部常委会议的有关文件(1–15次) (Relevant Documents of the Standing Committee Meetings of the Division of Technological Sciences, No. 1–15). Beijing: Zhongguo kexueyuan danganguan中国科学院档案馆(Archives of Chinese Academy of Sciences), 1956-17-001.

[36]Zhongkeyuan guanyu tuixuan xuebu weiyuan jingguo qingkuang xiang zhongyang de baogao 中科院关于推选学部委员经过情况向中央的报告(Report of the Chinese Academy of Sciences to the Central Committee on the Selection of Academic Division Members). Beijing: Zhongguo kexueyuan danganguan中国科学院档案馆(Archives of Chinese Academy of Sciences), 1955-01-005.

[37]Jishu kexuebu dishiwu zhi shibaci changwei hui wenjian技术科学部第15–18次常委会文件(Documents of the 15th–18th Meetings of Standing Committee of the Division of Technological Sciences). Beijing: Zhongguo kexueyuan danganguan中国科学院档案馆(Archives of Chinese Academy of Sciences), 1956-17-003.

list of candidates and consulted with relevant units. It was initially thought that the number of additional recruitment was unlikely to be large enough and the candidates needed to be reduced. It was planned to convene the 15th Standing Committee meeting on December 7th, 1956 to agree on a preliminary list of 15 to 20 members and then solicit the opinions of all members.[38]

In fact, with the postponement of the 15th Standing Committee meeting, the Division of Technological Sciences did notact as planned. Director Ny Tsai-ze and the secretary drew up a preliminary list of 9 candidates before the meeting, including 1 for civil engineering and water and 1 for chemical fuel, 2 for machinery and metallurgy respectively, and 3 for electrical engineering.[39] On January 14, 1957, the 15th Standing Committee meeting was held in the form of an enlarged one chaired by Deputy Director Mao T. E.[40] After studying the academic investigation of the nine candidates, the participating members discussed the reference list. The main opinions were as follows: there were fewer members in the chemical fuel group, and there was only one petroleum professional candidate in the reference list, so consider adding one chemical professional candidate; judging from the academic achievements, only one of the three candidates in the electrical engineering could be considered for additional member, and another candidate was needed; metallurgical group may consider reducing 2 candidates to 1; due to the lack of mineral processing and coal majors, additional candidates should be considered. In view of the importance and meticulous nature of the additional recruitment, the meeting did not agree on specific candidates. It was decided to invite the Standing Committee members who did not attend the meeting to deliberate after the meeting and to solicit the opinions of all the members after putting forward a new reference list according to the opinions of all the Standing Committee members.[41]

Division of Philosophy and Social Sciences was the latest to start preparing preliminary candidates. On November 22, 1956, the division sent a letter to all the members referring to the *Opinions* to "draw attention to whether the academic division members should be added and to propose a list to be added".[42] By the end

---

[38]Jishu kexuebu changwei huiyi de youguan wenjian (1–15 ci) 技术科学部常委会议的有关文件(1–15次) (Relevant Documents of the Standing Committee Meetings of the Division of Technological Sciences, No. 1–15). Beijing: Zhongguo kexueyuan danganguan中国科学院档案馆(Archives of Chinese Academy of Sciences), 1956-17-001.

[39]Jishu kexuebu dishiwu zhi shibaci changwei hui wenjian技术科学部第15–18次常委会文件(Documents of the 15th–18th Meetings of Standing Committee of the Division of Technological Sciences). Beijing: Zhongguo kexueyuan danganguan中国科学院档案馆(Archives of Chinese Academy of Sciences), 1956-17-003.

[40]The participants included members of the Standing Committee, Li Wencai, Wu Xuelin, Hou T. P., Huang Wenxi and Qian Zhidao and members in Beijing, Shi Zhiren, Tsai Fang-yin, Zhang Guangdou and Zhang Wei.

[41]Jishu kexuebu dishiwu zhi shibaci changwei hui wenjian技术科学部第15–18次常委会文件(Documents of the 15th–18th Meetings of Standing Committee of the Division of Technological Sciences). Beijing: Zhongguo kexueyuan danganguan中国科学院档案馆(Archives of Chinese Academy of Sciences), 1956-17-003.

[42]Guanyu zhaokai dierci xuebu weiyuan dahui he zengbu xuebu weiyuan wenti xiang zhongyang de qingshi baogao关于召开第二次学部委员大会和增补学部委员问题向中央的请示报告(Report

of December, 16 of the 60 members and the Wuhan branch had recommended 37 candidates from 11 disciplines:

History: Ji Wenfu, Zhou Yiliang, Tang Changru, Ku Chieh-kan and Li Jiannong;

International Relations: Wang Tieh-yai, Meng Yongqian, Liu Simu, Jin Zhonghua, Mei Ju-Ao and Qiao Guanhua;

Linguistics: Fang Guangtao, Lu Zhiwei, Cen Qixiang, Fu Maoji and Tang Lan;

Literature: You Guo'en, Qian Zhongshu and Li Erzhong;

Pedagogy: Liao Shicheng, Chen Heqin, Cao Fu and Dong Weichuan;

Law: Tsien T. S., Zhang Zhirang and Chou Keng-Shang;

Economics: Wu Ching-chao and Sun Xiaocun;

Ethnology: Fei Xiaotong, Woo Ting-liang and Xia Kangnong;

Art: Cai Yi;

Philosophy: Ma Yifu, Xiong Shili, Lü Cheng and Li Fanfu;

Archaeology: Chen Meng-chia.[43]

No one among them returned to China after 1955, and only 2 persons, Fu Maoji and Cao Fu, returned to China after 1949. Judging from the academic background of the candidates, most of the 16 members recommended candidates based on academic standards. It is because apart from the "old revolutionaries" Meng Yongqian and Li Fanfu, most of the candidates were experts with academic achievements. It is also worth noting that this time candidates for international relations, pedagogy, law, ethnology and other specialties were recommended, including Tsien T. S., Chou Keng-Shang and Woo Ting-liang, the first academicians of Academia Sinica. These specialties were considered in the selection of academic division members in 1955, but were eventually deleted as a whole, so that no one was selected as an academic division member.

On the choice of candidates, the Philosophy and Social Sciences Division adopted the method of studying one by one together with the heads of their affiliated research institutes, and held a meeting on January 7, 1957. The meeting believed: "When adding members to the Academic Division, the division members of the four disciplines of law, ethnology, pedagogy and international relations should be added, and some outstanding scientists should be appropriately added to the existing disciplines."[44] In the end, 20 preliminary candidates were prepared: (history) Ji Wenfu;

to the Central Committee for Instructions on Convening the Second General Assembly of Members of the Academic Divisions and the Problems of Supplementing Academic Division Members). Beijing: Zhongguo kexueyuan danganguan中国科学院档案馆(Archives of Chinese Academy of Sciences), 1957-02-020.

[43]Guanyu zhaokai dierci xuebu weiyuan dahui he zengbu xuebu weiyuan wenti xiang zhongyang de qingshi baogao关于召开第二次学部委员大会和增补学部委员问题向中央的请示报告(Report to the Central Committee for Instructions on Convening the Second General Assembly of Members of the Academic Divisions and the Problems of Supplementing Academic Division Members). Beijing: Zhongguo kexueyuan danganguan中国科学院档案馆(Archives of Chinese Academy of Sciences), 1957-02-020.

[44]Guanyu zhaokai dierci xuebu weiyuan dahui he zengbu xuebu weiyuan wenti xiang zhongyang de qingshi baogao关于召开第二次学部委员大会和增补学部委员问题向中央的请示报告(Report

(international relations) Qiao Guanhua, Mei Ju-Ao and Zhang Xiruo; (linguistics) Fang Guangtao and Lu Zhiwei; (literature) Feng Xuefeng and Lu Kanru; (pedagogy) Liao Shicheng, Meng Xiancheng, Lin Liru and Cheng Fangwu; (law) Tsien T. S., Zhang Zhirang and Chou Keng-Shang; (economics) Ji Chaoding and Sun Xiaocun; (ethnology) Liu Chun, Fei Xiaotong and Xia Kangnong.[45] Among them, 12 were on the initial list of nominees, and 8, Zhang Xiruo, Feng Xuefeng, Lu Kanru, Meng Xiancheng, Lin Liru, Cheng Fangwu, Ji Chaoding and Liu Chun were added at the meeting. In addition to these 20 people, the meeting thought thatKu Chieh-kan could be considered as a candidate in history. Gu is the founder of the "School of Distinguishing Ancient History" in the field of historiography. He was elected as the first academician of Academia Sinica in 1948, but was not elected as an academic division member in 1955.

(4) Proposal of the Principle of Recruitment and the Re-consideration of Candidates

When each division was preparing the initial candidates, the total number of recommended candidates exceeded 120. By the middle of January 1957, the number, after being cut down by the divisions themselves, still reached 70 or 80. The Party group of the CAS felt that "it is very difficult to balance among different divisions and disciplines".[46] After several discussions within and outside the Party and after asking for instructions from Zhou Yang, vice ministers of the Publicity Department of the CPC Central Committee, and Nie Rongzhen, the principles of additional recruitment were put forward:

> Supplement strictly; better to be deficient than to be indiscriminate; make up for the leak without asking for completeness; consider mainly from the academic level rather than from disciplines, regions and departments. Because the shortcomings of incompleteness can be overcome by augmenting later. If the candidate is not suitable, it will lead to passivity and increase the difficulty in unity. The selection of candidates should be based on the results of

---

to the Central Committee for Instructions on Convening the Second General Assembly of Members of the Academic Divisions and the Problems of Supplementing Academic Division Members). Beijing: Zhongguo kexueyuan danganguan中国科学院档案馆(Archives of Chinese Academy of Sciences), 1957-02-020.

[45]Guanyu zhaokai dierci xuebu weiyuan dahui he zengbu xuebu weiyuan wenti xiang zhongyang de qingshi baogao关于召开第二次学部委员大会和增补学部委员问题向中央的请示报告(Report to the Central Committee for Instructions on Convening the Second General Assembly of Members of the Academic Divisions and the Problems of Supplementing Academic Division Members). Beijing: Zhongguo kexueyuan danganguan中国科学院档案馆(Archives of Chinese Academy of Sciences), 1957-02-020.

[46]Guanyu zhaokai dierci xuebu weiyuan dahui he zengbu xuebu weiyuan wenti xiang zhongyang de qingshi baogao关于召开第二次学部委员大会和增补学部委员问题向中央的请示报告(Report to the Central Committee for Instructions on Convening the Second General Assembly of Members of the Academic Divisions and the Problems of Supplementing Academic Division Members). Beijing: Zhongguo kexueyuan danganguan中国科学院档案馆(Archives of Chinese Academy of Sciences), 1957-02-020.

> the consultation. Those approved by most people should be considered as candidates; those only agreed by some people and disapproved by others would be considered next time.[47]

The main purpose of these principles was to increase the number of candidates strictly rather than fill the vacancy, and to obtain the majority approval, at the same time, taking academic level as the main basis for considering candidates to downplay the influence of other factors on the increase in recruitment. The aim was to recruit more scholars with higher academic level, improve the authority of the Academic Divisions, and strengthen the academic leadership of the Divisions, but also in order not to affect the unity of the academic community.

On January 29, 1957, Zhang Jinfu, Vice-president of the CAS and Party secretary of the CAS Party group, presided over a joint meeting of directors of the Academic Divisions to discuss the increase in the number of academic division members and the second conference of academic division members. He stressed the principles of increasing the number of division members and pointed out that the candidates should be selected from three aspects: first, scientists who were not selected due to ideological reasons in 1955, such as Hu Hsen-hsu, Liu Sichi and Tang F. F.; second, scientists newly returned from abroad, such as Chang Hsiang-tung and Chang Wen-yu; third, the "new forces", such as Wu Wen-Tsün. In addition, the number of additional recruitment required should not be too large, with a total of 20 to 30.[48]

There was a difference between the new scope of candidates determined at this meeting and the scope of candidates approved at the General Assembly meeting in 1956. First of all, it lowered the political standard and "made an exception" for the so-called reactionary scientists. Secondly, it emphasized the recruitment of "new forces" instead of the disciplines newly developing or needing development urgently with insufficient force. Each division responded positively to the principles of increasing recruitment and the new scope of candidates. The directors of the Academic Divisions agreed, and the Standing Committee members and the members in Beijing also basically agreed and supported.[49] As a result, each Academic Division, in accordance with the spirit of the joint meeting of directors, re-considered the preliminary list of candidates.

---

[47] Guanyu zhaokai dierci xuebu weiyuan dahui he zengbu xuebu weiyuan wenti xiang zhongyang de qingshi baogao关于召开第二次学部委员大会和增补学部委员问题向中央的请示报告(Report to the Central Committee for Instructions on Convening the Second General Assembly of Members of the Academic Divisions and the Problems of Supplementing Academic Division Members). Beijing: Zhongguo kexueyuan danganguan中国科学院档案馆(Archives of Chinese Academy of Sciences), 1957-02-020.

[48] Chu Coching竺可桢(2008). *Chu Coching quanji*竺可桢全集(The Complete Works of Coching Chu), Vol. 14. Shanghai: Shanghai keji jiaoyu chubanshe上海科技教育出版社.505.

[49] Guanyu zhaokai dierci xuebu weiyuan dahui he zengbu xuebu weiyuan wenti xiang zhongyang de qingshi baogao关于召开第二次学部委员大会和增补学部委员问题向中央的请示报告(Report to the Central Committee for Instructions on Convening the Second General Assembly of Members of the Academic Divisions and the Problems of Supplementing Academic Division Members). Beijing: Zhongguo kexueyuan danganguan中国科学院档案馆(Archives of Chinese Academy of Sciences), 1957-02-020.

The Division of Mathematics, Physics and Chemistry adopted the method of first proposing a preliminary list of candidates, then soliciting the opinions of all members, and finally voting by secret ballot at the expanded meeting of the Standing Committee. On February 4th, 1957, based on the results of the deliberation at the enlarged Standing Committee meeting on January 17th, 8 candidates were proposed: Wu Wen-Tsün,Tsien H. S., Kuo Y. H., Wang De-zhao, Chang T. S., Chang Wen-yu, Chu Tze-tsin and Tsai Liu-sheng. Wu Wen-Tsün,Tsien H. S., Kuo Y. H., Chang T. S. and Chang Wen-yu were all the first or second in their disciplines in Table 8.1. In terms of chemistry, Gao Ji-yu and Sun Cheng-e, who ranked highest in Table 8.1, were eliminated and replaced by Chu Tze-tsin and Tsai Liu-sheng. In astronomy, there was no reservation. When soliciting opinions from all the members this time, the Division of Mathematics, Physics and Chemistry provided the resumes and thesis titles of the candidates for reference. 32 members replied and all agreed thatTsien H. S. and Chang Wen-yu should be selected. The number of people who agreed to the other six were: 31 for Wu Wen-Tsün, 31 for Kuo Y. H., 30 for Chang T. S., 28 for Chu Tze-tsin, 27 for Wang De-zhao and 26 for Tsai Liu-sheng.[50]

On February 23, 1957, the Division of Mathematics, Physics and Chemistry held an enlarged meeting of the Standing Committee to decide on the eight persons. Group discussions went first. The chemical group had a big difference of opinion on Chu Tze-tsin. Organic chemists generally believed that "there are still some people whose academic level is equal to or even higher than that of Chu Tze-tsin, such as Xing Qi-yi and Wang Pao-jen"; and that all three should be selected, or none of them should be selected. In the end, all the members present decided by secret ballot to select 7 people, Wu Wen-Tsün,Tsien H. S., Kuo Y. H., Chang Wen-yu, Wang De-zhao, Chang T. S. and Tsai Liu-sheng as the formal reference candidates for the selection of the Academy. Wu Wen-Tsün,Tsien H. S. and Chang Wen-yu got the most votes, 22 each. Kuo Y. H. and Tsai Liu-sheng came second with 21 votes each, while Chang T. S. and Wang De-zhao came third with 20 votes each. Chu Tze-tsin had only 6 votes and was defeated because his number of votes did not exceed 2/3 of the total.[51]

The Division of Biology and Earth Sciences adopted the method of consulting with members in Beijing and soliciting opinions from members outside Beijing through correspondence based on the original candidates. By March 26, the preliminary biological candidates Tseng C. K., Li Ju-chi, Ma Jung-chih and Xiong Yi had been

---

[50] 1957 nian xuebu changwei huiyi ji xueke xiaozu huiyi youguan wenjian一九五七年学部常委会议及学科小组会议有关文件(Relevant Documents of the 1957 Meetings of the Standing Committees of the Academic Divisions and the Meetings of the Discipline Teams). Beijing: Zhongguo kexueyuan danganguan中国科学院档案馆(Archives of Chinese Academy of Sciences), 1957-15-001.

[51] 1957 nian xuebu changwei huiyi ji xueke xiaozu huiyi youguan wenjian一九五七年学部常委会议及学科小组会议有关文件(Relevant Documents of the 1957 Meetings of the Standing Committees of the Academic Divisions and the Meetings of the Discipline Teams). Beijing: Zhongguo kexueyuan danganguan中国科学院档案馆(Archives of Chinese Academy of Sciences), 1957-15-001.

deleted, leaving only 5 persons[52] including Hu Hsen-hsu, Tang F. F., Liu Sichi, Chang Hsiang-tung and Feng Lan-chou. By April 1, earth sciences retained 4 persons, Feng Jinglan, Wang C. C., Fu C. Y. and Lee S. P.[53] However, there were objections to the addition of Hu Hsen-hsu, including Peking University professors Li Tsi-tung and Chang Chin-yueh, who were members of the Division of Biology and Earth Sciences, and Chun W. Y., director of the South China Institute of Botany of the CAS, as well as Wang Fuxiong and Zhong Buqiu of the Institute of Botany of the CAS where Hu was working.[54] On April 1, when the CAS held a meeting to discuss the list of additional recruits, Hu Hsen-hsu was deleted. Wang Shanyuan, a newly returned researcher in the Institute of Epidemiology of the Ministry of Health, was added.[55]

The method adopted by the Division of Technological Sciences was: first, put forward a preliminary proposal for the candidates based on the decision of the 15th enlarged meeting of the Standing Committee and the opinions of the Standing Committee members who did not attend the meeting; then, solicited the opinions of all the members; after that, proposed a new primary list; and finally the decision was made by discussions and secret ballot at the enlarged meeting of the Standing Committee. On February 7, 1957, the division sent a letter to all the members to solicit opinions on the preliminary proposal for the candidates in which 8 people were included: (civil engineering and water) Liu Hui-xian and Wang Juqian; (machinery) Wu Chung-hua; (fuel chemical industry) Wang Dexi and Zhao Zongyu; (mining and metallurgy) Wang Dezi and Huang Peiyun; (electrical engineering) Bao Guobao.[56] Among them, Wang Dezi was not among the candidates originally recommended by the various discipline teams.

After soliciting the opinions of all the members and many studies, the Division of Technological Sciences put forward a new primary list of 13 persons for consideration at the enlarged meeting of the Standing Committee held on April 5, 1957. The list included (civil engineering and water) Liu Hui-xian and Wang Juqian; (machinery) Wu Chung-hua; (fuel chemical industry) Wang Dexi, Zhao Zongyu, Zhang Hongyuan and Tsao Pen-hsi; (mining and metallurgy) Wang Dezi, Huang Peiyun and T. Ko; (electrical nngineering) Bao Guobao, Ming Nai-Ta and Tsai Chin-tao. The candidates in the preliminary proposal were all on the list. Tsai Chin-tao, T. Ko, Zhang Hongyuan, Ming Nai-Ta and Tsao Pen-hsi were included in the list

---

52Chu Coching竺可桢(2008). *Chu Coching quanji*竺可桢全集(The Complete Works of Coching Chu), Vol. 14. Shanghai: Shanghai keji jiaoyu chubanshe上海科技教育出版社.545.

53Chu Coching竺可桢(2008). *Chu Coching quanji*竺可桢全集(The Complete Works of Coching Chu), Vol. 14. Shanghai: Shanghai keji jiaoyu chubanshe上海科技教育出版社.548–549.

54Chu Coching竺可桢(2008). *Chu Coching quanji*竺可桢全集(The Complete Works of Coching Chu), Vol. 14. Shanghai: Shanghai keji jiaoyu chubanshe上海科技教育出版社.545, 549.

55Chu Coching竺可桢(2008). *Chu Coching quanji*竺可桢全集(The Complete Works of Coching Chu), Vol. 14. Shanghai: Shanghai keji jiaoyu chubanshe上海科技教育出版社.549.

56Jishu kexuebu dishiwu zhi shibaci changwei hui wenjian技术科学部第15–18次常委会文件(Documents of the 15th–18th Meetings of Standing Committee of the Division of Technological Sciences). Beijing: Zhongguo kexueyuan danganguan中国科学院档案馆(Archives of Chinese Academy of Sciences), 1956-17-003.

but not in the preliminary one. The last four were originally among the candidates originally recommended by the various discipline teams.

On April 5, the enlarged meeting of the Standing Committee of the Division of Technological Sciences discussed the new primary list first, then proposed two additional candidates, Luo Ying and Mao Henian, and basically reached the following consensus: (1) Because of the close combination of technological sciences and practice, the members of the Division of Technological Sciences should not only have a fairly theoretical level, but also have considerable production and practical experience. In 1955, when selecting and appointing academic division members, it was not given enough consideration, and attention should be paid to in this additional recruitment. (2) Technological sciences have a wide range of fields and categories and are directly related to the national industrialization task. Besides the existing members, there are still many categories missing in the Division of Technological Sciences, but the number of additional places is limited. When the number of additional places is to be increased again, the Academy should consider increasing the number of places appropriately. (3) Luo Ying is unable to work due to illness, but has considerable level and achievements. He is a rare talent in the bridge engineering field and should be considered for selection. (4) Although Ming Nai-Ta has numerous works, they have not been verified in practice. It is better not to take them for consideration.[57]

After that, the participants were divided into three groups: metallurgy, machinery and electrical engineering, fuel chemical, and civil engineering and water. They studied 15 preliminary candidates and put forward a compressed list. The metallurgical, mechanical and electrical group proposed Wang Dezi, Huang Peiyun, Zhang Hongyuan, Wu Chung-hua, Wang Juqian, Mao Henian and Zhao Zongyu. The fuel chemical group proposed Zhang Hongyuan, Bao Guobao, Zhao Zongyu, Wang Dezi, Wu Chung-hua, Wang Juqian or Luo Ying. The civil engineering and water group proposed Wu Chung-hua, Zhao Zongyu, T. Ko, Wang Juqian, Liu Hui-xian, Zhang Hongyuan, Tsai Chin-tao, Bao Guobao or Mao Henian. Wang Dexi, Tsao Pen-hsi and Ming Nai-Ta were eliminated.

After summarizing the list according to the candidates recommended by the groups, the meeting put forward a list of nine people: Zhao Zongyu, Wu Chung-hua, Zhang Hongyuan, Wang Juqian, Wang Dezi, Bao Guobao, Mao Henian, Huang Peiyun and T. Ko. In this way, Liu Hui-xian, director of the Institute of Civil Engineering and Architecture of the CAS, who was previously elected with high voice, was eliminated along with Luo Ying and Tsai Chin-tao. Then the meeting held that 7 people could be elected from the list by secret ballot, and it was better to elect one from Bao Guobao and Mao Henian, one from Huang Peiyun and T. Ko. As a result, Wu Chung-hua, Zhao Zongyu, Zhang Hongyuan, Wang Dezi and Wang Juqian all received a unanimous vote (15 votes), while Huang Peiyun, Mao Henian, Bao Guobao and T. Ko successively received a decreasing number of votes. Since

[57]Jishu kexuebu dishiwu zhi shibaci changwei hui wenjian技术科学部第15–18次常委会文件(Documents of the 15th–18th Meetings of Standing Committee of the Division of Technological Sciences). Beijing: Zhongguo kexueyuan danganguan中国科学院档案馆(Archives of Chinese Academy of Sciences), 1956-17-003.

Bao Guobao had only 2 votes less than Mao Henian, the meeting thought that if an additional place could be added, Bao should be included.

The Division of Philosophy and Social Sciences adopted the method of consulting with the members in Beijing and soliciting opinions from the members outside Beijing through correspondence. The Party group of the CAS also solicited the opinions of the heads of relevant departments on some candidates. On February 20, 1957, the Party group of the Academy sent a letter to Zhang Wentian, First Deputy Minister of Foreign Affairs: "In terms of international relations, Chen Han-seng is a member of the Division of Philosophy and Social Sciences, and now it is proposed to add Zhang Xiruo and Qiao Guanhua. Is it appropriate to do so considering academic prestige and the need for Party comrades? Or who should be better to add?"[58] On the same day, a letter was also sent to Dong Biwu, Secretary of the Central Supervisory Committee: "As for the law, the Division of Philosophy and Social Sciences intends to nominate Tsien T. S., Chou Keng-Shang and Zhang Zhirang as academic division members. We wonder if it is appropriate. At the same time, we consider that a Party member should be appointed as a member of the law discipline. We don't know who is better to add."[59] Visibly, because of the need for work or close ties with ideology, the Party group required that some of the candidates need to be Party members in the division.

By April 1, the Division of Philosophy and Social Sciences had drawn up a list of 10 people, including Chou Keng-Shang, OuyangYuqian, Tsien T. S., Zhang Jinfu, Pei Lisheng and Liu Daosheng.[60] The list changed a lot from the primary 20-member list which was proposed after brewing on January 7, 1957. Among them, Ouyang Yuqian was the president of the Central Academy of Drama, and Liu Daosheng was the deputy director of the division. They and Zhang Jinfu and Pei Lisheng were all experts or cadres in the Party, and were not included in the original 20-member list.

(5) **Appointment of the Final Candidates**

On April 22, 1957, the CAS held its Seventh Executive Meeting to discuss the list of candidates to be recruited by each division. The meeting agreed to the candidates proposed by each division and decided to submit the list to the State Council for

---

[58]Guanyu zhaokai dierci xuebu weiyuan dahui he zengbu xuebu weiyuan wenti xiang zhongyang de qingshi baogao关于召开第二次学部委员大会和增补学部委员问题向中央的请示报告(Report to the Central Committee for Instructions on Convening the Second General Assembly of Members of the Academic Divisions and the Problems of Supplementing Academic Division Members). Beijing: Zhongguo kexueyuan danganguan中国科学院档案馆(Archives of Chinese Academy of Sciences), 1957-02-020.

[59]Guanyu zhaokai dierci xuebu weiyuan dahui he zengbu xuebu weiyuan wenti xiang zhongyang de qingshi baogao关于召开第二次学部委员大会和增补学部委员问题向中央的请示报告(Report to the Central Committee for Instructions on Convening the Second General Assembly of Members of the Academic Divisions and the Problems of Supplementing Academic Division Members). Beijing: Zhongguo kexueyuan danganguan中国科学院档案馆(Archives of Chinese Academy of Sciences), 1957-02-020.

[60]Chu Coching竺可桢(2008). *Chu Coching quanji*竺可桢全集(The Complete Works of Coching Chu), Vol. 14. Shanghai: Shanghai keji jiaoyu chubanshe上海科技教育出版社.549.

approval.[61] A total of 34 candidates were proposed by all divisions: 7 people from the Division of Mathematics, Physics and Chemistry including Wu Wen-Tsün,Tsien H. S. and Chang Wen-yu; 9 from the Division of Biology and Earth Sciences, Tang F. F., Liu Sichi, Chang Hsiang-tung, Feng Lan-chou, Wang Shanyuan, Feng Jinglan, Wang C. C., Fu C. Y. and Lee S. P.; 6 from the Division of Technological Sciences, including Wang Juqian, Bao Guobao and Zhang Hongyuan; 12 from the Division of Philosophy and Social Sciences, including Ouyang Yuqian, Chou Keng-Shang, Tsien T. S., Lu Zhiwei, Zhang Jinfu, Pei Lisheng and Liu Daosheng.[62] So far, the number of candidates for the Philosophy and Social Sciences Division had increased by 2. The Division of Technological Sciences did not decide the candidates strictly according to the secret ballot result, but chose Bao Guobao.

As for candidates with political and historical problems or "reactionary thoughts" such as Tang F. F., Bao Guobao, Zhao Zongyu and Liu Sichi, the Party group of the CAS asked Nie Rongzhen and Zhou Yang for instructions and put forward suggestions: "After several years of examination, some of the political problems in history have been found out, and some of them are only suspicious and cannot be confirmed before and after liberation. If their academic level is adequate, we can consider adding them this time. In case of political problems in the future, we can also deal with them according to the law. We are not passive. The scientific community has changed a lot over the past few years on the issue of reactionary ideology, and they themselves have also changed somewhat. If their academic level is high enough, they can also be considered as candidates." Nie and Zhou agreed to these suggestions.[63]

Zhang Jinfu, Pei Lisheng and Liu Daosheng were Party members and administrative cadres. Their academic level was not high. Their selection undoubtedly went against the principles of increasing recruitment. At that time, the Party group of the CAS knew that scientists "expressed their approval, but it is estimated that many people still have opinions".[64] After the executive meeting of the Academy

---

[61]Zhongkeyuan zhaokai diwuci zhi dijiuci yuanwu changwu huiyi de tongji jilu jiyao ji youguan cailiao中科院召开第五次至第九次院务常务会议的通知、记录、纪要及有关材料(Notices, Minutes, Summaries and Related Materials of the Fifth to Ninth Executive Meetings of the Chinese Academy of Sciences). Beijing: Zhongguo kexueyuan danganguan中国科学院档案馆(Archives of Chinese Academy of Sciences), 1957-02-031.

[62]Chu Coching竺可桢(2008). *Chu Coching quanji*竺可桢全集(The Complete Works of Coching Chu), Vol. 14. Shanghai: Shanghai keji jiaoyu chubanshe上海科技教育出版社.562.

[63]Guanyu zhaokai dierci xuebu weiyuan dahui he zengbu xuebu weiyuan wenti xiang zhongyang de qingshi baogao关于召开第二次学部委员大会和增补学部委员问题向中央的请示报告(Report to the Central Committee for Instructions on Convening the Second General Assembly of Members of the Academic Divisions and the Problems of Supplementing Academic Division Members). Beijing: Zhongguo kexueyuan danganguan中国科学院档案馆(Archives of Chinese Academy of Sciences), 1957-02-020.

[64]Guanyu zhaokai dierci xuebu weiyuan dahui he zengbu xuebu weiyuan wenti xiang zhongyang de qingshi baogao关于召开第二次学部委员大会和增补学部委员问题向中央的请示报告(Report to the Central Committee for Instructions on Convening the Second General Assembly of Members of the Academic Divisions and the Problems of Supplementing Academic Division Members). Beijing: Zhongguo kexueyuan danganguan中国科学院档案馆(Archives of Chinese Academy of Sciences), 1957-02-020.

on April 22, according to the instructions of the Publicity Department of the CPC Central Committee, by studying and weighing, the Party group of the CAS decided to remove Zhang Jinfu, Pei Lisheng and Liu Daosheng from the list of candidates.[65] This shows that the Party group of the CAS, especially Zhang Jinfu and Pei Lisheng, who were the main leaders, were quite liberal and considerate of the overall situation.

However, according to the results of the consultation, the Party group of the CAS proposed not to recruit more candidates for law, ethnology, pedagogy or art. On May 3, 1957, the Party group of the Academy said in the above-mentioned report for instructions: "On the basis of the results of soliciting opinions from all sides, it is considered that there are few problems in these disciplines that are not supplemented at present. The scientific community has different opinions on the list of candidates, and there may be more problems if supplemented. Therefore, it is better to recruit them after the research work is carried out and when things are mature."[66]

According to the above decision of the Party group of the CAS, the list of 34 people approved by the executive meeting of the CAS on April 22 was soon cut to 25 people: (7 from the Division of Mathematics, Physics and Chemistry) Wu Wen-Tsün,Tsien H. S., Chang Wen-yu, Kuo Y. H., Tsai Liu-sheng, Chang T. S. and Wang De-zhao; (5 in Biology Division[67]) Tang F. F., Liu Sichi, Chang Hsiang-tung, Feng Lan-chou and Wang Shanyuan; (4 in Division of Earth Sciences) Feng Jinglan, Wang C. C., Fu C. Y. and Lee S. P.; (6 in Division of Technological Sciences) Wu Chung-hua, Zhao Zongyu, Zhang Hongyuan, Wang Dezi, Wang Juqian and Bao Guobao; (3 in Division of Philosophy and Social Sciences) Ji Wenfu, Lü Cheng and Lu Zhiwei. The candidates deleted all from the Division of Philosophy and Social Sciences, including Chou Keng-Shang and Tsien T. S., the first academicians of Academia Sinica.

On May 3, the Party group of the CAS reported the list of the 25 members in a request for instructions submitted to the Publicity Department of the CPC Central Committee, Nie Rongzhen and the Central Committee on the recruitment of additional members of the Academic Divisions. However, young people from the Institute of Geophysics of the CAS opposed Lee S. P.'s selection, saying he could not be tied with Fu C. Y. Some people also thought if Li was selected, Weng Wen-bo should be

---

[65]Guanyu zhaokai dierci xuebu weiyuan dahui he zengbu xuebu weiyuan wenti xiang zhongyang de qingshi baogao关于召开第二次学部委员大会和增补学部委员问题向中央的请示报告(Report to the Central Committee for Instructions on Convening the Second General Assembly of Members of the Academic Divisions and the Problems of Supplementing Academic Division Members). Beijing: Zhongguo kexueyuan danganguan中国科学院档案馆(Archives of Chinese Academy of Sciences), 1957-02-020.

[66]Guanyu zhaokai dierci xuebu weiyuan dahui he zengbu xuebu weiyuan wenti xiang zhongyang de qingshi baogao关于召开第二次学部委员大会和增补学部委员问题向中央的请示报告(Report to the Central Committee for Instructions on Convening the Second General Assembly of Members of the Academic Divisions and the Problems of Supplementing Academic Division Members). Beijing: Zhongguo kexueyuan danganguan中国科学院档案馆(Archives of Chinese Academy of Sciences), 1957-02-020.

[67]On May 3rd, 1957, the Division of Biology and Earth Sciences was divided into the Division of Biology and the Division of Earth Sciences.

selected.[68] Later, Li was deleted and the formal preliminary candidates to be hired became 24.

In the request report, the Party group of the CAS also hoped that the Central Committee would delegate power and the CAS would appoint its own candidates: "If approved by the Central Government, the new Academic Division Members can be announced through the method of' 'being recommended by the Chinese scientific community and selected by the CAS', which is stipulated in the regulations of the Academic Divisions, instead of the method of appointment approved by the State Council. In the future, it is planned to consider further democratic election by the scientific community. As for the election method, the current divisions can be entrusted to organize the scientific community to discuss and propose it."[69]

On May 20, 1957, the preparatory meeting for the Second General Assembly of CASAD Members was held. President Kuo Mo-jo stressed at the meeting that recruiting more academic division members was one of the main tasks of the conference.[70] In the evening, each division held a meeting of all its members. Vice President Woo Y. H. said when presiding over the meeting of the Division of Mathematics, Physics and Chemistry: This time the government handed over the power to scientists to make our own decisions and promote democracy. The candidates "will be approved by Old Guo, not by the State Council".[71] From this, we can see that the Central Government had agreed to delegate power, and the CAS could independently approve the academic division members.

At the meeting of each division that night, the members of the Biology Division, the Earth Science Division and the Technology Division all exchanged views on the issue of increasing the academic division members. The focus of the opinions was on the criteria, scope and principles of recruitment. Maa Dah-you suggested that candidates should be considered solely from academic achievements, not from the history of candidates as officials. Zhang Wen-you said: It is inappropriate to recruit

---

[68]Chu Coching竺可桢(2008). *Chu Coching quanji*竺可桢全集(The Complete Works of Coching Chu), Vol. 14. Shanghai: Shanghai keji jiaoyu chubanshe上海科技教育出版社.569.

[69]Guanyu zhaokai dierci xuebu weiyuan dahui he zengbu xuebu weiyuan wenti xiang zhongyang de qingshi baogao关于召开第二次学部委员大会和增补学部委员问题向中央的请示报告(Report to the Central Committee for Instructions on Convening the Second General Assembly of Members of the Academic Divisions and the Problems of Supplementing Academic Division Members). Beijing: Zhongguo kexueyuan danganguan中国科学院档案馆(Archives of Chinese Academy of Sciences), 1957-02-020.

[70]Zhongkeyuan 1957 nian zhaokai dierci xuebu dahui shulihua xuebu xiaozu de huiyi jilu中科院1957年召开第二次学部大会数理化学部小组的会议记录(Minutes of the Meetings of Group of the Division of Mathematics, Physics and Chemistry at the Second General Assembly of Members of Academic Divisions of the Chinese Academy of Sciences held in 1957). Beijing: Zhongguo kexueyuan danganguan中国科学院档案馆(Archives of Chinese Academy of Sciences), 1957-2-018.

[71]Zhongkeyuan 1957 nian zhaokai dierci xuebu dahui shulihua xuebu xiaozu de huiyi jilu中科院1957年召开第二次学部大会数理化学部小组的会议记录(Minutes of the Meetings of Group of the Division of Mathematics, Physics and Chemistry at the Second General Assembly of Members of Academic Divisions of the Chinese Academy of Sciences held in 1957). Beijing: Zhongguo kexueyuan danganguan中国科学院档案馆(Archives of Chinese Academy of Sciences), 1957-2-018.

Fig. 8.1 Venue of the Second General Assembly of the CASAD Members

new returnees as academic division members except a few. Because great achievements made abroad does not necessarily mean that they will be made at home. Too much attention has been paid to new returnees, those who returned in previous years would have their own opinions; academic division members should be supported by the domestic masses. Huang Bingwei and Tien C. C. believed that increasing the number of academic division members should take care of the balance of disciplines, "because the mission of the Academic Division is to promote national scientific work, otherwise the blank and weak disciplines will never occupy an important position."[72] In addition, some academic division members put forward opinions on specific candidates. For example, Chu Coching, Jaw Jeou-Jang, Hsia Chien-pei and Zhang Wen-you all suspected that the Division of Technological Sciences not only took care of the distribution of disciplines but also possibly took care of administrative positions when preparing the initial candidates (Fig. 8.1).[73]

[72]Dui dierci xuebu weiyuanhui qingkuang jianbao ji gebie weiyuan fayan zhailu对第二次学部委员会情况简报及个别委员发言摘录(Bulletins on the Second General Assembly of Members of Academic Divisions and Excerpts from Speeches of Individual Members). Beijing: Zhongguo kexueyuan danganguan中国科学院档案馆(Archives of Chinese Academy of Sciences), 1957-02-021.

[73]Dui dierci xuebu weiyuanhui qingkuang jianbao ji gebie weiyuan fayan zhailu对第二次学部委员会情况简报及个别委员发言摘录(Bulletins on the Second General Assembly of Members of Academic Divisions and Excerpts from Speeches of Individual Members). Beijing: Zhongguo kexueyuan danganguan中国科学院档案馆(Archives of Chinese Academy of Sciences), 1957-02-021.

Before the meeting, when Yu Wen, deputy secretary-general of the CAS, went to Hangzhou to report to Vice President Lee J. S. on his work, Lee J. S. put forward opinions on the recruitment of additional members of the Division of Philosophy and Social Sciences. Li was not in favor of excluding accomplished non-Marxist social scientists from the division. He said: The division includes many subjects. It covers widely and has many difficulties in work. It cannot be compared with the natural science divisions, "to have only three special people" as its members is still worthy of consideration. Although some people from the scientific community in the old society do not have Marxist-Leninist views, they have abundant historical data, and it is better to consider them. If there are fewer recruits, it will affect not only work but also political unity. If the list cannot be changed, it should also be properly explained when the list is announced so as to leave room to avoid being trapped in passivity.[74]

Most of these opinions did not have any substantial impact on the results of the supplement. On May 22, the final day of the preparatory meeting, each division decided on its formal preliminary candidates. Before the decision was made, the Academy issued resumes of the candidates and catalogues of their works as reference materials. Wu Chung-hua, Zhao Zongyu and Wang Juqian were elected by the Division of Technological Sciences, while Zhang Hongyuan, Wang Dezi and Bao Guobao lost the election. The official candidates originally considered by the other four divisions were all selected finally after discussion. When the decision was made, the Biology Division adopted the method of secret ballot. Chang Hsiang-tung won the most votes (37 votes), while Feng Lan-chou, Tang F. F. and Liu Sichi all lost 1 vote to Chang Hsiang-tung, and Wang Shanyuan got 26 votes. When the Division of Earth Sciences decided, with the exception of Feng Jinglan, Wang C. C. and Fu C. Y., there was a strong call for Ren Mei-e and Wang Zhizhuo to be elected, but "because of the need to balance the five divisions, everyone thought it was appropriate not to include them".[75] In the end, 21 members were elected as the academic division members. The list is as follows:

7 people from Division of Mathematics, Physics and Chemistry: Wu Wen-Tsün, Kuo Y. H.,Tsien H. S., Wang De-zhao, Chang Wen-yu, Chang T. S. and Tsai Liu-sheng; 5 from Division of Biology: Wang Shanyuan, Chang Hsiang-tung, Feng Lan-chou, Tang F. F. and Liu Sichi; 3 from Division of Earth Sciences: Wang C. C., Feng Jinglan and Fu C. Y.; 3 from Division of Technological Sciences: Wu Chung-hua, Zhao Zongyu and Wang Juqian; 3 people from Division of Philosophy and Social Sciences: Ji Wenfu, Lü Cheng and Lu Zhiwei.

On May 29, the 12th Executive Meeting of the CAS approved the appointment of additional academic division members according to the resolution of each Academic

---

[74]Yu Wen tongzhi fu hang xiang Li fuyuanzhang huibao gongzuo qingkuang ji Li fuyuanzhang de yijian郁文同志赴杭向李副院长汇报工作情况及李副院长的意见(Comrade Yu Wen Went to Hangzhou to Report to Vice-president Li on his Work and Vice-president Li's Opinions). Beijing: Zhongguo kexueyuan danganguan中国科学院档案馆(Archives of Chinese Academy of Sciences), 1957-01-019.

[75]Chu Coching竺可桢(2008). *Chu Coching quanji*竺可桢全集(The Complete Works of Coching Chu), Vol. 14. Shanghai: Shanghai keji jiaoyu chubanshe上海科技教育出版社.580.

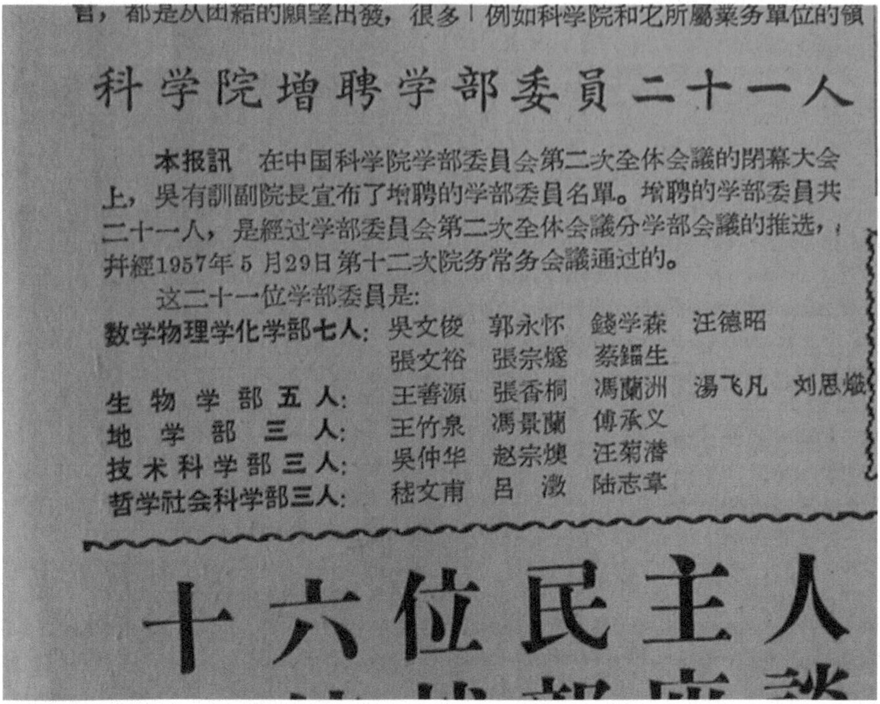

都是从团結的願望出發，很多 | 例如科学院和它所屬業务單位的領

科学院增聘学部委員二十一人

本报訊 在中国科学院学部委員会第二次全体会議的閉幕大会上，吳有訓副院長宣布了增聘的学部委員名單。增聘的学部委員共二十一人，是經过学部委員会第二次全体会議分学部会議的推选，并經1957年5月29日第十二次院务常务会議通过的。

这二十一位学部委員是：

数学物理学化学部七人： 吳文俊 郭永怀 錢学森 汪德昭 張文裕 張宗燧 蔡鎦生

生物学部五人： 王善源 張香桐 馮蘭洲 湯飞凡 刘思職

地学部三人： 王竹泉 馮景蘭 傅承义

技术科学部三人： 吳仲华 赵宗燠 汪菊潜

哲学社会科学部三人： 嵇文甫 吕 澂 陆志韋

十六位民主人

Fig. 8.2 The result of the supplement announced by *Guangming Daily* on May 31, 1957

Division.[76] The next day, it was read out by Woo Y. H. at the closing ceremony of the Second General Assembly of CASAD Members. On May 31, it was published by *People's Daily*, *Guangming Daily*, etc. (Fig. 8.2).[77]

There were no full-time Party members and administrative cadres among the 21 selected. They had basically received systematic scientific training in the West. Among them, 20 had studied abroad, 12 had got doctor's degrees (1 got it in China) and 3 had got master's degrees. Except Wang Juqian, all the candidates had delivered or published a certain number of academic papers or works. The one who published most was Feng Lan-chou, about 115 articles; the others, between 16 and more than 50 articles. Some works of the selected had certain international influence, including the

---

[76]Zhongguo kexueyuan zhaokai dishici zhi dishiliuci yuanwu changwu huiyi de tongzhi jilu jiyao ji youguan cailiao中国科学院召开第十次至第十六次院务常务会议的通知、记录、纪要及有关材料(Notices, Minutes, Summaries and Related Materials of the 10th to 16th Executive Meetings of the Chinese Academy of Sciences). Beijing: Zhongguo kexueyuan danganguan中国科学院档案馆(Archives of Chinese Academy of Sciences), 1957-02-032.

[77]Zhongguo kexueyuan zengpin xuebu weiyuan mingdan中国科学院增聘学部委员名单(The List of the Supplementary Academic Division Members of the Chinese Academy of Sciences). *Renmin ribao*人民日报(People's Daily), 1957-5-31 (1); kexueyuan zengpin xuebu weiyuan 21 ren科学院增聘学部委员二十一人(21 Supplementary Academic Division Members of the Chinese Academy of Sciences). *Guangming ribao*光明日报(Guangming Daily), 1957-5-31 (1).

English monograph *Engineering Cybernetics* published byTsien H. S. in the United States in 1954.[78]

Judging from the academic experience of the selected, 9 people, or about 42.9% had worked as professors or researchers in western academic institutions, namely, Wu Wen-Tsün,Tsien H. S., Kuo Y. H., Wang De-zhao, Chang Wen-yu, Tang F. F., Liu Sichi, Wang Shanyuan and Wu Chung-hua, while most of the other selected candidates had worked or were just working in important academic institutions in China.

Judging from the age of the selected, most of them were in their 40s and 50s. Wu Wen-Tsün and Wu Chung-hua were only 38 and 39 years old, and both won the Science Prize of the CAS in January 1957, four months before they were selected as academic division members. Wu Wen-Tsün won the first prize for his research on characteristic class and imbedding class; Wu Chung-hua won the second prize for research on gas turbine. All these reflected that the increase in recruitment was mainly based on academic standards.

Judging from the number of selected candidates in each division, the Division of Mathematics, Physics and Chemistry had the largest number, followed by the Division of Biology, and each of the other three divisions had 3 candidates. To some extent, it shows that the CAS paid more attention to basic disciplines in the fields of natural science such as mathematics, physics and chemistry. Nearly one-third of those selected returned to China after the founding of the CAS in 1955, includingTsien H. S., Kuo Y. H., Wang De-zhao, Chang Wen-yu, Wang Shanyuan and Chang Hsiang-tung, demonstrating that the new "returnees" were indeed favored.

8.2 Co-optation of Academic Division Members in 1980

(1) From the Impact of Political Movements to the Resumption of Activities of Academic Divisions

On June 8, 1957, that was shortly after the end of the increase in the number of academic division members in 1957, New China launched a large-scaled anti-rightist movement. In this political movement, many academic division members were attacked and criticized. 11 academic division members, Tseng Chao-lun, Chien Wei-zang, Meng Chao-ying, Lei Tianjue, Hsieh Chia-yung, Yu Ruihuang, Liu Sichi, Yuan Hanqing, Sheng Tongsheng, Xiang Da and Shen Zhiyuan were classified as rightists and were removed from their posts.[79] After the anti-rightist movement, under the continued influence of the "left" ideology, the "criticizing bourgeois" movement

[78]H. S. Tsien(1954). *Engineering Cybernetics*. New York: McGraw-Hill Book Company, Inc.

[79]Zhongguo kexueyuan zhaokai diwuci zhi dibaci yuanwu changwu huiyi de tongzhi jilu jiyao ji youguan cailiao中科院召开第5次至第8次院务常务会议的通知、记录、纪要及有关材料(Notices, Minutes, Summaries and Related Materials of the 5th to 8th Executive Meetings of the Chinese Academy of Sciences). Beijing: Zhongguo kexueyuan danganguan中国科学院档案馆(Archives of Chinese Academy of Sciences), 1958-02-005.

took place. Intellectuals bore the brunt and became the main target of criticism. Moreover, since 1958, the movement of "planting the red flag and pulling out the white flag" was rapidly developing in institutions of higher learning, intellectuals were criticized for taking the so-called "white professional road"; the wind of "The Great Leap Forward" spread throughout the country. In 1963, the National Socialist Education Movement was launched. In the stormy waves of the political movements, the Academic Divisions of the CAS rose and fell several times and stumbled.

After the "Cultural Revolution" broke out, the Academic Divisions struggled even more. On January 24, 1967, the "Revolutionary Rebellion Group" and other organizations in the Beijing area of the CAS held a joint power-taking conference, established the "Joint Power-taking Committee of Revolutionary Rebels of the CAS (Beijing Area)" and seized the leadership of the CAS. In Article 7 of Circular No. 1 after seizing power, they announced the abolition of the Academic Divisions, accusing the Academic Divisions as a product of the implementation of the "counter-revolutionary revisionist scientific research line" and the "expert line" by the CAS.[80] As a result, the divisions were "smashed". During the "Cultural Revolution", most of the academic division members were criticized and denounced as "reactionary academic authorities" and many were brutally persecuted.[81] The membership system survived in name only.

After Deng Xiaoping presided over the daily work of the Central Committee and the State Council in 1975, he sent Hu Yaobang and Li Chang to work in the CAS. According to Deng Xiaoping's directive spirit of rectifying the CAS, Hu Yaobang, after going to the CAS, made a thorough investigation and drafted a report outline for recovering losses, preserving the team and developing science and technology—*Several Issues Concerning Science and Technology Work* (renamed as *Report Outline on the Work of the Chinese Academy of Sciences* after revision). This outline put forward a plan to "set up academic divisions" in terms of organizational rectification. Deng Xiaoping fully affirmed the outline.[82] However, "to fight back the rightward overturn" soon began, the *Report Outline on the Work of the Chinese Academy of Sciences* was criticized by the Gang of Four. On February 16, 1976, the core group of

---

[80] Xue Pangao薛攀皋. Zhongguo kexueyuan shouci minzhu xuanju xuebu weiyuan (yuanshi) 中国科学院首次民主选举学部委员 (院士) (The First Democratic Election of the Chinese Academy of Sciences for Academic Division Members (Academicians)). In Xue Pangao (2011). *Keyuan qianchen wangshi*科苑前尘往事(Past Events of the Scientific Centre). Beijing: Kexue chubanshe 科学出版社.42.

[81] Guanyu Zhongguo kexueyuan xuebu dangqian jixiang gongzuo de qingshi baogao关于中国科学院学部当前几项工作的请示报告(Report for Instructions on Several Current Tasks of the Chinese Academy of Sciences). In Zhongguo kexueyuan bangongting中国科学院办公厅(General Office of the Chinese Academy of Sciences) (Eds.). *Zhongguo kexueyuan nianbao*中国科学院年报(The Annals of the Chinese Academy of Sciences) 1979: 211.

[82] 40 nian de zhuiqiu yu tansuo: Jinian Zhongguo kexueyuan jianyuan 40 zhounian四十年的追求与探索——纪念中国科学院建院四十周年(Forty Years of Pursuit and Exploration: Commemorating the 40th Anniversary of the Founding of the Chinese Academy Sciences). In Zhongguo kexueyuan bangongting中国科学院办公厅(General Office of the Chinese Academy of Sciences) (Eds.). *Zhongguo kexueyuan nianbao*中国科学院年报(The Annals of the Chinese Academy of Sciences) 1990: 5.

the CAS made a decision that "the establishment of the Academic Divisions is wrong and should be revoked".[83] Therefore, the reconstruction of the Academic Divisions plan aborted.

After the end of the "Cultural Revolution", the wrong line of "extreme left" was corrected, and China's scientific cause began to recover. On March 18, 1978, the National Science Congress opened in Beijing. Deng Xiaoping made an important speech, clearly pointing out that "the key to the four modernizations is the modernization of science and technology", "science and technology are productive forces", "mental workers serving socialism are part of the working people".[84] This congress fully mobilized the enthusiasm of the vast number of intellectuals, making the pursuit of knowledge and science a social trend and raising science and technology to an unprecedented position.[85] In December, the Third Plenary Session of the 11th CPC Central Committee was held in Beijing. The meeting clearly put forward the guiding principle of "emancipating the mind, using our brains, seeking truth from facts and looking forward in unity"; resolutely stopped using the slogan of "taking class struggle as the key link" and made the strategic decision to shift the focus of the whole Party's work to socialist modernization from 1979.[86]

In accordance with the spirit of the two conferences, the CAS began to implement the policies on science and technology and the intellectuals, and gradually shifted the work center of the whole Academy to scientific research. On January 15, 1979, the Central Government approved the resumption ofthe Academic Division activities by the CAS.[87] On January 24, the CAS issued the *Notice on Resuming the Work of the Academic Divisions* to all academic division members and their units and other relevant departments. The "Notice" stipulated: "Upon the approval of the Central Government on January 15, the Academic Divisions began to resume the activities. Therefore, it is decided to set up an office of the CAS Academic Divisions to handle the daily work under the leadership of the president, vice-president and the General

[83]Xue Pangao薛攀皋. Guanyu Zhongguo kexueyuan xuebu huifu yu chongjian gongzuo de huiyi关于中国科学院学部恢复与重建工作的回忆(The Memories of Restoration and Reconstruction of the Academic Divisions of Chinese Academy of Sciences). In Xue Pangao (2011). *Keyuan qianchen wangshi*科苑前尘往事(Past Events of the Scientific Centre). Beijing: Kexue chubanshe科学出版社.46.

[84]Deng Xiaoping邓小平(1994). Zai quanguo kexue dahui kaimushi shang de jianghua在全国科学大会开幕式上的讲话(Speech Delivered at the Opening Ceremony of the National Science Congress). In Deng Xiaoping. *Deng Xiaoping wenxuan* (II)邓小平文选(第2卷) (Selected Works of Deng Xiaoping), Vol. 2. Beijing: Renmin chubanshe人民出版社.85–100.

[85]Interviewed and organized by Sun Weilin孙伟林and Meng Wei孟玮(2008). Yi quanguo kexue dahui qianqian houhou: Yuan guojia kewei fuzhuren Wu Mingyu yixitan忆全国科学大会前前后后——原国家科委副主任吴明瑜一席谈(Before and After the National Science Conference: A Talk by Former Deputy Director of the State Science and Technology Commission Wu Mingyu). *Minzhu yu kexue*民主与科学(Democracy and Science) (1): 2–8.

[86]Mao Lirui毛礼锐and Shen Guanqun沈灌群(Eds.) (1989). *Zhongguo jiaoyu tongshi*(VI) 中国教育通史(第6卷) (General History of Chinese Education), Vol. 6. Jinan: Shandong jiaoyu chubanshe 山东教育出版社.252.

[87]The Division of Philosophy and Social Sciences was renamed the Chinese Academy of Social Sciences in May 1977.

Assembly."[88] On January 26, a tea party for academic division members of the CAS in Beijing was held in the Great Hall of the People, marking the official resumption of activities 12 years after the Academic Divisions were "smashed". The next day, the *People's Daily* reported on the tea party and announced the names of the academic division members attending the meeting.[89]

(2) **Brewing and Preparations for the Co-optation of Academic Division Members**

On March 23, 1979, the Party group of the CAS put forward "Several Opinions on Transferring the Work Center of the Whole Academy to Scientific Research" (hereinafter referred to as "*Opinions*"). The *Opinions* believed that "the institutions of the Academy should be reformed to strengthen academic leadership and scientific management", and "in order to strengthen academic leadership, the organization and activities of the Academic Divisions should be resumed." At the same time, it was emphasized that "the Academic Divisions are an academic leading organization under the leadership of the General Assembly. The number of academic division members should be expanded, and scientists who have made important contributions over the past 20 years should be selected, with special attention paid to the selection of middle-aged outstanding scientists."[90] On March 29, the CAS put forward the following considerations for the work of the Academic Divisions in its *Request for Instructions on the Work of the CAS Academic Divisions and Related Issues Such as the President and Vice President* (hereinafter referred to as "Request for Instructions") submitted to the State Council:

(1) Supplement the academic division members and enrich and perfect the institutions of the Academic Divisions. The "Request for Instructions" pointed out: There are 115 original academic division members of natural science still alive, with an average age of 73. Therefore, the Academic Divisions urgently need to supplement the new forces and expand the number of their members. In order to make the divisions vigorous, it is proposed to select a group of outstanding scientists with high academic level. The total number of members in the planning divisions was about 300. As for the scope of candidates for academic division members, besides mainland scientists, the CAS also considered Taiwan and

[88]Guanyu huifu xuebu gongzuo de tongzhi关于恢复学部工作的通知(Notice on Resuming the Work of the Academic Divisions). In Zhongguo kexueyuan bangongting中国科学院办公厅(General Office of the Chinese Academy of Sciences) (Eds.). *Zhongguo kexueyuan nianbao* 中国科学院年报(The Annals of the Chinese Academy of Sciences) 1979: 198.

[89]Tuanjie quanguo kexuejia jinkuai shixian sihua团结全国科学家尽快实现四化(Unite Scientists Across the Country to Realize the Four Modernizations as Soon as Possible). *Renmin ribao*人民日报(People's Daily), 1979-01-27 (1).

[90]Zhongguo kexueyuan dangzu guanyu ba quanyuan gongzuo de zhongxin zhuanyi dao keyan shang lai de jidian yijian中国科学院党组关于把全院工作的中心转移到科研上来的几点意见(Several Opinions of the Party Group of Chinese Academy of Sciences on Transferring the Work Center of the Whole Academy to Scientific Research). In Zhongguo kexueyuan bangongting中国科学院办公厅(General Office of the Chinese Academy of Sciences) (Eds.). *Zhongguo kexueyuan nianbao*中国科学院年报(The Annals of the Chinese Academy of Sciences) 1979: 9.

foreign scientists. The plan was to reserve part of the quota for famous scientists in Taiwan, and to consider and put forward specific plans when conditions were ripe. A small number of foreign scientists (including ethnic Chinese) with international academic status who had made contributions to China's scientific cause would be selected as foreign academic division members, with the aim of "keeping close ties with the scientific communities of all countries in the world". With regard to additional selection, it was stressed "to pay attention to the full development of democracy and agree in advance on the additional selection method for the academic division members."

(2) Revise the Regulations of the Academic Divisions to give full play to the role of Academic Divisions. The "Request for Instructions" said: "The CAS Academic Divisions have absorbed outstanding scientists from all over the country. They are actually a national authoritative academic institution and represent the scientific and technological level of our country. According to the new situation, we will revise the original Regulations of the Academic Divisions, clarify their nature and tasks, make their work rule-based, give full play to their academic leadership role under the leadership of General Assembly, and unite scientists from all over the country and promote the contacts and cooperation between the CAS and colleges, universities and various industrial departments."

(3) Prepare to hold the Fourth General Assembly of CASAD Members. After about half a year's preparation, the meeting will be held in November 1979, the time for the 30th anniversary of the founding of the CAS. The meeting will announce the result of co-optation of the academic division members, elect the directors, deputy directors and Standing Committee members of the Academic Divisions to form a new leadership of the Academic Divisions, and discuss and revise the Regulations of the Academic Divisions, etc.[91]

Soon, the report was approved by the State Council. Later, the CAS held a joint meeting of the Standing Committee members of the academic divisions from May 17 to 21, 1979, mainly to discuss issues related to the co-optation of academic

[91]Guanyu Zhongguo kexueyuan xuebu dangqian jixiang gongzuo de qingshi baogao关于中国科学院学部当前几项工作的请示报告(Report for Instructions on Several Current Tasks of the Chinese Academy of Sciences). In Zhongguo kexueyuan bangongting中国科学院办公厅(General Office of the Chinese Academy of Sciences) (Eds.). *Zhongguo kexueyuan nianbao*中国科学院年报(The Annals of the Chinese Academy of Sciences) 1979: 210–212; Xue Pangao薛攀皋. Guanyu Zhongguo kexueyuan xuebu huifu yu chongjian gongzuo de huiyi关于中国科学院学部恢复与重建工作的回忆(The Memories of Restoration and Reconstruction of the Academic Divisions of Chinese Academy of Sciences). In Xue Pangao (2011). *Keyuan qianchen wangshi*科苑前尘往事(Past Events of the Scientific Centre). Beijing: Kexue chubanshe科学出版社.48–50.

division members.[92] Members of the Standing Committees at the meeting unanimously agreed that the urgent task for each Academic Division was to supplement its members and enrich and improve its institutions. At the same time, they suggested that the co-optation of the academic division members should be based mainly on their achievements and contributions in scientific research and their ideological and political performance, and attention should be paid to the selection of younger outstanding talents and female scientists; the supplementary work should ensure the quality and proceed from reality, neither to simply fill up the positions nor to restrict the number. As for the total number of academic division members, the initial consideration was about 300. The meeting seriously discussed the steps of the work to supplement the academic division members and decided to adopt two ways for recommending candidates: one was candidates recommended directly by the existing academic division members; the other was candidates recommended by research institutions, colleges, universities and societies according to organizational systems. The Standing Committee of the Academic Division shall be responsible for the evaluation of the academic level, achievements and contributions of the candidates and the agreement on the list of candidates. As for the final candidates, it was decided to be elected by voting of the existing division members, the result of which would be reported to the State Council for approval by the CAS. The meeting drafted the "Measures for Supplementing Academic Division Members of the CAS" and was subsequently discussed and approved by the CAS General Assembly.[93]

The "Measures for Supplementing Academic Division Members of the CAS" stipulated first the criteria for the selection of academic division members: "The CAS Academic Divisions are composed of outstanding scientists from China. Researchers, professors, senior engineering and technical personnel, etc. or those who are of comparable level, have made important achievements and contributions in scientific and technological research may be elected as members of the Academic Divisions."[94] This standard emphasized that the candidate was an "outstanding scientist" and was

---

[92]Guanyu Zengbu Zhongguo kexueyuan xuebu weiyuan de tongzhi关于增补中国科学院学部委员的通知(Notice on Supplementing Academic Division Members of the Chinese Academy of Sciences). In Zhongguo kexueyuan bangongting中国科学院办公厅(General Office of the Chinese Academy of Sciences) (Eds.). *Zhongguo kexueyuan nianbao*中国科学院年报(The Annals of the Chinese Academy of Sciences) 1979: 210.

[93]Guanyu tuijian he linxuan Zhongguo kexueyuan xuebu weiyuan houxuanren de tongzhi he jige juti wenti de yijian关于推荐和遴选中国科学院学部委员候选人的通知和几个具体问题的意见(Notice on Recommending and Selecting Candidates for Academic Division Members of the Chinese Academy of Sciences and Opinions on Several Specific Issues). Beijing: Zhongguo kexueyuan danganguan中国科学院档案馆(Archives of Chinese Academy of Sciences), 1980-36-008; Guanyu Zengbu Zhongguo kexueyuan xuebu weiyuan de tongzhi关于增补中国科学院学部委员的通知(Notice on Supplementing Academic Division Members of the Chinese Academy of Sciences). In Zhongguo kexueyuan bangongting中国科学院办公厅(General Office of the Chinese Academy of Sciences) (Eds.). *Zhongguo kexueyuan nianbao*中国科学院年报(The Annals of the Chinese Academy of Sciences) 1979: 210.

[94]Guanyu tuijian he linxuan Zhongguo kexueyuan xuebu weiyuan houxuanren de tongzhi he jige juti wenti de yijian关于推荐和遴选中国科学院学部委员候选人的通知和几个具体问题的意见(Notice on Recommending and Selecting Candidates for Academic Division Members of the Chinese Academy of Sciences and Opinions on Several Specific Issues). Beijing: Zhongguo

based on considerable academic level, important scientific and technological achievements and contributions. It is close to the international academician standard that requires outstanding academic achievements and contributions. After deciding on the selection criteria, the "Measures for Supplementing Academic Division Members of the CAS" stipulated four steps for the addition, namely, recommendation, selection, evaluation and election. Among them, the recommendation referred to the recommendation of candidates for academic division members. There were two ways: one was jointly recommended by at least two academic division members who were required to send the recommendation to the "Office for Supplement of Academic Division Members" of the CAS; the other was recommended by ministries and commissions of the State Council, the General Political Department, the National Defense Science and Technology Commission, research institutes and institutions of higher learning affiliated to the office of National Defense Industry, branches, institutes and institutions of higher learning affiliated to the CAS, research institutes and institutions of higher learning directly affiliated to provinces, municipalities and autonomous regions, and societies directly affiliated to the China Association for Science and Technology. Candidates in relevant disciplines of the system, the region and the unit were mainly recommended. After each unit recommended, the recommendation form would be sent to its competent department for selection.

The reason why the CAS adopted the above two ways for recommendation was mainly due to the consideration that people in Scientific Community had been in a state of isolation during the ten-year "Cultural Revolution" and they lacked the knowledge of the dynamics of their specialized disciplines and fields. At the same time, researchers working in the national defense and military systems could not communicate with the local scientists or publish research reports or papers publicly because of their long-term anonymity due to the confidentiality of their work. Their achievements and contributions were difficult for the outside world to understand. However, these two recommendation ways could make up for the defects.

According to the provisions of the "Measures for Supplementing Academic Division Members of the CAS", the evaluation was conducted in two steps. First, the Standing Committees of the Academic Divisions reviewed the list of candidates recommended by the Academic Division members and selected by various departments and agreed on a preliminary list of candidates which was, then, announced to all academic division members for comments, and after full consultation, the list of candidates was determined. The election was held by secret ballot at the meeting of all academic division members. If a member of the Academic Division was unable to attend the meeting for some reason, "a sealed letter can be used to vote." The results of the election would be submitted by the CAS to the State Council for approval.[95]

---

kexueyuan danganguan中国科学院档案馆(Archives of Chinese Academy of Sciences), 1980-36-008.

[95]Guanyu tuijian he linxuan Zhongguo kexueyuan xuebu weiyuan houxuanren de tongzhi he jige juti wenti de yijian关于推荐和遴选中国科学院学部委员候选人的通知和几个具体问题的意见(Notice on Recommending and Selecting Candidates for Academic Division Members of the Chinese Academy of Sciences and Opinions on Several Specific Issues). Beijing: Zhongguo

At that time, there were some differences as to whether the election should be divided into divisions or a unified election of all the academic division members. At first, it was suggested that candidates in any discipline or professional field should not be elected by the division, but by all academic division members. Only in this way could they be truly "academic division members of the CAS". Finally, after a unified understanding, it was decided to elect by division, because "difference in profession makes one feel worlds apart", it is difficult for members of this discipline to objectively judge the academic work of new members of another discipline, it's so even in the same discipline with more branches. If a unified election had been carried out, it would have inevitably resulted in the dispersal of votes.[96]

At that time, there were also different opinions on how to stipulate the number of approval votes required for candidates to be electedas academic division members. The majority believed that strict control should be exercised and only those who obtained more than two-thirds of the votes could be elected. However, many people thought that it was appropriate for those who obtained more than half of the votes to be elected. The reason was that the scientific and technological community had been disconnected for more than a decade because of the "Cultural Revolution" and other reasons. Despite the recommended materials, there still was a lack of adequate knowledge about many candidates. Due to the relatively large number of people who agreed with the former opinion and the principle that "the minority is subordinate to the majority", the former opinion was written into the "Measures for Supplementing Academic Division Members of the CAS" (preliminary draft).[97] After the joint meeting of all divisions, Tsien San-Tsiang presided over the joint meeting of directors, deputy directors and acting directors of all divisions. The meeting held that the actual situation should be taken into consideration, and that strict control should be exercised without going to extremes so that the election results would not be affected by undue influence. It was suggested that the number of votes needed for the elected academic division members should be changed from more than two thirds to more than half.[98]

---

kexueyuan danganguan中国科学院档案馆(Archives of Chinese Academy of Sciences), 1980-36-008.

[96]Xue Pangao薛攀皋. Guanyu Zhongguo kexueyuan xuebu huifu yu chongjian gongzuo de huiyi关于中国科学院学部恢复与重建工作的回忆(The Memories of Restoration and Reconstruction of the Academic Divisions of Chinese Academy of Sciences). In Xue Pangao (2011). *Keyuan qianchen wangshi*科苑前尘往事(Past Events of the Scientific Centre). Beijing: Kexue chubanshe科学出版社.51–52.

[97]Zhongguo kexueyuan xuebu weiyuan zengbu banfa中国科学院学部委员增补办法(Measures for Supplementing Academic Division Members of the Chinese Academy of Sciences). In Zhongguo kexueyuan bangongting中国科学院办公厅(General Office of the Chinese Academy of Sciences) (Eds.). *Zhongguo kexueyuan nianbao*中国科学院年报(The Annals of the Chinese Academy of Sciences) 1979: 212–213.

[98]Xue Pangao薛攀皋. Guanyu Zhongguo kexueyuan xuebu huifu yu chongjian gongzuo de huiyi关于中国科学院学部恢复与重建工作的回忆(The Memories of Restoration and Reconstruction of the Academic Divisions of Chinese Academy of Sciences). In Xue Pangao (2011). *Keyuan qianchen wangshi*科苑前尘往事(Past Events of the Scientific Centre). Beijing: Kexue chubanshe科学出版社.52–53.

On June 20, 1979, the CAS submitted to the State Council the "Report on the Work of Supplementing the CAS Academic Division Members".[99] In July 10, the State Council approved it and forwarded it to the Revolutionary Committees of all provinces, municipalities and autonomous regions, ministries and commissions of the State Council, agencies directly under the State Council, the General Political Department, the National Defense Science and Technology Commission, the National Defense Industry Office, and the China Association for Science and Technology for research and implementation.[100]

(3) Official Commencement and Results of the Co-optation of the Academic Division Members

On July 23, 1979, the CAS issued official letters and notices to relevant departments, units and academic division members, inviting them to recommend candidates for academic division members. The supplement of the members of the Academic Divisions immediately began. Both the official letter and the notice stated that the number was about 180 and the election was to be completed by the end of 1979. It was hoped that all grass-roots units completed the recommendation work by the end of August and all provinces, municipalities, autonomous regions and departments completed the selection by the end of September. Before October 5th, the recommendation letters would be sent to the Office for Supplement of Academic Division Members of the CAS. The notice required the academic division members to send the recommendation letter before August 31.[101]

After the State Council approved the report of the CAS on the work of supplementing academic division members on July 10, many departments and units began to study and deploy the work, and raised questions at the same time. After the official start of the supplement work on July 23, the current (acting) and deputy directors of various divisions of the CAS listened to the report of the Office for Supplement of Academic Division Members and discussed several specific issues in the work of

[99] Guanyu tuijian he linxuan Zhongguo kexueyuan xuebu weiyuan houxuanren de tongzhi he jige juti wenti de yijian关于推荐和遴选中国科学院学部委员候选人的通知和几个具体问题的意见(Notice on Recommending and Selecting Candidates for Academic Division Members of the Chinese Academy of Sciences and Opinions on Several Specific Issues). Beijing: Zhongguo kexueyuan danganguan中国科学院档案馆(Archives of Chinese Academy of Sciences), 1980-36-008.

[100] Guanyu tuijian he linxuan Zhongguo kexueyuan xuebu weiyuan houxuanren de tongzhi he jige juti wenti de yijian关于推荐和遴选中国科学院学部委员候选人的通知和几个具体问题的意见(Notice on Recommending and Selecting Candidates for Academic Division Members of the Chinese Academy of Sciences and Opinions on Several Specific Issues). Beijing: Zhongguo kexueyuan danganguan中国科学院档案馆(Archives of Chinese Academy of Sciences), 1980-36-008.

[101] Guanyu tuijian he linxuan Zhongguo kexueyuan xuebu weiyuan houxuanren de tongzhi he jige juti wenti de yijian关于推荐和遴选中国科学院学部委员候选人的通知和几个具体问题的意见(Notice on Recommending and Selecting Candidates for Academic Division Members of the Chinese Academy of Sciences and Opinions on Several Specific Issues). Beijing: Zhongguo kexueyuan danganguan中国科学院档案馆(Archives of Chinese Academy of Sciences), 1980-36-008.

recommending and selecting candidates for academic division members. On August 16, the Office for Supplement of Academic Division Members drafted "Opinions on Several Specific Issues in the Work of Recommending and Selecting Candidates for the CAS Academic Division Members", giving clear opinions on the allocation of additional number of academic division members, the selection of various kinds of talents, the specific specialties, achievements and contributions of the recommended candidates for academic division members, as well as papers and other issues.[102]

Regarding the allocation of additional number of academic division members, the "Opinions" considered it inconvenient to assign. The reason was that candidates initially recommended by various parties had to go through the Standing Committee of the Academic Division or a certain range of peer reviews to solicit opinions, from which a preliminary list of candidates could be agreed upon, and then a formal list of candidates would be determined after re-examination and approval, with the existing members of the Academic Divisions holding unequal elections. The "Opinions" also clearly stated the additional number of academic division members for various disciplines asked by some departments and units: "Based on recommendation, selection and preliminary examination, the general scope of supplement will be negotiated according to specific conditions and needs, and no specific provisions will be made at present." Regarding the selection of various talents, the "Opinions" emphasized that attention should be paid to the selection of talents from new or weak disciplines when recommending and selecting, "but their achievements, contributions and level should be considered first, and the standards should not be lowered." At the same time, it was also possible to select talents who were not researchers, professors, or senior engineering and technical personnel, but who "have a comparable level". The "Opinions" also put forward that under the premise of giving top priority to "achievements, contributions and academic level", attention should be paid to the selection of younger outstanding talents and outstanding female scientists as candidates. As for the selection of the former, it was because the academic division members would take part in certain activities and undertake some work.[103]

Regarding the achievements and contributions of the recommended candidates for academic division members, the "Opinions" pointed out that it was the main basis for the academic division members to be added. At the same time, when recommending, it was required to specify as much as possible the research results obtained by the recommended person and their actual effects on the national economic construction

---

[102]Guanyu tuijian he linxuan Zhongguo kexueyuan xuebu weiyuan houxuanren de tongzhi he jige juti wenti de yijian关于推荐和遴选中国科学院学部委员候选人的通知和几个具体问题的意见(Notice on Recommending and Selecting Candidates for Academic Division Members of the Chinese Academy of Sciences and Opinions on Several Specific Issues). Beijing: Zhongguo kexueyuan danganguan中国科学院档案馆(Archives of Chinese Academy of Sciences), 1980-36-008.

[103]Guanyu tuijian he linxuan Zhongguo kexueyuan xuebu weiyuan houxuanren de tongzhi he jige juti wenti de yijian关于推荐和遴选中国科学院学部委员候选人的通知和几个具体问题的意见(Notice on Recommending and Selecting Candidates for Academic Division Members of the Chinese Academy of Sciences and Opinions on Several Specific Issues). Beijing: Zhongguo kexueyuan danganguan中国科学院档案馆(Archives of Chinese Academy of Sciences), 1980-36-008.

and national defense construction, or their role in promoting the development of certain disciplines and fields. In view of the outstanding achievements and contributions made by some scientific and technological workers who published no papers or few papers, the "Opinions" also pointed out: "when recommending, the value and level of the results are mainly considered, and the presence or absence of papers or the number of papers is not the determining factor."[104]

In the process of recommending candidates for academic division members, some members said that the requirement of more than two members jointly recommending candidates had certain difficulties in implementation. On this issue, the joint meeting of the academic divisions' current (acting) and deputy directors decided that "the original regulations will remain unchanged. If there are difficulties, it is also possible to recommend separately."[105] This left room for flexibility in the original regulations. In addition, according to the opinions of some regions and departments, the CAS extended the deadline for sending the materials recommending candidates for academic division members to October 20, no later than the end of October.[106]

By the end of the recommendation and selection process, more than 1100 candidates had been directly recommended by the academic division members and recommended by the competent departments responsible for the selection. There were 28 provinces, municipalities and autonomous regions, 38 ministries and commissions and directly affiliated units of the State Council, and 58 societies of the China Association for Science and Technology that had made the recommendations.[107] According to the relevant provisions of the "Measures for Supplementing Academic Division

---

[104]Guanyu tuijian he linxuan Zhongguo kexueyuan xuebu weiyuan houxuanren de tongzhi he jige juti wenti de yijian关于推荐和遴选中国科学院学部委员候选人的通知和几个具体问题的意见(Notice on Recommending and Selecting Candidates for Academic Division Members of the Chinese Academy of Sciences and Opinions on Several Specific Issues). Beijing: Zhongguo kexueyuan danganguan中国科学院档案馆(Archives of Chinese Academy of Sciences), 1980-36-008.

[105]Guanyu tuijian he linxuan Zhongguo kexueyuan xuebu weiyuan houxuanren de tongzhi he jige juti wenti de yijian关于推荐和遴选中国科学院学部委员候选人的通知和几个具体问题的意见(Notice on Recommending and Selecting Candidates for Academic Division Members of the Chinese Academy of Sciences and Opinions on Several Specific Issues). Beijing: Zhongguo kexueyuan danganguan中国科学院档案馆(Archives of Chinese Academy of Sciences), 1980-36-008.

[106]Guanyu tuijian he linxuan Zhongguo kexueyuan xuebu weiyuan houxuanren de tongzhi he jige juti wenti de yijian关于推荐和遴选中国科学院学部委员候选人的通知和几个具体问题的意见(Notice on Recommending and Selecting Candidates for Academic Division Members of the Chinese Academy of Sciences and Opinions on Several Specific Issues). Beijing: Zhongguo kexueyuan danganguan中国科学院档案馆(Archives of Chinese Academy of Sciences), 1980-36-008.

[107]Zhongguo kexueyuan xuebu weiyuan quanti huiyi(1980 nian 3 yue 27 ri zhi 4 yue 2 ri) (1) (Li Chang Qin Lisheng deng tongzhi zai bimuhui shang jianghua)中国科学院学部委员全体会议(1980.3.27–4.2) (1) (李昌、秦力生等同志在闭幕会上讲话) (The Plenary Meeting of Academic Division Members of Chinese Academy of Sciences (March 27–April 2, 1980) (1) (Speeches made by Comrades Li Chang, Qin Lisheng and Others at the Closing Meeting). Beijing: Zhongguo kexueyuan danganguan中国科学院档案馆(Archives of Chinese Academy of Sciences), 1980-36-005.

Members of the CAS", 996 candidates were confirmed as actually and validly recommended by the joint meeting of the directors, deputy directors and acting directors of the Academic Divisions.[108] From the end of 1979 to February 1980, 196 candidates were initially selected after evaluation by all divisions.[109] Most of them were old scientists over 60 years old.[110]

From March 28 to April 2, 1980, the CAS held a plenary meeting of the academic division members in Beijing. Ny Tsai-ze presided over the meeting with 91 academic division members present.[111] One of the agenda of the meeting was to determine the formal candidates as the academic division members. The meeting held that if the 196 candidates were added, "it is difficult for the Academic Divisions to reflect the current situation of vigorous development of various disciplines and the continuous emergence of outstanding young and middle-aged scientists in our country. The natural aging problem of the academic division members cannot be solved (the average age of the existing 118 members is over 74 years old), and the Academic Divisions are also having difficulties in meeting the needs of rapid progress in the modernization of science and technology. In order to fully reflect the development of scientific cause in our country, it is planned to expand the number of supplementary personnel from the original 180 to 350 so as to care for some emerging disciplines

---

[108]Xue Pangao薛攀皋. Zhongguo kexueyuan shouci minzhu xuanju xuebu weiyuan (yuanshi) 中国科学院首次民主选举学部委员 (院士) (The First Democratic Election of the Chinese Academy of Sciences for Academic Division Members (Academicians)). In Xue Pangao (2011). *Keyuan qianchen wangshi*科苑前尘往事(Past Events of the Scientific Centre). Beijing: Kexue chubanshe 科学出版社.43.

[109]Guanyu xuebu jige wenti de qingshi baogao关于学部几个问题的请示报告(Report for Instructions on Several Issues Concerning the Academic Divisions). In Zhongguo kexueyuan bangongting 中国科学院办公厅(General Office of the Chinese Academy of Sciences)(Eds.). *Zhongguo kexueyuan nianbao*中国科学院年报(The Annals of the Chinese Academy of Sciences) 1980: 181.

[110]Zhongguo kexueyuan xuebu weiyuan quanti huiyi(1980 nian 3 yue 27 ri zhi 4 yue 2 ri) (1) (Li Chang Qin Lisheng deng tongzhi zai bimuhui shang jianghua)中国科学院学部委员全体会议(1980.3.27–4.2) (1) (李昌、秦力生等同志在闭幕会上讲话) (The Plenary Meeting of Academic Division Members of Chinese Academy of Sciences (March 27–April 2, 1980) (1) (Speeches made by Comrades Li Chang, Qin Lisheng and Others at the Closing Meeting). Beijing: Zhongguo kexueyuan danganguan中国科学院档案馆(Archives of Chinese Academy of Sciences), 1980-36-005.

[111]Zhongguo kexueyuan xuebu weiyuan quanti huiyi(1980 nian 3 yue 27 ri zhi 4 yue 2 ri) (1) (Li Chang Qin Lisheng deng tongzhi zai bimuhui shang jianghua)中国科学院学部委员全体会议(1980.3.27–4.2) (1) (李昌、秦力生等同志在闭幕会上讲话) (The Plenary Meeting of Academic Division Members of Chinese Academy of Sciences (March 27–April 2, 1980) (1) (Speeches made by Comrades Li Chang, Qin Lisheng and Others at the Closing Meeting). Beijing: Zhongguo kexueyuan danganguan中国科学院档案馆(Archives of Chinese Academy of Sciences), 1980-36-005.

and to enable some outstanding middle-aged scientists to enter the Academic Divisions."[112] Among them, 20 places were reserved for the members of the "interdisciplinary subjects of natural science and philosophy and social sciences", and the actual number of additional places was 330.[113]

As the meeting had decided to expand the number ofadditional academic division members, the Academic Divisions did not limit the evaluation to 196 people who passed the preliminary examination. At last, through group discussion and negotiation, 378 candidates in total were determined by all divisions, including 128 in the Division of Mathematics, Physics and Chemistry, 65 in the Division of Biology, 71 in the Division of Earth Sciences and 114 in the Division of Technological Science (76 in the first plan).[114] After a review by all the academic division members, the Academic Divisions finally agreed on 367 formal candidates.[115]

On November 26, 1980, 283 new academic division members were elected by secret ballot in the Academic Divisions. Among them, there were 51 in the Division of Mathematics and Physics, 51 in the Division of Chemistry, 53 in the Division of Biology, 64 in the Division of Earth Sciences and 64 in the Division of Technological Sciences.[116] The number of people elected was 47 fewer than the expected additional 330. The average age of those elected was 62.8 years old, with more than 40 middle-aged scientists under 55 years old and the youngest was 41 years old. Most newly-added academic division members were researchers, professors or senior engineering and technical personnel. However, due to the selection of talents "with comparable level", candidates who were associate researchers and associate professors were elected, for example, Jiang Boju, Tang Xiaowei, Ni Jiazuan, Zou

---

[112]Guanyu xuebu jige wenti de qingshi baogao关于学部几个问题的请示报告(Report for Instructions on Several Issues Concerning the Academic Divisions). In Zhongguo kexueyuan bangongting 中国科学院办公厅(General Office of the Chinese Academy of Sciences)(Eds.). *Zhongguo kexueyuan nianbao*中国科学院年报(The Annals of the Chinese Academy of Sciences) 1980: 181.

[113]Guanyu xuebu jige wenti de qingshi baogao关于学部几个问题的请示报告(Report for Instructions on Several Issues Concerning the Academic Divisions). In Zhongguo kexueyuan bangongting 中国科学院办公厅(General Office of the Chinese Academy of Sciences)(Eds.). *Zhongguo kexueyuan nianbao*中国科学院年报(The Annals of the Chinese Academy of Sciences) 1980: 181–182.

[114]Zhongguo kexueyuan xuebu weiyuan quanti huiyi(1980 nian 3 yue 27 ri zhi 4 yue 2 ri) (1) (Li Chang Qin Lisheng deng tongzhi zai bimuhui shang jianghua)中国科学院学部委员全体会议(1980.3.27–4.2) (1) (李昌、秦力生等同志在闭幕会上讲话) (The Plenary Meeting of Academic Division Members of Chinese Academy of Sciences (March 27–April 2, 1980) (1) (Speeches made by Comrades Li Chang, Qin Lisheng and Others at the Closing Meeting). Beijing: Zhongguo kexueyuan danganguan中国科学院档案馆(Archives of Chinese Academy of Sciences), 1980-36-005.

[115]Xue Pangao薛攀皋. Zhongguo kexueyuan shouci minzhu xuanju xuebu weiyuan (yuanshi) 中国科学院首次民主选举学部委员 (院士) (The First Democratic Election of the Chinese Academy of Sciences for Academic Division Members (Academicians)). In Xue Pangao (2011). *Keyuan qianchen wangshi*科苑前尘往事(Past Events of the Scientific Centre). Beijing: Kexue chubanshe 科学出版社.43.

[116]In October 1980, the Division of Mathematics, Physics and Chemistry became the Division of Mathematics and Physics and the Division of Chemistry.

Gang and Liang Dongcai. This time, Taiwan and foreign scientists were not added as academic division members.

After this supplement, the total numberof academic division members of the CAS reached 400, with the average age falling from 73 to 65.8, but still on the high side. The number of female academic division members increased from 1 to 15.[117] All the academic division members were distributed in 24 provinces, cities and autonomous regions, 27 departments, 43 universities and 114 research units.[118] One third of them were working in the CAS system.[119]

After the completion of this supplement, the CPC Secretariat, when discussing the report of the CAS on the results of the supplement of the academic division members on January 29th, 1981, instructed that there should be more middle-aged scientific workers with true skill and genuine knowledge, and the number of academic division members could be increased.[120] In response, the CAS proposed: "According to the procedures stipulated in the supplementary measures approved by the State Council, it takes a long time to co-opt the academic division members. Moreover, it is difficult to concentrate opinions without a consultation meeting of all the academic division members. Therefore, we plan to discuss the issue of adding some middle-aged scientists to the Academic Divisions at the conference of academic division members to be held in April this year, and to implement it earnestly so as to gradually rejuvenate the Academic Divisions.[121] Since then, the Secretariat of the Central Committee has respected the opinions of the CAS and has not given any administrative intervention.

---

[117]Guanyu chengqing shenpi Zhongguo kexueyuan xuebu weiyuan zengbu mingdan de baogao 关于呈请审批中国科学院学部委员增补名单的报告(Report on Petition for Approval of the List of Supplementary Academic Division Members of the Chinese Academy Sciences). In Zhongguo kexueyuan bangongting中国科学院办公厅(General Office of the Chinese Academy of Sciences)(Eds.). *Zhongguo kexueyuan nianbao*中国科学院年报(The Annals of the Chinese Academy of Sciences) 1981: 355–356.

[118]Ny Tsai-ze严济慈(1981). Zhongguo kexueyuan disici xuebu weiyuan dahui kaimuci中国科学院第四次学部委员大会开幕词(Opening Speech of the Fourth General Assembly of Members of Academic Divisions of the Chinese Academy of Sciences). In Zhongguo kexueyuan bangongting中国科学院办公厅(General Office of the Chinese Academy of Sciences) (Eds.). *Zhongguo kexueyuan nianbao*中国科学院年报(The Annals of the Chinese Academy of Sciences) 1981: 57.

[119]Guowuyuan zhongkeyuan xuebuban youguan pizhuan zengxuan de qingshi zengxuan tongzhi tuijian qingkuang pingshen banfa xinwen fabuhui deng国务院、中科院、学部办有关批转增选的请示、增选通知、推荐情况、评审办法、新闻发布会等(Requests for Instructions on Supplement Approved and Transferred, Supplement Notices, Recommendation Information, Evaluation Methods and Press Conference, etc. Handled by the State Council, the Chinese Academy of Sciences and Its Office of Academic Divisions Respectively). Beijing: Zhongguo kexueyuan danganguan中国科学院档案馆(Archives of Chinese Academy of Sciences), 1991-36-001.

[120]Xue Pangao薛攀皋. Guanyu Zhongguo kexueyuan xuebu huifu yu chongjian gongzuo de huiyi 关于中国科学院学部恢复与重建工作的回忆(The Memories of Restoration and Reconstruction of the Academic Divisions of Chinese Academy of Sciences). In Xue Pangao (2011). *Keyuan qianchen wangshi*科苑前尘往事(Past Events of the Scientific Centre). Beijing: Kexue chubanshe 科学出版社.54.

[121]Guanyu chengqing shenpi Zhongguo kexueyuan xuebu weiyuan zengbu mingdan de baogao 关于呈请审批中国科学院学部委员增补名单的报告(Report on Petition for Approval of the List of Supplementary Academic Division Members of the Chinese Academy Sciences). In Zhongguo kexueyuan bangongting中国科学院办公厅(General Office of the Chinese Academy

On March 23, 1981, the State Council approved 283 additional academic division members of the CAS.[122]

8.3 Summary

The supplement of academic division members in 1957 was the only supplement of members to the Academic Divisions of the CAS before the "Cultural Revolution". It started formally in May 1956 and ended in May 1957. It lasted about one year and finally selected 21 members. Compared with the selection of academic division members in 1955, the increase was an obvious improvement. First of all, although it nominally used the 1955 selection criteria, it actually mainly considered the academic level and downplayed political criteria. Secondly, it implemented the principle of "better to be insufficient than to be indiscriminate." The number of recruits did not exceed the maximum of 30, and almost all the recruits were experts with higher academic level. Third, it was basically run independently by the CAS and the final candidates were also decided independently by the CAS without the approval of the State Council. It received less external negative interference. Fourth, the selection of candidates in the field of natural science was more serious in this increase. The Division of Mathematics, Physics and Chemistry and the Division of Technological Sciences both adopted a secret ballot when preparing the initial candidates and the Division of Biology did so in determining the final candidates. The final choice in the field of natural science was basically based on the outcome of the deliberation and resolution of each division. These improvements reflected the new atmosphere of the CAS and even the academia during the "Indian Spring". It was precisely because of these improvements that, through the additional recruitment, the academic division members, although still in name and job title, had actually begun to evolve into academicians with academic honor in nature.

Objectively speaking, there were still some shortcomings in the increase. First of all, like that in 1955, there was no strict and institutionalized selection procedure for the selection of the academic division members. Secondly, this increase in recruitment was also affected by administrative interference and political factors. In addition, although the Party group of the CAS had let go of the natural science divisions, it intervened more in the Division of Philosophy and Social Sciences. Moreover, this increase in recruitment in favor of recent returnees actually intended to strengthen academic unity, which reflected the political orientation.

The co-optation of members to the Academic Divisions in 1980 was the first time in history of the CAS that academic division members were independently

---

of Sciences)(Eds.). *Zhongguo kexueyuan nianbao*中国科学院年报(The Annals of the Chinese Academy of Sciences) 1981: 355–356.

[122]Zhongguo kexueyuan 1981 nian dashiji中国科学院一九八一年大事记(Major Events of the Chinese Academy of Sciences in 1981). In Zhongguo kexueyuan bangongting中国科学院办公厅(General Office of the Chinese Academy of Sciences) (Eds.). *Zhongguo kexueyuan nianbao*中国科学院年报(The Annals of the Chinese Academy of Sciences) 1981: 517.

elected through a more rigorous and institutionalized democratic selection process. Compared with the increase in the number of academic division members in 1957, this co-optation emphasized achievements, contributions and academic level, while downplaying the political standards of the candidates. Combining two ways of recommendation, by recommending candidates through joint names of academic division members and through organizational systems, the source range of candidates was expanded, and the probability of "leaving pearls" was reduced. Through the four steps, recommendation, selection, evaluation and election, the degree of procedure and democratization in the selection of candidates was improved. This further promoted the nature of the academic division members to the title of academician with academic honor. However, the selection criteria were relatively lower than those of international academicians, and the size of the elected was relatively large, which limited the overall level of this group of academic division members.

Chapter 9
Preparation and Suspension of Academician System Between 1980 and 1981

The Chinese Academy of Sciences made legislative preparations for its academician system after it set up the academic divisions in 1955. It also organized personnel to discuss academician system in January 1962.[1] However, the CAS failed to establish the system due to various reasons. After the "Cultural Revolution", Hu Qiaomu, president of the Chinese Academy of Social Sciences (CASS), suggested establishing an academician system in 1979. In 1980, the CAS, the CASS and the State Scientific and Technological Commission jointly carried out preparations for the academician system under the impetus of Hu Qiaomu and Fang Yi, president of the CAS and director of the State Scientific and Technological Commission. On October 21st, the executive meeting of the State Council approved the "Request for Instructions on Establishing Academician System" submitted by the three units. However, the plan to establish the academician system finally ran aground. This chapter examines how a joint academician system nearly came about. It also looks at why this plan was suspended, and how China's academician system was reconstructed between 1980 and 1981.

9.1 Preparatory Work and Formulation of Academician Regulations

Hu Qiaomu was an important leader in the ideological, cultural, publicity and education fronts of the Communist Party of China. In the 1940s, he was secretary to Mao Tse-tung and secretary of the Political Bureau of the CPC central Committee. After the founding of New China, he served successively as president of Xinhua

[1] Youguan yuanshi zhidu cankao ziliao guanyu xuewei xuexian tiaoli deng wenti de baogao有关院士制度参考资料，关于学位、学衔条例等问题的报告(Reference Materials on Academician System, Reports on Regulations of Academic Degrees, Academic Titles, etc.). Beijing: Zhongguo kexueyuan danganguan中国科学院档案馆(Archives of Chinese Academy of Sciences), 1964–02–041.

J. Guo, *The Establishment and Reconstruction of the Academician System in China*,
https://doi.org/10.1007/978-981-15-7208-1_9

Fig. 9.1 Hu Qiaomu, president of the CASS

News Agency, Director of the General Administration of Information, executive vice-minister of the Publicity Department of the CPC Central Committee, and the secretary-general of the Culture and Education Committee of the State Council. He was selected as a CASAD member in 1955. In November 1977, he became the first president and secretary of the Party group of the CASS. He was elected a member of the Central Committee in 1978. He was elected a secretary of the Central Secretariat in 1980 and resigned as the president of the CASS in December. In 1979, he raised the issue of establishing the academician system, which was related to his emphasis on academic organization and system construction.[2] The preparation for academician system of the CAS, the CASS and the State Scientific and Technological Commission was initiated by Hu Qiaomu in 1979, but it still had profound international and domestic backgrounds (Fig. 9.1).

[2]In the CASS, Hu Qiaomu attached great importance to the establishment and improvement of academic organizations and systems at the two levels of the Academy and Institutes. See Hu Qiaomu zhuan bianxiezu «胡乔木传» 编写组(Compilation Team of "Biography of Hu Qiaomu")(Eds.) (2007). *Hu Qiaomu yu Zhongguo shehui kexueyuan* 胡乔木与中国社会科学院(Hu Qiaomu and the Chinese Academy of Social Sciences). Beijing: Renmin chubanshe人民出版社. 3–85.

At that time, the academician system, as an important institution of the National Academy of Sciences to promote academic development, had been in use in the world for many years. It had been for 31 years in Academia Sinica since its establishment. Moreover, New China was then in a period of social transformation after the "Cultural Revolution" when science and technology were beginning to recover and their status had never been better, and the realization of the modernization of science and technology was strongly advocated. In this social environment, the reform of the national science and technology system began to be pushed forward with increasing international academic exchanges. The establishment of the academician system became the necessary choice and the demand of the times for the Academy of Sciences of New China. In March 1979, at the suggestion of Hu Qiaomu and Fang Yi, the Ministry of Education and the Bureau of Science and Technology Cadres of the State Council began to set up a degree system. At that time, in science developed countries such as Europe, America and Japan, the academician system and the degree system were parallel in the national academic system. It was the general trend for China to establish the academician system to improve its national academic system and make it more cohesive with the world.

It was before November 28, 1979 that Hu Qiaomu raised the issue of establishing the academician system. At that time, the expansion of the CASAD members had officially begun. Fang Yi, President of the CAS, who was also the director of the State Scientific and Technological Commission, supported Hu Qiaomu's proposal. For this, Fang Yi solicited the opinions of the head responsible for science and education in the State Scientific and Technological Commission. After obtaining approval, he appointed Li Chang, vice-president and Secretary of the Party group of the CAS, Tsien San-Tsiang, vice-president of the CAS, and Wu Heng, vice-secretary of the Party group of the CAS and deputy director of the State Scientific and Technological Commission, to work together on the establishment of the academician system. In the CASS, the vice-president Yu Guang-yuan was specifically responsible for preparing the academician system. On November 28, Li Chang wrote to Hu Qiaomu explaining Fang Yi's arrangement and pointing out:

> There are some issues that need to be studied. They are, what is the relationship between the academicians and the academic division members, the latter are currently being supplemented (nearly 1,000 candidates have been put forward by various systems and a lot of work has been done, with the approval of the State Council for the by-election)? Is academician also a kind of academic title like PhD? I have asked Comrade Sanqiang and some (referring to a few) scientists to make preparations. Shall we have a talk about it after collecting the opinions?[3]

---

[3]Guojia kewei kexueyuan shehui kexueyuan gunayu jianli yuanshi zhidu wenti de qingshi yuanshi tiaoli (caoan) ji guowuyuan changwu huiyi taolun jiyao国家科委、科学院、社会科学院关于建立院士制度问题的请示、院士条例(草案)及国务院常务会议讨论纪要(Request for Instructions by the State Scientific and Technological Commission, the Chinese Academy of Sciences and the Chinese Academy of Social Sciences on the Establishment of Academician System, *Academician Regulations (Draft)* and Summary of Discussion at the Executive Meeting of the State Council). Beijing: Zhongguo kexueyuan danganguan中国科学院档案馆(Archives of Chinese Academy of Sciences), 1980–02–024.

Judging from this letter, Li Chang did not know the relationship between the CAS members and the CASAD members, nor did he know that the former was an honorary title, not an academic title. Before officially starting the preparation for the academician system, Tsien San-Tsiang and Wu Heng did literature research work and collected relevant materials of the academician systems of eight countries, Romania, Hungary, Poland, the Soviet Union, the United States, Britain, France and Japan, which were sent to Li Chang on January 17, 1980 "for research reference and hope to prepare opinions".[4] After studying, Yu Guang-yuan, Wu Heng and Tsien San-Tsiang all believed that the proposal to establish the academician system in China was very important and should be settled.[5]

After that, according to the arrangement of Yu Guang-yuan, Tsien San-Tsiang took the responsibility to draft the *Proposal on Setting up the Academician System of the Chinese Academy of Sciences* and the *Academician Regulations of the Chinese Academy of Sciences (Draft)*, which were handed to Yu Guang-yuan on February 29, 1980 for further discussion. These two documents were formed on the basis of investigation and research, and represented the preliminary suggestions and opinions of the CAS on the academician system. "Proposal on Setting up the Academician System of the Chinese Academy of Sciences" introduced Kuo Mo-jo's speech on academician system at the Inaugural Assembly of the Academic Divisions in 1955 and its implementation, Deng Xiaoping's speech at the National Science Congress in 1978 and the ongoing co-optation of CASAD members, and then emphasized:

> The increase of new forces will certainly strengthen the academic leadership of our Academy. The academic division members come from outstanding scientists, professors and senior engineers in all aspects of the country, many of whom are not only first-class in China, but also famous in international academic circles. In order to give full play to the role of these scientists in the construction of the four modernizations and to strengthen international academic exchanges, it should be put on the agenda to further establish the academician system of the CAS on the basis of the academic division members. Because it is objectively necessary and possible. For example, after establishing the academician system, some CAS members could also hold positions of CASAD members. CAS members mainly focus on the national level, while CASAD members are mainly academic leaders in the Academy.

---

[4]Guojia kewei kexueyuan shehui kexueyuan gunayu jianli yuanshi zhidu wenti de qingshi yuanshi tiaoli (caoan) ji guowuyuan changwu huiyi taolun jiyao国家科委、科学院、社会科学院关于建立院士制度问题的请示、院士条例(草案)及国务院常务会议讨论纪要(Request for Instructions by the State Scientific and Technological Commission, the Chinese Academy of Sciences and the Chinese Academy of Social Sciences on the Establishment of Academician System, *Academician Regulations (Draft)* and Summary of Discussion at the Executive Meeting of the State Council). Beijing: Zhongguo kexueyuan danganguan中国科学院档案馆(Archives of Chinese Academy of Sciences), 1980–02–024.

[5]Guojia kewei kexueyuan shehui kexueyuan gunayu jianli yuanshi zhidu wenti de qingshi yuanshi tiaoli (caoan) ji guowuyuan changwu huiyi taolun jiyao国家科委、科学院、社会科学院关于建立院士制度问题的请示、院士条例(草案)及国务院常务会议讨论纪要(Request for Instructions by the State Scientific and Technological Commission, the Chinese Academy of Sciences and the Chinese Academy of Social Sciences on the Establishment of Academician System, *Academician Regulations (Draft)* and Summary of Discussion at the Executive Meeting of the State Council). Beijing: Zhongguo kexueyuan danganguan中国科学院档案馆(Archives of Chinese Academy of Sciences), 1980–02–024.

Academicians are of lifelong tenure, while academic division members have their term of office.[6]

According to this proposal, the both memberships were in parallel; there was no term limit for CAS members and there was the term limit for CASAD members. This was in line with the plan put forward by the CAS on September 20, 1955 in its *Summary Report on the Work of the Inaugural Assembly of the Academic Divisions* submitted to Chen Yi and the State Council to postpone the establishment of the academician system which was compatible with the academic division membership system for one to three years, and a stipulation in the *First Draft of Draft Constitution of the Chinese Academy of Sciences* (the second revision) that CASAD members had terms of office.[7]

The *Proposal on Setting up the Academician System of the Chinese Academy of Sciences* had higher requirements on the conditions for candidates of academicians. "It is suggested to give priority to academic level, candidates should be innovative academically, or be the founder in some fields of science and technology in our country." The "Proposal" held: "Only in this way can a real academic authority be formed, which will be a good advisor to major issues in China's development of science and technology and become a highly honorary organization in China's scientific community." With regard to the number of academicians and the method for their selection, the "Proposal" put forward: "It is appropriate to have about 250 academicians (including social sciences) throughout the country. The method for selecting academicians is suggested to be first conceived among a small number of scientists with high academic status and representativeness (see the attached page for the list of the CAS part), submitted to the State Council for approval, and then appointed by the Standing Committee of the National People's Congress."[8] "The

[6]Guojia kewei kexueyuan shehui kexueyuan gunayu jianli yuanshi zhidu wenti de qingshi yuanshi tiaoli (caoan) ji guowuyuan changwu huiyi taolun jiyao国家科委、科学院、社会科学院关于建立院士制度问题的请示、院士条例(草案)及国务院常务会议讨论纪要(Request for Instructions by the State Scientific and Technological Commission, the Chinese Academy of Sciences and the Chinese Academy of Social Sciences on the Establishment of Academician System, *Academician Regulations (Draft)* and Summary of Discussion at the Executive Meeting of the State Council). Beijing: Zhongguo kexueyuan danganguan中国科学院档案馆(Archives of Chinese Academy of Sciences), 1980–02–024.

[7]Xue bu chengli dahui gongzuo zongjie baogao学部成立大会工作总结报告(Summary Report on the Work of the Inaugural Assembly of the Academic Divisions). In Zhongguo kexueyuan xueshu mishuchu中国科学院学术秘书处(Academic Secretariat of the Chinese Academy of Sciences)(Eds.). *Zhongguo kexueyuan nianbao*中国科学院年报(The Annals of the Chinese Academy of Sciences) 1955: 53–54; Youguan yuanshi zhidu cankao ziliao guanyu xuewei xuexian tiaoli deng wenti de baogao有关院士制度参考资料, 关于学位、学衔条例等问题的报告(Reference Materials on Academician System, Reports on Regulations of Academic Degrees, Academic Titles, etc.). Beijing: Zhongguo kexueyuan danganguan中国科学院档案馆(Archives of Chinese Academy of Sciences), 1964–02–041.

[8]Guojia kewei kexueyuan shehui kexueyuan gunayu jianli yuanshi zhidu wenti de qingshi yuanshi tiaoli (caoan) ji guowuyuan changwu huiyi taolun jiyao国家科委、科学院、社会科学院关于建立院士制度问题的请示、院士条例(草案)及国务院常务会议讨论纪要(Request for Instructions by the State Scientific and Technological Commission, the Chinese Academy of Sciences and

list of the CAS part" included in the "Proposal" amounted to 23 persons: Hua Loo-Keng, Su Buchin, Chou Pei-yuan, Tsien H. S., Ny Tsai-ze, Tsien San-Tsiang, Wang Ganchang, Yu Guang-yuan, Yang Shixian, Lu Jia-xi, Tang Au-chin, Feng Teh-Pei, PaiSitsan, Wang Ying-Lai, Huang C. S., Jin Shanbao, Wu Heng, Yin T. H., Huang T. K., Mao T. E., Zhang Guangdou, Li Xun and Wang Daheng. Later Kiang Tsai-han was added in.[9] These candidates were all the CAS Academic Division members, and most of them were scientists with high prestige in China (Fig. 9.2).

There were seven articles in the *Academician Regulations of the Chinese Academy of Sciences (Draft)*, which stipulated the purpose of establishing the academician system of the CAS, the nature, main activities, election and appointment of the members, and the conditions for honorary and foreign members, etc.

1. Since the founding of the People's Republic of China, a group of excellent scientific and technological talents with outstanding achievements and contributions have emerged in the socialist construction. In order to rely on and give full play to their positive role in building a modern and powerful socialist country, the academician system of the CAS is specially established.
2. The CAS members are the staff officers of the Party and the State leading the people of the whole country to climb the world's scientific and technological summit, playing an important advisory role, enjoying high honors and academic authority.
3. The main activities of the CAS members are:
   (1) Make academic comments and suggestions on major scientific and technological issues in the construction of China's four modernizations;
   (2) Participate in major international academic exchanges and cooperation, keep close contact with well-known foreign scientists, and recommend honorary CAS members and foreign members;
   (3) Make academic appraisals of the special awards for natural science inventions and creations of the people's Republic of China;
   (4) Participate in major activities of the CASAD and strengthen the academic leadership of the Academy.
4. The first group of the CAS members shall be selected by the CAS from among distinguished scientists, professors and senior engineers with high prestige, most

the Chinese Academy of Social Sciences on the Establishment of Academician System, *Academician Regulations (Draft)* and Summary of Discussion at the Executive Meeting of the State Council). Beijing: Zhongguo kexueyuan danganguan中国科学院档案馆(Archives of Chinese Academy of Sciences), 1980–02–024.

[9]Guojia kewei kexueyuan shehui kexueyuan gunayu jianli yuanshi zhidu wenti de qingshi yuanshi tiaoli (caoan) ji guowuyuan changwu huiyi taolun jiyao国家科委、科学院、社会科学院关于建立院士制度问题的请示、院士条例(草案)及国务院常务会议讨论纪要(Request for Instructions by the State Scientific and Technological Commission, the Chinese Academy of Sciences and the Chinese Academy of Social Sciences on the Establishment of Academician System, *Academician Regulations (Draft)* and Summary of Discussion at the Executive Meeting of the State Council). Beijing: Zhongguo kexueyuan danganguan中国科学院档案馆(Archives of Chinese Academy of Sciences), 1980–02–024.

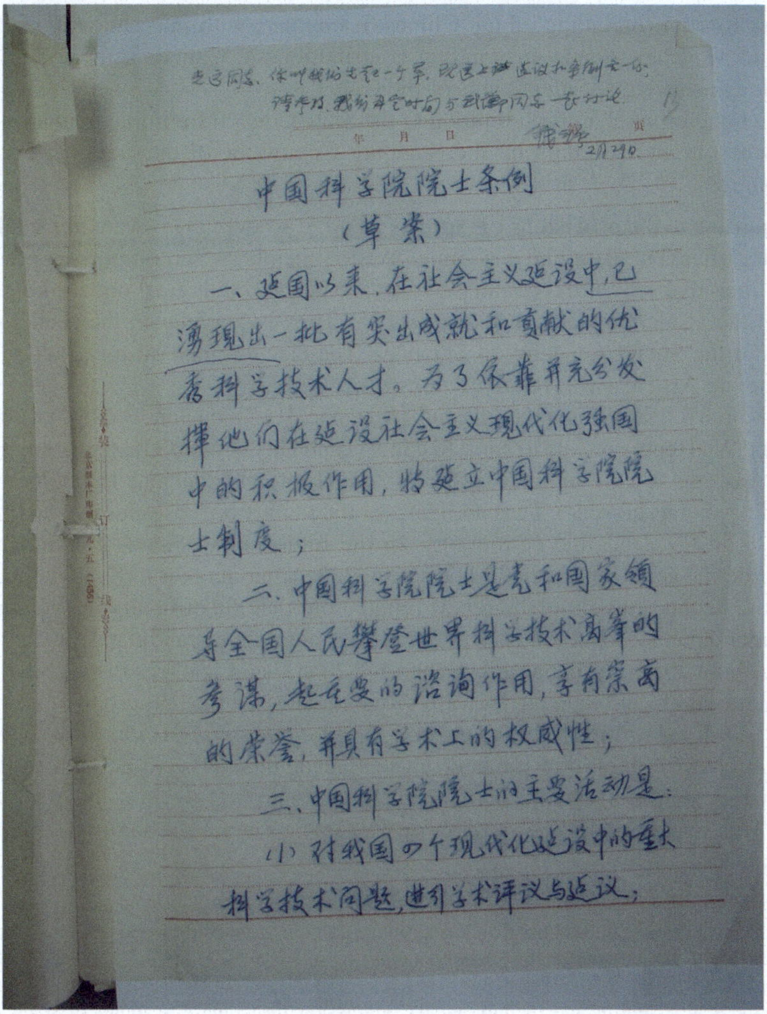
中国科学院院士条例
（草案）
一、建国以来，在社会主义建设中，已涌现出一批有突出成就和贡献的优秀科学技术人才。为了依靠并充分发挥他们在建设社会主义现代化强国中的积极作用，特建立中国科学院院士制度；
二、中国科学院院士是党和国家领导全国人民攀登世界科学技术高峰的参谋，起重要的咨询作用，享有崇高的荣誉，并具有学术上的权威性；
三、中国科学院院士的主要活动是：
（1）对我国四个现代化建设中的重大科学技术问题，进行学术评议与建议；

Fig. 9.2 *Academician Regulations of the Chinese Academy of Sciences (Draft)* (Part) drafted by Tsien San-Tsiang

achievements and contributions, and reported to the State Council for appointment by the Standing Committee of the National People's Congress. After that, the CAS members will be elected by the existing academicians every five years, discussed by the CAS and reported to the State Council for appointment by the Standing Committee of the National People's Congress.
5. Elect leaders who have made significant contributions to China's scientific and technological undertakings as honorary academicians.

6. Hire foreign (including ethnic Chinese) scientists with international academic status who have made outstanding contributions to China's science and technology as foreign academicians of the CAS.
7. The regulations shall be submitted to the State Council for implementation after discussion and adoption by the Standing Committee of the National People's Congress."[10]

According to the provisions of Articles 2 and 3, academicians were not academic honorary titles, but also, like academic division members, should play an advisory and deliberative role in the country's scientific and technological undertakings and take on the academic leadership of the CAS. This was similar to the academician system of the Soviet Academy of Sciences. The election interval for academicians stipulated in Article 4 was relatively long, once 5 years.

On March 5, 1980, Tsien San-Tsiang, in the name of Tsien San-Tsiang, Yu Guang-yuan and Wu Heng, drew up a report to Fang Yi and Hu Qiaomu. The report suggested that it was suis for the number of the first batch of academicians to be about 200 (including social sciences). The candidates could mainly be those elderly and famous scientists in the academic community. In the future, new scientists with remarkable academic achievements could be selected one after another through democratic methods. At the same time, the report pointed out that the difficulty in establishing the academician system mainly lied in the determination of candidates. It suggested to have about 20 people in natural science and 20 in social sciences to make deliberation and research, and then put forward the candidates for the first group of academicians.[11] Different from the *Academician Regulations of the Chinese Academy of Sciences (Draft)*, the report pointed out that academicians were honorary titles of scholarship, unlike academic division members who had to do a lot of practical work. This was to adopt the opinion of Yu Guang-yuan.[12] In addition, the report mentioned:

---

[10]Guojia kewei kexueyuan shehui kexueyuan gunayu jianli yuanshi zhidu wenti de qingshi yuanshi tiaoli (caoan) ji guowuyuan changwu huiyi taolun jiyao国家科委、科学院、社会科学院关于建立院士制度问题的请示、院士条例(草案)及国务院常务会议讨论纪要(Request for Instructions by the State Scientific and Technological Commission, the Chinese Academy of Sciences and the Chinese Academy of Social Sciences on the Establishment of Academician System, *Academician Regulations (Draft)* and Summary of Discussion at the Executive Meeting of the State Council). Beijing: Zhongguo kexueyuan danganguan中国科学院档案馆(Archives of Chinese Academy of Sciences), 1980–02–024.

[11]Guojia kewei kexueyuan shehui kexueyuan gunayu jianli yuanshi zhidu wenti de qingshi yuanshi tiaoli (caoan) ji guowuyuan changwu huiyi taolun jiyao国家科委、科学院、社会科学院关于建立院士制度问题的请示、院士条例(草案)及国务院常务会议讨论纪要(Request for Instructions by the State Scientific and Technological Commission, the Chinese Academy of Sciences and the Chinese Academy of Social Sciences on the Establishment of Academician System, *Academician Regulations (Draft)* and Summary of Discussion at the Executive Meeting of the State Council). Beijing: Zhongguo kexueyuan danganguan中国科学院档案馆(Archives of Chinese Academy of Sciences), 1980–02–024.

[12]Guojia kewei kexueyuan shehui kexueyuan gunayu jianli yuanshi zhidu wenti de qingshi yuanshi tiaoli (caoan) ji guowuyuan changwu huiyi taolun jiyao国家科委、科学院、社会科学院关于建立院士制度问题的请示、院士条例(草案)及国务院常务会议讨论纪要(Request for Instructions by the State Scientific and Technological Commission, the Chinese Academy of Sciences and

> Not long ago, the degree system adopted by the Standing Committee of the National People's Congress is also of your concern. However, we think it is not enough to have the degree system alone. We should also establish the academician system. Academician is different from the current academic leadership titles (such as academic division members and academic committee members) of various academic institutions, it is an honorary title. Because of this, we have drafted the *Academician Regulations of the Chinese Academy of Sciences*.[13]

On March 12, 1980, after deliberation by Wu Heng, Yu Guang-yuan and Tsien San-Tsiang, Wu Heng hosted a symposium on the establishment of the academician system at the State Scientific and Technological Commission. Ny Tsai-ze and Chou Pei-yuan, vice-presidents of the CAS; Shiah Nae, Director of the Institute of Archaeology of the CASS; Feng Zhi, Director of the Institute of Foreign Literature of the CASS; Huang C. S., President of the Chinese Academy of Medical Sciences and Lin Shixiao, Secretary of its Party Committee; Jin Shanbao, President of the Chinese Academy of Agricultural Sciences and Xu Yuanquan, Secretary of its Party Committee; Wu Benxia from the academic degree group of the Ministry of Education, etc. attended the meeting. At the meeting, Yu Guang-yuan and Tsien San-Tsiang first put forward a preliminary idea of setting up the academician system. The participants expressed their support for the establishment of the academician system, but they hoped to be cautious and not rush into action.[14] Then, the meeting discussed the purpose of establishing the academician system, the main tasks of academicians, the generation and quota of the first group of academicians, and the treatment of academicians.

On the purpose of establishing the academician system, the participants had different understandings. Feng Zhi believed that the academician system was established to promote the academic development of China and not to consider anything else. Academicians must have real talent, practical knowledge and contributions, and

---

the Chinese Academy of Social Sciences on the Establishment of Academician System, *Academician Regulations (Draft)* and Summary of Discussion at the Executive Meeting of the State Council). Beijing: Zhongguo kexueyuan danganguan中国科学院档案馆(Archives of Chinese Academy of Sciences), 1980–02–024.

[13]Guojia kewei kexueyuan shehui kexueyuan gunayu jianli yuanshi zhidu wenti de qingshi yuanshi tiaoli (caoan) ji guowuyuan changwu huiyi taolun jiyao国家科委、科学院、社会科学院关于建立院士制度问题的请示、院士条例(草案)及国务院常务会议讨论纪要(Request for Instructions by the State Scientific and Technological Commission, the Chinese Academy of Sciences and the Chinese Academy of Social Sciences on the Establishment of Academician System, *Academician Regulations (Draft)* and Summary of Discussion at the Executive Meeting of the State Council). Beijing: Zhongguo kexueyuan danganguan中国科学院档案馆(Archives of Chinese Academy of Sciences), 1980–02–024.

[14]Guojia kewei kexueyuan shehui kexueyuan gunayu jianli yuanshi zhidu wenti de qingshi yuanshi tiaoli (caoan) ji guowuyuan changwu huiyi taolun jiyao国家科委、科学院、社会科学院关于建立院士制度问题的请示、院士条例(草案)及国务院常务会议讨论纪要(Request for Instructions by the State Scientific and Technological Commission, the Chinese Academy of Sciences and the Chinese Academy of Social Sciences on the Establishment of Academician System, *Academician Regulations (Draft)* and Summary of Discussion at the Executive Meeting of the State Council). Beijing: Zhongguo kexueyuan danganguan中国科学院档案馆(Archives of Chinese Academy of Sciences), 1980–02–024.

could not only consider their age and status. Lin Shixiao believed that the establishment of the academician system should not only promote academic development, but also consider promoting international exchanges and the reunification of the motherland.[15]

There were two opinions on the main tasks of academicians. One agreed with what Yu Guang-yuan and Tsien San-Tsiang said, “Academician is honorary”. For example, Shiah Nae said: “Academicians do not do specific work, but do their jobs well.” Huang C. S. said: “If academic division members are of a working nature, academicians can be honorary and permanent.” Another view was that academicians should not only be honorary, but should undertake certain work as well.[16]

There were two opinions on the main tasks of academicians. One agreed with what Yu Guang-yuan and Tsien San-Tsiang said, “Academician is honorary”. For example, Shiah Nae said: “Academicians do not do specific work, but do their jobs well.” Huang C. S. said: “If academic division members are of a working nature, academicians can be honorary and permanent.” Another view was that academicians should not only be honorary, but should undertake certain work as well.[17]

Regarding the treatment of academicians, part of the participants believed that being elected as academicians was an honor and special material treatment could not be considered. For example, Shiah Nae said: “Our intellectuals do not want money, but honor. We cannot adopt the Soviet approach and give preferential treatment to academicians in terms of salary and treatment. Because Soviet scientific work cannot be pushed forward, it depends on material stimulation. But material considerations

---

[15]Guojia kewei kexueyuan shehui kexueyuan gunayu jianli yuanshi zhidu wenti de qingshi yuanshi tiaoli (caoan) ji guowuyuan changwu huiyi taolun jiyao国家科委、科学院、社会科学院关于建立院士制度问题的请示、院士条例(草案)及国务院常务会议讨论纪要(Request for Instructions by the State Scientific and Technological Commission, the Chinese Academy of Sciences and the Chinese Academy of Social Sciences on the Establishment of Academician System, *Academician Regulations (Draft)* and Summary of Discussion at the Executive Meeting of the State Council). Beijing: Zhongguo kexueyuan danganguan中国科学院档案馆(Archives of Chinese Academy of Sciences), 1980–02–024.

[16]Guojia kewei kexueyuan shehui kexueyuan gunayu jianli yuanshi zhidu wenti de qingshi yuanshi tiaoli (caoan) ji guowuyuan changwu huiyi taolun jiyao国家科委、科学院、社会科学院关于建立院士制度问题的请示、院士条例(草案)及国务院常务会议讨论纪要(Request for Instructions by the State Scientific and Technological Commission, the Chinese Academy of Sciences and the Chinese Academy of Social Sciences on the Establishment of Academician System, *Academician Regulations (Draft)* and Summary of Discussion at the Executive Meeting of the State Council). Beijing: Zhongguo kexueyuan danganguan中国科学院档案馆(Archives of Chinese Academy of Sciences), 1980–02–024.

[17]Guojia kewei kexueyuan shehui kexueyuan gunayu jianli yuanshi zhidu wenti de qingshi yuanshi tiaoli (caoan) ji guowuyuan changwu huiyi taolun jiyao国家科委、科学院、社会科学院关于建立院士制度问题的请示、院士条例(草案)及国务院常务会议讨论纪要(Request for Instructions by the State Scientific and Technological Commission, the Chinese Academy of Sciences and the Chinese Academy of Social Sciences on the Establishment of Academician System, *Academician Regulations (Draft)* and Summary of Discussion at the Executive Meeting of the State Council). Beijing: Zhongguo kexueyuan danganguan中国科学院档案馆(Archives of Chinese Academy of Sciences), 1980–02–024.

should also be taken into account." Lin Shixiao believed that the issue of material benefits could be ignored then until the well-off society was built in 2000.[18]

Although there were differences among participants on the above issues due to different perspectives of consideration, there was unanimous support for the establishment of the academician system. After the symposium, Yu Guang-yuan wrote to Wu Heng and Tsien San-Tsiang on April 30, 1980, pointing out: "It is not enough to say only 'Chinese Academy of Sciences' in the academician regulations."[19] Thus, the original *Academician Regulations of the Chinese Academy of Sciences (Draft)* was changed to *Academician Regulations of the Chinese Academy of Sciences and the Chinese Academy of Social Sciences (Draft)* (hereinafter referred to as *Academician Regulations (Draft)*). The *Academician Regulations (Draft)* was finally drawn up by Yu Guang-yuan (hereinafter referred to as "Draft by Yu Guang-yuan"). Yu Guang-yuan drafted a total of 10 articles, the full text of which is shown in the left column of Table 9.1, which was quite different from the *Academician Regulations of the Chinese Academy of Sciences (Draft)* drawn up by Tsien San-Tsiang. Among them, the biggest change was that the Draft by Yu Guang-yuan stipulated that academician was the country's highest academic honorary title and was a permanent title, deleting the main activities of academicians stipulated in Article 3 of the *Academician Regulations of the Chinese Academy of Sciences (Draft).*

Before July 1, the Chinese Academy of Social Sciences solicited opinions from experts in the academy, including Chen Han-seng, Qian Zhongshu, Fei Xiaotong, Weng Dujian, Yu Guanying, Bao Zhenghu and Shiah Nae, on the Draft by Yu Guang-yuan. Chen Han-seng and Qian Zhongshu agreed, and Fei Xiaotong, Weng Dujian, Yu Guanying, Bao Zhenghu and Shiah Nae put forward written opinions. As for the second article, everyone agreed that academicians should be "those with outstanding academic achievements", the contents followed were unnecessary and should be deleted. Shiah Nae believed that those with outstanding academic achievements should have academic works. Bao Zhenghu proposed that the Chinese nationality of academicians should be emphasized, saying that "it should be stipulated that only those of Chinese nationality and who often live in China can be elected as Chinese academicians." Regarding Article 4, Bao Zhenghu raised the question: "Doesn't the

[18]Guojia kewei kexueyuan shehui kexueyuan gunayu jianli yuanshi zhidu wenti de qingshi yuanshi tiaoli (caoan) ji guowuyuan changwu huiyi taolun jiyao国家科委、科学院、社会科学院关于建立院士制度问题的请示、院士条例(草案)及国务院常务会议讨论纪要(Request for Instructions by the State Scientific and Technological Commission, the Chinese Academy of Sciences and the on the Establishment of Academician System, *Academician Regulations (Draft)* and Summary of Discussion at the Executive Meeting of the State Council). Beijing: Zhongguo kexueyuan danganguan中国科学院档案馆(Archives of Chinese Academy of Sciences), 1980–02–024.

[19]Guojia kewei kexueyuan shehui kexueyuan gunayu jianli yuanshi zhidu wenti de qingshi yuanshi tiaoli (caoan) ji guowuyuan changwu huiyi taolun jiyao国家科委、科学院、社会科学院关于建立院士制度问题的请示、院士条例(草案)及国务院常务会议讨论纪要(Request for Instructions by the State Scientific and Technological Commission, the Chinese Academy of Sciences and the Chinese Academy of Social Sciences on the Establishment of Academician System, *Academician Regulations (Draft)* and Summary of Discussion at the Executive Meeting of the State Council). Beijing: Zhongguo kexueyuan danganguan中国科学院档案馆(Archives of Chinese Academy of Sciences), 1980–02–024.

Table 9.1 Comparison of the Draft by Yu Guang-yuan and finalized *Academician Regulations (Draft)*

Draft by Yu Guang-yuan	finalized *Academician Regulations* (Draft)
Article 1 The purpose of establishing the academician system in our country is to give outstanding scientists a high degree of honor, to show the State's emphasis on science and to encourage scientific workers, and to promote the development and prosperity of science in our country Article 2 Academicians shall be those who have outstanding academic achievements, academic works and have made important contributions in certain academic fields Article 3 Academician is the highest academic honorary title in our country Article 4 Academician shall have the title for lifetime Article 5 The title of academician shall be frequently used when publishing important academic works and participating in international and domestic major academic activities Article 6 Foreigners who have made outstanding contributions to China's scientific development and prosperity and enjoy high academic status in the world may be awarded the title of foreign academicians Article 7 The first number of academicians shall be submitted by the CAS and the CASS to the Standing Committee of the National People's Congress for confirmation Article 8 The admission of new academicians shall be recommended and nominated by the existing academicians. After the election, the new academicians shall be reported to the Standing Committee of the National People's Congress for examination and approval, and then the issuance of certificates Article 9 Admission of new members shall be conducted every two years Article 10 These Regulations shall come into force after being submitted to the Standing Committee of the National People's Congress for approval	Article 1 The purpose of establishing the academician system in our country and conferring the title of academician of the CAS or academician of the CASS on scientists is to give outstanding scientists the highest honor, to express the State's emphasis on science and to encourage scientific workers, and to promote the development and prosperity of science in our country Article 2 Academicians shall be Chinese scholars who have made outstanding academic achievements and important contributions in a certain academic field Article 3 Academician is the highest academic honorary title in our country Article 4 Academician shall have the title for lifetime Article 5 The first batch of academicians shall be submitted by the CAS and the CASS to the Standing Committee of the National People's Congress for approval and issuance of certificates. Article 6 The admission of new academicians shall be recommended and nominated by the existing academicians of both houses respectively, and elected by the Academician Assembly of both houses, and submitted to the Standing Committee of the National People's Congress for approval and issuance of certificates Article 7 Admission of new academicians shall be conducted every two years. The General Assembly of Academicians is held every two years Article 8 Foreigners who have made outstanding contributions to China's scientific development and prosperity and enjoy high academic status in the world may be conferred the title of foreign academicians by the two houses respectively Article 9 These Regulations shall come into force after being submitted by the House Affairs Committees of both houses to the Standing Committee of the National People's Congress for approval

Sources Guojia kewei kexueyuan shehui kexueyuan gunayu jianli yuanshi zhidu wenti de qingshi yuanshi tiaoli (caoan) ji guowuyuan changwu huiyi taolun jiyao国家科委、科学院、社会科学院关于建立院士制度问题的请示、院士条例 (草案) 及国务院常务会议讨论纪要(Request for Instructions by the State Scientific and Technological Commission, the Chinese Academy of Sciences and the Chinese Academy of Social Sciences on the Establishment of Academician System, *Academician Regulations (Draft)* and Summary of Discussion at the Executive Meeting of the State Council). Beijing: Zhongguo kexueyuan danganguan中国科学院档案馆(Archives of Chinese Academy of Sciences), 1980–02–024.

lifelong professional title exclude 'delisting'? What happens if a political situation arises?" "How to deal with academicians who voluntarily renounce their titles?" With regard to Article 5, Shiah Nae said: "This Article is unnecessary to be a regulation. As an international practice, academicians naturally have the right to use this title but are not obligated to use it." Regarding Article 6, Weng Dujian thought that "honorary" academicians could be awarded to foreigners; Fei Xiaotong suggested that "honorary fellows or honorary academicians" could be awarded; Bao Zhenghu believed that "foreign scholars can only be corresponding or contacting academicians". For Article 8, Weng Dujian suggested changing it to "an academician's task is to recommend, nominate and elect new academicians. The new academician must be reported to the Standing Committee of the National People's Congress for examination, approval and issuance of a certificate." Shiah Nae said:" It will take effect after the election in the General Assembly of academicians, and need not be reviewed by the Standing Committee of the National People's Congress. " In addition to the articles in the Draft by Yu Guang-yuan, Shiah Nae proposed to add three articles: first, the annual or biennial General Assembly of Academicians will be held to elect new academicians, and to formulate, revise and approve various detailed rules related to the academician system; second, for work outside the period of the General Assembly of Academicians, there should be an organization of the nature of a standing committee, composed of elected personnel from among the academicians (the number is 12 to 15, and one third is re-elected every year); third, the General Assembly of Academicians (and its standing committee) can put forward suggestions on measures to promote China's scientific development and submit them to relevant departments for consideration.[20] These three articles were all related to the General Assembly of Academicians, and the proposal that the General Assembly of Academicians should be held every two years was adopted.

On July 1, Liu Yuren, secretary of Yu Guang-yuan sorted out the opinions of the above experts' and sent them to Wu Heng and Tsien San-Tsiang for a look.[21] After that, the Draft by Yu Guang-yuan was revised. For example, In Article 4, after "academician is a lifelong professional title", added "special circumstantial treatment

[20]Guojia kewei kexueyuan shehui kexueyuan gunayu jianli yuanshi zhidu wenti de qingshi yuanshi tiaoli (caoan) ji guowuyuan changwu huiyi taolun jiyao国家科委、科学院、社会科学院关于建立院士制度问题的请示、院士条例(草案)及国务院常务会议讨论纪要(Request for Instructions by the State Scientific and Technological Commission, the Chinese Academy of Sciences and the Chinese Academy of Social Sciences on the Establishment of Academician System, *Academician Regulations (Draft)* and Summary of Discussion at the Executive Meeting of the State Council). Beijing: Zhongguo kexueyuan danganguan中国科学院档案馆(Archives of Chinese Academy of Sciences), 1980–02–024.

[21]Guojia kewei kexueyuan shehui kexueyuan gunayu jianli yuanshi zhidu wenti de qingshi yuanshi tiaoli (caoan) ji guowuyuan changwu huiyi taolun jiyao国家科委、科学院、社会科学院关于建立院士制度问题的请示、院士条例(草案)及国务院常务会议讨论纪要(Request for Instructions by the State Scientific and Technological Commission, the Chinese Academy of Sciences and the Chinese Academy of Social Sciences on the Establishment of Academician System, *Academician Regulations (Draft)* and Summary of Discussion at the Executive Meeting of the State Council). Beijing: Zhongguo kexueyuan danganguan中国科学院档案馆(Archives of Chinese Academy of Sciences), 1980–02–024.

measures shall be determined separately"; for the "certificate issuance" proposed in Article 8, it clearly stipulated that the certificate should be issued by the Standing Committee of the National People's Congress. On July 25, the President's Meeting of the CAS discussed the revised version of the Draft by Yu Guang-yuan, and mainly put forward the following suggestions: In natural science, 25 people should definitely be chosen, and in social sciences, 20 people should be chosen to form a group for brewing academicians; the functions and powers of foreign academicians may not be mentioned; the article "special circumstantial treatment measures shall be determined separately" can be deleted and shall be formulated when necessary in the future.[22] After the draft was revised again, the *Academician Regulations (Draft)* was finalized. The Draft by Yu Guang-yuan and the finalized *Academician Regulations (Draft)* are listed in Table 9.1 for comparison.

There are 9 articles in the finalized *Academician Regulations (Draft)*. Its first article has the same purpose as the establishment of the academician system stipulated in Article 1 of the Draft by Yu Guang-yuan, but the final draft clearly defined that academicians were divided into two types, namely, academicians of the CAS and of the CASS. Article 2 of the final draft absorbed the opinions of Shiah Nae, Bao Zhenghu and others. Article 5 of the final draft was similar to Article 7 of the Draft by Yu Guang-yuan, except that the final draft stipulated that the first batch of academicians should be approved and issued certificates by the Standing Committee of the National People's Congress, while the Draft by Yu Guang-yuan stipulated that the Standing Committee of the National People's Congress should determine them. Article 6 of the final draft was revised from Article 8 of the Draft by Yu Guang-yuan, which specified the responsibilities of the two academies, namely, the CAS and the CASS, in recommending and nominating new academicians and the election of new academicians by the two academies' assemblies, and the academician certificate was issued by the Standing Committee of the National People's Congress. Article 7 of the final draft was to add "The General Assembly of Academicians is held every two years" to Article 9 of the Draft by Yu Guang-yuan. Article 8 of the final draft basically copied Article 6 of the Draft by Yu Guang-yuan, with the difference that the final draft made it clear that the title of foreign academicians could be conferred by both houses. Article 9 of the final draft was similar to Article 10 of the Draft by Yu Guang-yuan, but it clearly stipulated that the department that submitted the regulations to the Standing Committee of the National People's Congress for approval was the House Affairs Committee of both houses. Article 5 of the Draft by Yu Guang-yuan was not included in the final draft, which adopted the opinion of Shiah Nae's.

---

[22]Guojia kewei kexueyuan shehui kexueyuan gunayu jianli yuanshi zhidu wenti de qingshi yuanshi tiaoli (caoan) ji guowuyuan changwu huiyi taolun jiyao国家科委、科学院、社会科学院关于建立院士制度问题的请示、院士条例(草案)及国务院常务会议讨论纪要(Request for Instructions by the State Scientific and Technological Commission, the Chinese Academy of Sciences and the Chinese Academy of Social Sciences on the Establishment of Academician System, *Academician Regulations (Draft)* and Summary of Discussion at the Executive Meeting of the State Council). Beijing: Zhongguo kexueyuan danganguan中国科学院档案馆(Archives of Chinese Academy of Sciences), 1980–02–024.

On the whole, the finalized *Academician Regulations (Draft)* stipulated the academician system of pure academic honor for life. Apart from recommending, nominating and electing new academicians, academicians had no other actual work. The conditions for selecting academicians were based on the academic achievements and contributions of the candidates and were in line with international standards.

9.2 From Reporting the "Request for Instructions" to the Emergence of Differences

On August 7, 1980, the CAS, the CASS and the State Scientific and Technological Commission jointly submitted to the State Council the *Request for Instructions on the Establishment of Academician System* and the final version of the attached *Academician Regulations (Draft)*. The "Request for Instructions on the Establishment of Academician System" was drafted mainly on the basis of Yu Guang-yuan's opinions. It was revised by Mei Yi, Deputy Secretary-General of the CASS, and revised by Tsien San-Tsiang. It was discussed and finalized at the president's office meeting of the CAS on July 25 and adopted by the CAS, the CASS and the State Scientific and Technological Commission (Fig. 9.3).[23]

The *Request for Instructions on the Establishment of Academician System* first stated that "The academician system is set up in more developed countries (including some developing countries) in the world. There were also academicians of Academia Sinica of the former Kuomintang government." Then, it introduced the consideration of the academician system by the CAS when its Academic Divisions were founded, and the joint preparation for the academician system of the CAS, the CASS and the State Scientific and Technological Commission. It pointed out: "It is necessary and it is in place today to establish the academician system in our country to give outstanding scientists the highest honor, to further indicate the highest level achieved in our scientific research, and to express that the State attaches importance to science and encourages scientific workers." Then, it put forward the opinions on the nature and duties of academicians, the selection conditions for candidates, the generation of the first batch of academicians, the granting of foreign academicians and honorary academicians, and the number of years for electing new academicians. Its main spirit

---

[23]Guojia kewei kexueyuan shehui kexueyuan gunayu jianli yuanshi zhidu wenti de qingshi yuanshi tiaoli (caoan) ji guowuyuan changwu huiyi taolun jiyao国家科委、科学院、社会科学院关于建立院士制度问题的请示、院士条例 (草案) 及国务院常务会议讨论纪要(Request for Instructions by the State Scientific and Technological Commission, the Chinese Academy of Sciences and the Chinese Academy of Social Sciences on the Establishment of Academician System, *Academician Regulations (Draft)* and Summary of Discussion at the Executive Meeting of the State Council). Beijing: Zhongguo kexueyuan danganguan中国科学院档案馆(Archives of Chinese Academy of Sciences), 1980–02–024; Kexueyuan dangzu shuji Li Chang gei Fang Yi deng yuanlingdao tonghzi de xin科学院党组书记李昌同志给方毅等院领导同志的信(Comrade Li Chang, Secretary of the Party Group of the Chinese Academy of Sciences, Addressed a Letter to Fang Yi and Other Leaders of the Academy). Beijing: Zhongguo kexueyuan danganguan中国科学院档案馆(Archives of Chinese Academy of Sciences), 1981–02–035.

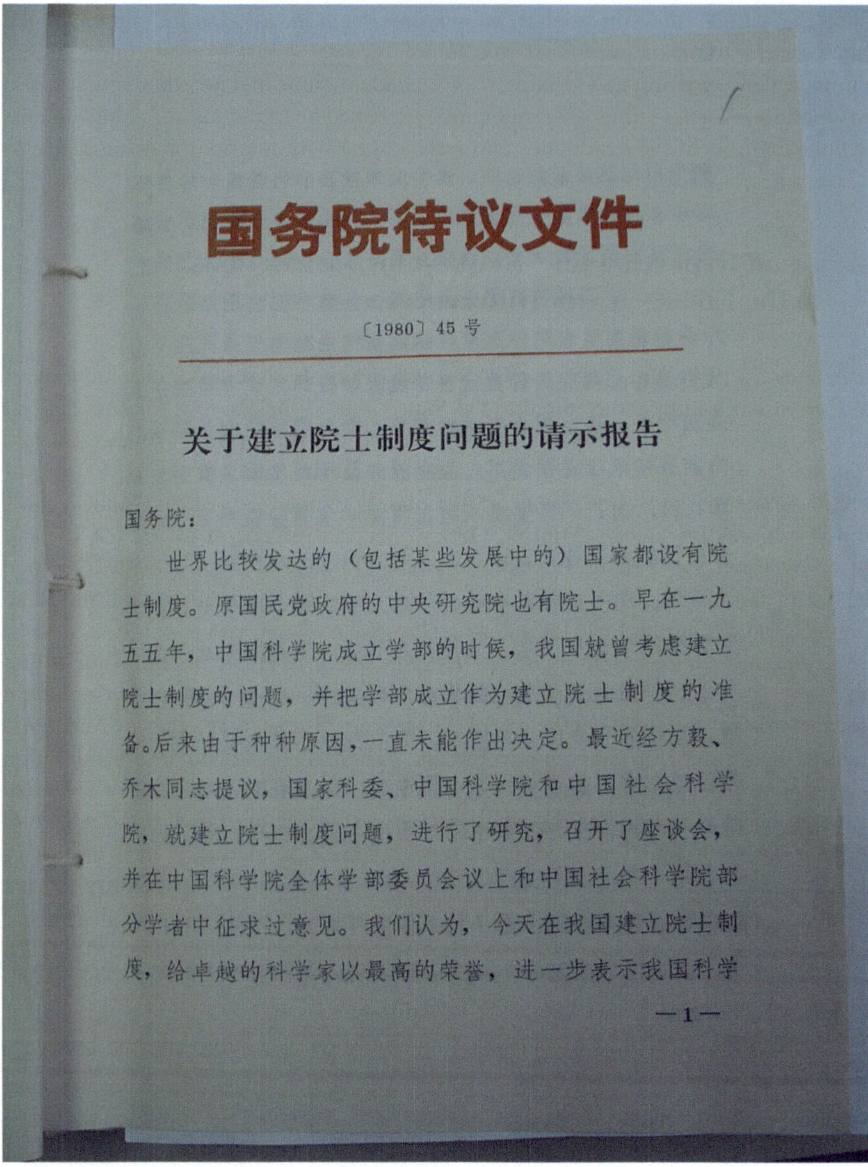

国务院待议文件

〔1980〕45号

关于建立院士制度问题的请示报告

国务院：

世界比较发达的（包括某些发展中的）国家都设有院士制度。原国民党政府的中央研究院也有院士。早在一九五五年，中国科学院成立学部的时候，我国就曾考虑建立院士制度的问题，并把学部成立作为建立院士制度的准备。后来由于种种原因，一直未能作出决定。最近经方毅、乔木同志提议，国家科委、中国科学院和中国社会科学院，就建立院士制度问题，进行了研究，召开了座谈会，并在中国科学院全体学部委员会议上和中国社会科学院部分学者中征求过意见。我们认为，今天在我国建立院士制度，给卓越的科学家以最高的荣誉，进一步表示我国科学

—1—

Fig. 9.3 *Request for Instructions on the Establishment of Academician System* (Page 1)

was consistent with the finalized *Academician Regulations (Draft)*, but its content was more specific. For instance, the selection conditions for candidates required "having outstanding academic achievements and making significant contributions in the academic field", it also stated that "as long as the conditions are met, such an honorary title shall be given, and no seniority or criteria other than scientific standard shall be proposed." For the first group of academicians, it explained that "in view of the current situation in our country, the first batch of academicians can only be selected from among the scholars who have already enjoyed a high reputation in academia", the quota is about 200.[24]

On October 21, 1980, Wan Li presided over the executive meeting of the State Council. The meeting discussed the issues of the CASAD and the establishment of the academician system, and held: "the academic division membership is a professional title, and academician is the highest honorary title for life. Both can exist simultaneously in China. This is conducive to strengthening academic leadership and promoting the development of science and technology. It is also an expression of the State's emphasis on science and encouragement to scientific workers." The meeting agreed to the *Request for Instructions on the Establishment of Academician System* and the *Report for Instructions on Several Issues Concerning the Academic Divisions* submitted by the CAS on May 31. It agreed in principle to the finalized *Academician Regulations (Draft)*, requiring the CAS and the CASS to revise them according to the discussion opinions and report them directly to the Standing Committee of the National People's Congress for deliberation.[25]

After the meeting, the final draft of the *Academician Regulations (Draft)* was revised and Article 5 was deleted.[26] It could be inferred from this that during the discussion at the executive meeting of the State Council on October 21, there were objections to the approval and issuance of certificates by the Standing Committee of the National People's Congress for the first batch of academicians. However, this had little impact on the establishment of the academician system.

---

[24]Guojia kewei kexueyuan shehui kexueyuan gunayu jianli yuanshi zhidu wenti de qingshi yuanshi tiaoli (caoan) ji guowuyuan changwu huiyi taolun jiyao国家科委、科学院、社会科学院关于建立院士制度问题的请示、院士条例(草案)及国务院常务会议讨论纪要(Request for Instructions by the State Scientific and Technological Commission, the Chinese Academy of Sciences and the Chinese Academy of Social Sciences on the Establishment of Academician System, *Academician Regulations (Draft)* and Summary of Discussion at the Executive Meeting of the State Council). Beijing: Zhongguo kexueyuan danganguan中国科学院档案馆(Archives of Chinese Academy of Sciences), 1980–02–024; Kexueyuan dangzu shuji Li Chang gei Fang Yi deng yuanlingdao tonghzi de xin科学院党组书记李昌同志给方毅等院领导同志的信(Comrade Li Chang, Secretary of the Party Group of the Chinese Academy of Sciences, Addressed a Letter to Fang Yi and Other Leaders of the Academy). Beijing: Zhongguo kexueyuan danganguan中国科学院档案馆(Archives of Chinese Academy of Sciences), 1981–02–035.

[25]Guojia kewei kexueyuan shehui kexueyuan gunayu jianli yuanshi zhidu wenti de qingshi yuanshi tiaoli (caoan) ji guowuyuan changwu huiyi taolun jiyao国家科委、科学院、社会科学院关于建立院士制度问题的请示、院士条例(草案)及国务院常务会议讨论纪要(Request for Instructions by the State Scientific and Technological Commission, the Chinese Academy of Sciences and the Chinese Academy of Social Sciences on the Establishment of Academician System, *Academician Regulations (Draft)* and Summary of Discussion at the Executive Meeting of the State Council). Beijing: Zhongguo kexueyuan danganguan中国科学院档案馆(Archives of Chinese Academy of Sciences), 1980–02–024.

[26]Zhongguo kexueyuan Zhongguo shehui kexueyuan yuanshi tiaoli (caoan) 中国科学院、中国社会科学院院士条例(草案) (Academician Regulations of the Chinese Academy of Sciences and the Chinese Academy of Social Sciences (Draft)). In Zhongguo kexueyuan bangongting中国科学院办公厅(General Office of the Chinese Academy of Sciences) (Eds.). *Zhongguo kexueyuan nianbao* 中国科学院年报(The Annals of the Chinese Academy of Sciences) 1980: 73.

No later than March 1981, CAS had planned to elect academicians within the year. On March 19, Li Chang said at the working meeting of the CAS: "In order to express the respect of the country and the people for science and scientists, and to enable our scientists to meet friends in literature and be well-matched in international academic activities, the State Council has approved the establishment of the academician system in our Academy (and the Academy of Social Sciences), that is, high-level scientists who have made significant contributions to our science are awarded lifetime honorary titles—academicians of the CAS. We plan to recommend a group of academicians within the year (the first group will be smaller, for example, dozens of people) and submit them to the Standing Committee of the National People's Congress for approval, and then academicians will elect new academicians."[27]

However, less than a month after the working meeting, there was disagreement within the Party group of the CAS over the establishment ofthe academician system. On the morning of April 14, 1981, Li Chang presided over the Secretary's office meeting, attended by Hu Keshi and Qin Lisheng, deputy party secretaries of the CAS, and Liu Chun, member of the Party group and deputy secretary-general of the Academy; Huang Shulin, Zhu Jianming and Wang Minxi attended the meeting.[28] According to Li Chang's plan, the "academician issue" would be discussed at the Fourth General Assembly of CASAD members in May 1981. Hu Keshi suggested postponing consideration of the issue until later. There were two main reasons for this: one was that the assembly should focus on the formation of the presidium, president and vice-presidents of the CAS, as well as the directors, deputy directors and standing committee members of all academic divisions. "The workload is very heavy and there is a lot of detailed work to be done"; the other was that Hu Qiaomu suggested taking it slowly and calmly in establishing the academician system. At the same time, Hu Keshi also put forward another two reasons: one was that the CASAD members have some opinions on establishing the academician system, "we can sum up", if "we vote for CAS members now", "CASAD members have no ideological preparation"; the other was that after the executive meeting of the State Council approved the "Request for Instructions on the Establishment of Academician System," "the three of us did not meet or report it to the Standing Committee of the National People's Congress."[29] He thought that the methods and steps of establishing the academician system between the CAS and the CASS were "better to be consistent".[30]

---

[27]Li Chang tongzhi zai Zhongguo kexueyuan gongzuo huiyi shang de zongjie yaodian李昌同志在中国科学院工作会议上的总结要点(Comrade Li Chang's Summary Points at the Working Meeting of the Chinese Academy of Sciences). In Zhongguo kexueyuan bangongting中国科学院办公厅(General Office of the Chinese Academy of Sciences) (Eds.). *Zhongguo kexueyuan nianbao* 中国科学院年报(The Annals of the Chinese Academy of Sciences) 1981: 33.

[28]Yu Wen, secretary-general of the CAS, and Tsien San-Tsiang, vice-president of the CAS, were supposed to be present, but were absent due to business trips and hospitalization respectively.

[29]The "three" are the Chinese Academy of Sciences, the Chinese Academy of Social Sciences and the State Scientific and Technological Commission. "report it to the Standing Committee of the National People's Congress" refers to submitting the finalized draft of the *Academician Regulations (Draft)* to the Standing Committee of the National People's Congress for deliberation after revision.

[30]1981 nian dangzu changwei huiyi zhenggongkou pengtou huiyi jilu—九八一年党组常委会议、政工口碰头会议记录(Minutes of the 1981 Meetings of the Standing Committee of the Party Group

Liu Chun agreed with Hu Keshi's suggestion, saying: Comrade Qiao Mu's opinion is not to rush into action and the choice of candidates should be cautious. Moreover, Liu Chun believed that there was still much to be studied in the academician system itself which required a high level for academicians, "not lower than that of foreign countries, not to mention lower than that of Taiwan". It was better to be cautious in selecting the first group of academicians. Huang Shulin, Zhu Jianming and Wang Minxi basically supported postponing the consideration of the "academician issue" until later. Huang Shulin said: "We are pressed for time. Preparing for the establishment of academician system is likely to affect the conference of the CASAD members." Zhu Jianming expressed his "support for stability" on the establishment of academician system. Wang Minxi said: "Mei Yi said at first that we also considered the matter to reach agreement on the steps. Later, he said that you should go first. Later Comrade Qiaomu expressed the opinion that it was better for both houses to reach agreement on the steps and be cautious."[31] It can be seen from this that the CASS was initially active in establishing the academician system, but later became somewhat passive, and decided to let the CAS "take a step first ". However, Hu Qiaomu advocated that the CAS and the CASS should take the same steps and be cautious in establishing the academician system.

Qin Lisheng had no comment on when to "propose academicians", but said "don't make academician honor absolute." He believed that both memberships had honor and should coexist. They are not on the same track, they should be on two tracks. CAS members could not be elected by CASAD members.[32]

Faced with the support of most people to consider the "academician issue" later, Li Chang said: "It is good to be cautious. However, it should be prudent and active. It is not the way to delay." However, he made concessions and said: "Academician issue is mentioned in the Academy Constitution, it can be discussed and some hints can be made, but it will not involve the choice of candidates." He advocated that the "academician issue" should still be discussed at the conference of CASAD members, but the selection of candidates "will be decided by the presidium later."[33] The "Academy Constitution" said by Li Chang referred to the *Draft Constitution of the Chinese Academy of Sciences* being formulated by the CAS.

---

and Brief Meetings of the Political Work Section). Beijing: Zhongguo kexueyuan danganguan中国科学院档案馆(Archives of Chinese Academy of Sciences), 1981–01–005.

[31] 1981 nian dangzu changwei huiyi zhenggongkou pengtou huiyi jilu—九八—年党组常委会议、政工口碰头会议记录(Minutes of the 1981 Meetings of the Standing Committee of the Party Group and Brief Meetings of the Political Work Section). Beijing: Zhongguo kexueyuan danganguan中国科学院档案馆(Archives of Chinese Academy of Sciences), 1981–01–005.

[32] 1981 nian dangzu changwei huiyi zhenggongkou pengtou huiyi jilu一九八一年党组常委会议、政工口碰头会议记录(Minutes of the 1981 Meetings of the Standing Committee of the Party Group and Brief Meetings of the Political Work Section). Beijing: Zhongguo kexueyuan danganguan中国科学院档案馆(Archives of Chinese Academy of Sciences), 1981–01–005.

[33] 1981 nian dangzu changwei huiyi zhenggongkou pengtou huiyi jilu一九八一年党组常委会议、政工口碰头会议记录(Minutes of the 1981 Meetings of the Standing Committee of the Party Group and Brief Meetings of the Political Work Section). Beijing: Zhongguo kexueyuan danganguan中国科学院档案馆(Archives of Chinese Academy of Sciences), 1981–01–005.

9.3 The Fourth General Assembly of the CASAD Members and the Plan to Establish the Academician System Ran Aground

From May 11 to 20, 1981, the Fourth General Assembly of the CASAD Members was held. The agreement of the "Academy Constitution" was one of the agenda of the assembly. However, around the agreement on the "Academy Constitution", academic division members had great differences on the "academician issue". Before the assembly, the Preparatory Committee of the assembly sent the *Draft Constitution of the Chinese Academy of Sciences* (draft for comments) to all academic division members, the directly affiliated institutions under the CAS and the dual leadership units on April 21, 1981, for comments. The draft was divided into seven chapters: "General Provisions", "Leading Organizations", "CAS Members", "CASAD Members, Academic Divisions", "Research Institutes and Related Organizations", "Funds" and "Supplementary Provisions".[34] In the chapter of "Leading Organizations", it was stipulated that "the General Assembly of academic division members is the highest decision-making body of the CAS". Chapter 3, "CAS Members", consisted of four articles. Based on the finalized *Academician Regulations (Draft)* and the "Request for Instructions on the Establishment of Academician System", it respectively stipulated the nature and standards of academicians, the seats of academicians, the approval of the first batch of academicians and admission procedures for the new academicians, the conditions and methods for the formation of foreign academicians, and the activities and responsibilities of academicians. There are six articles in Chapter 4, "CASAD Members, Academic Divisions." Its Article 1 stipulated that "CASAD members are academic leaders, selected from among the outstanding scientists in the country, and elected by the conference of CASAD members in each Academic Division. The term of office of the academic division members is four years, and they can be re-elected once."[35]

During the discussion of the *Draft Constitution of the Chinese Academy of Sciences* on May 12 and 13 at the assembly, the academic division members held different views on the implementation of the division member system and the academician system at the same time, that is, the "double track system". At that time, the members were clearly divided into two groups: supporters and opponents. The supporters were represented by Su Buchin and Hu Shihua, members of the Physics and Mathematics Division, and Chen Zongji, member of the Division of Technological Sciences. The opponents were represented by Liu Hui-xian, Zhang Guangdou, Guo Kexin, Yan Kai, Qian Ning and Huang Wenxi, members of the Division of Technological Sciences, and Li Guoping and Tan Gaosheng, members of

[34]Zhongguo kexueyuan shixing zhangcheng中国科学院试行章程(The Tentative Constitution of the Chinese Academy of Sciences). Beijing: Zhongguo kexueyuan danganguan中国科学院档案馆(Archives of Chinese Academy of Sciences), 1981–02–092.

[35]Zhongguo kexueyuan shixing zhangcheng中国科学院试行章程(The Tentative Constitution of the Chinese Academy of Sciences). Beijing: Zhongguo kexueyuan danganguan中国科学院档案馆(Archives of Chinese Academy of Sciences), 1981–02–092.

the Division of Physics and Mathematics, and Yin T. H., member of the Division of Earth Sciences.

Among the supporters, Su Buchin said: "China should have an academician system, which was repeatedly proposed many times before the 'Cultural Revolution', but was reversed every time. It is not easy to bring it up this time." Hu Shihua suggested: "There should also be the General Assembly of academicians to decide on major issues of principle, and the Academic Divisions implement the resolutions of the General Assembly of academicians on academic issues."[36] Chen Zongji said: "Academician system should be established. This is a matter of national honor." He advocated publishing the academic achievements of academician candidates in newspapers for everyone to select.[37]

In opposition, Liu Hui-xian said that there was no precedent in the world to implement the two systems together, "and to do so is to lower the status of the current academic division members." Zhang Guangdou, Guo Kexin, Yan Kai and Qian Ning all said that the coexistence of the two systems in the world would also cause chaos.[38] Li Guoping believed, "The academician system was put forward on the basis of lowering the academic level of the members of the Academic Divisions, which is an impolite expression to the academic division members." In the opponents, some members of the Academic Divisions also said that in history, there was the saying that academic division members were equal to academicians. In foreign countries and in foreign affairs activities, the academic title of equivalent academician has been used, so the establishment of academician system deserved consideration again.[39]

As for the term of office of the academic division members, the academic division members also had different opinions. Chen Zongji said: "It is better for the academic division members to hold office for four years a term, so as to supplement the newly emerged scientific research talents. At present in our country, it is still the old one who has the final say, never the younger one. This will affect the development of our country.[40] Some members of Physics and Mathematics Division shared similar

[36]Disici xuebu weiyuan dahui jianbao(21–40 qi) 第四次学部委员大会简报(21–40期) (Bulletins on the Fourth General Assembly of Members of the Academic Divisions, Issue 21–40). Beijing: Zhongguo kexueyuan danganguan中国科学院档案馆(Archives of Chinese Academy of Sciences), 1981–02–112.

[37]Dahui choubei gongzuo huiyi jianbao (1–6) ji dahui jianbao (1–20) 大会筹备工作会议简报(1–6)及大会简报(1–20) (Bulletins on the Meetings of the Preparatory Work for the Assembly, Issue1–6, and Bulletins on the Assembly, Issue1–20). Beijing: Zhongguo kexueyuan danganguan中国科学院档案馆(Archives of Chinese Academy of Sciences), 1981–02–111.

[38]Dahui choubei gongzuo huiyi jianbao (1–6) ji dahui jianbao (1–20) 大会筹备工作会议简报(1–6)及大会简报(1–20) (Bulletins on the Meetings of the Preparatory Work for the Assembly, Issue 1–6, and Bulletins on the Assembly, Issue1–20). Beijing: Zhongguo kexueyuan danganguan中国科学院档案馆(Archives of Chinese Academy of Sciences), 1981–02–111.

[39]Disici xuebu weiyuan dahui jianbao(21–40 qi) 第四次学部委员大会简报(21–40期) (Bulletins on the Fourth General Assembly of Members of the Academic Divisions, Issue 21–40). Beijing: Zhongguo kexueyuan danganguan中国科学院档案馆(Archives of Chinese Academy of Sciences), 1981–02–112.

[40]Dahui choubei gongzuo huiyi jianbao (1–6) ji dahui jianbao (1–20) 大会筹备工作会议简报(1–6)及大会简报(1–20) (Bulletins on the Meetings of the Preparatory Work for the Assembly, Issue

views with Chen Zongji, saying that "the members of the Academic Divisions are representatives of the scientific community and are re-elected every four years, which is beneficial for more young people to come out to work."[41] But the opposition was stronger. For example, Tan Gaosheng said: "The members of the Academic Divisions are outstanding representatives of the scientific community. If they don't make mistakes, how can they be elected out?" Cheng Min-Teh said: "Academic division members should not talk about their term of office. They can be re-elected but not for life. Don't replace some members by re-election. It is important that new members should be added continuously. A member can propose not to go on serving as a member due to old age and other reasons." Many members of the Division of Physics and Mathematics agreed with Cheng Min-Teh and considered that even if they were re-elected for another term, when the expiration of their two terms of eight years came, "all the present members will retire, which is very disadvantageous". Wang Xianghao, member of the Physics and Mathematics Division, also said: "Members can be re-elected, especially those working in the Academy of Sciences. However, members must be elected and one-third of them must be replaced each time."[42] 11 members of the Division of Technological Sciences disagreed with the selection of members every four years, saying that "this is too frequent, the workload is too heavy, and the big shift is not conducive to stability."[43] Huang Wenxi advocated "the implementation of lifelong tenure system for academic division members." He said: "As long as the members of the Academic Divisions do not leave their research posts, their standards should be improved day by day. For these outstanding scientists, they should be given full play to their role. A larger number is also beneficial. Don't use a 'special-grade chef' who can already cook famous dishes as an 'ordinary cook' ".[44] There were also "compromise" suggestions from academic division members. For example, Wang H. C., a member of the Division of Technological Sciences, said that a 70-year-old limit could be set for the members, and those who exceeded the limit would retire. Li Minhua, member of the Division of Technological Sciences, thought

---

1–6, and Bulletins on the Assembly, Issue 1–20). Beijing: Zhongguo kexueyuan danganguan中国科学院档案馆(Archives of Chinese Academy of Sciences), 1981–02–111.

[41]Disici xuebu weiyuan dahui jianbao(21–40 qi) 第四次学部委员大会简报(21–40期) (Bulletins on the Fourth General Assembly of Members of the Academic Divisions, Issue 21–40). Beijing: Zhongguo kexueyuan danganguan中国科学院档案馆(Archives of Chinese Academy of Sciences), 1981–02–112.

[42]Disici xuebu weiyuan dahui jianbao(21–40 qi) 第四次学部委员大会简报(21–40期) (Bulletins on the Fourth General Assembly of Members of the Academic Divisions, Issue 21–40). Beijing: Zhongguo kexueyuan danganguan中国科学院档案馆(Archives of Chinese Academy of Sciences), 1981–02–112.

[43]Dahui choubei gongzuo huiyi jianbao (1–6) ji dahui jianbao (1–20) 大会筹备工作会议简报(1–6)及大会简报(1–20) (Bulletins on the Meetings of the Preparatory Work for the Assembly, Issue 1–6, and Bulletins on the Assembly, Issue 1–20). Beijing: Zhongguo kexueyuan danganguan中国科学院档案馆(Archives of Chinese Academy of Sciences), 1981–02–111.

[44]Dahui choubei gongzuo huiyi jianbao (1–6) ji dahui jianbao (1–20) 大会筹备工作会议简报(1–6)及大会简报(1–20) (Bulletins on the Meetings of the Preparatory Work for the Assembly, Issue 1–6, and Bulletins on the Assembly, Issue 1–20). Beijing: Zhongguo kexueyuan danganguan中国科学院档案馆(Archives of Chinese Academy of Sciences), 1981–02–111.

that it was not necessary to have an age limit, but it should depend on whether the member was competent for the job. Hu Haichang, a member from the same division, suggested that some academic division members could be replaced every four years.[45]

On the evening of May 14, 1981, the CAS held a Party group meeting to discuss the *Draft Constitution of the Chinese Academy of Sciences*. The meeting was chaired by Li Chang and attended by Hu Keshi, Yu Wen, Qin Lisheng, Liu Chun, Li Su, Zhang Wensong, Huang Shulin, Qi Kechang, Zhu Jianming and Zhong Bingchang. Ny Tsai-ze, Lu Jia-xi, Yan Dongsheng, Li Xun, Ji Bo, Shi Shan, Zhao Beike, Wu Xuezhen, Luo Wei and Xue Pangao attended. Hu Keshi said: "Seventeen groups reported last night. There were more opinions on the issue of Academicians and academic division members. We phoned Comrade Fang Yi and reported. Comrade Fang Yi said that the chapter of academicians could not be written this time, and it needed more deliberation among scientists. Academic division members are elected once every four years, and it is difficult for them to choose themselves. We'll study the problem again ourselves." Li Chang said: Regarding the two memberships, "since there is no agreement, the chapter of academicians will not be written. It is not good to pass it reluctantly without solving the problem of understanding." Yu Wen interjected, "Just say that the chapter of academicians will not be written for the time being because of differences in opinions. Let academic division members be academicians." In response, Li Chang said: "The Party group should discuss whether to set up the academician system in the future. If we say that it's better not to do it, we won't submit the report to the Central Government, but we can't go on like this without understanding. I would agree if you all decide that an academic division member equals an academician, but we also need to explain it clearly."[46]

At this Party group meeting, Qin Lisheng, Liu Chun, Zhang Wensong, Huang Shulin and Lu Jia-xi all agreed not to write the chapter of academicians or not to mention academicians for the time being. However, Liu Chun believed that the academician system "will still be implemented" later. Qin Lisheng said: Not writing the chapter of academicians "doesn't mean not to engage in it".[47]

With regard to the term of office of academic division members, Qin Lisheng proposed "it needs discussion to change the lifelong tenure to the term system". Zhang Wensong thought not to write too specifically. In addition, Qin Lisheng suggested to write feasible items into the constitution first, and renamed it as a "trial constitution". Yu Wen, Liu Chun and Zhang Wensong agreed, and Li Chang finally decided to adopt

[45]Dahui choubei gongzuo huiyi jianbao (1–6) ji dahui jianbao (1–20) 大会筹备工作会议简报(1–6)及大会简报(1–20) (Bulletins on the Meetings of the Preparatory Work for the Assembly, Issue 1–6, and Bulletins on the Assembly, Issue 1–20). Beijing: Zhongguo kexueyuan danganguan中国科学院档案馆(Archives of Chinese Academy of Sciences), 1981–02–111.

[46]1981 nian dangzu huiyi jiluben一九八一年党组会议记录本(Notebooks of the Party Group Meetings in 1981). Beijing: Zhongguo kexueyuan danganguan中国科学院档案馆(Archives of Chinese Academy of Sciences), 1981–01–004.

[47]1981 nian dangzu huiyi jiluben一九八一年党组会议记录本(Notebooks of the Party Group Meetings in 1981). Beijing: Zhongguo kexueyuan danganguan中国科学院档案馆(Archives of Chinese Academy of Sciences), 1981–01–004.

them.[48] As a result, on May 15, 1981, the *Draft Constitution of the Chinese Academy of Sciences* was officially renamed the *Tentative Draft Constitution of the Chinese Academy of Sciences*.

After this Party group meeting on the night of May 14, the revised draft of the *Tentative Draft Constitution of the Chinese Academy of Sciences* deleted the chapter of "CAS Members" and noted "this part shall be withdrawn and kept temporarily".[49] On the morning of May 15, Ny Tsai-ze presided over the Second Plenary Meeting of the Presidium of the Fourth General Assembly of CASAD Members. At the meeting, Hu Keshi explained the revision of the academy chapter, pointing out that "according to the opinions discussed by everyone, the original Chapter 3— "'CAS Members' will be completely removed. The issue of academicians will be discussed separately." The Presidium Meeting of the Fourth General Assembly basically agreed to the amendment.[50] On the morning of May 17, the executive meeting of the Presidium of the Fourth General Assembly was held to discuss several issues concerning the revision of the *Tentative Draft Constitution of the Chinese Academy of Sciences* (the second revised draft). It was decided that "the chapter on CAS Members in the original discussion draft is not included in the trial constitution as there are different views and at the same time, this issue also involves the CASS. Therefore, it has yet to be discussed."[51]

Regarding the term of office of academic division members, the executive meeting of the Presidium considered that "the Academy constitution is of a trial nature, and the term of office may be changed to 'tentative four years' until the next assembly of the academic division members to discuss appropriate measures."[52] As some members at the division meeting on May 17, 1981 strongly suggested that "since the academicians have been abolished, the term of office of the members of the Academic

---

[48] 1981 nian dangzu huiyi jiluben一九八一年党组会议记录本(Notebooks of the Party Group Meetings in 1981). Beijing: Zhongguo kexueyuan danganguan中国科学院档案馆(Archives of Chinese Academy of Sciences), 1981–01–004.

[49] Zhongguo kexueyuan shixing zhangcheng中国科学院试行章程(The Tentative Constitution of the Chinese Academy of Sciences). Beijing: Zhongguo kexueyuan danganguan中国科学院档案馆(Archives of Chinese Academy of Sciences), 1981–02–092.

[50] Yuan sici xuebu dahui zhuxituan huiyi jiyao院四次学部大会主席团会议纪要(Summaries of the Presidium Meeting of the Fourth General Assembly of Members of the Academic Divisions of Chinese Academy of Sciences). Beijing: Zhongguo kexueyuan danganguan中国科学院档案馆(Archives of Chinese Academy of Sciences), 1981–2–98.

[51] Yuan sici xuebu dahui zhuxituan huiyi jiyao院四次学部大会主席团会议纪要(Summaries of the Presidium Meeting of the Fourth General Assembly of Members of the Academic Divisions of Chinese Academy of Sciences). Beijing: Zhongguo kexueyuan danganguan中国科学院档案馆(Archives of Chinese Academy of Sciences), 1981–2–98.

[52] Yuan sici xuebu dahui zhuxituan huiyi jiyao院四次学部大会主席团会议纪要(Summaries of the Presidium Meeting of the Fourth General Assembly of Members of the Academic Divisions of Chinese Academy of Sciences). Beijing: Zhongguo kexueyuan danganguan中国科学院档案馆(Archives of Chinese Academy of Sciences), 1981–2–98.

Divisions should not be specified",[53] when the assembly passed the *Tentative Draft Constitution of the Chinese Academy of Sciences* on May 18, the provision that "the term of office of the academic division members is tentatively four years and they can be re-elected once" was deleted.[54] In addition, the stipulation that "academic division members are academic leaders" was also deleted from the *Tentative Draft Constitution of the Chinese Academy of Sciences* (the second revised version, May 15, 1981).[55]

After these deletions and the retention of the stipulation that "the General Assembly of academic division members is the highest decision-making body of the CAS", the academician system was likely to be replaced by the membership system of academic divisions. On May 19, 1981, Hu Qiaomu, who was already a secretary of the Secretariat of the CPC Central Committee, wrote to Fang Yi, president of the CAS, and Li Chang, vice President of the CAS, stressing (Fig. 9.4):

> I hope the General Assembly of the academic division members will never deny the academician system. Academician system is a system that has been in use in the world for hundreds of years, and it cannot be replaced by the academic division members in any case. The nature of the two is very different. If the Chinese Academy of Sciences vetoed it down, it would not be possible for the Chinese Academy of Social Sciences to implement it alone. This is indeed a national issue, and be careful to consider it.[56]

However, the letter failed to urge the CAS not to postpone the "academician issue". The plan to establish the academician system ran aground in the CAS after the Fourth General Assembly of the CASAD Members. Subsequently, Hu Qiaomu expressed that the establishment of the academician system in the CASS could be temporarily postponed.[57] The plan to establish the academician system also ran aground in the CASS.

---

[53]Zhongguo kexueyuan shixing zhangcheng中国科学院试行章程(The Tentative Constitution of the Chinese Academy of Sciences). Beijing: Zhongguo kexueyuan danganguan中国科学院档案馆(Archives of Chinese Academy of Sciences), 1981–02–092.

[54]Zhongguo kexueyuan shixing zhangcheng中国科学院试行章程(The Tentative Constitution of the Chinese Academy of Sciences). Beijing: Zhongguo kexueyuan danganguan中国科学院档案馆(Archives of Chinese Academy of Sciences), 1981–02–092.

[55]Zhongguo kexueyuan shixing zhangcheng中国科学院试行章程(The Tentative Constitution of the Chinese Academy of Sciences). Beijing: Zhongguo kexueyuan danganguan中国科学院档案馆(Archives of Chinese Academy of Sciences), 1981–02–092.

[56]Guojia kewei kexueyuan shehui kexueyuan gunayu jianli yuanshi zhidu wenti de qingshi yuanshi tiaoli (caoan) ji guowuyuan changwu huiyi taolun jiyao国家科委、科学院、社会科学院关于建立院士制度问题的请示、院士条例(草案)及国务院常务会议讨论纪要(Request for Instructions by the State Scientific and Technological Commission, the Chinese Academy of Sciences and the Chinese Academy of Social Sciences on the Establishment of Academician System, *Academician Regulations (Draft)* and Summary of Discussion at the Executive Meeting of the State Council). Beijing: Zhongguo kexueyuan danganguan中国科学院档案馆(Archives of Chinese Academy of Sciences), 1980–02–024.

[57]Wei Lili韦莉莉(2009). Zhongguo shehui kexueyuan xuebu weiyuan zhidu de xiangguan wenti yanjiu中国社会科学院学部委员制度的相关问题研究(Studies on Related Problems of the CASS Membership System). *Shehui kexue guanli yu pinglun*社会科学管理与评论(Management and Review of Social Sciences) (1): 57.

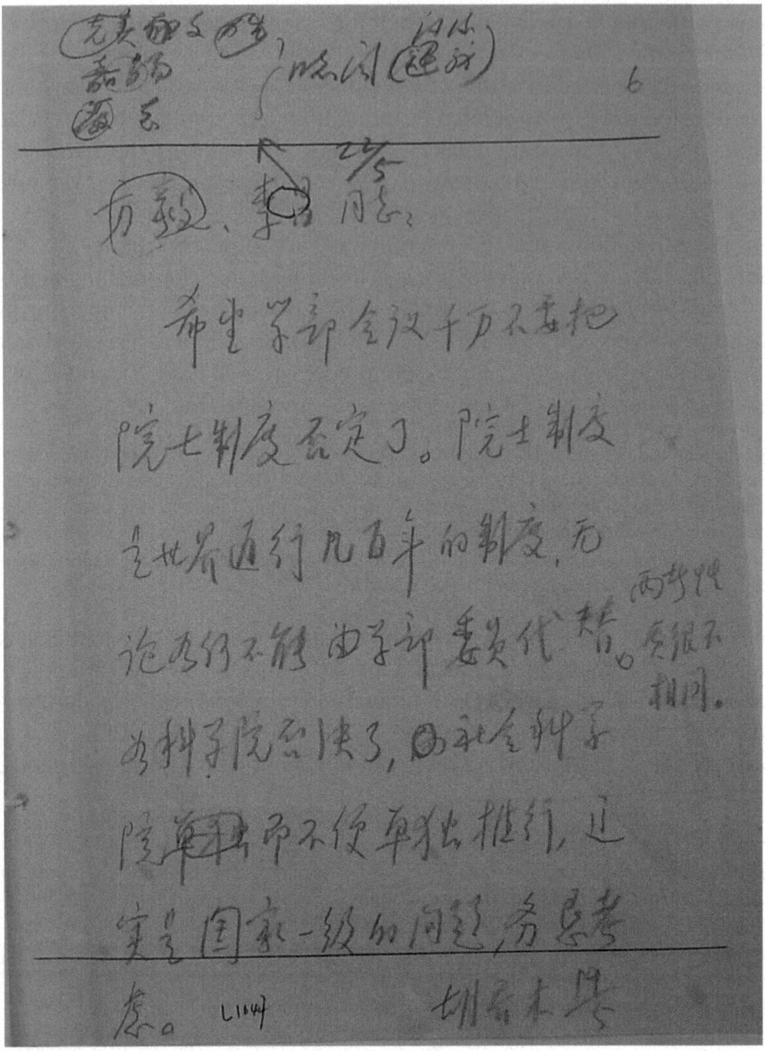

方毅、李昌同志：

希望学部会议千万不要把院士制度否定了。院士制度是世界通行几百年的制度，无论如何不能由学部委员代替。（两者性质很不相同。）如科学院否决了，由社会科学院单独推行，这实是国家一级的问题，务恳考虑。

胡乔木 19/5

Fig. 9.4 Letter to Fang Yi and Li Chang by Hu Qiaomu on May 19, 1981

9.4 Summary

In 1980, the CAS, the CASS and the State Scientific and Technological Commission jointly prepared for the academician system, which was an important activity to rebuild the academician system in Chinese history after 1949. According to the plan, the CAS and the CASS would establish a lifelong and purely honorary academician system by electing outstanding scientists as academicians to form a real academic authority in China. At the same time, the system of academic division members went hand in hand with the academician system, but the former is positioned as an academic leader with a term of four years and could only be re-elected once.

Although the voice of approval and support dominated the preparatory process and the executive meeting of the State Council approved the "Request for Instructions on the Establishment of the Academician System", the plan to establish the academician system ran aground in 1981. The grounding was mainly due to the strong opposition of many CASAD members to the implementation of the academician system at the Fourth General Assembly of the CASAD Members in May 1981. Moreover, many academic division members advocated to implement lifetime tenure. These directly led to the decision by the Party group of the CAS to delete the "CAS Members" chapter in the revised draft of the *Draft Constitution of the Chinese Academy of Sciences*. Since then, the provisions on the nature of professional title and the term system of the CASAD members were deleted, and the provision that the "the General Assembly of academic division members is the highest decision-making body of the CAS" was retained. This made the academician system likely to be replaced by academic division membership. At the same time, the plan to set up the academician system was also affected to some extent by the differences among the Party group members of the CAS on the establishment of the academician system and the slow action of the CASS on the establishment of the system. To be cautious or to keep the steps consistent with the CASS was one of the reasons that most people in the Party group of the CAS advocated or supported postponing the consideration of the "academician issue".

The reason for many academic division members to oppose the academician system was that the academician system formulated in this preparatory work was radical and idealistic to them, and also conflicted with the system of academic division members of the CAS. Because according to the academician system drawn up in the preparatory work, only a few academic division members would be selected as academicians. However, no matter in the history of the CAS or at that time, the members of the academic divisions actually had the academic honor of academicians to some degree. In foreign countries, there were also cases of recognition of CASAD members as academicians. Not only that, regarding the system in the *Draft Constitution of the Chinese Academy of Sciences* (draft for comments) dated April 21, 1981, academic division members could only serve two terms (8 years) at most, while deep in the hearts of most academic division members, they actually attached great importance to academic honor and were not willing to lose the position. What Shiah Nae said that "our intellectuals do not want money, but honor",[58] in fact, reflected the psychological thoughts of the members who value academic honor.

[58]Guojia kewei kexueyuan shehui kexueyuan gunayu jianli yuanshi zhidu wenti de qingshi yuanshi tiaoli (caoan) ji guowuyuan changwu huiyi taolun jiyao国家科委、科学院、社会科学院关于建立院士制度问题的请示、院士条例 (草案) 及国务院常务会议讨论纪要(Request for Instructions by the State Scientific and Technological Commission, the Chinese Academy of Sciences and the Chinese Academy of Social Sciences on the Establishment of Academician System, *Academician Regulations (Draft)* and Summary of Discussion at the Executive Meeting of the State Council). Beijing: Zhongguo kexueyuan danganguan中国科学院档案馆(Archives of Chinese Academy of Sciences), 1980–02–024.

Under these circumstances, it was inevitable that many CASAD members objected the academician system.

From a deeper level, the stranding of the plan to establish the academician system not only shows the faltering reconstruction of the academician system in New China, but also profoundly reflects the abortion of the genuine intention of establishing the academic system of the national Academy of Sciences in line with the international standards caused by the serious differences within the academic circle. It is of great reflective significance in the history of contemporary science society of China.

Chapter 10 The Establishment of the Academician System in New China

After a plan to establish the academician system ran aground in 1981, the call for the system did not disappear. The Chinese Academy of Sciences (CAS) finally established the membership system in 1994 by converting all academic division members into CAS Members, on the basis of the 1983 conversion of academic division members into the country's highest academic honorary title in science and technology and the 1991 co-optation of its members. When the Chinese Academy of Engineering (CAE) was established in 1994, its membership system was also established. As a result, an academician system was finally established and rebuilt in new China. However, since the Chinese Academy of Social Sciences (CASS) had not established a membership system yet, despite making many efforts to do so since 1980, the reconstruction of the academician system in China still had some defects.

10.1 From Academic Division Members to CAS Members: The Membership System Established by the CAS

Although a plan to establish the academician system was shelved at the Fourth General Assembly of the CASAD Members, the Assembly and the leading party of the CAS did not draw a final conclusion on whether to establish the membership system or not. Since then, some academic division members still advocated for the plan. On June 14, 1983, Ge Tingsui, an academic division member, deputy to the National People's Congress, vice president of the CAS' Hefei Branch, and director of the Institute of Solid State Physics, put forth a proposal during the first session of the Sixth National People's Congress, suggesting that "the titles of 'Member of the CAS' and 'Associate Member of the CAS' should be established immediately and the current system for the academic division members of CAS should be abolished." He said: "The academic division members seem to be both honorary and professional. Their significance is not clear and their tasks are not specific, resulting in many insurmountable contradictions. Therefore, it is suggested that we

J. Guo, *The Establishment and Reconstruction of the Academician System in China*,
https://doi.org/10.1007/978-981-15-7208-1_10

abolish the system of academic division members and establish full honorary titles for the Members and Associate Members of the CAS. In the proposal, Ge Tingsui also suggested: "The Members and Associate Members of CAS should be selected nationwide according to their academic attainments or contributions to the development of China's scientific cause. There should be no term of office, no quota, and no age limit at all."[1] Ge Tingsui's proposal represented one of the voices in the academic circle. In the proposal, Members and Associate Members coexisted together, reflecting the membership system of the Soviet Academy of Sciences. In 1955, when CAS made legislative preparations for the establishment of its membership system after the founding conference of the Academic Divisions, it imitated the membership system of the Soviet Academy of Sciences and had the idea of setting up Members and Associate Members in the CAS. *First Draft of Draft Constitution of the Chinese Academy of Sciences* (the second revision) had the relevant provisions.[2]

On June 15, the General Office of the Standing Committee of the National People's Congress sent the proposal to the CAS for research. On May 16, 1984, the second day of the Second Session of the Sixth National People's Congress, the CAS responded to the proposal:

> 1. In terms of the nature of the academic division members of the CAS, it has been reported to the Secretariat of the CPC Central Committee in November 1983 for discussion and was confirmed as the highest academic accolade title in our country.
>
> 2. On whether to co-opt academic division members or abolish the system of academic division members in order to establish the system of "Members and Associate Members of the CAS", due to the many problems involved, it will be studied and opinions will be put forth respectively, and will be submitted to the Central Committee and the State Council for evaluation and decision.
>
> 3. Other suggestions and opinions will be taken as reference when we amend the *Constitution of the Chinese Academy of Sciences*.[3]

Judging by the later situation, Ge Tingsui's suggestion in the proposal was not adopted. As for the nature of the academic division members of the CAS mentioned in the above reply, "it has been clearly recognized as China's highest academic

[1]Woyuan dui quanguo renda liujie erci huiyi daibiao jianyi piping yijian banli qingkuang我院对全国人大六届二次会议代表建议、批评、意见办理情况(Our Academy's Handling of Suggestions, Criticisms and Opinions of Deputies at Two Sessions of the Sixth National People's Congress). Beijing: Zhongguo kexueyuan danganguan中国科学院档案馆(Archives of Chinese Academy of Sciences), 1984-02-020.

[2]Youguan yuanshi zhidu cankao ziliao guanyu xuewei xuexian tiaoli deng wenti de baogao有关院士制度参考资料, 关于学位、学衔条例等问题的报告(Reference Materials on Academician System, Reports on Regulations of Academic Degrees, Academic Titles, etc.). Beijing: Zhongguo kexueyuan danganguan中国科学院档案馆(Archives of Chinese Academy of Sciences), 1964-02-041.

[3]Woyuan dui quanguo renda liujie erci huiyi daibiao jianyi piping yijian banli qingkuang我院对全国人大六届二次会议代表建议、批评、意见办理情况(Our Academy's Handling of Suggestions, Criticisms and Opinions of Deputies at Two Sessions of the Sixth National People's Congress). Beijing: Zhongguo kexueyuan danganguan中国科学院档案馆(Archives of Chinese Academy of Sciences), 1984-02-020.

accolade title," which was determined after the group working on the institutional reform of the central and state organs and the leading science and technology group of the State Council sent investigation teams to the CAS in 1983. On November 5, 1983, the investigation team of the science and technology leading group of the State Council and the leading party group of the CAS proposed in their request report to the Secretariat of the Central Committee: "In the future, the General Assembly of academic division members will be changed from the highest decision-making body of the CAS to the supreme academic appraisal body of the Academy of Sciences and the highest national advisory body in science and technology. Academic division membership will still be the country's highest academic accolade in science and technology."[4] On November 15, the Secretariat of the Central Committee agreed to this in its *Notice on the Decisions of the Central Committee Meeting*.[5]

From January 5 to 12, 1984, the Fifth General Assembly of CASAD Members was held. At the assembly, President Lu Jia-xi clearly stated: "The investigation team organized by the office of the science and technology leading group of the State Council and the leading party group of the CAS together proposed to make it clear that the General Assembly and the Presidium of the CAS would no longer be the decision-making bodies of CAS, while academic division membership would still be the country's highest accolade in terms of science and technology."[6] As a result, the nature and status of academic division members became completely consistent with those of academicians from other countries in the world; however, academic division members were not CAS Members by name. After the General Assembly of academic division members was no longer the highest decision-making body of the CAS, the system obviously became different from the academician system with the academician assembly as its core. This left room for the establishment of a membership system to be raised again.

On December 17, 1984, the Secretariat of the Central Committee held a meeting to discuss the issue of the membership system and decided to establish one in our

[4]Guanyu zhaokai Zhongguo kexueyuan diwuci xuebu weiyuan dahui wenti de qingshi baogao关于召开中国科学院第五次学部委员大会问题的请示报告(Request for Instructions on Convening the Fifth General Assembly of Members of the Academic Divisions of Chinese Academy of Sciences). In Zhongguo kexueyuan bangongting中国科学院办公厅(General Office of the Chinese Academy of Sciences) (Eds.). *Zhongguo kexueyuan nianbao*中国科学院年报(The Annals of the Chinese Academy of Sciences) 1983: 428–429.

[5]Li Jishi李吉士(2012). Diwuci xuebu weiyuan dahui buxuanchu zhuxituan wuwei zhixing zhuxi hou wuwei zhixing zhuxi zhuyao huodong jishi第五次学部委员大会补选出主席团五位执行主席后，五位执行主席主要活动纪事(The Chronicle of Main Activities of the Five Executive Chairmen of the Presidium after They Were Re-elected at the Fifth General Assembly of Members of the Academic Divisions of Chinese Academy of Sciences). *Yuanshi ziliao yu yanjiu*院史资料与研究(Historical Materials and Research of the Chinese Academy of Sciences) (2): 1.

[6]Lu Jia-xi卢嘉锡(1984). Zai Zhongguo kexueyuan diwuci xuebu weiyuan dahui shang de gongzuo baogao在中国科学院第五次学部委员大会上的工作报告(Work Report at the Fifth General Assembly of Members of the Academic Divisions of Chinese Academy of Sciences). In Zhongguo kexueyuan bangongting中国科学院办公厅(General Office of the Chinese Academy of Sciences) (Eds.). *Zhongguo kexueyuan nianbao*中国科学院年报(The Annals of the Chinese Academy of Sciences) 1984: 29.

country. It clearly stated: "We may consider converting academic division members into CAS Members. The number of CAS Members can be expanded from several hundred to about one thousand, to promote a group of young and middle-aged scientific and technological personnel as CAS Members; the senior and respected academic division members could be honorary members of the CAS. Being a CAS Member only indicates academic status and is not linked with administrative function. Please ask the State Science and Technology Commission to take lead in discussing this with the CAS, CASS, and other relevant divisions and then putting forth a plan for approval."[7] As for the decision of the Secretariat of the Central Committee, the five executive chairmen of the Presidium of CAS—Ny Tsai-ze, Wu Chung-hua, Lu Jia-xi, Wu Heng, andTsien H. S.—held special discussions. On January 13, 1985, Ny Tsai-ze also listened to opinions from the thirteen members of the Standing Committee of the National People's Congress. On January 17, the five of them jointly wrote to the central leaders Fang Yi, Hu Qili, Zhao Ziyang, and Hu Yaobang, saying they supported the establishment of the membership system and to change "all academic division members (about two-thirds of the members outside the Academy) into CAS Members." At the same time, they hoped to regard outstanding academic achievements as the only condition for electing Members. In order to distinguish CAS Members inside and outside the academy, it was suggested that the CAS be renamed the Chinese Academy of Natural Sciences and be tied with CASS.[8]

Fang Yi agreed with the opinions of the five executive chairmen of the Presidium of the CAS. After reading the letter he forwarded it to Song Jian, director of the State Science and Technology Commission, asking him to "consider the opinions of Comrade Old Ny[9] and the other comrades and to submit it to the Central Committee for further instruction."[10] On March 25, the five executive chairmen felt it necessary to reflect their opinions to Zhao Ziyang after reading the *Resolution on Granting Noble Honorary Titles to Experts with Particularly Significant Contributions and Outstanding Achievements (Draft),* drafted by Hu Qiaomu. On March 26, according to the opinions of Ny Tsai-ze and Wu Heng, the academic secretariat of the Presidium

[7]Li Jishi李吉士(2012). Diwuci xuebu weiyuan dahui buxuanchu zhuxituan wuwei zhixing zhuxi hou wuwei zhixing zhuxi zhuyao huodong jishi第五次学部委员大会补选出主席团五位执行主席后, 五位执行主席主要活动纪事(The Chronicle of Main Activities of the Five Executive Chairmen of the Presidium after They Were Re-elected at the Fifth General Assembly of Members of the Academic Divisions of Chinese Academy of Sciences). *Yuanshi ziliao yu yanjiu*院史资料与研究(Historical Materials and Research of the Chinese Academy of Sciences) (2): 5.

[8]Zhongyang lingdao dui Zhongguo kexueyuan youguan wenti de pishi中央领导对中国科学院有关问题的批示(Instructions of the Central Leaders on Relevant Issues of Chinese Academy of Sciences), Beijing: Zhongguo kexueyuan danganguan中国科学院档案馆(Archives of Chinese Academy of Sciences), 1985-02-072.

[9]Old Ny refers to Ny Tsai-ze.

[10]Li Jishi李吉士(2012). Diwuci xuebu weiyuan dahui buxuanchu zhuxituan wuwei zhixing zhuxi hou wuwei zhixing zhuxi zhuyao huodong jishi第五次学部委员大会补选出主席团五位执行主席后, 五位执行主席主要活动纪事(The Chronicle of Main Activities of the Five Executive Chairmen of the Presidium after They Were Re-elected at the Fifth General Assembly of Members of the Academic Divisions of Chinese Academy of Sciences) . *Yuanshi ziliao yu yanjiu*院史资料与研究(Historical Materials and Research of the Chinese Academy of Sciences) (2): 7.

of the CAS sent letters to Zhao Ziyang, Hu Qili, and Hu Yaobang respectively, and distributed copies of the joint letter of January 17 from the five executive chairmen. This time, the Central Government responded quickly. On April 1, Zhao Ziyang instructed, "Forward it to Comrade Qi Li and Qiao Mu to read." Hu Qiaomu, who was then a member of the Political Bureau of the Central Committee, did not give his comments until April 8.[11]

Although Hu Qiaomu urged the establishment of the membership system, he was unwilling to lower the level of CAS Members and opposed changing it. Therefore, he pointed out in the instructions, "This proposal is not desirable. Many current academic division members do not have high academic achievements, and few are administrative personnel. If they were all appointed to be CAS Members, they will not be as good as Taiwan's." He thought that it was unnecessary to change the name of the CAS just to distinguish between its Members inside and outside the Academy. In the instructions, he said: "Members and the CAS have no practical work connection except to regularly convene existing Members to elect new Members. For example, many Taiwan academicians are in the United States and have nothing to do with Taiwan's Academia Sinica. Therefore, there is no distinction between the Members inside and outside of the Academy (the Soviet Union calls those outside the academy "corresponding members", and we need not follow suit). It is not necessary for the Academy to change its name; changing its name cannot solve the question of why scientists outside the Academy are called members of the CAS."[12] After Hu Qiaomu's instructions, Hu Yaobang, General Secretary of the Central Committee, instructed, "it is necessary to study this carefully before a decision is made."[13] Therefore, the conversion did not proceed smoothly.

Hu Qiaomu's objection was pertinent. According to international practice, academician was the highest academic accolade given to outstanding scientists who have made important academic achievements. Under the circumstance of the low academic achievements of many of the CAS Members, the conversion of all academic division members into CAS Members would inevitably lower the overall academic level of CAS Members, and would also undermine their academic accolade title.

---

[11]Zhongyang lingdao dui Zhongguo kexueyuan youguan wenti de pishi中央领导对中国科学院有关问题的批示(Instructions of the Central Leaders on Relevant Issues of Chinese Academy of Sciences), Beijing: Zhongguo kexueyuan danganguan中国科学院档案馆(Archives of Chinese Academy of Sciences), 1985-02-072.

[12]Li Jishi李吉士(2012). Diwuci xuebu weiyuan dahui buxuanchu zhuxituan wuwei zhixing zhuxi hou wuwei zhixing zhuxi zhuyao huodong jishi第五次学部委员大会补选出主席团五位执行主席后，五位执行主席主要活动纪事(The Chronicle of Main Activities of the Five Executive Chairmen of the Presidium after They Were Re-elected at the Fifth General Assembly of Members of the Academic Divisions of Chinese Academy of Sciences). *Yuanshi ziliao yu yanjiu*院史资料与研究(Historical Materials and Research of the Chinese Academy of Sciences) (2): 10.

[13]Li Jishi李吉士(2012). Diwuci xuebu weiyuan dahui buxuanchu zhuxituan wuwei zhixing zhuxi hou wuwei zhixing zhuxi zhuyao huodong jishi第五次学部委员大会补选出主席团五位执行主席后，五位执行主席主要活动纪事(The Chronicle of Main Activities of the Five Executive Chairmen of the Presidium after They Were Re-elected at the Fifth General Assembly of Members of the Academic Divisions of Chinese Academy of Sciences). *Yuanshi ziliao yu yanjiu*院史资料与研究(Historical Materials and Research of the Chinese Academy of Sciences) (2): 10.

In February 1986, in view of the possibility of discussing the establishment of the membership system and other issues at the Committee of the Division of Technological Sciences (to be held at the end of August), Wu Heng instructed that "these issues should be handled only after consulting the party organization and putting forward an opinion, or after the party organization solicits the opinions of the executive chairman; otherwise the executive chairmen cannot consider them."[14] Later, Yan Dongsheng, Party Secretary of the CAS, Hu Yongchang, Deputy Secretary General, Li Jishi, Academic Secretary of the Presidium of the Academic Divisions, Zhao Fushan, Vice President of the CASS, and Ru Xin exchanged ideas and opinions on the establishment of the membership system. From the end of July to the beginning of August, Song Jian—director of the State Science and Technology Commission—and Yan Dongsheng agreed to organize a group to draft a report to the Central Committee on the establishment of the membership system. Somehow, after the report was drafted in early September, the establishment of the membership system was shelved.[15]

However, in March 1988, Hu Keshi and 41 other delegates proposed the establishment of the membership system at the First Session of the Seventh National People's Congress.[16] On November 8, the Fourth Session of the Congress approved of the proposal, saying that "at present, countries such as the Soviet Union, Britain, the United States, France, Japan, and India have implemented a membership system, most of which have a history of more than 100 years. Other countries, including third world countries, have also implemented an academician system, awarding outstanding scientists with the title of academician, reflecting the country's emphasis on science and technology, and promoting the progress of science and technology. Implementing the membership system in our country is of great significance for promotion of the whole nation to respect knowledge and talents."[17] At that time, the State Science and Technology Commission had brought the proposal into the 1988

---

[14]Li Jishi李吉士(2012). Diwuci xuebu weiyuan dahui buxuanchu zhuxituan wuwei zhixing zhuxi hou wuwei zhixing zhuxi zhuyao huodong jishi第五次学部委员大会补选出主席团五位执行主席后，五位执行主席主要活动纪事(The Chronicle of Main Activities of the Five Executive Chairmen of the Presidium after They Were Re-elected at the Fifth General Assembly of Members of the Academic Divisions of Chinese Academy of Sciences). *Yuanshi ziliao yu yanjiu*院史资料与研究(Historical Materials and Research of the Chinese Academy of Sciences) (2): 13–14.

[15]Li Jishi李吉士(2012). Diwuci xuebu weiyuan dahui buxuanchu zhuxituan wuwei zhixing zhuxi hou wuwei zhixing zhuxi zhuyao huodong jishi第五次学部委员大会补选出主席团五位执行主席后，五位执行主席主要活动纪事(The Chronicle of Main Activities of the Five Executive Chairmen of the Presidium after They Were Re-elected at the Fifth General Assembly of Members of the Academic Divisions of Chinese Academy of Sciences). *Yuanshi ziliao yu yanjiu*院史资料与研究(Historical Materials and Research of the Chinese Academy of Sciences) (2): 14–15.

[16]Zhonghua renmin gongheguo diqijie quanguo renmin daibiao dahui diyici huiyi mishuchu中华人民共和国第七届全国人民代表大会第一次会议秘书处(The Secretariat of the First Session of the Seventh National People's Congress of the People's Republic of China) (Eds.) (1988). *Zhonghua renmin gongheguo diqijie quanguo renmin daibiao dahui diyici huiyi huikan*中华人民共和国第七届全国人民代表大会第一次会议会(Proceedings of the First Session of the Seventh National People's Congress of the People's Republic of China). Beijing: Renmin chubanshe人民出版社.253.

[17]Quanguo renmin daibiao dahui changwu weiyuanhui bangongting全国人民代表大会常务委员会办公厅(Office of the Standing Committee of the National People's Congress) (Eds.) (2004). Zhonghua renmin gongheguo quanguo renmin daibiao dahui changwu weiyuanhui

soft science research topic on science policy and scientific management. According to the instructions of Song Jian, director of the State Science and Technology Commission, the CAS, CASS, State Education Commission and other institutes and departments, led by the State Science and Technology Commission, formed a "joint research group" to conduct research on the membership system. This prompted the CAS to put the establishment of a membership system on the agenda. On November 16, Ny Tsai-ze chaired a meeting of the executive chairman of the Presidium of the CAS to discuss the establishment of a membership system. He believed that the CAS had the responsibility of adopting a positive attitude to promote its realization as soon as possible.[18] However, with the sudden death of the head of the joint research group, the issue of establishing a membership system was shelved once again.[19]

What was more unfortunate was that the CAS did not co-opt academic division members for ten years after 1980. By 1990, the problem of aging members had become very serious and there was an obvious "fault" in talents. In May 1990, the total number had been reduced from 400 in 1980 to 322, with an average age of over 75. Of these 322 people, only ten were under 60 years old and none were under 50 years old. This situation caused the Academic Divisions to lose the vitality of work to a large extent; many members were anxious about this. In view of this, on May 7, 1990, Tsien San-Tsiang, who had retired from the position of vice-president of the CAS, wrote to Premier Li Peng with the name of "a veteran of the scientific and technological field", proposing to resume the co-optation of academic division members and to lower the average age:

> It is understood that the average age of existing academic division members is over 75 years old. Among the 322 members none are under 50 years old, and only a dozen are under 60 years old. Although everyone has a strong patriotic enthusiasm, they all want more to be contributed to the prosperity of science and technology and the prosperity of the motherland. After all, most of them are too old to do their best. We ourselves often feel anxious about this and hope to change this "fault" of a situation as soon as possible. With the consideration of many aspects, it is most feasible to change the present situation by co-opting the members. Although there are some different opinions within the academic circle on the implementation and establishment of the membership system (relevant parties have also organized special investigations), in view of the fact that the conditions and timing for the implementation of the membership system are not yet ready, everyone believes that it is an urgent matter to

gongbao (1988–1989 nian juan)中华人民共和国全国人民代表大会常务委员会公报(1988–1989年卷) (Communiqué of the Standing Committee of the National People's Congress of the People's Republic of China), Vol. 1988–1989. Beijing: Zhongguo fazhi chubanshe中国法制出版社.360.

[18]Zhongguo kexueyuan zhuxituan zhixing zhuxi huiyi jiyao中国科学院主席团执行主席会议纪要(Summary of the Executive Chairmen's Meeting of the Presidium of the Chinese Academy of Sciences). In Zhongguo kexueyuan bangongting中国科学院办公厅(General Office of the Chinese Academy of Sciences) (Eds.). *Zhongguo kexueyuan nianbao*中国科学院年报(The Annals of the Chinese Academy of Sciences) 1988: 408.

[19]Wei Lili韦莉莉(2009). Zhongguo shehui kexueyuan xuebu weiyuan zhidu de xiangguan wenti yanjiu中国社会科学院学部委员制度的相关问题研究(Studies on Related Problems of the CASS Membership System). *Shehui kexue guanli yu pinglun*社会科学管理与评论(Management and Review of Social Sciences) (1): 58.

immediately start selecting more young CAS Members. At the same time, it is also possible to carry out research on the membership system.[20]

Li Peng gave principled instructions on Tsien San-Tsiang's proposal.[21] In order to listen to a wider spectrum of opinions, all academic divisions of the CAS held separate joint meetings with academic division members, the Standing Committee, and the directors and deputy directors of the Academic Divisions in Beijing. The participants agreed that it was absolutely necessary and timely to co-opt the members. On June 20, 1990, the CAS and the State Science and Technology Commission jointly submitted to the State Council the *Request for the Co-optation of Academic Division Members of Chinese Academy of Sciences* and its annex *Measures for the Co-optation of Academic Division Members of Chinese Academy of Sciences*. Regarding the co-optation, the *Request for the Co-optation of Academic Division Members of Chinese Academy of Sciences* mainly put forward the following suggestions.

(1) New members should be elected "mainly based on their achievements and contributions to science and technology, their academic level, and their ideological and political performance. In the process of co-optation, special attention should be paid to the selection of qualified, outstanding young and middle-aged scientists."[22]Among these, "ideological and political performance" referred to two aspects: one was to love the motherland with dedication, and the other was to have a good style of studying.[23] (2) The number of co-opted candidates was about 200. The co-opted candidates must ensure quality, adhere to the standards and strict requirements, and never increase quantity blindly. (3) In the absence of a clear definition of the membership system, future co-optation would be held biennially, about 55 people at a time (fifteen in the Division of Technological Sciences and ten in other divisions). By 2000, the total number would not exceed 750, and co-optation work would gradually be institutionalized and standardized. (4) In order to give full play to the overall role of the Academic Divisions and enhance the vitality of the division's

---

[20]1991 nian xuebu weiyuan zengxuan cailiao1991年学部委员增选材料(Materials on Co-optation of Academic Division Members in 1991). Beijing: Zhongguo kexueyuan danganguan中国科学院档案馆(Archives of Chinese Academy of Sciences), 1991-36-035.

[21]Dashiji大事记(Chronicle of Events). In Zhongguo kexueyuan bangongting中国科学院办公厅(General Office of the Chinese Academy of Sciences) (Eds.). *Zhongguo kexueyuan nianbao* 中国科学院年报(The Annals of the Chinese Academy of Sciences) 1990: 632.

[22]1991 nian xuebu weiyuan zengxuan cailiao1991年学部委员增选材料(Materials on Co-optation of Academic Division Members in 1991). Beijing: Zhongguo kexueyuan danganguan中国科学院档案馆(Archives of Chinese Academy of Sciences), 1991-36-035.

[23]Guowuyuan zhongkeyuan xuebuban youguan pizhuan zengxuan de qingshi zengxuan tongzhi tuijian qingkuang pingshen banfa xinwen fabuhui deng国务院、中科院、学部办有关批转增选的请示、增选通知、推荐情况、评审办法、新闻发布会等(Requests for Instructions on Supplement Approved and Transferred, Supplement Notices, Recommendation Information, Evaluation Methods and Press Conference, etc. Handled by the State Council, the Chinese Academy of Sciences and Its Office of Academic Divisions Respectively). Beijing: Zhongguo kexueyuan danganguan中国科学院档案馆(Archives of Chinese Academy of Sciences), 1991-36-001.

work, some senior academic division members might be considered as honorary members at an appropriate time after this co-optation.[24]

The *Measures for the Co-optation of Academic Division Members of Chinese Academy of Sciences* stipulated specific methods for co-optation, which were divided into four steps: recommendation, primary election, review, and election. The operation methods and requirements of these four steps were similar to the four steps of recommendation, selection, review, and election in the co-optation of academic division members in 1980, but there were also some differences. For example, the recommendation of candidates for academic division membership was still divided into two channels: a recommendation from the academic division members and a recommendation from the organizational system; the recommendation from the academic division members was changed to at least two members jointly or separately recommending each candidate, and each member could recommend up to three candidates. In the election step, while still adopting the stipulation that more than half of the votes must be approved in order to be elected as an academic division member, it was also stipulated that "the number of voters should be equal to or greater than two-thirds of the members in each existing academic division, and the election is valid that way."[25] These differences increased the stringency of co-optation.

The *Measures for the Co-optation of Academic Division Members of Chinese Academy of Sciences* also explained the composition of the academic divisions and the nature of its members, i.e. "Academic Divisions of the Chinese Academy of Sciences is composed of outstanding scientists from China. CASAD membership is the country's highest academic accolade in science and technology." It also stipulated the criteria for selecting academic division members: "Researchers, professors, or senior scientific and technological experts with an outstanding scientific and technological level, who have made systematic and creative scientific achievements and significant contributions in a certain scientific and technological field may be co-opted as academic division members."[26] This was stricter than the selection criteria of the co-optation in 1980, which was basically equivalent to the international academician criteria. The increase in selection criteria was the result of a complete change in the nature of academic division members.

On October 26, 1990, Li Peng and Yao Yilin presided over the 69th Executive Meeting of the State Council to discuss the aforementioned *Request for the Co-optation of Academic Division Members of Chinese Academy of Sciences* and *Measures for the Co-optation of Academic Division Members of Chinese Academy of Sciences*. The meeting agreed to co-opt the members and agreed in principle to the

[24] 1991 nian xuebu weiyuan zengxuan cailiao1991年学部委员增选材料(Materials on Co-optation of Academic Division Members in 1991). Beijing: Zhongguo kexueyuan danganguan中国科学院档案馆(Archives of Chinese Academy of Sciences), 1991-36-035.

[25] 1991 nian xuebu weiyuan zengxuan cailiao1991年学部委员增选材料(Materials on Co-optation of Academic Division Members in 1991). Beijing: Zhongguo kexueyuan danganguan中国科学院档案馆(Archives of Chinese Academy of Sciences), 1991-36-035.

[26] 1991 nian xuebu weiyuan zengxuan cailiao1991年学部委员增选材料(Materials on Co-optation of Academic Division Members in 1991). Beijing: Zhongguo kexueyuan danganguan中国科学院档案馆(Archives of Chinese Academy of Sciences), 1991-36-035.

Measures for the Co-optation of Academic Division Members of Chinese Academy of Sciences. The meeting held that co-opted members should be considered not only for their academic level, but also their actual contribution to economic and national defense construction. Necessary restrictions should be imposed on the measures for co-optation, so that outstanding young and middle-aged scientists and scientists engaged in technical science should make up an appropriate proportion of the total number.[27] On November 16, the State Council issued a notice to the People's Governments in all provinces, autonomous regions, and municipalities directly under the Central Government, and all ministries and agencies directly under the State Council to forward the *Request for the Co-optation of Academic Division Members of Chinese Academy of Sciences*, pointing out that:

> The co-optation of academic division members is an important event for China's scientific and technological field. It is of great significance to promote the development of China's scientific and technological undertakings, reflect respect for knowledge and talents, and mobilize the enthusiasm of scientific and technological personnel to serve the socialist modernization of the motherland. Therefore, the co-optation must adhere to high standards and strict requirements in order to ensure good quality. First of all, when recommending candidates, we must recommend outstanding scientific and technological experts who have truly made systematic and creative scientific achievements and major contributions, have outstanding academic standards, love the motherland and have a decent style of studying, and we must carefully select them. Because of the current situation in which the members are too old, special attention should be paid to and necessary measures should be taken in the process of co-optation so that qualified young and middle-aged scientific and technological experts will make up a certain proportion. At the same time, the co-optation should be gradually institutionalized in order to promote the continuous growth of outstanding scientific and technological talents. All regions and departments should cooperate closely and carry out this task conscientiously and responsibly.[28]

Prior to this, Zhou Guangzhao, the CAS president, presided over a joint meeting of the directors of all academic divisions on October 31, 1990, conveying the spirit of the 69th Executive Meeting of the State Council and agreeing on additional seats for all academic divisions. Among them, 33 were added to the Divisions of Mathematics and Physics, Chemistry, Geology, and Biology, and 66 were added to the Division of Technological Sciences.[29] The increase in the number of places in the Division of Technological Sciences was mainly due to the suggestions made by Tsien San-Tsiang and other academic division members in a letter to Li Peng on May 7, 1990. In Qian's view, "according to the development trend of world science and technology

---

[27] 1991 nian xuebu weiyuan zengxuan cailiao1991年学部委员增选材料(Materials on Co-optation of Academic Division Members in 1991). Beijing: Zhongguo kexueyuan danganguan中国科学院档案馆(Archives of Chinese Academy of Sciences), 1991-36-035.

[28] 1991 nian xuebu weiyuan zengxuan cailiao1991年学部委员增选材料(Materials on Co-optation of Academic Division Members in 1991). Beijing: Zhongguo kexueyuan danganguan中国科学院档案馆(Archives of Chinese Academy of Sciences), 1991-36-035.

[29] Dashiji大事记(Chronicle of Events). In Zhongguo kexueyuan bangongting中国科学院办公厅(General Office of the Chinese Academy of Sciences) (Eds.). *Zhongguo kexueyuan nianbao* 中国科学院年报(The Annals of the Chinese Academy of Sciences) 1990: 639.

and the urgent need of China's national economic development, it is very necessary to emphasize the importance of science and technology."[30]

On November 30, 1990, the CAS issued a notice to the People's Governments in all provinces, autonomous regions, municipalities directly under the Central Government, all ministries and agencies directly under the State Council, all headquarters of the Military Commission, all services and arms, and the China Association for Science and Technology, stating the number of members to be co-opted and the criteria and conditions for co-optation, and stipulating that the nomination and primary selection would be closed on May 15, 1991.[31] On December 10, the CAS issued another notice to all affiliated units of the Academy, explaining "any relevant issues concerning the recommendation of candidates for academic division membership."[32]

The CAS and any relevant units, divisions, and all the academic division members generally attached great importance to the co-optation. In the primary election, the State Education Commission and the China Association for Science and Technology employed dozens of top 100 members to form the primary election expert committee and evaluated candidates from their respective divisions.[33] In order to do a good job in the recommendation and primary selection of candidates, the CAS specially set up a leading group for recommendation and primary selection. Zhang Yungang and Ge Nengquan served as team leader and deputy team leader, respectively. On December 17, 1990, the leading group held a meeting. The meeting decided that the recommendation work of each unit should be carried out under the leadership

---

[30] 1991 nian xuebu weiyuan zengxuan cailiao1991年学部委员增选材料(Materials on Co-optation of Academic Division Members in 1991). Beijing: Zhongguo kexueyuan danganguan中国科学院档案馆(Archives of Chinese Academy of Sciences), 1991-36-035; Guowuyuan zhongkeyuan xuebuban youguan pizhuan zengxuan de qingshi zengxuan tongzhi tuijian qingkuang pingshen banfa xinwen fabuhui deng国务院、中科院、学部办有关批转增选的请示、增选通知、推荐情况、评审办法、新闻发布会等(Requests for Instructions on Supplement Approved and Transferred, Supplement Notices, Recommendation Information, Evaluation Methods and Press Conference, etc. Handled by the State Council, the Chinese Academy of Sciences and Its Office of Academic Divisions Respectively). Beijing: Zhongguo kexueyuan danganguan中国科学院档案馆(Archives of Chinese Academy of Sciences), 1991-36-001.

[31] 1991 nian xuebu weiyuan zengxuan cailiao1991年学部委员增选材料(Materials on Co-optation of Academic Division Members in 1991). Beijing: Zhongguo kexueyuan danganguan中国科学院档案馆(Archives of Chinese Academy of Sciences), 1991-36-035.

[32] Guanyu tuijian zhongguo keuxeyuan xuebu weiyuan houxuanren de tongzhi关于推荐中国科学院学部委员候选人的通知(Notice on Recommending Candidates of Academic Division Members of Chinese Academy of Sciences). In Zhongguo kexueyuan bangongting中国科学院办公厅(General Office of the Chinese Academy of Sciences) (Eds.). *Zhongguo kexueyuan nianbao* 中国科学院年报(The Annals of the Chinese Academy of Sciences) 1990: 292–293.

[33] Guowuyuan zhongkeyuan xuebuban youguan pizhuan zengxuan de qingshi zengxuan tongzhi tuijian qingkuang pingshen banfa xinwen fabuhui deng国务院、中科院、学部办有关批转增选的请示、增选通知、推荐情况、评审办法、新闻发布会等(Requests for Instructions on Supplement Approved and Transferred, Supplement Notices, Recommendation Information, Evaluation Methods and Press Conference, etc. Handled by the State Council, the Chinese Academy of Sciences and Its Office of Academic Divisions Respectively). Beijing: Zhongguo kexueyuan danganguan中国科学院档案馆(Archives of Chinese Academy of Sciences), 1991-36-001.

of the directors, principals, or factory managers, and clarified the organization and procedures for each unit of the Academy to recommend candidates.

(1) Each unit shall set up a candidate recommendation team with academic division members serving as ex officio members, and the other members shall be selected by the director from among the research fellows of the unit, on the basis of soliciting opinions from relevant parties; (2) The director convened all the researchers of the unit to convey the documents concerning the co-optation of academic division members, and in particular to specify the criteria and conditions of candidates for academic division membership. After full deliberation, all research fellows should propose candidates for academic division membership of this unit through appropriate means. According to the recommendations, the recommendation team would propose the first round of candidates; (3) The recommendation team would select candidates for academic division membership of a unit through a democratic review of the candidates in the first round (when a member of the recommendation team is reviewed, the members should withdraw him) and through secret ballot. Candidates must obtain more than half of the votes of all members on the recommendation team; (4) The personnel department should be responsible for filling in the recommendation materials (including votes) of candidates for academic division membership according to the comments of the recommendation team. The leader (or deputy leader) of the recommendation team should sign and affix the seal of the particular unit; (5) The list of persons who were approved of by more than half of the recommendation team's secret ballot should be submitted in a sealed manner for storage.[34]

On December 31, the leading group held its second meeting. After discussion, the meeting decided that the number of candidates recommended by each unit should be open-ended, but they must adhere to the standards and conditions and recommend the best scientific and technological experts. The people who were approved by more than half of the recommendation team were candidates for academic division membership, and were required to formally sort out their materials and report to the CAS as required.[35]

[34]Zhongguo kexueyuan yuannei xuebu weiyuan houxuanren tuijian he chuxuan gongzuo lingdao xiaozu huiyi jiyao中国科学院院内学部委员候选人推荐和初选工作领导小组会议纪要(Summary of the Meeting of the Leading Group on Candidates Recommendation and Primary Selection for Academic Division Members of Chinese Academy of Sciences). In Zhongguo kexueyuan bangongting中国科学院办公厅(General Office of the Chinese Academy of Sciences) (Eds.). *Zhongguo kexueyuan nianbao*中国科学院年报(The Annals of the Chinese Academy of Sciences) 1990: 293–294.

[35]Zhongguo kexueyuan yuannei xuebu weiyuan houxuanren tuijian he chuxuan gongzuo lingdao xiaozu dierci huiyi jiyao中国科学院院内学部委员候选人推荐和初选工作领导小组第二次会议纪要(Summary of the Second Meeting of the Leading Group on Candidates Recommendation and Primary Selection for Academic Division Members of Chinese Academy of Sciences). In Zhongguo kexueyuan bangongting中国科学院办公厅(General Office of the Chinese Academy of Sciences) (Eds.). *Zhongguo kexueyuan nianbao*中国科学院年报(The Annals of the Chinese Academy of Sciences) 1991: 224.

By the end of May 1991, more than 300 members and 43 ministries and agencies directly under the State Council, 26 provinces, autonomous regions, and municipalities directly under the Central Government, as well as the headquarters of the Chinese People's Liberation Army, various services and arms, the China Association for Science and Technology, etc. had recommended and initially elected 1,079 science and technology experts as valid candidates for academic division membership. The candidates belonged to 169 different research institutes, 181 institutions of higher learning, and nearly 100 first-class societies.[36]

For co-optation, review work was the key point. In order to do a good review job, the CAS held two joint meetings with the directors of the Academic Divisions to discuss and approve of the two rounds of review methods of academic division member candidates, the arrangement of materials submitted to the review meeting, the confirmation of effective recommendations, the determination of the academic divisions to be recommended, the principles, procedures, and methods for handling complaints, etc. Among them, the first round of review stipulated that each division was responsible for reviewing all valid candidates and their attached materials. The review process could be divided into two stages: the first stage—review of materials (including investigation materials and complaint materials), group review, exchange of information, preliminary voting, and putting forward a preliminary list of candidates; the second stage—organizing reconsiderations, verification, discussion, consultation (or a second preliminary vote) on the preliminary list of candidates to determine an official list of candidates; for the members who were unable to attend the review meeting for some reason, they would not send candidate materials for review outside of the review meeting. The written evaluations of the academic division members may be read out loud during the introduction and review of the concerned candidates. The second round of review stipulated that, during the review, each division should carefully review, objectively introduce, democratically discuss, and fully communicate and negotiate with each candidate on the preliminary list of candidates for academic division membership. At the same time, for some candidates, various methods (such as inviting them to introduce themselves, individual investigation, organizing a discussion, etc.) could be adopted to solicit the opinions of other concerned members. In the second round of review, especially when selecting the official candidate list, each division should pay special attention to the selection of outstanding young or middle-aged scientific and technological experts, on the premise of strictly adopting the unified standard conditions, so that candidates under 60 years old should not be fewer than one-third of the official candidate list for each division.[37]

---

[36]Guanyu chengqing shenpi Zhongguo kexueyuan xuebu weiyuan zengxuan mingdan de baogao 关于呈请审批中国科学院学部委员增选名单的报告(Report on Petition for Approval the List of Co-optation of Academic Division Members of Chinese Academy of Sciences). In Zhongguo kexueyuan bangongting中国科学院办公厅(General Office of the Chinese Academy of Sciences) (Eds.). *Zhongguo kexueyuan nianbao*中国科学院年报(The Annals of the Chinese Academy of Sciences 1991: 218–219.

[37]Guowuyuan zhongkeyuan xuebuban youguan pizhuan zengxuan de qingshi zengxuan tongzhi tuijian qingkuang pingshen banfa xinwen fabuhui deng国务院、中科院、学部办有关批转

Before and after confirming the effective recommendation of candidates, the CAS received 32 formal letters of complaint, eighteen of which had entered the list of effective candidates. On June 18, 1991, the joint meeting of the directors of the Academic Divisions decided that a very serious and prudent attitude must be adopted when handling the complaint letters, and that it should be handled in strict accordance with the relevant disciplines and procedures. If the respondent had entered the list of valid candidates, a letter would be sent in the name of the Joint Office of the Academic Divisions, asking the competent division in charge of the primary election to organize an investigation and verification and submit a written investigation report within the prescribed time. It was required that the official seals of the competent department and the provincial and municipal personnel departments would be affixed when responsibly sending the investigation materials. In order to have more time for investigation and verification and to avoid any possible deviation, the review materials of the effective candidates who complained must be, in addition to the original recommendation letter, accompanied by complaint letters, investigation materials, summary tables, attached papers, and works when they entered the first round of the pre-selection list for the second round of review.[38]

In order to ensure the smooth progress of co-optation under the principle of "strictness, rightness, impartiality, and objectivity", the Joint Office of the Ministry of Education also formulated relevant confidentiality rules. Among them, it was stipulated that candidate materials should not be taken out of the prescribed storage place and should not be extracted or copied without approval. The materials sent to the academic division members for review should not be browsed by irrelevant personnel. Personnel who needed access to any materials due to work-related needs should not disclose the situation in the process of recommendation, primary selection, review, and election to any unit or individual in any way; in the review process, all records on co-optation should use the uniformly distributed record book. The minutes of the meeting should be kept by the recorder. After the meeting finishes, it should be uniformly filed by each academic division office. Records of other personnel should be submitted to the Joint Office of Academic Divisions for unified handling after co-optation is completed. In order to create an environment for the members to speak freely and express their opinions fully, when the academic division members were discussing and evaluating the candidates, other irrelevant personnel would not participate in the meeting except the designated recorder, the heads of the Joint

---

增选的请示、增选通知、推荐情况、评审办法、新闻发布会等(Requests for Instructions on Supplement Approved and Transferred, Supplement Notices, Recommendation Information, Evaluation Methods and Press Conference, etc. Handled by the State Council, the Chinese Academy of Sciences and Its Office of Academic Divisions Respectively). Beijing: Zhongguo kexueyuan danganguan中国科学院档案馆(Archives of Chinese Academy of Sciences), 1991-36-001.

[38]Guowuyuan zhongkeyuan xuebuban youguan pizhuan zengxuan de qingshi zengxuan tongzhi tuijian qingkuang pingshen banfa xinwen fabuhui deng国务院、中科院、学部办有关批转增选的请示、增选通知、推荐情况、评审办法、新闻发布会等(Requests for Instructions on Supplement Approved and Transferred, Supplement Notices, Recommendation Information, Evaluation Methods and Press Conference, etc. Handled by the State Council, the Chinese Academy of Sciences and Its Office of Academic Divisions Respectively). Beijing: Zhongguo kexueyuan danganguan中国科学院档案馆(Archives of Chinese Academy of Sciences), 1991-36-001.

Office, and the academic division office. All the voting materials of the candidates for the academic members should be sorted out and properly kept by designated filing personnel for future reference.[39]

The first round of review was conducted from June 24 to July 30, 1991. After group deliberation, conference introduction, exchange of information, pre-voting, and voting, each division first produced a 480-member list twice as many as the official candidates. After that, 472 candidates were selected from 1079 validly recommended candidates by secret ballot.[40] In order to facilitate the evaluation of the candidate materials on the preliminary list by the academic division members and to judge the candidates fairly and accurately, the Joint Office of Academic Divisions decided in July to set up material-consulting rooms in Beijing, Shanghai, and Nanjing to specially receive the academic division members and consult the candidate materials.[41] The second round of review was held from October 25 to November 2, 1991, together with the election. After the review and preliminary selection by various divisions, 239 people passed this round of review and became formal candidates. Among them, 129 were under 60 years old, accounting for 54% of the total number of formal candidates. The results of these two rounds of review have all passed at least three votes: sub-discipline review and voting to produce the list of candidates, introduced at the General Assembly; a preliminary vote conducted by all members to analyze the preliminary vote situation; and a formal vote held to determine the results of this round of review.[42]

The election was held centrally from November 1 to November 2, 1991, with a margin of 20% and by secret ballot. Immediately after the vote, the ballot box was

---

[39]Guowuyuan zhongkeyuan xuebuban youguan pizhuan zengxuan de qingshi zengxuan tongzhi tuijian qingkuang pingshen banfa xinwen fabuhui deng国务院、中科院、学部办有关批转增选的请示、增选通知、推荐情况、评审办法、新闻发布会等(Requests for Instructions on Supplement Approved and Transferred, Supplement Notices, Recommendation Information, Evaluation Methods and Press Conference, etc. Handled by the State Council, the Chinese Academy of Sciences and Its Office of Academic Divisions Respectively). Beijing: Zhongguo kexueyuan danganguan中国科学院档案馆(Archives of Chinese Academy of Sciences), 1991-36-001.

[40]Guowuyuan zhongkeyuan xuebuban youguan pizhuan zengxuan de qingshi zengxuan tongzhi tuijian qingkuang pingshen banfa xinwen fabuhui deng国务院、中科院、学部办有关批转增选的请示、增选通知、推荐情况、评审办法、新闻发布会等(Requests for Instructions on Supplement Approved and Transferred, Supplement Notices, Recommendation Information, Evaluation Methods and Press Conference, etc. Handled by the State Council, the Chinese Academy of Sciences and Its Office of Academic Divisions Respectively). Beijing: Zhongguo kexueyuan danganguan中国科学院档案馆(Archives of Chinese Academy of Sciences), 1991-36-001.

[41]1991 nian xuebu weiyuan zengxuan cailiao1991年学部委员增选材料(Materials on Co-optation of Academic Division Members in 1991). Beijing: Zhongguo kexueyuan danganguan中国科学院档案馆(Archives of Chinese Academy of Sciences), 1991-36-035.

[42]Guowuyuan zhongkeyuan xuebuban youguan pizhuan zengxuan de qingshi zengxuan tongzhi tuijian qingkuang pingshen banfa xinwen fabuhui deng国务院、中科院、学部办有关批转增选的请示、增选通知、推荐情况、评审办法、新闻发布会等(Requests for Instructions on Supplement Approved and Transferred, Supplement Notices, Recommendation Information, Evaluation Methods and Press Conference, etc. Handled by the State Council, the Chinese Academy of Sciences and Its Office of Academic Divisions Respectively). Beijing: Zhongguo kexueyuan danganguan中国科学院档案馆(Archives of Chinese Academy of Sciences), 1991-36-001.

sealed. Members who were unable to attend the meeting for some reason voted by sealed letter and the voting was closed on November 20. By the end of the voting date, 289 out of 306 members had participated in the voting, accounting for 94% of the total. This far exceeded the legal number of valid voters. From November 27 to December 2, each division opened the box to check and count the votes under the supervision of the chief scrutineer and his workers. As a result, 210 people received more than half of the votes. The election situation and results were checked and confirmed by either the standing committee of each division or the director of the division, then reviewed by the CAS and submitted to the State Council for approval.[43] On December 26, the State Council approved the co-optation list.[44]

Of the 210 new members, 38 were from the Division of Mathematics and Physics, 35 were from the Division of Chemistry, 34 were from the Division of Biology, 35 were from the Division of Earth Sciences, and 68 were from the Division of Technological Sciences—all of which were similar to the number of places agreed upon before co-optation. Based on the units and work areas of the elected people, they were distributed into 30 departments and 20 provinces, autonomous regions, and municipalities directly under the central government. From the perspective of discipline and specialty, the scope was wider than before. A number of outstanding scientific and technological experts were also elected in some previously blank, weak disciplines and emerging fields of science and technology. Their average age was 61 years old, of which 111 were under 60 years old, accounting for 52.9% of the elected people. There were twelve people under 50 and the youngest was only 42. Through this co-optation, the average age of the members was further made younger.[45]

[43]Guowuyuan zhongkeyuan xuebuban youguan pizhuan zengxuan de qingshi zengxuan tongzhi tuijian qingkuang pingshen banfa xinwen fabuhui deng国务院、中科院、学部办有关批转增选的请示、增选通知、推荐情况、评审办法、新闻发布会等(Requests for Instructions on Supplement Approved and Transferred, Supplement Notices, Recommendation Information, Evaluation Methods and Press Conference, etc. Handled by the State Council, the Chinese Academy of Sciences and Its Office of Academic Divisions Respectively). Beijing: Zhongguo kexueyuan danganguan中国科学院档案馆(Archives of Chinese Academy of Sciences), 1991-36-001.

[44]Dashiji大事记(Chronicle of Events). In Zhongguo kexueyuan bangongting中国科学院办公厅(General Office of the Chinese Academy of Sciences) (Eds.). *Zhongguo kexueyuan nianbao* 中国科学院年报(The Annals of the Chinese Academy of Sciences) 1991: 467.

[45]Guowuyuan zhongkeyuan xuebuban youguan pizhuan zengxuan de qingshi zengxuan tongzhi tuijian qingkuang pingshen banfa xinwen fabuhui deng国务院、中科院、学部办有关批转增选的请示、增选通知、推荐情况、评审办法、新闻发布会等(Requests for Instructions on Supplement Approved and Transferred, Supplement Notices, Recommendation Information, Evaluation Methods and Press Conference, etc. Handled by the State Council, the Chinese Academy of Sciences and Its Office of Academic Divisions Respectively). Beijing: Zhongguo kexueyuan danganguan中国科学院档案馆(Archives of Chinese Academy of Sciences), 1991-36-001.

Although there were some shortcomings[46] in the co-optation of academic division members in 1991—such as complicated procedures, too much time spent by the members, and pressure exerted by the leaders of some divisions on the members in many aspects[47]—the academic standards of the candidates in the 1991 co-optation of academic division members were higher than those in 1980, which were basically equivalent to international academician standards, and the selection method was more standardized, rigorous, and democratic. After the co-optation, the CAS entered a normal, two-year development track.

From April 20 to 25, 1992, the Sixth General Assembly of CASAD Members was held. On April 23, the assembly passed the *Regulations for the Academic Division Members of Chinese Academy of Sciences (Trial). The Regulations (Trial)* were changed to the *Constitution for the Academic Division Members of Chinese Academy of Sciences (Trial)* in August 1992, after being recommended by the Legislative Affairs Bureau of the State Council and approved by the CAS. Before December 7, the *Constitution for the Academic Division Members of Chinese Academy of Sciences (Trial)* were promulgated and implemented with the approval of the State Council.[48] It was the first fundamental regulation on the work of the Academic Divisions.

The *Constitution for the Academic Division Members of Chinese Academy of Sciences (Trial)* consisted of eight chapters and 28 articles. Among them, Chapter 1 was titled *Academic Division Members,* and there were two articles that stipulated the nature, obligations, and rights of the academic division members. With regard to the nature of the members, it was stipulated that "it is the highest academic title in science and technology that is established by the state, and it is a lifelong honor."[49] The word "lifelong" was added to the nature of the academic division members stipulated in the previous relevant documents, which clearly defined the lifelong honor of the title. Chapter 2 was titled *Procedure for the Election of Academic Division Members* and consisted of five articles. Article 1 stipulated the criteria for selecting academic division members: "researchers, professors, or scholars and experts with the same title who have made systematic and creative achievements and significant contributions in the field of science and technology, love the motherland,

---

[46]Guowuyuan zhongkeyuan xuebuban youguan pizhuan zengxuan de qingshi zengxuan tongzhi tuijian qingkuang pingshen banfa xinwen fabuhui deng国务院、中科院、学部办有关批转增选的请示、增选通知、推荐情况、评审办法、新闻发布会等(Requests for Instructions on Supplement Approved and Transferred, Supplement Notices, Recommendation Information, Evaluation Methods and Press Conference, etc. Handled by the State Council, the Chinese Academy of Sciences and Its Office of Academic Divisions Respectively). Beijing: Zhongguo kexueyuan danganguan中国科学院档案馆(Archives of Chinese Academy of Sciences), 1991-36-001.

[47]Xuebu zhuxituan huiyi cailiao学部主席团会议材料(Meeting Materials of the Presidium of the Academic Divisions). Beijing: Zhongguo kexueyuan danganguan中国科学院档案馆(Archives of Chinese Academy of Sciences), 1994-36-017.

[48]Yuanshi tiaoli (Zhangcheng) 院士条例(章程) (Statutes for Membership of the Chinese Academy of Sciences). Beijing: Zhongguo kexueyuan danganguan中国科学院档案馆(Archives of Chinese Academy of Sciences), 1994-36-018.

[49]Yuanshi tiaoli (Zhangcheng) 院士条例(章程) (Statutes for Membership of the Chinese Academy of Sciences). Beijing: Zhongguo kexueyuan danganguan中国科学院档案馆(Archives of Chinese Academy of Sciences), 1994-36-018.

and have a decent style of studying may be recommended for election as academic division members of the CAS."[50] Compared to the criteria in the *Measures for the Co-optation of Academic Division Members of Chinese Academy of Sciences*, which was agreed on in principle by the 69th Executive Meeting of the State Council on October 26, 1990, this criteria removed the point "outstanding scientific and technological level," but specified the requirements for the candidates' ideological and political performance; namely, "love the motherland and have a decent style of studying." Article 2 stipulated: "The election of academic division members shall be held biennially, and the total number of co-opted members should not exceed sixty. The number of co-opted members for each division should be determined through discussion by the Presidium."[51] This has provided a legal basis for the normalization and stricter co-optation of academic division members since then. Article 3 stipulated the channels and requirements for "recommending candidates". Compared to the related provisions of the aforementioned *Measures for the Co-optation of Academic Division Members of Chinese Academy of Sciences*, there was little change. Article 4 stipulated the requirements for the evaluation of candidates and the election of academic division members. Regarding the election of academic division members, it was stipulated that:

> For the election of academic division members, a secret ballot was adopted according to each division, which accounted for 40 percent. The number of voters in each division was no less than two-thirds of the number of members in the existing division, and the election was valid. Candidates who had obtained more than one-half of the votes cast would be selected according to the number of places to be elected and votes to be obtained until the number of places was full. If more than one-half of the candidates had equal votes and could not be selected, the candidate with equal votes would be re-voted, and the candidate with more votes would be selected. The election results should be reported to the Presidium.[52]

This was more stringent than the relevant provisions of the aforementioned *Measures for the Co-optation of Academic Division Members of Chinese Academy of Sciences* and fully reflected the principle of "margin election".

Chapter 3 of the *Constitution for the Academic Division Members of Chinese Academy of Sciences (Trial)* is titled *Foreign Academic Division Members and Their Election Procedures*, which stipulated the selection criteria and election procedures for foreign members. According to the regulations, "foreign scholars who have made great contributions to China's scientific and technological undertakings and enjoy high academic status in the world can be recommended to be elected as foreign Academic division members. Candidates for foreign academic division members

[50] Yuanshi tiaoli (Zhangcheng) 院士条例(章程) (Statutes for Membership of the Chinese Academy of Sciences). Beijing: Zhongguo kexueyuan danganguan中国科学院档案馆(Archives of Chinese Academy of Sciences), 1994-36-018.

[51] Yuanshi tiaoli (Zhangcheng) 院士条例(章程) (Statutes for Membership of the Chinese Academy of Sciences). Beijing: Zhongguo kexueyuan danganguan中国科学院档案馆(Archives of Chinese Academy of Sciences), 1994-36-018.

[52] Yuanshi tiaoli (Zhangcheng) 院士条例(章程) (Statutes for Membership of the Chinese Academy of Sciences). Beijing: Zhongguo kexueyuan danganguan中国科学院档案馆(Archives of Chinese Academy of Sciences), 1994-36-018.

shall be recommended by at least five academic division members and reviewed by the relevant standing committees of the Academic Division. The Presidium meeting should determine the official candidates, and the General Assembly of academic division members should elect them by secret ballot. The number of voters should be no less than two-thirds of the current members. The election was valid and the number of votes won was no less than two-thirds of the number of voters."[53]

Chapters 4 to 8 of the *Constitution for the Academic Division Members of Chinese Academy of Sciences (Trial)* were successively the *Academic Divisions*, the *General Assembly of Academic Division Members*, *The Standing Leadership Organization*, *Publications and Funds*, and *Other*. They successively stipulated the nature, organization, and leadership composition of the school, the functions and tasks of the Academic Divisions, the nature, frequency, and functions of the General Assembly, the nature, organization, personnel composition, and functions of the Presidium, publications and the divisions' funding for activities, the quorum of the General Assembly for academic division members, the Plenary Meetings of all divisions, the meetings of the Standing Committee, the meetings of the Presidium and Executive Committee, how to resign or revoke the title of academic division member, and so on.[54]

The promulgation and implementation of the *Constitution for the Academic Division Members of Chinese Academy of Sciences (Trial)* laid a stable, institutional foundation for the more rigorous and standardized co-optation of the academic division members of the CAS and its development. In order to prepare for the co-optation of academic division members in 1993, the Joint Office of the CAS drafted *Detailed Rules for the Co-optation of Academic Division Members of Chinese Academy of Sciences (draft for comments)* according to the *Constitution for the Academic Division Members of Chinese Academy of Sciences (Trial)* and the aforementioned *Measures for the Co-optation of Academic Division Members of Chinese Academy of Sciences,* as well as the six documents discussed and adopted by the joint meeting of the directors of the CAS in 1991. The *Detailed Rules for the Co-optation of Academic Division Members of Chinese Academy of Sciences (draft for comments)* was submitted to the standing committees of the various divisions for comments on September 2, 1992. On December 4, the Presidium Meeting of the CASAD discussed and approved the *Detailed Rules for Implementation of the Co-optation of Academic Division Members of Chinese Academy of Sciences*.[55] The four steps of the *Detailed Rules for Implementation of the Co-optation of Academic Division Members*

---

[53] Yuanshi tiaoli (Zhangcheng) 院士条例(章程) (Statutes for Membership of the Chinese Academy of Sciences). Beijing: Zhongguo kexueyuan danganguan中国科学院档案馆(Archives of Chinese Academy of Sciences), 1994-36-018.

[54] Yuanshi tiaoli (Zhangcheng) 院士条例(章程) (Statutes for Membership of the Chinese Academy of Sciences). Beijing: Zhongguo kexueyuan danganguan中国科学院档案馆(Archives of Chinese Academy of Sciences), 1994-36-018.

[55] Xuebu changwei dui zengxuan gongzuo xize tichu de xiugai buchong yijian学部常委对 "增选工作细则"提出的修改、补充意见(Revision and Supplemental Opinions of the Standing Committee of the Academic Divisions on the *Detailed Rules for Co-optation*). Beijing: Zhongguo kexueyuan danganguan中国科学院档案馆(Archives of Chinese Academy of Sciences), 1994-36-016.

of Chinese Academy of Sciences—recommendation, primary selection, review, and election—made relatively detailed provisions. This would provide further institutional guarantee for guiding and standardizing the co-optation of the members of the CASAD.

According to the provisions of the *Constitution for the Academic Division Members of Chinese Academy of Sciences (Trial)* and the *Detailed Rules for Implementation of the Co-optation of Academic Division Members of Chinese Academy of Sciences*, an additional 59 members were elected in November 1993.[56] Although the average age of those elected members was nearly two years older than that of those elected in 1991,[57] the limited number of co-opted candidates and the stricter criteria for selecting candidates increased the bargaining chip for the final transition of academic division members to CAS Members.

On February 4, 1993, in preparation for the CAE, the CAS and the State Science and Technology Commission jointly submitted to the Central Committee and State Council *Request for Instructions on Issues Concerning the Establishment of Chinese Academy of Engineering*, which raised the issue of "the appellation of CAE Members":

> In the process of discussion and soliciting opinions, it is better that the members of the CAE are called Members (academicians), which is not only consistent with its honorary nature but also convenient for international contact and exchange. At the same time, it is also suggested that academic division members should also be renamed Members. Judging from the current situation of China's scientific and technological strength and its external influence, the academic division members are renamed as Members while the conditions are ripe and the time is favorable, and while it was also a long-standing common voice of the scientific and technological circles in the country.[58]

"The conditions are ripe" obviously referred to the enhancement of China's scientific and technological strength and its external influence, but it also meant that the divergences within the academic circle on the establishment of the membership system were reduced. On August 31, the State Science and Technology Commission, jointly with the CAS, submitted to the State Council and the Central Committee a revised *Request for Instructions on Issues Concerning the Establishment of Chinese Academy of Engineering*. The revised *Request for Instructions on Issues Concerning the Establishment of Chinese Academy of Engineering* not only raised the issue of the designation of members of the CAE and the change of the name of academic

---

[56]Xuebu zhuxituan huiyi cailiao学部主席团会议材料(Meeting Materials of the Presidium of the Academic Divisions). Beijing: Zhongguo kexueyuan danganguan中国科学院档案馆(Archives of Chinese Academy of Sciences), 1994-36-017.

[57]Xuebu zhuxituan huiyi cailiao学部主席团会议材料(Meeting Materials of the Presidium of the Academic Divisions). Beijing: Zhongguo kexueyuan danganguan中国科学院档案馆(Archives of Chinese Academy of Sciences), 1994-36-017.

[58]Guanyu jianli Zhongguo gongchengyuan youguan wenti de qingshi关于建立中国工程院有关问题的请示(Request for Instructions on Issues Concerning the Establishment of Chinese Academy of Engineering). In Zhongguo kexueyuan bangongting中国科学院办公厅(General Office of the Chinese Academy of Sciences) (Eds.). *Zhongguo kexueyuan nianbao*中国科学院年报(The Annals of the Chinese Academy of Sciences) 1993: 142–143.

division members, but it also raised the issue of the honorary membership system that was to be implemented in both academies.

On October 19, 1993, the 11th Executive Meeting of the State Council discussed the revised *Request for Instructions on Issues Concerning the Establishment of Chinese Academy of Engineering* and agreed on principle to establish the CAE.[59] On January 6, 1994, Jiang Zemin presided over the 46th Meeting of the 14th Standing Committee of the Political Bureau of the Central Committee.[60] The meeting discussed and agreed on principle to the revised *Request for Instructions on Issues Concerning the Establishment of Chinese Academy of Engineering*. They also decided to invite the State Science and Technology Commission to revise this *Request for Instructions on Issues Concerning the Establishment of Chinese Academy of Engineering* and organize its implementation, according to the opinions discussed at the meeting.[61] At the same time, it was approved that the academic division members of the CAS should be renamed as CAS Members.[62] On January 14, the CAS issued a notice to all academic division members: "The Party Central Committee and the State Council have decided that the academic division members of the CAS shall be renamed as CAS Members."[63] As a result, the General Assembly of academic division members was renamed the General Assembly of CAS Members. From June 3 to June 8, the Seventh General Assembly of the CAS Members[64] and its establishment

[59]Zhongkeyuan bangongting guanyu jianli Zhongguo gongchengyuan timing Qian Zhengying tongzhi wei 1995 nian qiaola fasi ziran kexuejiang houxuanren Zhou Guangzhao yuanzhang zai Zhongguo kexuebao gongzuo huiyi shang de jianghua tongzhi qingshi中科院、办公厅关于建立中国工程院、提名钱正英同志为1995年乔拉法斯自然科学奖候选人、周光召院长在中国科学报工作会议上的讲话、通知、请示(Remarks, Notices and Requests of Chinese Academy of Sciences and Its General Office on Establishing the Chinese Academy of Engineering, Nominating Comrade Qian Zhengying as Candidate for the 1995 Chorafas Natural Sciences Award, and President Zhou Guangzhao at the Work Conference of China Science Daily). Beijing: Zhongguo kexueyuan danganguan中国科学院档案馆(Archives of Chinese Academy of Sciences), 1993-02-055.

[60]1994 nian dashiji 1994年大事记(Chronicle of Events in 1994). In Zhongguo gongchengyuan中国工程院(Chinese Academy of Engineering) (Eds.) (2004). *Zhongguo gongchengyuan nianjian* (1994–1997) 中国工程院年鉴(1994–1997) (Yearbook of Chinese Academy of Engineering, 1994–1997). Beijing: Zhongguo gongchengyuan中国工程院.155.

[61]Zhongyang huiyi jueding shixiang tongzhi (Zhaideng) 中央会议决定事项通知(摘登) (Notice of Decisions of the Central Committee Meeting) (Excerpts). In Zhongguo gongchengyuan中国工程院(Chinese Academy of Engineering) (Eds.) (2004). *Zhongguo gongchengyuan nianjian* (1994–1997) 中国工程院年鉴(1994–1997) (Yearbook of Chinese Academy of Engineering, 1994–1997). Beijing: Zhongguo gongchengyuan中国工程院.13.

[62]Xuebu zhuxituan huiyi cailiao学部主席团会议材料(Meeting Materials of the Presidium of the Academic Divisions). Beijing: Zhongguo kexueyuan danganguan中国科学院档案馆(Archives of Chinese Academy of Sciences), 1994-36-017.

[63]Zhongguo kexueyuan 1994 nian dashiji中国科学院一九九四年大事记(Chronicle of Events of Chinese Academy of Sciences in 1994). In Zhongguo kexueyuan bangongting中国科学院办公厅(General Office of the Chinese Academy of Sciences) (Eds.). *Zhongguo kexueyuan nianbao*中国科学院年报(The Annals of the Chinese Academy of Sciences) 1994: 427.

[64]Considering the inheritance relationship between the membership system of CAS and the past membership system of academic divisions, the First General Assembly of CAS Members was actually called as the Seventh General Assembly. See Li Peng李鹏. Zai Zhongguo gongchengyuan

and the First General Assembly of the CAE were held simultaneously in Beijing. This meant the completion of the transition, marking the establishment of the CAS' membership system and the CAE, and the reconstruction of the academician system in China.

The Seventh General Assembly of CAS Members revised and approved the *Constitution for the Members of Chinese Academy of Sciences* on June 8, based on the *Constitution for the Academic Division Members of Chinese Academy of Sciences (Trial)*. According to the *Constitution for the Members of Chinese Academy of Sciences*, membership in the CAS was the highest academic title in science and technology established by the state and is a lifelong honor. If they have systematic and creative achievements and major contributions in the field of science and technology, love the motherland, and have a decent style of study, researchers, professors with Chinese nationality, or scholars and experts with the same title (including Chinese scholars and experts living in Taiwan Province, Hong Kong, and Macao and residing in other countries) could be recommended for election as CAS Members. The co-optation of Members should be held biennially, and the total number of co-opted Members should not exceed sixty. Candidates for Membership could be recommended through two channels: first, they could be directly recommended by Members; second, they could be recommended by relevant domestic scientific and technological research institutes, universities, and first-level societies affiliated with the China Association for Science and Technology, according to their organizational systems; the review of candidates for Membership would be conducted according to the academic division. The number of Members voting in each division should not be less than two-thirds of the number of Members in the division. Candidates who had won more than one-half of the votes cast would be elected according to the number of places to be elected by the division and the number of votes won, until the number was full. Foreign scholars and experts who had made important contributions to China's science and technology and have high academic status in the world can be recommended as foreign Members of the CAS. The election of foreign Members would be held biennially. Candidates for foreign Membership should not have the recommendation of less than five Members. They shall be elected by secret ballot through the General Assembly of CAS Members. The number of Members participating in the voting should not be less than two-thirds of all Members. Candidates with no less than two-thirds of the votes cast can be elected.[65]

---

chengli dahui he Zhongguo kexueyuan diqici yuanshi dahui shang de jianghua在中国工程院成立大会和中国科学院第七次院士大会上的讲话(Speech at the Founding Conference of Chinese Academy of Engineering and the Seventh General Assembly of Members of Chinese Academy of Sciences). In Zhongguo gongchengyuan中国工程院(Chinese Academy of Engineering) (Eds.) (1995). *Zhongguo gongchengyuan shoujie yuanshi dahui wenjian jinkan*中国工程院首届院士大会文件集刊(Collection of Documents of the First General Assembly of Members of Chinese Academy of Engineering). Beijing: Zhongguo gongchengyuan中国工程院.5.

[65]Xuebuban youguan yuanshi zhangcheng renzhi (Wang Chen Qian Yingjie Ai Xinyuan) 学部办有关《院士章程》、任职(王晨、钱莹洁、艾新源) (Relevant Statutes for Membership of the Chinese Academy of Sciences by the Office of CASAD, Position Held (Wang Chen, Qian Yingjie, Ai Xinyuan). Beijing: Zhongguo kexueyuan danganguan中国科学院档案馆(Archives of Chinese Academy of Sciences), 1994-36-007; Zhongguo kexueyuan yuanhsi zhangcheng中国科学院院士

The Seventh General Assembly of CAS Members also elected the first batch of foreign Members on June 8 with a total of fourteen: Derek Harold Barton, Shiing-Shen Chern, Samuel C. C. Ting, Yuan-Cheng B. Fung, P. H. Raven, Joseph Needham, Tsung-Dao Lee, Chia-Chiao Lin, Shing-Tung Yau, Herbert A. Simon, Chang-Lin Tien, Chien-Shiung Wu, Chen-Ning Yang and Leroy L. Chang.[66] They were selected from 25 valid candidates recommended by academic division members. Excluding Barton, Joseph Needham, P. H. Raven, and Herbert A. Simon, the other ten were all Chinese scholars. Among them, twelve were American and two were British. There were twelve people working in the United States, one working in Britain, and one working in Hong Kong. The fourteen foreign academicians all had high academic status in the world, and were scientists who were friendly to China and had made important contributions to China's scientific and technological undertakings. For example, Shiing-Shen Chern, a member of the first batch of members of Academia Sinica and a foreign Member of the CAS recommended by nineteen academic division members, was an international leader in differential geometry and an academician of the American Academy of Sciences. Before he went to the United States at the end of 1948, he trained a group of topology talents for China at the Institute of Mathematics of Academia Sinica. Since 1972, he has given lectures in China many times, and in the 1980s he contributed an important part to the establishment of the Nankai Institute of Mathematics. Under his cultivation and promotion, a group of outstanding young scholars emerged in China. Shing-Tung Yau was an academician of the American Academy of Sciences and was the first Chinese mathematician to win the Fields Medal. Before winning the Fields Medal, he gave lectures at the CAS and introduced his best work, including unpublished ones, to the Chinese. Since then, he has given lectures in China many times, recommended Chinese mathematicians to famous research institutes and universities for exchanges, and personally guided many doctoral students in China, some of whom became internationally renowned mathematicians. And there was Chen-Ning Yang and Tsung-Dao Lee as well, winners of the Nobel Prize in Physics (1957). Since 1971, Chen-Ning Yang had visited and lectured in China many times, promoted scientific and technological exchanges and cooperation between China and the United States, put forth many important suggestions on China's scientific and technological policies. In order to promote the development of education and science in China, Tsung-Dao Lee had done a lot of work in personnel training, construction of high-energy accelerators, and the development of high-energy physics since the 1970s. In 1979, he founded the "CUSPEA" program and sent 915 graduate students to the United States by 1989. Barton was a famous international organic chemist and Nobel Prize winner in chemistry (1969). He had always been friendly to China, and cared for and supported

---

章程(Statutes for Membership of the Chinese Academy of Sciences). In Zhongguo kexueyuan bangongting中国科学院办公厅(General Office of the Chinese Academy of Sciences) (Eds.). *Zhongguo kexueyuan nianbao*中国科学院年报(The Annals of the Chinese Academy of Sciences) 1994: 225–228.

[66]Fan Hongye樊洪业(Eds.) (1999). *Zhongguo kexueyuan biannianshi: 1949–1999*中国科学院编年史: 1949–1999 (Chronicle of Chinese Academy of Sciences, 1949–1999). Shanghai: Shanghai keji jiaoyu chubanshe上海科技教育出版社.361.

the development of China's organic chemistry department. In 1974, at the invitation of the CAS, he visited it as well as the Shanghai Institute of Organic Chemistry. He once proposed that Chinese organic chemists be Honorary Members of the Royal Society of England. He nominated and recommended Chinese organic chemists several times as consultants and editors of the famous international organic chemistry publications *Tetrahedron* and *Tetrahedron Letters*. He made a lot of contributions to the promotion of academic exchanges between China and the international organic chemistry community.[67] These reflected China's efforts to integrate into the international membership system, in terms of membership system construction.

After the Seventh General Assembly, following the *Constitution for the Members of Chinese Academy of Sciences* passed by the General Assembly, the Joint Office of the CAS revised the *Detailed Rules for Implementation of the Co-optation of Academic Division Members of Chinese Academy of Sciences* discussed and approved by the Presidium Meeting of the CASAD on December 4, 1992, formed the *Detailed Rules for Implementation of the Co-optation of Members of Chinese Academy of Sciences* (draft for comments), and solicited opinions from all Members on September 24, 1994. Important amendments to the draft included: changing "academic division members" to "Members". In the *General Provisions*, a provision was added, saying "special attention should be paid to the selection of qualified, outstanding young and middle-aged scientific and technological experts, and in the official list for each division, those sixty years old and under should not be fewer than one-third of the candidates." According to Article 5 of Chapter 2 of the *Statutes for Membership in the Chinese Academy of Sciences*, the limit of recommending candidates by each Member was changed from three to two. From the original stipulation that the recommendation of two Members was valid, the recommendation of three academicians was also valid. In order to reduce the meeting and shorten the review time of the meeting, review work was divided into two stages: correspondent review conducted by the discipline group and meeting review conducted by the divisions. With the change of the review method, corresponding changes were made at the operational level, such as material distribution and the generation of the preliminary candidate list. The draft was approved on principle by the Executive Committee of the Presidium of the CAS on September 23, 1994.[68]

On December 16, 1994, the Presidium Meeting of the CASAD discussed and approved the *Detailed Rules for Implementation of the Co-optation of Members of Chinese Academy of Sciences*. The *Detailed Rules for Implementation of the Co-optation of Members of Chinese Academy of Sciences* mainly revised the "review" part of the draft for comments and did not adopt the correspondence review done by any relevant discipline review teams. Instead, it stipulated that both the first-step

[67] Xuebu zhuxituan huiyi cailiao学部主席团会议材料(Meeting Materials of the Presidium of the Academic Divisions). Beijing: Zhongguo kexueyuan danganguan中国科学院档案馆(Archives of Chinese Academy of Sciences), 1994-36-017.

[68] Xuebu changwei dui zengxuan gongzuo xize tichu de xiugai buchong yijian学部常委对"增选工作细则"提出的修改、补充意见(Revision and Supplemental Opinions of the Standing Committee of the Academic Divisions on the *Detailed Rules for Co-optation*). Beijing: Zhongguo kexueyuan danganguan中国科学院档案馆(Archives of Chinese Academy of Sciences), 1994-36-016.

review and the second-step review should be conducted in a meeting. In addition, the "review" section added provisions for the implementation of the challenge system.[69] These amendments were beneficial to improving the fairness and scientificity of the evaluation work. The *Detailed Rules for Implementation of the Co-optation of Members of Chinese Academy of Sciences* mainly provided a concrete and guiding institutional basis for the co-optation of CAS Members.

After the establishment of the membership system of the CAS in June 1994, the academic divisions were positioned as the country's highest advisory body in science and technology. Members participated in the activities of one division, according to their discipline areas. The General Assembly was the highest organization form of the CAS and was held biennially. The permanent leading body of the General Assembly when it was not in session was the Presidium of the CASAD. The president was the executive chairman of the presidium and presided over the presidium meeting.[70] This kind of institutional arrangement has continued to today.

10.2 The Preparations and Establishment of the CAE and Its Membership System

The CAE is the highest honorary and advisory academic institute in the national engineering science and technology (hereinafter referred to as "engineering technology") community that is composed of Members. It has research institutes, design institutes, and development entities. Membership in the CAE is the highest academic title in engineering technology established by the state and is a lifelong honor.[71] Both Academia Sinica and the CAS established the membership system several years later than the CAE. It had established its membership system when the academy itself was established, which was the same process as that of the western countries when they

[69]Xuebu changwei dui zengxuan gongzuo xize tichu de xiugai buchong yijian学部常委对"增选工作细则"提出的修改、补充意见(Revision and Supplemental Opinions of the Standing Committee of the Academic Divisions on the *Detailed Rules for Co-optation*). Beijing: Zhongguo kexueyuan danganguan中国科学院档案馆(Archives of Chinese Academy of Sciences), 1994-36-016.

[70]Xuebuban youguan yuanshi zhangcheng renzhi (Wang Chen Qian Yingjie Ai Xinyuan) 学部办有关《院士章程》、任职(王晨、钱莹洁、艾新源) (Relevant Statutes for Membership of the Chinese Academy of Sciences by the Office of CASAD, Position Held (Wang Chen, Qian Yingjie, Ai Xinyuan). Beijing: Zhongguo kexueyuan danganguan中国科学院档案馆(Archives of Chinese Academy of Sciences), 1994-36-007; Zhongguo kexueyuan yuanhsi zhangcheng中国科学院院士章程(Statutes for Membership of the Chinese Academy of Sciences). In Zhongguo kexueyuan bangongting中国科学院办公厅(General Office of the Chinese Academy of Sciences) (Eds.). *Zhongguo kexueyuan nianbao*中国科学院年报(The Annals of the Chinese Academy of Sciences) 1994: 225–228.

[71]Zhongguo gongchengyuan Zhangcheng中国工程院章程(Statutes for the Chinese Academy of Engineering). In Zhongguo gongchengyuan中国工程院(Chinese Academy of Engineering) (Eds.) (1995). *Zhongguo gongchengyuan shoujie yuanshi dahui wenjian jinkan*中国工程院首届院士大会文件集刊(Collection of Documents of the First General Assembly of Members of Chinese Academy of Engineering). Beijing: Zhongguo gongchengyuan中国工程院.29–30.

established their academies of sciences. The CAE was also determined by its "virtual entity" organizational form. As a milestone in the process of institutionalizing China's engineering technology, the establishment of the CAE's membership system was an integral part of the reconstruction of the membership system in China.

(1) **The Background of the Establishment of the CAE and Its Membership System**

Before the establishment of the CAE, Chinese scientists, engineering and technical experts, and relevant personages had repeatedly put forth active proposals for the establishment of the CAE. In 1992, at the fifth meeting of the Seventh National Committee of the Chinese People's Political Consultative Conference, and during the Sixth General Assembly of CASAD Members, many members of the Chinese People's Political Consultative Conference and the CASAD offered different proposals and suggestions, hoping to establish the CAE as soon as possible.[72] In particular, the *Proposals on the Early Establishment of the Chinese Academy of Engineering and Technological Sciences* (hereinafter referred to as the *Proposals*),[73] put forth by six academic divisionmembers—Zhang Guangdou, Wang Daheng, Shi Changxu, Zhang Wei, Hou Hsiong-Lin and Luo Peilin—received the attention of the Central Committee and the State Council leaders[74] and played a vital role in the establishment of the CAE and its membership system.

The establishment of the CAE and its membership system was also closely related to the following international and domestic backgrounds. Since the twentieth century, engineering technology, which used to rely mainly on accumulated experience, had gained theoretical support and guidance due to the needs of industrial production and the rapid development of technological science. It had developed rapidly at an unprecedented speed and scale, and was playing an increasingly important role in promoting industrial and agricultural production and economic and social development. Therefore, the scientific and technological circles and decision-makers of

[72]Zhu Guangya朱光亚. Zai Zhongguo gongchengyuan chengli dahui shang de jianghua在中国工程院成立大会上的讲话(Speech at the Founding Conference of Chinese Academy of Engineering). In Zhongguo gongchengyuan中国工程院(Chinese Academy of Engineering) (Eds.) (1995). *Zhongguo gongchengyuan shoujie yuanshi dahui wenjian jinkan*中国工程院首届院士大会文件集刊(Collection of Documents of the First General Assembly of Members of Chinese Academy of Engineering). Beijing: Zhongguo gongchengyuan中国工程院.10.

[73]Liu wei xuebu weiyuan guanyu zaori jianli Zhongguo gongcheng yu jishu kexueyuan de jianyi六位学部委员关于早日建立中国工程与技术科学院的建议(Six Academic Division Members' Proposals on the Early Establishment of the Chinese Academy of Engineering and Technological Sciences). In Zhongguo gongchengyuan中国工程院(Chinese Academy of Engineering) (Eds.) (2004). *Zhongguo gongchengyuan nianjian* (1994–1997) 中国工程院年鉴(1994–1997) (Yearbook of Chinese Academy of Engineering, 1994–1997). Beijing: Zhongguo gongchengyuan中国工程院.33–35.

[74]Guanyu jianli Zhongguo gongchengyuan youguan wenti de qingshi关于建立中国工程院有关问题的请示(Request for Instructions on Issues Concerning the Establishment of Chinese Academy of Engineering). In Zhongguo kexueyuan bangongting中国科学院办公厅(General Office of the Chinese Academy of Sciences) (Eds.). *Zhongguo kexueyuan nianbao*中国科学院年报(The Annals of the Chinese Academy of Sciences) 1993: 141.

many countries around the world began to pay more and more attention to engineering technology, and had taken measures to strengthen its status and role. In 1919, Sweden took the lead in establishing the Royal Academy of Engineering, followed by Denmark, Norway, the United States, Mexico, Australia, Britain, Belgium, Switzerland, France, Japan, Finland, Uruguay, Argentina, Canada, India, Hungary, Brazil, the Soviet Union, and other countries in establishing similar academic institutes.[75] By 1992, almost all industrially developed countries and some still-developing countries including India, Mexico, and Hungary had established their own Academy of Engineering Sciences and its membership system.[76]

In 1978, the Council of the Academies of Engineering and Technological Sciences (CAETS) was established by the Swedish, American, Mexican, Australian, and British Academies of Engineering or like organizations.[77] This was an international academic institute with national engineering academies from all over the world as group members. Its goal was to take on the role of an international forum that would discuss the important engineering and technical issues in the world, encourage cooperative, international engineering undertakings, assist countries that did not yet have an engineering academy to establish one, and be committed to strengthening engineering and technical cooperation in order to promote world economic and social welfare.[78] China was invited to send members of the Division of Technological

[75]Guanyu jianli Zhongguo gongchengyuan youguan wenti de qingshi关于建立中国工程院有关问题的请示(Request for Instructions on Issues Concerning the Establishment of Chinese Academy of Engineering). In Zhongguo gongchengyuan中国工程院(Chinese Academy of Engineering) (Eds.) (2004). *Zhongguo gongchengyuan nianjian* (1994–1997) 中国工程院年鉴(1994–1997) (Yearbook of Chinese Academy of Engineering, 1994–1997). Beijing: Zhongguo gongchengyuan中国工程院.38; Zhu Guangya朱光亚. Guanyu Zhongguo gongchengyuan de chengli关于中国工程院的成立(About the Establishment of Chinese Academy of Engineering) . In Zhongguo gongchengyuan中国工程院(Chinese Academy of Engineering) (Eds.) (1995). *Zhongguo gongchengyuan shoujie yuanshi dahui wenjian jinkan*中国工程院首届院士大会文件集刊(Collection of Documents of the First General Assembly of Members of Chinese Academy of Engineering). Beijing: Zhongguo gongchengyuan中国工程院.69.

[76]Guowai gongchengyuan ji leisi jigou de gaikuang国外工程院及类似机构的概况(Overview of Foreign Academies of Engineering and Similar Institutes). In Zhongguo gongchengyuan中国工程院(Chinese Academy of Engineering) (Eds.) (1995). *Zhongguo gongchengyuan shoujie yuanshi dahui wenjian jinkan*中国工程院首届院士大会文件集刊(Collection of Documents of the First General Assembly of Members of Chinese Academy of Engineering). Beijing: Zhongguo gongchengyuan中国工程院.76–87.

[77]The Council was established as the Convocation of Engineering Academies and Like Organizations. In 1983, it was renamed as Convocation of Academies of Engineering and Technological Sciences. It was finally renamed the Council of the Academy of Engineering and Technological Sciences. See Guoji zuzhi: Geguo gongcheng yu jishu kexueyuan lishihui国际组织——各国工程与技术科学院理事会(International Organization: the Councils of National Academies of Engineering and Technological Sciences). In Zhongguo gongchengyuan中国工程院(Chinese Academy of Engineering) (Eds.) (1995). *Zhongguo gongchengyuan shoujie yuanshi dahui wenjian jinkan* 中国工程院首届院士大会文件集刊(Collection of Documents of the First General Assembly of Members of Chinese Academy of Engineering). Beijing: Zhongguo gongchengyuan中国工程院.88.

[78]Guoji zuzhi: Geguo gongcheng yu jishu kexueyuan lishihui国际组织——各国工程与技术科学院理事会(International Organization: the Councils of National Academies of Engineering and

Sciences of the CAS to attend the annual meeting of the CAETS six times. In 1990, China formally applied to join the CAETS in the name of the Division of Technological Sciences. However, due to the fact that the Division of Technological Sciences was not an independent academic institute, China was not accepted. If China had not set up an independent engineering academy, it would have definitely created an unfavorable situation for the exchanges between China and the international engineering and technology community.[79]

In other words, entering the 1990s, the establishment of the CAE and its membership system not only became the general trend for China's engineering technology system to conform to international standard, but had also become an important way to ensure and strengthen exchanges between China and the international engineering technology community.

In the 30 to 40 years since 1949, a large number of outstanding engineering and technical experts and engineers who had made noteworthy contributions to national economic development and national defense construction had grown up in China. However, due to the limitation in the total number of seats and the different characteristics and requirements of science and engineering, many outstanding experts among them were not elected as academic division members.[80] The establishment of the CAE and its membership system would solve this problem on a large scale. After the reform and opening up of China, the Central Government paid special attention to economic construction, emphasizing it as the center. The role of engineering technology in national economic construction was also clearly reflected. However, China's engineering technology level and research, design, and construction capabilities were still unable to meet the great demands of the country's economic construction. In order to solve this problem, it was necessary to establish a national-level engineering academy—with engineering and technical experts as the main body—and the membership system, so as to improve the social status of China's engineering and technical circles, mobilize the enthusiasm of engineering and technical personnel, and give full play to their overall role.

Technological Sciences). In Zhongguo gongchengyuan中国工程院(Chinese Academy of Engineering) (Eds.) (1995). *Zhongguo gongchengyuan shoujie yuanshi dahui wenjian jinkan*中国工程院首届院士大会文件集刊(Collection of Documents of the First General Assembly of Members of Chinese Academy of Engineering). Beijing: Zhongguo gongchengyuan中国工程院.88–93.

[79]Guanyu jianli Zhongguo gongchengyuan youguan wenti de qingshi关于建立中国工程院有关问题的请示(Request for Instructions on Issues Concerning the Establishment of Chinese Academy of Engineering). In Zhongguo gongchengyuan中国工程院(Chinese Academy of Engineering) (Eds.) (2004). *Zhongguo gongchengyuan nianjian* (1994–1997) 中国工程院年鉴(1994–1997) (Yearbook of Chinese Academy of Engineering, 1994–1997). Beijing: Zhongguo gongchengyuan中国工程院.38.

[80]Guanyu jianli Zhongguo gongchengyuan youguan wenti de qingshi关于建立中国工程院有关问题的请示(Request for Instructions on Issues Concerning the Establishment of Chinese Academy of Engineering). In Zhongguo gongchengyuan中国工程院(Chinese Academy of Engineering) (Eds.) (2004). *Zhongguo gongchengyuan nianjian* (1994–1997) 中国工程院年鉴(1994–1997) (Yearbook of Chinese Academy of Engineering, 1994–1997). Beijing: Zhongguo gongchengyuan中国工程院.38–39.

More importantly, the Central Committee and the State Council were vigorously implementing Deng Xiaoping's guiding ideology that "science and technology are the first productive forces" when the six members Zhang Guangdou, Wang Daheng, Shi Changxu, Zhang Wei, Hou Hsiong-Lin, and Luo Peilin put forward the *Proposals*.[81] As engineering technology was a direct productive force, the Central Government and the State Council attached unprecedented importance to it, in order to implement this guiding ideology and improve the level of engineering technology and research, design, and construction capabilities. In 1991, when Jiang Zemin, General Secretary of the CPC Central Committee, spoke on behalf of the CPC Central Committee and the State Council at the Fourth National Congress of the China Association for Science and Technology, he pointed out:

> According to certain needs and possibilities, we will concentrate the country's material and financial resources on some major projects related to the national economy, the people's livelihood, and national security, and jointly tackle key problems so as to achieve achievements with great practical application value and breakthroughs with advanced world standards. … In deepening the reform of the science and technology system and ensuring the development of science and technology in the 1990s, the CPC Central Committee and the State Council have taken and will further take on a number of major measures to better implement various scientific and technological tasks.[82]

Major projects related to the international people's livelihood and national security mentioned here mainly refers to those in the field of engineering and technology. The *Proposals* of the six members received the attention of the Central Committee and State Council leaders, and was also directly related to this background.

(2) The *Proposals* of the Six Academic Division Members, the Instructions of the Central Committee, and the Opinions of the CAS

The six academic division members put forth the *Proposals* on April 21, 1992. The *Proposals* was drafted by Luo Peilin and jointly submitted to the Central Committee by the six members through different channels.[83] In the *Proposals*, they introduced the development of engineering technology in the world as well as the establishment

[81]Jiang Zemin江泽民. Zai Zhongguo kexue jishu xiehui disici quanguo daibiao dahui shang de jianghua在中国科学技术协会第四次全国代表大会上的讲话(Speech at the Fourth National Congress of the China Association of Science and Technology). In Guojia kexue jishu weiyuanhui国家科学技术委员会(State Scientific and Technological Commission) et al. (Eds.) (1991). *Kexue jishu shi diyi shengchanli*科学技术是第一生产力(Science and Technology is the First Productive Force). Jinan: Shandong kexue jishu chubanshe山东科学技术出版社.1–2.

[82]Jiang Zemin江泽民. Zai Zhongguo kexue jishu xiehui disici quanguo daibiao dahui shang de jianghua在中国科学技术协会第四次全国代表大会上的讲话(Speech at the Fourth National Congress of the China Association of Science and Technology). In Guojia kexue jishu weiyuanhui国家科学技术委员会(State Scientific and Technological Commission) et al. (Eds.) (1991). *Kexue jishu shi diyi shengchanli*科学技术是第一生产力(Science and Technology is the First Productive Force). Jinan: Shandong kexue jishu chubanshe山东科学技术出版社.4–5.

[83]Shi Changxu师昌绪(2011). *Zai rensheng daolu shang: Shi Changxu zizhuan*在人生道路上: 师昌绪自传(On the Road of Life: Autobiography of Shi Changxu). Beijing: Kexue chubanshe科学出版社.186.

of the national academy of engineering and similar institutions, and proposed: "Since China is still a developing country, its development of engineering technology and technological science is still not enough and the level of industrial technology is still very poor. Starting from the implementation of the principle that science and technology are the first productive forces, and implementing the policies of 'subordinating to the economy' and 'serving the economy', we propose to establish the Chinese Academy of Engineering and Technological Sciences as soon as possible, in order to promote the development of economic construction and national defense construction."[84] Then, around the central task of the "Chinese Academy of Engineering and Technological Sciences," they proposed the following suggestions:

> The central task of this academy should be to provide authoritative consultation, demonstration, and review for the country and the government on major engineering, technological, and scientific decisions as well as technical and economic issues, and to identify especially important engineering, technological, and scientific achievements. It naturally transcends departmental and regional limitations. In order to accomplish such a central task, its members should be selected among the engineering and technical personnel at the national level and those who have made significant contributions to the development of engineering and technology. Of course, this is also the highest honor for personnel elected in engineering science and technology. We suggest that the CAS be immediately instructed to undertake specific preparations. On the basis of some members of the Division of Technological Sciences and other divisions, the preparatory committee will be composed of personnel who are highly inventive in engineering and technology outside of the CAS.[85]

These proposals had already involved the selection criteria for the Members of "the Chinese Academy of Engineering and Technological Sciences" and the preliminary plan for the preparation of the Academy.

In the *Proposals*, the six members also put forth suggestions on the organizational form of the institute: "The Academy of Engineering and Technological Sciences should be a 'virtual body'; that is, it does not have jurisdiction over any research institute, school, or factory. This is consistent with the existing divisions of the CAS, and there would be no duplication or contradiction with the existing organizations."[86] This organizational form was different from that of Academia Sinica and the CAS,

[84]Guanyu zaori jianli Zhongguo gongcheng yu jishu kexueyuan de jianyi关于早日建立中国工程与技术科学院的建议(Proposal on the Establishment of the Chinese Academy of Engineering and Technology as Early as Possible). In Zhongguo kexueyuan jinshu yanjiusuo中国科学院金属研究所(Institute of Metal Research, Chinese Academy of Sciences) (Eds.) (2000). *Shi Changxu keji huodong shengya*师昌绪科技活动生涯(Shi Changxu's Career in Science and Technology). Beijing: Kexue chubanshe科学出版社.97.

[85]Guanyu zaori jianli Zhongguo gongcheng yu jishu kexueyuan de jianyi关于早日建立中国工程与技术科学院的建议(Proposal on the Establishment of the Chinese Academy of Engineering and Technology as Early as Possible). In Zhongguo kexueyuan jinshu yanjiusuo中国科学院金属研究所(Institute of Metal Research, Chinese Academy of Sciences) (Eds.) (2000). *Shi Changxu keji huodong shengya*师昌绪科技活动生涯(Shi Changxu's Career in Science and Technology). Beijing: Kexue chubanshe科学出版社.97.

[86]Guanyu zaori jianli Zhongguo gongcheng yu jishu kexueyuan de jianyi关于早日建立中国工程与技术科学院的建议(Proposal on the Establishment of the Chinese Academy of Engineering and Technology as Early as Possible). In Zhongguo kexueyuan jinshu yanjiusuo中国科学院金属研究所(Institute of Metal Research, Chinese Academy of Sciences) (Eds.) (2000). *Shi Changxu*

and was based on the internationally accepted organizational form of the Academy of Engineering Sciences.

The six members also put forward suggestions on the relationship between the CAS and the Academy of Engineering and Technological Sciences: "The two academies independently conduct the election of their members, set up their own presidiums and presidents, and independently determine their own rules, regulations, principles, policies, and work plans."[87] At the same time, they proposed: "The CAS still retains the division of technological sciences and the existing members of the division of technological sciences would be ex-officio members of the Academy of Engineering and Technological Sciences. An expert can be elected as a member of academic division of both academies."[88] Based on the later situation, only the proposal that "the existing members of the division of technological sciences would be ex-officio members of the Academy of Engineering and Technological Sciences" was implemented. The proposal that "both academies should organize and coordinate with each other" was generally implemented; all other suggestions were fully implemented.

After twenty days—that is, on May 11, 1992—Jiang Zemin gave instructions to the *Proposals* published in the 54th issue of the general office of the Central Committee's *Synthesis and Summary*: "Comrade Jia Bao: this matter has been mentioned many times. It seems that we need to exchange opinions with various parties to study it and make decisions. Please consider it."[89] "Jia Bao," also known as Wen Jiabao, was then the Alternate Secretary of the Secretariat of the Central Committee and Director of the Central General Office. The next day, upon Wen Jiabao's instructions and with the support of State Councilor and Director of the State Science and Technology Commission Song Jian and Secretary-General of the State Council Luo Gan, the CAS took the lead and President Zhou Guangzhao convened any relevant personnel

*keji huodong shengya*师昌绪科技活动生涯(Shi Changxu's Career in Science and Technology). Beijing: Kexue chubanshe科学出版社.97.

[87]Guanyu zaori jianli Zhongguo gongcheng yu jishu kexueyuan de jianyi关于早日建立中国工程与技术科学院的建议(Proposal on the Establishment of the Chinese Academy of Engineering and Technology as Early as Possible). In Zhongguo kexueyuan jinshu yanjiusuo中国科学院金属研究所(Institute of Metal Research, Chinese Academy of Sciences) (Eds.) (2000). *Shi Changxu keji huodong shengya*师昌绪科技活动生涯(Shi Changxu's Career in Science and Technology). Beijing: Kexue chubanshe科学出版社.97–98.

[88]Guanyu zaori jianli Zhongguo gongcheng yu jishu kexueyuan de jianyi关于早日建立中国工程与技术科学院的建议(Proposal on the Establishment of the Chinese Academy of Engineering and Technology as Early as Possible). In Zhongguo kexueyuan jinshu yanjiusuo中国科学院金属研究所(Institute of Metal Research, Chinese Academy of Sciences) (Eds.) (2000). *Shi Changxu keji huodong shengya*师昌绪科技活动生涯(Shi Changxu's Career in Science and Technology). Beijing: Kexue chubanshe科学出版社.98.

[89]Zhonggong zhongyang guowuyuan lingdao tongzhi guanyu jianli Zhongguo gongchengyuan de pishi中共中央、国务院领导同志关于建立中国工程院的批示(Instructions of Leaders of the CPC Central Committee and the State Council on the Establishment of Chinese Academy of Engineering). In Zhongguo gongchengyuan中国工程院(Chinese Academy of Engineering) (Eds.) (2004). *Zhongguo gongchengyuan nianjian* (1994–1997) 中国工程院年鉴(1994–1997) (Yearbook of Chinese Academy of Engineering, 1994–1997). Beijing: Zhongguo gongchengyuan中国工程院.33.

to study and discuss the issue of the Chinese Academy of Engineering and Technological Sciences. The "Research Group on the Issues of the CAE" was formed by the head, deputy head, and six advisors of the Division of Technological Sciences of the CAS to conduct a special investigation and study.[90] On this basis, on July 18, the CAS, in the name of President Zhou Guangzhao, put forward five principled opinions on the proposed Chinese Academy of Engineering and Technological Sciences and submitted them to Luo Gan, Jiang Zemin, Li Peng, Wen Jiabao and Song Jian for instructions.[91] The five principles are as follows:

> First, the Chinese Academy of Engineering and Technological Sciences (temporarily named) should be an academic institute composed of scientists and engineers who have made great contributions in engineering technology, and should not be an administrative body.
>
> Second, the plan for the establishment of the Chinese Academy of Engineering and Technological Sciences should be prepared, discussed, and put forward mainly by scientists and technical experts working in the field of engineering and technology, and should be submitted to the Central Committee and the State Council for a decision after widely listening to the opinions of relevant parties.
>
> Third, the Chinese Academy of Engineering and Technological Sciences should be a "virtual entity"; that is, it should only consist of Members (or academic division members) and not establish or have jurisdiction over entities such as research and development.
>
> Fourth, an important feature of the development of modern science is the mutual infiltration of and the comprehensive crossing of multiple disciplines. In order to avoid the situation in which science and technology are completely separated, which is not conducive to the development of science and technology, the Chinese Academy of Engineering and Technological Sciences and the CAS should establish inseparable organic links through various means, such as referring to the American model, acknowledging that the Chinese Academy of Engineering and Technological Sciences is a group member of the CAS, and establishing that leaders of the two academies are appropriately cross-appointed.
>
> Fifth, in order to streamline administration, improve efficiency, and keep the two academies in close touch, it is proposed that, after the establishment of the Academy of Engineering

[90]Guanyu jianli Zhongguo gongchengyuan youguan wenti de qingshi关于建立中国工程院有关问题的请示(Request for Instructions on Issues Concerning the Establishment of Chinese Academy of Engineering). In Zhongguo kexueyuan bangongting中国科学院办公厅(General Office of the Chinese Academy of Sciences) (Eds.). *Zhongguo kexueyuan nianbao*中国科学院年报(The Annals of the Chinese Academy of Sciences) 1993: 141; Zhonggong zhongyang guowuyuan lingdao tongzhi guanyu jianli Zhongguo gongchengyuan de pishi中共中央、国务院领导同志关于建立中国工程院的批示(Instructions of Leaders of the CPC Central Committee and the State Council on the Establishment of Chinese Academy of Engineering). In Zhongguo gongchengyuan中国工程院(Chinese Academy of Engineering) (Eds.) (2004). *Zhongguo gongchengyuan nianjian* (1994–1997) 中国工程院年鉴(1994–1997) (Yearbook of Chinese Academy of Engineering, 1994–1997). Beijing: Zhongguo gongchengyuan中国工程院.33.

[91]Zhonggong zhongyang guowuyuan lingdao tongzhi guanyu jianli Zhongguo gongchengyuan de pishi中共中央、国务院领导同志关于建立中国工程院的批示(Instructions of Leaders of the CPC Central Committee and the State Council on the Establishment of Chinese Academy of Engineering). In Zhongguo gongchengyuan中国工程院(Chinese Academy of Engineering) (Eds.) (2004). *Zhongguo gongchengyuan nianjian* (1994–1997) 中国工程院年鉴(1994–1997) (Yearbook of Chinese Academy of Engineering, 1994–1997). Beijing: Zhongguo gongchengyuan中国工程院.33.

> and Technological Sciences, there will not be an additional administrative office. Its administrative functions will be undertaken by the administrative office of the CASAD, and a set of two brands of institutions will be implemented externally.[92]

Luo Gan, Song Jian, Wen Jiabao, Li Peng, and Jiang Zemin successively agreed to these five principles.[93] These opinions represented the basic views of the CAS on its nature, establishment plan, administrative organization, and its relationship with the Chinese Academy of Engineering and Technological Sciences. On August 31, 1992, after receiving the original instructions of the central leadership on the five principles, Zhou Guangzhao said that "the matter can be proceeded."[94] As a result, the preparatory work for the CAE was officially launched under the leadership of the CAS.

(3) **The First Phase of Preparatory Work Led by the CAS**

After taking lead in the preparatory work, the CAS conducted research and discussions on the establishment of the CAE within the scope of over 200 members of the Division of Technological Sciences and the standing committees of the Division of Mathematics and Physics, the Division of Chemistry, the Division of Biology, and the Division of Earth Sciences. Moreover, opinions were solicited from the heads of the relevant ministries and academic organizations under the State Council, including the State Education Commission, the Commission of Science, Technology, and Industry for National Defense, the Ministry of Aeronautics and Astronautics, the Ministry of Machinery and Electricity, the Ministry of Metallurgy, the Ministry of Chemical Industry, the Ministry of Agriculture, the General Corporation of Nuclear Industry, the General Corporation of Oil and Gas, the China Association for Science and Technology, and the National Natural Science Foundation of China. Since then, the CAS and the State Science and Technology Commission jointly submitted to the Central

---

[92]Zhou Guangzhao guanyu jianli Zhongguo gongchengyuan de wutiao yuanze周光召关于建立中国工程院的五条原则(Zhou Guangzhao's Five Principles for Establishing the Chinese Academy of Engineering). In Zhongguo gongchengyuan中国工程院(Chinese Academy of Engineering) (Eds.) (2004). *Zhongguo gongchengyuan nianjian* (1994–1997) 中国工程院年鉴(1994–1997) (Yearbook of Chinese Academy of Engineering, 1994–1997). Beijing: Zhongguo gongchengyuan中国工程院.35–36.

[93]Zhonggong zhongyang guowuyuan lingdao tongzhi guanyu jianli Zhongguo gongchengyuan de pishi中共中央、国务院领导同志关于建立中国工程院的批示(Instructions of Leaders of the CPC Central Committee and the State Council on the Establishment of Chinese Academy of Engineering). In Zhongguo gongchengyuan中国工程院(Chinese Academy of Engineering) (Eds.) (2004). *Zhongguo gongchengyuan nianjian* (1994–1997) 中国工程院年鉴(1994–1997) (Yearbook of Chinese Academy of Engineering, 1994–1997). Beijing: Zhongguo gongchengyuan中国工程院.33.

[94]Zhonggong zhongyang guowuyuan lingdao tongzhi guanyu jianli Zhongguo gongchengyuan de pishi中共中央、国务院领导同志关于建立中国工程院的批示(Instructions of Leaders of the CPC Central Committee and the State Council on the Establishment of Chinese Academy of Engineering). In Zhongguo gongchengyuan中国工程院(Chinese Academy of Engineering) (Eds.) (2004). *Zhongguo gongchengyuan nianjian* (1994–1997) 中国工程院年鉴(1994–1997) (Yearbook of Chinese Academy of Engineering, 1994–1997). Beijing: Zhongguo gongchengyuan中国工程院.33.

Committee and the State Council on February 4, 1993 the *Request for Instructions on Issues Concerning the Establishment of Chinese Academy of Engineering*.[95]

The *Request for Instructions on Issues Concerning the Establishment of Chinese Academy of Engineering* was a relatively complete plan for the construction of the CAE. It pointed out that "most scientific and technological experts and departments" agreed that "it is very necessary and timely to establish an independent academic institution of highest honor and consultation with scientific and technological engineering experts as the main body." Then, around the "request for instructions on several issues", the *Request for Instructions on Issues Concerning the Establishment of Chinese Academy of Engineering* successively put forward seven requests on "the question of the title," "the nature and function of the CAE," "the relationship between the CAE and the CAS", "the appellation of CAE Members", "the criteria and conditions for CAE Members", "the quota of the first batch of CAE Members and the method for their generation", and "the leadership system and daily administrative organization of CAE".[96]

"On the issue of title," the *Request for Instructions on Issues Concerning the Establishment of Chinese Academy of Engineering* suggested that the proposed Chinese Academy of Engineering and Technological Sciences adopt the name "CAE". The reason was: "this title can not only clearly and concisely highlight the dominant position of engineering science and technology, but also reflect the characteristics of the division of labor among China's major scientific and technological institutes." "The CAE is also easily understood with international contacts."[97]

"On the nature and role of the CAE," the *Request for Instructions on Issues Concerning the Establishment of Chinese Academy of Engineering* stated: "In accordance with international practices and China's national conditions, the CAE should be the highest honorary and advisory academic institute in the engineering science and technology field, composed of China's outstanding engineering science and technology experts and engineers; not as an administrative institute, nor should it set up or have jurisdiction over research, teaching, development, design, production, and other entities. The main task of the CAE is to provide consultation on the decision-making and planning of major national engineering science and technology advances from a

[95]Guanyu jianli Zhongguo gongchengyuan youguan wenti de qingshi关于建立中国工程院有关问题的请示(Request for Instructions on Issues Concerning the Establishment of Chinese Academy of Engineering). In Zhongguo kexueyuan bangongting中国科学院办公厅(General Office of the Chinese Academy of Sciences) (Eds.). *Zhongguo kexueyuan nianbao*中国科学院年报(The Annals of the Chinese Academy of Sciences) 1993: 141–144.

[96]Guanyu jianli Zhongguo gongchengyuan youguan wenti de qingshi关于建立中国工程院有关问题的请示(Request for Instructions on Issues Concerning the Establishment of Chinese Academy of Engineering). In Zhongguo kexueyuan bangongting中国科学院办公厅(General Office of the Chinese Academy of Sciences) (Eds.). *Zhongguo kexueyuan nianbao*中国科学院年报(The Annals of the Chinese Academy of Sciences) 1993: 142–143.

[97]Guanyu jianli Zhongguo gongchengyuan youguan wenti de qingshi关于建立中国工程院有关问题的请示(Request for Instructions on Issues Concerning the Establishment of Chinese Academy of Engineering). In Zhongguo kexueyuan bangongting中国科学院办公厅(General Office of the Chinese Academy of Sciences) (Eds.). *Zhongguo kexueyuan nianbao*中国科学院年报(The Annals of the Chinese Academy of Sciences) 1993: 142.

global perspective; to study and discuss major engineering and technical issues and put forward suggestions for them; to carry out academic activities and participate in the activities of corresponding international organizations on behalf of China's engineering science and technology sector, and to promote extensive exchanges and cooperation in the field of engineering science and technology."[98] Among them, the nature of the CAE is better than that the one stated in the five principles of the proposed Chinese Academy of Engineering and Technology, put forward by the CAS in the name of Zhou Guangzhao on July 18, 1992. It is clear that the CAE is the highest honorary and advisory academic institute in the Chinese engineering science and technology field. The main task of the CAE, compared to the central task of the Chinese Academy of Engineering and Technological Sciences in the *Proposals,* put forth by the six members Zhang Guangdou, Wang Daheng, Shi Changxu, Zhang Wei, Hou Hsiong-Lin and Luo Peilin of the academic divisions, was "to carry out academic activities and participate in the activities of corresponding international organizations on behalf of China's engineering science and technology sector; to promote extensive exchanges and cooperation in the field of engineering science and technology."

"On the relationship between the CAE and the CAS", the *Request for Instructions on Issues Concerning the Establishment of Chinese Academy of Engineering* suggested: "After the CAE is established, it, along with the CAS, would be the two highest academic institutes that are independent of each other and have their own priorities in their work." However, "it is necessary to maintain the organic connection between the two institutions." It was suggested that academic division members of the CAE and CAE Members" be cross-elected according to their respective standards and prescribed procedures. In carrying out major consultation work and undertaking other tasks, the two academies could organize joint activities as needed to facilitate exchanges and complementation."[99] This changed the five principles put forward by the CAS in the name of Zhou Guangzhao on July 18, 1992, including "referring to the US model" and "implementing a set of two brands of institutions abroad."

"On the appellation of the members of the CAE", the *Request for Instructions on Issues Concerning the Establishment of Chinese Academy of Engineering* suggested: "In the process of discussion and soliciting opinions, everyone thinks that the appellation of 'Member of the CAE' would be better than academician, because it is not only in line with its honorary nature but is also convenient for international contact

---

[98]Guanyu jianli Zhongguo gongchengyuan youguan wenti de qingshi关于建立中国工程院有关问题的请示(Request for Instructions on Issues Concerning the Establishment of Chinese Academy of Engineering). In Zhongguo kexueyuan bangongting中国科学院办公厅(General Office of the Chinese Academy of Sciences) (Eds.). *Zhongguo kexueyuan nianbao*中国科学院年报(The Annals of the Chinese Academy of Sciences) 1993: 142.

[99]Guanyu jianli Zhongguo gongchengyuan youguan wenti de qingshi关于建立中国工程院有关问题的请示(Request for Instructions on Issues Concerning the Establishment of Chinese Academy of Engineering). In Zhongguo kexueyuan bangongting中国科学院办公厅(General Office of the Chinese Academy of Sciences) (Eds.). *Zhongguo kexueyuan nianbao*中国科学院年报(The Annals of the Chinese Academy of Sciences) 1993: 142.

and favorable for communication. At the same time, it is also suggested that the academic division members of the CAS should be renamed Members."[100]

"On the criteria and conditions for the academicians of the CAE", the *Request for Instructions on Issues Concerning the Establishment of Chinese Academy of Engineering* put forward the following suggestions:

> Members of the CAE, as the country's highest academic title in engineering science and technology, must be elected through strict procedures. Its business standards and conditions should be different from those of the Members of the CAS, and more emphasis and attention should be paid to the candidates' contribution and achievements in engineering science and technology.
>
> Senior engineers, researchers, professors, or engineering science and technology experts and scholars with equivalent titles who have made significant and creative achievements and contributions in the field of engineering science and technology, love the motherland, and have a decent style of studying can be recommended for election as Members of the CAE.
>
> Among them, 'significant and creative achievements and contributions' mainly referred to those in engineering design and implementation, having made significant contributions to creatively solve engineering science and technology problems, having significant inventions and creations in a certain field of engineering science and technology, being a founder and pioneer in an important field of engineering science and technology, or making outstanding contributions to education in an important field of engineering science and technology.[101]

This suggestion emphasized the strictness of the election procedure for academicians. The requirements for Membership were very high, and they were equivalent to the standards for Membership in the CAS. However, the criteria for selecting CAE Members was focused on the major, creative achievements and contributions of the candidates to the field of engineering technology, unlike the criteria for selecting Members of the CAS, which emphasized that the candidates must "make systematic and creative achievements and major contributions to the field of science and technology."[102]

"Regarding the number of the first batch of CAE Members and the method for their selection", the *Request for Instructions on Issues Concerning the Establishment of Chinese Academy of Engineering* stated: "Referring to the practice of the first batch of academic division members and the first batch of academicians from other countries, the first batch of CAE Members can be recommended and selected by

---

[100] Guanyu jianli Zhongguo gongchengyuan youguan wenti de qingshi关于建立中国工程院有关问题的请示(Request for Instructions on Issues Concerning the Establishment of Chinese Academy of Engineering). In Zhongguo kexueyuan bangongting中国科学院办公厅(General Office of the Chinese Academy of Sciences) (Eds.). *Zhongguo kexueyuan nianbao*中国科学院年报(The Annals of the Chinese Academy of Sciences) 1993: 142–143.

[101] Guanyu jianli Zhongguo gongchengyuan youguan wenti de qingshi关于建立中国工程院有关问题的请示(Request for Instructions on Issues Concerning the Establishment of Chinese Academy of Engineering). In Zhongguo kexueyuan bangongting中国科学院办公厅(General Office of the Chinese Academy of Sciences) (Eds.). *Zhongguo kexueyuan nianbao*中国科学院年报(The Annals of the Chinese Academy of Sciences) 1993: 143.

[102] Yuanshi tiaoli (Zhangcheng) 院士条例(章程) (Statutes for Membership of the Chinese Academy of Sciences). Beijing: Zhongguo kexueyuan danganguan中国科学院档案馆(Archives of Chinese Academy of Sciences), 1994-36-018.

relevant units and divisions according to the pre-determined standards and conditions. After careful discussion, negotiation, and selection by the preparatory leading group (with the assistance of the preparatory committee), the proposed list of Members can be presented and employed in the name of the CAE after being approved of by the State Council."[103] The selection of the first batch of Members to be employed would take place through two channels. One was to invite some academic division members from the CASAD who had been engaged in major national engineering and technical work to be included in the list of candidates. The other was to make recommendations and a primary selection according to the standards and conditions among personnel who were engaged in research, design, manufacturing, production, and teaching in various engineering and technical fields nationwide. On the basis of the lists put forward by these two channels, the first list of Members to be employed was selected. The *Request for Instructions* suggested that the number of people to be employed from the latter channel should account for the majority; the list of candidates for the first batch of Members to be employed should include no less than two-thirds from the industrial sector and no less than one-third under the age of sixty. The first batch of Members should have a total number of about three hundred. After that, the number of Members should increase by a certain number biennially, and the total number of Members should exceed the total number of CAS Members.[104]

"On the leadership system and daily work organization of the CAE", the *Request for Instructions on Issues Concerning the Establishment of Chinese Academy of Engineering* states: "The CAE shall have a president and several vice presidents (the personnel and their salary relations shall not be separated from the original unit) to form a collective leadership that leads the work of the whole academy, which shall be changed according to the procedures stipulated in the relevant articles of the *Statutes*." At the same time, "according to different professional fields, the CAE will be divided into several divisions (e.g. electromechanical energy, materials and chemicals, information and electronics, civil construction, agronomy, medicine, etc.), and the Members of each division will elect a division director, deputy director, and standing committee to lead the work of that division."[105]

---

[103] Guanyu jianli Zhongguo gongchengyuan youguan wenti de qingshi关于建立中国工程院有关问题的请示(Request for Instructions on Issues Concerning the Establishment of Chinese Academy of Engineering). In Zhongguo kexueyuan bangongting中国科学院办公厅(General Office of the Chinese Academy of Sciences) (Eds.). *Zhongguo kexueyuan nianbao*中国科学院年报(The Annals of the Chinese Academy of Sciences) 1993: 143.

[104] Guanyu jianli Zhongguo gongchengyuan youguan wenti de qingshi关于建立中国工程院有关问题的请示(Request for Instructions on Issues Concerning the Establishment of Chinese Academy of Engineering). In Zhongguo kexueyuan bangongting中国科学院办公厅(General Office of the Chinese Academy of Sciences) (Eds.). *Zhongguo kexueyuan nianbao*中国科学院年报(The Annals of the Chinese Academy of Sciences) 1993: 143.

[105] Guanyu jianli Zhongguo gongchengyuan youguan wenti de qingshi关于建立中国工程院有关问题的请示(Request for Instructions on Issues Concerning the Establishment of Chinese Academy of Engineering). In Zhongguo kexueyuan bangongting中国科学院办公厅(General Office of the Chinese Academy of Sciences) (Eds.). *Zhongguo kexueyuan nianbao*中国科学院年报(The Annals of the Chinese Academy of Sciences) 1993: 143.

Additionally, the *Request for Instructions on Issues Concerning the Establishment of Chinese Academy of Engineering* put forward a list of suggestions from the CAE's preparatory leading group, which was produced through consultation and discussion at the meeting of the Presidium of the CAS. The list consisted of seventeen people; the team leader was Song Jian, the deputy team leaders were Luo Gan and Zhou Guangzhao, and the group members were Wang Daheng, Shi Changxu, Zhu Kaixuan, Zhu Guangya, Zhu Lilan, Liu Jibin, Huai Guomo, Zhang Wei, Zhang Cunhao, Zhang Guangdou, Luo Peilin, Hou Hsiong-Lin, Xu Zhijian, and Sheng Shuren.[106] Apart from Luo Gan and Xu Zhijian, the other fifteen people had important direct or indirect relations with the CAS: one of them was the president of the CAS (Zhou Guangzhao), seven were advisors to the Presidium of the CAS, eight were academic division members, and two were advisors to the Presidium as well as academic division members. All six proposers of the *Proposal* were on the list.

According to the plan of the *Request for Instructions on Issues Concerning the Establishment of Chinese Academy of Engineering*, after the list of the preparatory leading group of the CAE was approved, the leading group invited forty or so representative members who were distributed in various divisions and who had made outstanding achievements and contributions in engineering science and technology to form a preparatory committee that would assist the leading group in organizing discussions, selecting the first batch of Members, drafting the *Constitution of Chinese Academy of Engineering*, and preparing for the First General Assembly of CAE Members.[107]

(4) The Second Phase of Preparatory Work, Led by the State Science and Technology Commission

1. From the Revision of the *Request for Instructions* to the Adjustment of the List of Preparatory Leading Groups

Due to the *Request for Instructions on Issues Concerning the Establishment of Chinese Academy of Engineering,* jointly submitted by the CAS and the State Science and Technology Commission, "there are some deficiencies, and some leading comrades of the industrial departments have opinions on it." Premier Li Peng of the State Council issued an instruction on April 9, 1993 that the *Request for Instructions on Issues Concerning the Establishment of Chinese Academy of Engineering* should be revised. Moreover, he clearly proposed that the preparatory work for the CAE

---

[106] Guanyu jianli Zhongguo gongchengyuan youguan wenti de qingshi关于建立中国工程院有关问题的请示(Request for Instructions on Issues Concerning the Establishment of Chinese Academy of Engineering). In Zhongguo kexueyuan bangongting中国科学院办公厅(General Office of the Chinese Academy of Sciences) (Eds.). *Zhongguo kexueyuan nianbao*中国科学院年报(The Annals of the Chinese Academy of Sciences) 1993: 143–145.

[107] Guanyu jianli Zhongguo gongchengyuan youguan wenti de qingshi关于建立中国工程院有关问题的请示(Request for Instructions on Issues Concerning the Establishment of Chinese Academy of Engineering). In Zhongguo kexueyuan bangongting中国科学院办公厅(General Office of the Chinese Academy of Sciences) (Eds.). *Zhongguo kexueyuan nianbao*中国科学院年报(The Annals of the Chinese Academy of Sciences) 1993: 143–144.

should be led by the State Science and Technology Commission, and that the preparatory leading group should be adjusted and expanded.[108] Song Jian conveyed the spirit of Li Peng's instructions to the leaders of the CAS. Subsequently, preparatory work for the second phase, led by the State Science and Technology Commission, began. Song Jian was in charge of the preparatory work for this stage and Zhu Lilan was in charge of the executive work. According to Song Jian's opinion, specific preparatory work should still be done by the Joint Office of the CAS so Ge Nengquan, who had been more involved in this work, became the director of the preparatory leading group office, and Feng Sijian, deputy director of the High Technology Department of the State Science and Technology Commission, became the deputy director of the office.[109] On August 31, the State Science and Technology Commission, jointly with the CAS, submitted to the State Council and the Central Committee a revised *Request for Instructions on Issues Concerning the Establishment of Chinese Academy of Engineering*.[110]

On October 19, the 11th Executive Meeting of the State Council agreed on principle to establish the CAE. On November 8, after learning of the decision, Zhang Guangdou, Zhang Wei, Wang Daheng, Hou Hsiong-Lin, and Shi Changxu sent letters to Song Jian, Luo Gan, and Wen Jiabao and forwarded them to Jiang Zemin, Li Peng, and Zhu Rongji, Vice Premier of the State Council, giving their opinions on the establishment of the CAE. They thought: "Since the CAE is the highest academic institute in the engineering science and technology field, its members enjoy the highest academic title in the national engineering technology field and must strictly follow the standard conditions when selecting."[111] They reiterated the criteria and

---

[108]Shi Changxu师昌绪(2011). *Zai rensheng daolu shang: Shi Changxu zizhuan*在人生道路上: 师昌绪自传(On the Road of Life: Autobiography of Shi Changxu). Beijing: Kexue chubanshe科学出版社.188.

[109]Shi Changxu师昌绪(2011). *Zai rensheng daolu shang: Shi Changxu zizhuan*在人生道路上: 师昌绪自传(On the Road of Life: Autobiography of Shi Changxu). Beijing: Kexue chubanshe科学出版社.188–189.

[110]Zhongkeyuan bangongting guanyu jianli Zhongguo gongchengyuan timing Qian Zhengying tongzhi wei 1995 nian qiaola fasi ziran kexuejiang houxuanrenZhouGuangzhao yuanzhang zai Zhongguo kexuebao gongzuo huiyi shang de jianghua tongzhi qingshi中科院、办公厅关于建立中国工程院、提名钱正英同志为1995年乔拉法斯自然科学奖候选人、周光召院长在中国科学报工作会议上的讲话、通知、请示(Remarks, Notices and Requests of Chinese Academy of Sciences and Its General Office on Establishing the Chinese Academy of Engineering, Nominating Comrade Qian Zhengying as Candidate for the 1995 Chorafas Natural Sciences Award, andPresident ZhouGuangzhao at the Work Conference of China Science Daily). Beijing: Zhongguo kexueyuan danganguan中国科学院档案馆(Archives of Chinese Academy of Sciences), 1993-02-055.

[111]Zhang Guangdou deng wuwei xuebu weiyuan jiu youguan zujian zhongguo gongchengyuan de yixie yijian zhi zhongyang lingdao xin张光斗等五位学部委员就有关组建中国工程院的一些意见致中央领导的信(A Letter Sent by Zhang Guangdou and Other Five Members of Academic Divisions to the Central Leadership on Some Opinions on the Establishment of Chinese Academy of Engineering). In Zhongguo gongchengyuan中国工程院(Chinese Academy of Engineering) (Eds.) (2004). *Zhongguo gongchengyuan nianjian* (1994–1997) 中国工程院年鉴(1994–1997) (Yearbook of Chinese Academy of Engineering, 1994–1997). Beijing: Zhongguo gongchengyuan中国工程院.36.

requirements for CAE Members in the revised *Request for Instructions on Issues Concerning the Establishment of Chinese Academy of Engineering,* issued by the State Science and Technology Commission and the CAS; namely that "any expert in engineering science and technology who has made significant and creative achievements and contributions in the field of engineering science and technology, loves the motherland, and has a decent style of studying may be recommended for election as a Member of the CAE." The specific requirements were: in a certain engineering science and technology field, to have obtained important research results and major inventions, made great contributions in the design and construction of major projects and creatively solved any engineering and technical problems, or to have been a founder or pioneer in an important field of engineering science and technology.[112] At the same time they said: "The CAE takes engineers as its main body and also absorbs some researchers and professors who have made outstanding contributions to engineering science and technology."[113] Furthermore, they pointed out that:

> What we are most concerned about is that the CAE cannot become an organization that arranges cadres. All members must meet the aforementioned criteria; otherwise it will damage the prestige of the CAE, fail to achieve the purpose of establishing the Academy, and cause difficulties in international exchanges.[114]

They did not agree that engineering technology and science should be completely separated. They suggested that "the technical division of the Chinese Academy of Science must be preserved after the establishment of the Chinese Academy of Engineering, and they cannot be replaced each other." They also suggested that "some

[112]Zhang Guangdou deng wuwei xuebu weiyuan jiu youguan zujian zhongguo gongchengyuan de yixie yijian zhi zhongyang lingdao xin张光斗等五位学部委员就有关组建中国工程院的一些意见致中央领导的信(A Letter Sent by Zhang Guangdou and Other Five Members of Academic Divisions to the Central Leadership on Some Opinions on the Establishment of Chinese Academy of Engineering). In Zhongguo gongchengyuan中国工程院(Chinese Academy of Engineering) (Eds.) (2004). *Zhongguo gongchengyuan nianjian* (1994–1997) 中国工程院年鉴(1994–1997) (Yearbook of Chinese Academy of Engineering, 1994–1997). Beijing: Zhongguo gongchengyuan中国工程院.36–37.

[113]Zhang Guangdou deng wuwei xuebu weiyuan jiu youguan zujian zhongguo gongchengyuan de yixie yijian zhi zhongyang lingdao xin张光斗等五位学部委员就有关组建中国工程院的一些意见致中央领导的信(A Letter Sent by Zhang Guangdou and Other Five Members of Academic Divisions to the Central Leadership on Some Opinions on the Establishment of Chinese Academy of Engineering). In Zhongguo gongchengyuan中国工程院(Chinese Academy of Engineering) (Eds.) (2004). *Zhongguo gongchengyuan nianjian* (1994–1997) 中国工程院年鉴(1994–1997) (Yearbook of Chinese Academy of Engineering, 1994–1997) . Beijing: Zhongguo gongchengyuan中国工程院.37.

[114]Zhang Guangdou deng wuwei xuebu weiyuan jiu youguan zujian zhongguo gongchengyuan de yixie yijian zhi zhongyang lingdao xin张光斗等五位学部委员就有关组建中国工程院的一些意见致中央领导的信(A Letter Sent by Zhang Guangdou and Other Five Members of Academic Divisions to the Central Leadership on Some Opinions on the Establishment of Chinese Academy of Engineering). In Zhongguo gongchengyuan中国工程院(Chinese Academy of Engineering) (Eds.) (2004). *Zhongguo gongchengyuan nianjian* (1994–1997) 中国工程院年鉴(1994–1997) (Yearbook of Chinese Academy of Engineering, 1994–1997) . Beijing: Zhongguo gongchengyuan中国工程院.37.

academic division members with strong engineering backgrounds become members of both academies. This would be conducive to interdisciplinary research and establish a closer relationship between science and engineering, mutual understanding, and common improvement."[115] These opinions were quite pertinent for maintaining the prestige of the CAE and for developing Chinese science and engineering technology.

On November 12, the State Science and Technology Commission and the CAS jointly submitted to the State Council and the Central Committee the *Supplementary Report on the Request for Instructions on Issues Concerning the Establishment of Chinese Academy of Engineering*. The *Supplementary Report* proposed that the CAE should be subordinate to the State Council as a directly affiliated institute, in terms of its management system. On October 19, the 11th Executive Meeting of the State Council raised the question of how to strictly control the criteria and requirements for selecting CAE Members, and pointed out that:

> Everyone agrees that it is very important and timely for the Executive Meeting of the State Council to bring it out. From the very beginning of the establishment of the CAE, the criteria, requirements, and necessary procedures for selecting Members must be strictly controlled. The discussion held that the criteria and requirements for the selection of CAE Members should focus more on their contributions and applicable achievements in the field of engineering technology than the current members of the CAS. The specific requirements are: to obtain important research results and have major inventions in a certain engineering and technical field, to have significant contributions and creative solutions to engineering and technical problems in the design and construction of major projects, to be the founder and pioneer of an important engineering technology field, or to have made outstanding achievements in the application and practice of engineering technology, and so forth. Only by strictly mastering the standard requirements can membership in the CAE be truly recognized as the highest academic title in the field of engineering technology in our country. [116]

This made the criteria and conditions for the selection of CAE Members and the emphasis on the criteria and requirements clearer. It also clarified the stance of the

[115]Zhang Guangdou deng wuwei xuebu weiyuan jiu youguan zujian zhongguo gongchengyuan de yixie yijian zhi zhongyang lingdao xin张光斗等五位学部委员就有关组建中国工程院的一些意见致中央领导的信(A Letter Sent by Zhang Guangdou and Other Five Members of Academic Divisions to the Central Leadership on Some Opinions on the Establishment of Chinese Academy of Engineering). In Zhongguo gongchengyuan中国工程院(Chinese Academy of Engineering) (Eds.) (2004). *Zhongguo gongchengyuan nianjian* (1994–1997) 中国工程院年鉴(1994–1997) (Yearbook of Chinese Academy of Engineering, 1994–1997). Beijing: Zhongguo gongchengyuan中国工程院.36–37.

[116]Zhongkeyuan bangongting guanyu jianli Zhongguo gongchengyuan timing Qian Zhengying tongzhi wei 1995 nian qiaola fasi ziran kexuejiang houxuanrenZhouGuangzhao yuanzhang zai Zhongguo kexuebao gongzuo huiyi shang de jianghua tongzhi qingshi中科院、办公厅关于建立中国工程院、提名钱正英同志为1995年乔拉法斯自然科学奖候选人、周光召院长在中国科学报工作会议上的讲话、通知、请示(Remarks, Notices and Requests of Chinese Academy of Sciences and Its General Office on Establishing the Chinese Academy of Engineering, Nominating Comrade Qian Zhengying as Candidate for the 1995 Chorafas Natural Sciences Award, andPresident ZhouGuangzhao at the Work Conference of China Science Daily). Beijing: Zhongguo kexueyuan danganguan中国科学院档案馆(Archives of Chinese Academy of Sciences), 1993-02-055.

State Science and Technology Commission and the CAS, that they would strictly control the criteria and requirements for the selection of CAE Members.

At the same time, the *Supplementary Report* stressed the importance of the selection of the first batch of Members; it revised the number of the first batch and put forth that "everyone believes that the selection of the first batch of CAE Members is of vital importance and will play a guiding role. For this reason, after the CAE's preparation plan is approved, the preparatory leading group will organize experts to fully discuss and negotiate the selection according to extensive nomination by engineering and technical circles, and propose a list of about 150 people (including a few experts selected from the academic division members of the CAS) to be employed, and submit it to the State Council for approval before appointment."[117] According to the plan, the number of the first batch of Members would be reduced from the original planned number of 300 by about 150.

The *Supplementary Report* also proposed a revised list of preparatory leading groups, totaling 45 members. The group leader was Song Jian, while the deputy team leaders were Qian Zhengying, Zhou Guangzhao, Ding Henggao, Zhu Lilan,[118] Qi Yuanjing, Lin Hanxiong, and Shi Changxu. The group members consisted of Wang Yue, Wang Daheng, Wang Zhibao, Wang Dianzuo, Wei Yu, Lu Liangshu, Zhu Erming, Zhu Guangya, Zhu Gaofeng, Min Enze, Shen Zhijie, Zhang Wei, Zhang Guangdou, Zhang Fengxiang, Zhang Bingxi, Lu Yuanjiu, Lin Hua, Luo Xibei, Luo Peilin, Ji Guobiao, Jin Yilian, Zhou Ganzhi, Zheng Guangdi, Hou Hsiong-Lin, Hu Qiheng, Zhao Renkai, Fan Weitang, Yao Fusheng, Qin Zhongda, Gu Songfen, Xu Zhijian, Yin Ruiyu, Guo Kong hui, Lu Yongxiang, Guan De, Pan Beilei, and Feng Binglin.[119] The list had 28 more people than the original list; mainly engineering and technical experts and responsible persons from the industrial sector were added.

[117] Zhongkeyuan bangongting guanyu jianli Zhongguo gongchengyuan timing Qian Zhengying tongzhi wei 1995 nian qiaola fasi ziran kexuejiang houxuanrenZhouGuangzhao yuanzhang zai Zhongguo kexuebao gongzuo huiyi shang de jianghua tongzhi qingshi中科院、办公厅关于建立中国工程院、提名钱正英同志为1995年乔拉法斯自然科学奖候选人、周光召院长在中国科学报工作会议上的讲话、通知、请示(Remarks, Notices and Requests of Chinese Academy of Sciences and Its General Office on Establishing the Chinese Academy of Engineering, Nominating Comrade Qian Zhengying as Candidate for the 1995 Chorafas Natural Sciences Award, andPresident ZhouGuangzhao at the Work Conference of China Science Daily). Beijing: Zhongguo kexueyuan danganguan中国科学院档案馆(Archives of Chinese Academy of Sciences), 1993-02-055.

[118] Zhu Lilan is the executive deputy head.

[119] Zhongkeyuan bangongting guanyu jianli Zhongguo gongchengyuan timing Qian Zhengying tongzhi wei 1995 nian qiaola fasi ziran kexuejiang houxuanrenZhouGuangzhao yuanzhang zai Zhongguo kexuebao gongzuo huiyi shang de jianghua tongzhi qingshi中科院、办公厅关于建立中国工程院、提名钱正英同志为1995年乔拉法斯自然科学奖候选人、周光召院长在中国科学报工作会议上的讲话、通知、请示(Remarks, Notices and Requests of Chinese Academy of Sciences and Its General Office on Establishing the Chinese Academy of Engineering, Nominating Comrade Qian Zhengying as Candidate for the 1995 Chorafas Natural Sciences Award, andPresident ZhouGuangzhao at the Work Conference of China Science Daily). Beijing: Zhongguo kexueyuan danganguan中国科学院档案馆(Archives of Chinese Academy of Sciences), 1993-02-055.

Of the 45 people, eighteen of them were academic division members of the CAS.[120] There was no change in the selection of team leader, and the number of deputy team leaders increased to seven. Luo Gan, Zhu Kaixuan, Liu Jibin, Huai Guomo, Zhang Cunhao, and Sheng Shuren were no longer on the list.

2. The Selection of the First Batch of CAE Members

On January 6, 1994, the 46th meeting of the 14th Standing Committee of the Political Bureau of the Central Committee discussed and agreed on principle to the revised *Request for Instructions on Issues Concerning the Establishment of Chinese Academy of Engineering,* which was jointly proposed by the State Science and Technology Commission and the CAS, and decided to invite the State Science and Technology Commission to revise it according to the opinions discussed at the meeting and organize its implementation.[121] On February 25, the State Council issued a notice to the people's governments in all provinces, autonomous regions, and municipalities directly under the Central Government and all ministries and agencies directly under the State Council to forward the *Request for Instructions on Issues Concerning the Establishment of Chinese Academy of Engineering*, which pointed out that:

> The Party Central Committee and the State Council have approved of the *Request for Instructions on Issues Concerning the Establishment of Chinese Academy of Engineering* that was issued by the State Science and Technology Commission and the CAS, which is hereby forwarded to you for your cooperation and implementation.
>
> The CAE will implement a membership system that will be the highest honorary and advisory academic institute in the engineering and technology field in China. Members of the CAE, the highest academic title in engineering technology established by the state, must be elected from outstanding engineering and technical experts who have made significant achievements and contributions. The first batch of CAE Members will be appointed by the CAE after a certain selection process and will be submitted to the State Council for approval, according to unified standards and conditions.
>
> At the same time, it was decided that the academic division members of the CAS should be renamed as CAS Members.
>
> With the approval of the CPC Central Committee and the State Council, a preparatory leading group for the CAE, headed by Comrade Song Jian, has been established to organize and coordinate the preparatory work for the establishment of the Academy. After its establishment, the CAE will be subordinate to the State Council and will be a directly affiliated institution of it. Its office is affiliated with the State Science and Technology Commission.

---

[120] The 18 academic division members of CAS were Song Jian, Zhou Guangzhao, Shi Changxu, Wang Yue, Wang Daheng, Wang Dianzuo, Zhu Guangya, Min Enze, Zhang Wei, Zhang Guangdou, Zhang Bingxi, Lu Yuanjiu, Luo Peilin, Zhou Ganzhi, Hou Hsiong-Lin, Zhao Renkai, Gu Songfen, and Lu Yongxiang.

[121] Zhongyang huiyi jueding shixiang tongzhi (Zhaideng) 中央会议决定事项通知(摘登) (Notice of Decisions of the Central Committee Meeting) (Excerpts). In Zhongguo gongchengyuan中国工程院(Chinese Academy of Engineering) (Eds.) (2004). *Zhongguo gongchengyuan nianjian* (1994–1997) 中国工程院年鉴(1994–1997) (Yearbook of Chinese Academy of Engineering, 1994–1997). Beijing: Zhongguo gongchengyuan中国工程院.13.

The staffing and operating expenses of the CAE shall be examined and agreed upon by the Central Organization Committee and the Ministry of Finance respectively.[122]

The notice showed the affirmation of the *Request for Instructions on Issues Concerning the Establishment of Chinese Academy of Engineering* by the Central Committee and the State Council and their support for the establishment of the CAE and its membership system.

Prior to this, the head of the preparatory leading group, Song Jian, chaired a meeting with the head and deputy head of the CAE's preparatory leading group on January 11, 1994. He informed the spirit of the meetings of the Central Committee and the State Council, determined the principles for the establishment of the CAE, and discussed the methods for the selection of the first batch of Members and the progress of the work involved. On January 15, Song Jian presided over the first plenary meeting of the CAE's preparatory leading group, discussed the nomination, selection, approval, and appointment methods of the candidates of the first batch of Members, and divided the discipline selection group and the corresponding discipline grouping of the members of the leading group.[123] Among them, the discipline selection group was divided into six groups according to major industry classifications: the civil and hydraulic engineering group, the electronics and information engineering group, the agriculture, textile, and environmental engineering group, the metallurgy and material chemistry engineering group, the machinery and transportation engineering group, and the mining and energy engineering group. The members of the leading group were divided into groups to form corresponding discipline groups, called "The Selection Group of the Preparatory Leading Group of the CAE."[124] The sub-group

[122]Guowuyuan pizhuan guojia kewei Zhongguo kexueyuan guanyu jianli Zhongguo gongchengyuan qingshi de tongzhi国务院批转国家科委、中国科学院关于建立中国工程院请示的通知(Notice of the State Council on Approving and Transferring to the Request for Instructions on Issues Concerning the Establishment of Chinese Academy of Engineering by the State Science and Technology Commission and the Chinese Academy of Sciences. In Zhongguo gongchengyuan中国工程院(Chinese Academy of Engineering) (Eds.) (2004). *Zhongguo gongchengyuan nianjian* (1994–1997) 中国工程院年鉴(1994–1997) (Yearbook of Chinese Academy of Engineering, 1994–1997). Beijing: Zhongguo gongchengyuan中国工程院.14.

[123]1994 nian dashiji 1994年大事记(Chronicle of Events in 1994). In Zhongguo gongchengyuan中国工程院(Chinese Academy of Engineering) (Eds.) (2004). *Zhongguo gongchengyuan nianjian* (1994–1997) 中国工程院年鉴(1994–1997) (Yearbook of Chinese Academy of Engineering, 1994–1997). Beijing: Zhongguo gongchengyuan中国工程院.155.

[124]Zhongguo gongchengyuan choubei lingdao xiaozu linxuanzu fenzu shuoming中国工程院筹备领导小组遴选组分组说明(Explanation for Grouping of Selection Group of Preparatory Leading Group of Chinese Academy of Engineering). In Zhongguo gongchengyuan中国工程院(Chinese Academy of Engineering) (Eds.) (2004). *Zhongguo gongchengyuan nianjian* (1994–1997) 中国工程院年鉴(1994–1997) (Yearbook of Chinese Academy of Engineering, 1994–1997). Beijing: Zhongguo gongchengyuan中国工程院.50–51; Zhongguo gongchengyuan choubei lingdao xiaozu linxuanzu fenzu mingdan中国工程院筹备领导小组遴选组分组名单(The Sub-group List of the Selection Group of the Preparatory Leading Group of Chinese Academy of Engineering). In Zhongguo gongchengyuan中国工程院(Chinese Academy of Engineering) (Eds.) (2004). *Zhongguo gongchengyuan nianjian* (1994–1997) 中国工程院年鉴(1994–1997) (Yearbook

list of the selection group of the CAE's preparatory leading group is shown in Table 10.1.

Both the head and deputy head of the preparatory leading group participated in each selection group. With grouping instructions from the selection group, the members of the preparatory leading group were assigned to each selection group according to their expertise or the type of industry or industrial work they were responsible for at that time. Under the leadership of the preparatory leading group,

Table 10.1 Sub-group list of selection group of the CAE's preparatory leading group

Group	Personnel
Civil and Hydraulic Engineering Group	Convenors: Zhang Wei (engineering mechanics), Zhou Ganzhi (urban planning), Zhu Erming hydraulic structure), Qian Zhengying (deputy head of the preparatory group), Lin Hanxiong (deputy head of the preparatory group), Shen Zhijie (railway engineering)
Electronics and Information Engineering Group	Convenors: Wang Daheng (applied optics), Luo Peilin (electronics), Hu Qiheng (automatic control and information science), Song Jian (head of preparatory team), Wang Yue (radar and communication system), Zhu Gaofeng (communication system), Jin Yilian (computer science), Wei Yu (bioelectronics)
Agronomy, Textile and Environmental Engineering Group	Convenor: Lu Liangshu (wheat breeding and cultivation), Wang Zhibao (forestry), Pan Beilei (light industry), Qi Yuanjing (deputy head of the preparatory group), Xu Zhijian (deputy secretary general of the State Council), Ji Guobiao (textile)
Metallurgy and Material Chemistry Engineering Group	Convenor: Shi Changxu (metal material), Hou Hsiong-Lin (chemical engineering), Min Enze (petrochemical), Zhu Lilan (deputy head of the preparatory group), Lin Hua (chemical engineering), Qin Zhongda (chemical engineering), Yin Ruiyu (metallurgy), Wang Dianzuo (mineral engineering)
Machinery and Transportation Engineering Group	Convenors: Gu Songfen (aircraft aerodynamics), Yao Fusheng (steam turbine), Lu Yuanjiu (inertial technology and precision machinery), Ding Henggao (deputy head of the preparatory group), Zheng Guangdi (transportation), Feng Binglin (underwater weapon), Lu Yongxiang (fluid transmission and control), Guan De (aircraft design), Guo Konghui (automobile design)

(continued)

of Chinese Academy of Engineering, 1994–1997). Beijing: Zhongguo gongchengyuan中国工程院.51–52.

Table 10.1 (continued)

Group	Personnel
Mining and Energy Engineering Group	Convenors: Zhang Guangdou (water and hydropower), Zhang Fengxiang (power system), Fan Weitang (mining engineering), Zhou Guangzhao (deputy head of the preparatory group), Zhu Guangya (nuclear physics), Zhao Renkai (nuclear power engineering), Luo Xibei (hydropower planning), Zhang Bingxi (ore deposit geology)

Sources Zhongguo gongchengyuan choubei lingdao xiaozu linxuanzu fenzu mingdan中国工程院筹备领导小组遴选组分组名单(The Sub-group List of the Selection Group of the Preparatory Leading Group of Chinese Academy of Engineering). In Zhongguo gongchengyuan中国工程院(Chinese Academy of Engineering) (Eds.) (2004). *Zhongguo gongchengyuan nianjian* (1994–1997) 中国工程院年鉴(1994–1997) (Yearbook of Chinese Academy of Engineering, 1994–1997). Beijing: Zhongguo gongchengyuan中国工程院.51–52

each selection group was responsible for reviewing and selecting the candidates of the first batch of CAE Members.[125]

On January 17, the preparatory leading group of the CAE issued a *Notice on Nominating Candidates for CAE Membership* to all relevant ministries and commissions under the State Council and the General Political Department of the Chinese People's Liberation Army. This marked the official start of the selection of the first batch of CAE Members. The *Notice* explained the nature of CAE Members and the criteria and requirements for their selection:

> Membership in the CAE is the highest academic title in engineering technology that is established by the state. Candidates nominated for CAS Membership must be senior engineers, researchers, professors, or other engineering and technical experts with equivalent titles who have made significant and creative achievements and contributions in the field of engineering and technology, love the motherland, and have a decent style of studying.
>
> Significant and creative achievements and contributions' mainly refers to: important research results and major inventions in a certain engineering and technical field, significant contributions and creative solutions to engineering and technical problems in the design and construction of major projects, being the founder and pioneer of an important engineering technology field, outstanding achievements in the application and practice of engineering technology, and so forth.[126]

---

[125]Zhongguo gongchengyuan choubei lingdao xiaozu linxuanzu fenzu shuoming中国工程院筹备领导小组遴选组分组说明(Explanation for Grouping of Selection Group of Preparatory Leading Group of Chinese Academy of Engineering). In Zhongguo gongchengyuan中国工程院(Chinese Academy of Engineering) (Eds.) (2004). *Zhongguo gongchengyuan nianjian* (1994–1997) 中国工程院年鉴(1994–1997) (Yearbook of Chinese Academy of Engineering, 1994–1997). Beijing: Zhongguo gongchengyuan中国工程院.50–51.

[126]Guanyu timing Zhongguo gongchengyuan yuanshi houxuanren de tongzhi关于提名中国工程院院士候选人的通知(Notice on Nominating Candidates for Members of Chinese Academy of Engineering). In Zhongguo gongchengyuan中国工程院(Chinese Academy of Engineering) (Eds.) (2004). *Zhongguo gongchengyuan nianjian* (1994–1997) 中国工程院年鉴(1994–1997) (Yearbook

Among these, the criteria and conditions for the selection of CAE Members were basically the same as those listed in the *Supplementary Report on the Request for Instructions on Issues Concerning the Establishment of Chinese Academy of Engineering*, and there was no essential difference. However, according to the provisions of the *Notice*, the first batch of Members was numbered at about one hundred,[127] which was about fifty fewer than the 150 or so proposed in the *Supplementary Report*." It had increased the strict control over the number of academicians to be recruited.

In order to let the selection of the first batch of CAE Members proceed smoothly and ensure the quality of the selection, the *Notice* required all relevant ministries, commissions, directly-affiliated institutions, and the General Political Department of the Chinese People's Liberation Army to "put forward one to two (no more than two) candidates for the first batch of CAE Members through necessary procedures alongside the professional conditions of the system, so as to make a uniform selection. Among the candidates mentioned, there should be no fewer than one candidate 65 years old or younger. The General Political Department would be responsible for nominating and submitting the corresponding materials for one to two candidates in the military system." The nominations would close on February 8, 1994.[128]

According to the results of the discussions at the first plenary meeting of the CAE's preparatory leading group, the candidates for the first batch of Members could also be directly nominated by members of the preparatory leading group. On January 18, 1994, the preparatory leading group issued the *Notice on Nominating Candidates for CAE Members* to all members of the group, requiring that "each person nominate up to two candidates, and the candidates nominated by two or more members of the leading group shall be valid." The nominations would close on February 18.[129]

The nomination of candidates for Membership was not carried out in all provinces, cities, or municipalities across the country. This was mainly because the preparatory

---

of Chinese Academy of Engineering, 1994–1997). Beijing: Zhongguo gongchengyuan中国工程院.48.

127Guanyu timing Zhongguo gongchengyuan yuanshi houxuanren de tongzhi关于提名中国工程院院士候选人的通知(Notice on Nominating Candidates for Members of Chinese Academy of Engineering). In Zhongguo gongchengyuan中国工程院(Chinese Academy of Engineering) (Eds.) (2004). *Zhongguo gongchengyuan nianjian* (1994–1997) 中国工程院年鉴(1994–1997) (Yearbook of Chinese Academy of Engineering, 1994–1997). Beijing: Zhongguo gongchengyuan中国工程院.48–49.

128Guanyu timing Zhongguo gongchengyuan yuanshi houxuanren de tongzhi关于提名中国工程院院士候选人的通知(Notice on Nominating Candidates for Members of Chinese Academy of Engineering). In Zhongguo gongchengyuan中国工程院(Chinese Academy of Engineering) (Eds.) (2004). *Zhongguo gongchengyuan nianjian* (1994–1997) 中国工程院年鉴(1994–1997) (Yearbook of Chinese Academy of Engineering, 1994–1997). Beijing: Zhongguo gongchengyuan中国工程院.48–49.

129Guanyu timing Zhongguo gongchengyuan yuanshi houxuanren de tongzhi关于提名中国工程院院士候选人的通知(Notice on Nominating Candidates for Members of Chinese Academy of Engineering). In Zhongguo gongchengyuan中国工程院(Chinese Academy of Engineering) (Eds.) (2004). *Zhongguo gongchengyuan nianjian* (1994–1997) 中国工程院年鉴(1994–1997) (Yearbook of Chinese Academy of Engineering, 1994–1997). Beijing: Zhongguo gongchengyuan中国工程院.49–50.

leading group considered that the selection of Members was mainly to facilitate the formation of the CAE and nationwide election immediately after the establishment of the Academy, and the nomination time was relatively tight.[130]

Until the nomination deadline, a total of 43 ministries and commissions, directly affiliated institutions, relevant scientific and technological organizations, the General Political Department of the Chinese People's Liberation Army, and members of the preparatory leading group had nominated 108 candidates for Membership through deliberation, negotiation, and primary selection in accordance with the relevant regulations.[131] Among them, there were 88 people from the industrial sector system, accounting for 82.2% of the total number. There were 82 people under 65 years old, accounting for 76.6% of the total number. The professional fields of the effective candidates included civil engineering, architecture, hydraulic engineering, electric power, agriculture, textile, disaster prevention, environment, electronics, information, machinery, transportation, metallurgy, materials, chemical industry, exploration, mining, energy, and so on.[132]

On February 23, the preparatory leading group held a second plenary meeting to discuss and adopt the *Measures for the Nomination, Selection, Approval, and Appointment of the First Batch of CAE Members* and the recommended list of candidates for the first batch of CAE Members. From March 1 to March 5, the preparatory leading group held its third plenary meeting and concentrated its time and energy on the selection of all candidates for CAE Membership. Apart from Zhang Bingxi's sick leave, the remaining 44 members of the preparatory leading group attended the meeting. The meeting specifically emphasized and stipulated the challenge system. Any candidate for CAE Membership who both was a member of the leading group and had immediate or major collateral relatives should be withdrawn during the group appraisal and general assembly introduction. In order to ensure smooth progress in the selection process, safeguard the unity of the scientific and technological community, and fully listen to different opinions, the meeting stipulated that no one should disseminate the opinions and discussions expressed at the meeting to evaders or the outside

---

[130] Guanyu chengqing shenpi Zhongguo gongchengyuan shoupi yuanshi nipin mingdan de qingshi 关于呈请审批中国工程院首批院士拟聘名单的请示(Request for Approval of the Candidate List of the First Batch of Members of Chinese Academy of Engineering). In Zhongguo gongchengyuan 中国工程院(Chinese Academy of Engineering) (Eds.) (2004). *Zhongguo gongchengyuan nianjian* (1994–1997) 中国工程院年鉴(1994–1997) (Yearbook of Chinese Academy of Engineering, 1994–1997). Beijing: Zhongguo gongchengyuan中国工程院.20–21.

[131] Guanyu chengqing shenpi Zhongguo gongchengyuan shoupi yuanshi nipin mingdan de qingshi 关于呈请审批中国工程院首批院士拟聘名单的请示(Request for Approval of the Candidate List of the First Batch of Members of Chinese Academy of Engineering). In Zhongguo gongchengyuan 中国工程院(Chinese Academy of Engineering) (Eds.) (2004). *Zhongguo gongchengyuan nianjian* (1994–1997) 中国工程院年鉴(1994–1997) (Yearbook of Chinese Academy of Engineering, 1994–1997). Beijing: Zhongguo gongchengyuan中国工程院.20.

[132] Shoupi yuanshi de chansheng首批院士的产生(The Generation of the First Batch of Members of Chinese Academy of Engineering). In Zhongguo gongchengyuan中国工程院(Chinese Academy of Engineering) (Eds.) (2004). *Zhongguo gongchengyuan nianjian* (1994–1997) 中国工程院年鉴(1994–1997) (Yearbook of Chinese Academy of Engineering, 1994–1997). Beijing: Zhongguo gongchengyuan中国工程院.47.

world, and the avoiders should not inquire about the meeting. The meeting divided 108 candidates into six selection groups according to their professional fields: civil, architecture and hydraulic engineering; electronics and information engineering; agriculture, textile, disaster prevention, and environmental engineering; metallurgy, materials, and chemical engineering, machinery and transportation engineering; and exploration, mining, and energy engineering. These groups generally corresponded to the preparatory leading group's selection groups.[133]

From March 1 to 2, the members of the preparatory leading group first reviewed the recommendation materials of the candidates, and held democratic discussions and comments within the selection group. Each selection group adopted a pre-voting method to generate a list of candidates for CAE Membership that would be presented to the General Assembly, according to the proportion of 80% of the candidates recommended by the group. A total of 86 candidates were presented to the conference by the six selection teams. From March 3 to March 4, representatives of each selection group introduced 86 candidates at the General Assembly. After inquiry, discussions, and voting, the General Assembly elected 84 of the 108 valid candidates as the official candidates of the first batch of CAE Members. On March 5, the preparatory leading group voted on the 84 official candidates of the first batch of CAE Members, 64 of whom won more than half of the votes and were included in the list of candidates for the first batch of Members.[134]

According to the scheduled plan, the number of the first batch of CAE Members was about 100, of which about seventy should be elected by the preparatory leading group. In order to "maintain close ties between the CAE and the CAS (AD), which would be conducive to the coordinated development of China's science and technology, economy, and society," the other thirty people were selected by the Presidium of the CASAD from the CAS Members entrusted by the preparatory leading group.[135] On March 24, the preparatory leading group held its fourth plenary meeting. The

---

[133]Zhongguo gongchengyuan choubei lingdao xiaozu disanci quanti huiyi jiyao中国工程院筹备领导小组第三次全体会议纪要(Summary of the Third Plenary Meeting of the Preparatory Leading Group of Chinese Academy of Engineering). In Zhongguo gongchengyuan中国工程院(Chinese Academy of Engineering) (Eds.) (2004). *Zhongguo gongchengyuan nianjian* (1994–1997) 中国工程院年鉴(1994–1997) (Yearbook of Chinese Academy of Engineering, 1994–1997). Beijing: Zhongguo gongchengyuan中国工程院.54–56.

[134]Zhongguo gongchengyuan choubei lingdao xiaozu disanci quanti huiyi jiyao中国工程院筹备领导小组第三次全体会议纪要(Summary of the Third Plenary Meeting of the Preparatory Leading Group of Chinese Academy of Engineering). In Zhongguo gongchengyuan中国工程院(Chinese Academy of Engineering) (Eds.) (2004). *Zhongguo gongchengyuan nianjian* (1994–1997) 中国工程院年鉴(1994–1997) (Yearbook of Chinese Academy of Engineering, 1994–1997). Beijing: Zhongguo gongchengyuan中国工程院.55.

[135]Zhongguo gongchengyuan choubei lingdao xiaozu disici quanti huiyi jiyao中国工程院筹备领导小组第四次全体会议纪要(Summary of the Fourth Plenary Meeting of the Preparatory Leading Group of Chinese Academy of Engineering). In Zhongguo gongchengyuan中国工程院(Chinese Academy of Engineering) (Eds.) (2004). *Zhongguo gongchengyuan nianjian* (1994–1997) 中国工程院年鉴(1994–1997) (Yearbook of Chinese Academy of Engineering, 1994–1997). Beijing: Zhongguo gongchengyuan中国工程院.56; Xuebu zhuxituan huiyi cailiao学部主席团会议材料(Meeting Materials of the Presidium of the Academic Divisions). Beijing: Zhongguo kexueyuan danganguan中国科学院档案馆(Archives of Chinese Academy of Sciences), 1994-36-017.

meeting was attended by 35 people, more than two-thirds of the quorum. In view of the large difference between the 64 Members to be selected in the first batch and the 70 candidates to be selected, and in light of the arduous and heavy task of preparing for the establishment of the CAE and the additional selection of Members after its establishment, and considering the representativeness and universality of the Members to be selected, the meeting passed the *Supplementary Provisions on the Selection of the First Batch of CAE Members* through discussion and voting. "It is agreed that the preparatory leading group of the CAE will vote again on the twenty other candidates on the original difference list."[136] The twenty candidates of the 84 from the first batch of official candidates were not elected on March 5. The vote was held at this meeting with Gu Songfen, Wang Yue and Wang Dianzuo serving as the scrutineers, and those who won more than half of the votes were still elected. In the end, two people were elected and included in the list of candidates for the first batch of CAE Members.[137]

In the selection process, members of the preparatory leading group were divided as to whether or not leading cadres with rich experience and contributions in the field of engineering technology but did not have academic papers or monographs should be selected. Among the six people who put forward the *Proposal*s, Hou Hsiong-Lin held a favorable attitude. He argued that "there must be such people," but "of course, not too many."[138] As the members of the preparatory leading group held different views on this issue, a dispute occurred when discussing the specific candidates. Lin Hua, a former vice-director of the State Science and Technology Commission and

[136]Zhongguo gongchengyuan choubei lingdao xiaozu disici quanti huiyi jiyao中国工程院筹备领导小组第四次全体会议纪要(Summary of the Fourth Plenary Meeting of the Preparatory Leading Group of Chinese Academy of Engineering). In Zhongguo gongchengyuan中国工程院(Chinese Academy of Engineering) (Eds.) (2004). *Zhongguo gongchengyuan nianjian* (1994–1997) 中国工程院年鉴(1994–1997) (Yearbook of Chinese Academy of Engineering, 1994–1997). Beijing: Zhongguo gongchengyuan中国工程院.56; Guanyu shoupi gongchengyuan yuanshi linxuan de buchong guiding关于首批工程院院士遴选的补充规定(Supplementary Provisions on the Selection of the First Batch of Members of Chinese Academy of Engineering). In Zhongguo gongchengyuan中国工程院(Chinese Academy of Engineering) (Eds.) (2004). *Zhongguo gongchengyuan nianjian* (1994–1997) 中国工程院年鉴(1994–1997) (Yearbook of Chinese Academy of Engineering, 1994–1997). Beijing: Zhongguo gongchengyuan中国工程院.52.

[137]Zhongguo gongchengyuan choubei lingdao xiaozu disici quanti huiyi jiyao中国工程院筹备领导小组第四次全体会议纪要(Summary of the Fourth Plenary Meeting of the Preparatory Leading Group of Chinese Academy of Engineering). In Zhongguo gongchengyuan中国工程院(Chinese Academy of Engineering) (Eds.) (2004). *Zhongguo gongchengyuan nianjian* (1994–1997) 中国工程院年鉴(1994–1997) (Yearbook of Chinese Academy of Engineering, 1994–1997). Beijing: Zhongguo gongchengyuan中国工程院.56.

[138]HouHsiong-Lin侯祥麟(2001). *Wo yu shiyou youyuan: Hou Hsiong-Lin zishu*我与石油有缘; 侯祥麟自述(I am Predestined with Oil: Autobiography of HouHsiong-Lin). Beijing: Shiyou gongye chubanshe石油工业出版社200.

vice-director of the State Planning Commission, was one of the figures at issue.[139] After voting, Lin Hua was finally elected.

With regard to the thirty other candidates for the first batch of CAE Members, the meeting of the group leader and deputy group leader, convened by the preparatory leading group on January 11, 1994, agreed to entrust the Presidium of the CAS to put forward a list of suggestions, which, after being confirmed by the preparatory leading group, would be directly included in the list of candidates for the first batch of CAE Members and submitted to the State Council for approval. At that time, it was considered inappropriate and unnecessary to go through the selection process, because time was very tight and membership in the CAS was already the country's highest academic title in science and technology. However, the Presidium of the CAS put forth four principles and two specific requirements for the candidates. The four principles were: (1) his/her work scope (including previous work), with a strong engineering and technical background; (2) he/she agreed to serve concurrently; (3) he/she has a certain representativeness, so as to expand the influence of the CAE and the mutual connection and exchange of information between major scientific and technological institutions; (4) Regarding the proposed thirty Members of the CAS, it is required to have a reasonable distribution in terms of departments, units, specialties, ages, etc. The two specific requirements were: (1) from the list of thirty people, no fewer than two-thirds should work in the industrial sector; (2) from the list of thirty people, those under 65 years old should not be less than half.[140]

In view of the fact that the eighteen CAS Members[141] in the preparatory leading group were more in line with the above principles and that many factors had been taken into account when they were members of the preparatory leading group, the CAS planned to put them all in the list of proposed candidates. As a result, the other seventeen were put on the preliminary list,[142] except for one who was not included because he expressed his unwillingness to serve concurrently as a member of the CAE.[143]

---

[139] HouHsiong-Lin侯祥麟(2001). *Wo yu shiyou youyuan: Hou Hsiong-Lin zishu*我与石油有缘: 侯祥麟自述(I am Predestined with Oil: Autobiography of HouHsiong-Lin). Beijing: Shiyou gongye chubanshe石油工业出版社200.

[140] Xuebu zhuxituan huiyi cailiao学部主席团会议材料(Meeting Materials of the Presidium of the Academic Divisions). Beijing: Zhongguo kexueyuan danganguan中国科学院档案馆(Archives of Chinese Academy of Sciences), 1994-36-017.

[141] The 18 Members are the 18 academic division members of Chinese Academy of Sciences in the above-mentioned preparatory leading group. At this time, academic division members of CAS have been renamed CAS Members.

[142] Xuebu zhuxituan huiyi cailiao学部主席团会议材料(Meeting Materials of the Presidium of the Academic Divisions). Beijing: Zhongguo kexueyuan danganguan中国科学院档案馆(Archives of Chinese Academy of Sciences), 1994-36-017.

[143] He is Zhang Bingxi. Because only Zhang Bingxi and Zhou Guangzhao, among 18 members of the CAS in the preparatory leading group, were not included in the list of 30 recommended candidates for the first batch of members of CAE finally determined by the presidium of CAS, while Zhou Guangzhao made concessions because the Division of Mathematics and Physics did not nominate Tsien H. S. when preparing the initial list of candidates.

Each division of the CAS also proposed a preliminary list of candidates. The list of candidates from the Division of Mathematics and Physics, the Division of Chemistry, and the Division of Earth Sciences was determined by the division office according to the aforementioned principles, after listening to the opinions of relevant Members and division directors, putting forward a reference list, making suggestions according to the overall situation by the division director, and soliciting the opinions of the deputy director of the division and any relevant academicians. The list of candidates for the Division of Biology was agreed upon by the Standing Committee through the nomination of relevant Members. Because there were too many people in the Division of Technological Sciences, an enlarged meeting of the Standing Committee was held to discuss and vote on the reference list proposed by the director and office of the division and to decide on the list of candidates.[144]

Both the Division of Mathematics and Physics and the Division of Technological Sciences suffered setbacks in the process of preparing their list of candidates for the preliminary proposal. In the Division of Mathematics and Physics,Tsien H. S. was not included in the preliminary list of recommended candidates for the division, although he was a scientist who had made great achievements and contributions to the field of engineering technology. The President of the CAS Zhou Guangzhao, who was included in the list, believed thatTsien H. S. should have been put in the list,[145] so he "voluntarily withdrew and gave way toTsien H. S.".[146] This incident reflected Zhou Guangzhao's demeanor of taking the overall situation into consideration at the expense of his personal reputation. However, it was no accident that Tsien H. S. was not included in the preliminary list of recommended candidates; although Tsien H. S.'s achievements and contributions in the field of engineering technology were quite high, he had a poor reputation in the Division of Mathematics and Physics of the CAS and had opposed the establishment of the CAE.[147]

Several principles were determined when the Division of Technical Sciences was preparing the preliminary list of recommended candidates: first, he or she must be an initiator of the CAE; second, he or she must be an engineer and technician closely related to the industrial sector; third, he or she must have won the title of academician of the Academy of Engineering internationally; fourth, he or she must be a young academic divisionmember of the CAS who had indeed made outstanding achievements in engineering and technology. At the same time, we should also consider the distribution of disciplines and the rejuvenation of young people. After the list was drawn up, the division's standing committee held a secret ballot to conduct a

[144]Xuebu zhuxituan huiyi cailiao学部主席团会议材料(Meeting Materials of the Presidium of the Academic Divisions). Beijing: Zhongguo kexueyuan danganguan中国科学院档案馆(Archives of Chinese Academy of Sciences), 1994-36-017.

[145]According to the conversation between Mr. Ge Nengquan and the author on October 11, 2013.

[146]Shi Changxu师昌绪(2011). *Zai rensheng daolu shang: Shi Changxu zizhuan*在人生道路上: 师昌绪自传(On the Road of Life: Autobiography of Shi Changxu). Beijing: Kexue chubanshe科学出版社.188–189.

[147]According to the conversation between Mr. Ge Nengquan and the author on October 11, 2013.

margin election.[148] As a result, civil engineer Pan Jiazheng, the backbone of the Three Gorges Project, was not selected. The leadership of the CAS believed that this was a problem, so the head of the Joint Office of the CAS mobilized Shi Changxu to withdraw, but to no avail. In the end, Pan Jiazheng was given the place of a selected academic division member who did not work in Beijing. As for the reason why he did not withdraw, Shi Changxu said in his memoir: "I am not worried about losing the title of CAE Member, but rather about the relationship between the two academies. If the two academies cannot coordinate at the beginning, it will be even more unmanageable in the future. It will not only damage the reputation of the Members, but will also be detrimental to the development and effect of consultation work."[149]

On February 28, 1994, the Presidium of the CASAD held a meeting and, after discussion, finalized the list of thirty recommended candidates: Wang Xuan, Wang Yue, Wang Daheng, Wang Dianzuo, Shi Yuanchun, Shi Changxu, Liu Yongtan, Zhu Guangya, Li Deren, Li Guohao, Yan Dongsheng, Min Guirong, Min Enze, Shen Zhiyun, Song Jian, Zhang Wei, Zhang Guangdou, Zhang Zongyou, Lu Yuanjiu, Chen Junliang, Luo Peilin, Zhou Ganzhi, Zheng Zhemin, Chang Yinfo, Zhao Renkai, Hou Hsiong-Lin,Tsien H. S., Gu Songfen, Lu Yongxiang, and Pan Jiazheng. Among them, there were 20 people working in the industrial sector, accounting for two-thirds, which met the regulations. 14 people under 65, accounting for 46.7%, one fewer than the required ratio of being no fewer than half.[150] Subsequently, the preparatory leading group of the CAE deliberated and voted on the list of recommended candidates.[151]

On March 31, the preparatory leading group of the CAE submitted to the State Council and the Central Committee a *Request for Approval of the Candidate List*

[148] Shi Changxu师昌绪. Resheng daolu 80 zai: Shi Changxu zishu人生道路八十载——师昌绪自述(80 Years on the Road of Life: Autobiography of Shi Changxu). In Zhongguo kexueyuan jinshu yanjiusuo中国科学院金属研究所(Institute of Metal Research, Chinese Academy of Sciences) (Eds.) (2000). *Shi Changxu keji huodong shengya*师昌绪科技活动生涯(Shi Changxu's Career in Science and Technology). Beijing: Kexue chubanshe科学出版社.564.

[149] Shi Changxu师昌绪. Resheng daolu 80 zai: Shi Changxu zishu人生道路八十载——师昌绪自述(80 Years on the Road of Life: Autobiography of Shi Changxu). In Zhongguo kexueyuan jinshu yanjiusuo中国科学院金属研究所(Institute of Metal Research, Chinese Academy of Sciences) (Eds.) (2000). *Shi Changxu keji huodong shengya*师昌绪科技活动生涯(Shi Changxu's Career in Science and Technology). Beijing: Kexue chubanshe科学出版社.564.

[150] Zhongguo kexueyuan guanyu Zhongguo kexueyuan yuanshi nipin wei Zhongguo gongchengyuan yuanshi de jianyi mingdan de han中国科学院关于中国科学院院士拟聘为中国工程院院士的建议名单的函(A Letter from Chinese Academy of Sciences on the Proposed List of CAS Members to be Appointed as CAE Members). In Zhongguo gongchengyuan中国工程院(Chinese Academy of Engineering) (Eds.) (2004). *Zhongguo gongchengyuan nianjian* (1994–1997) 中国工程院年鉴(1994–1997) (Yearbook of Chinese Academy of Engineering, 1994–1997). Beijing: Zhongguo gongchengyuan中国工程院.52–53.

[151] Guanyu chengqing shenpi Zhongguo gongchengyuan shoupi yuanshi nipin mingdan de qingshi 关于呈请审批中国工程院首批院士拟聘名单的请示(Request for Approval of the Candidate List of the First Batch of Members of Chinese Academy of Engineering). In Zhongguo gongchengyuan 中国工程院(Chinese Academy of Engineering) (Eds.) (2004). *Zhongguo gongchengyuan nianjian* (1994–1997) 中国工程院年鉴(1994–1997) (Yearbook of Chinese Academy of Engineering, 1994–1997). Beijing: Zhongguo gongchengyuan中国工程院.21.

of the First Batch of CAE Members.[152] The *Request for Approval* was accompanied by a list of 66 selected CAE Member candidates from the first bath and a list of thirty selected candidates from the first batch of CAS Members. On May 10, the State Council gave an official reply to the *Request for Approval*, agreeing to appoint the 96 members as the first batch of CAE Members in the name of the CAE, and requiring that "when the inaugural assembly of the CAE is held this year, the list of CAE Members shall be published and the appointment procedures shall be fulfilled."[153] The list of these 96 Members and their respective academic divisions after the establishment of the CAE are shown in Table 10.2.

Of these 96 people, 75 worked in the industrial sector, accounting for 78.1% (more than two-thirds) of the number. Sixty-three people were under 65, accounting for 65.6% (more than half) of the number. Their fields of expertise generally covered the major industries in China's national economy, national defense construction, and social development at that time, and were broadly representative. Most of the 96 people were researchers and senior engineers of research institutions, while a few were heads of research institutes, universities, and relevant committees. There were three officials at or above the national deputy ministerial level, namely State Councilor Song Jian, Deputy Minister of the Metallurgical Industry Yin Ruiyu, and Deputy Minister of the Coal Industry Fan Weitang. Apart from the fourteen members Qian Zhengying, Zhou Guangzhao, Zhu Lilan, Lin Hanxiong, Wang Zhibao, Zhu Erming, Shen Zhijie, Zhang Fengxiang, Zhang Bingxi, Luo Xibei, Zheng Guangdi, Qin Zhongda, Xu Zhijian, and Feng Binglin, the remaining 31 people were among the 96 Members.

After the first batch of Members was elected, the preparatory leading group also carried out the drafting of the *Constitution of Chinese Academy of Engineering*, worked out the principles, methods, and conditions for the selection of the CAE's president and vice-president, and discussed and determined the agenda of the inaugural assembly of the CAE and any related matters. From May 31 to June 1, the preparatory meeting for the founding meeting of the CAE was held. The meeting discussed the candidates for president and vice-president of the CAE, the proposed list of the Presidium and executive chairman of the conference, and the candidates

---

[152]Guanyu chengqing shenpi Zhongguo gongchengyuan shoupi yuanshi nipin mingdan de qingshi 关于呈请审批中国工程院首批院士拟聘名单的请示(Request for Approval of the Candidate List of the First Batch of Members of Chinese Academy of Engineering). In Zhongguo gongchengyuan 中国工程院(Chinese Academy of Engineering) (Eds.) (2004). *Zhongguo gongchengyuan nianjian* (1994–1997) 中国工程院年鉴(1994–1997) (Yearbook of Chinese Academy of Engineering, 1994–1997). Beijing: Zhongguo gongchengyuan中国工程院.20–21.

[153]Guowuyuan guanyu Zhongguo gongchengyuan pinren shoupi yuanshi wenti de pifu国务院关于中国工程院聘任首批院士问题的批复(The Reply from the State Council to the Issue of the First Batch of Members of Chinese Academy of Engineering). In Zhongguo gongchengyuan中国工程院(Chinese Academy of Engineering) (Eds.) (2004). *Zhongguo gongchengyuan nianjian* (1994–1997) 中国工程院年鉴(1994–1997) (Yearbook of Chinese Academy of Engineering, 1994–1997). Beijing: Zhongguo gongchengyuan中国工程院.19.

Table 10.2 A list of the first batch of CAE members (1994)

Subordinate division	Name of the members
Division of Mechanical and Transportation Engineering (19)	Ding Henggao, Wang Yongzhi, Guan Qiao, Ruan Xueyu, Li Hongzhi, Wu Yousheng, Min Guirong, Wang Yousheng, Shen Zhiyun, Lu Yuanjiu, Yao Fusheng, Gu Songfen, Tsien H. S., Guo Konghui, Huang Xuhua, Tu Shan-cheng, Xie Youbai, Lu Yongxiang, Guan De
Division of Information and Electronic Engineering (23)	Wang Xuan, Wang Yue, Wang Daheng, Wei Yu, Zhu Gaofeng, Liu Yongtan, Liu Xianlin, Su Junhong, Li Tongbao, Li Deren, He Dequan, Wang Chengwei, Song Jian, Zhang Zhizhong, Chen Liwei, Chen Junliang, Luo Peilin, Jin Guofan, Jin Yilian, Zhou Zhongyi, Hu Qiheng, Ni Guangnan, Jiang Xinsong
Division of Chemical, Metallurgical and Material Engineering (16)	Wang Dianzuo, Shi Changxu, Yan Dongsheng, Li Dadong, Li Hengde, Wu Zhongwei, Zou Jing, Min Enze, Lin Hua, Zhou Lian, Hou Hsiong-Lin, Xu Gengguang, Xu Chengen, Yin Ruiyu, Huang Peiyun, Qi Yuanjing
Division of Energy and Mining Engineering (13)	Ma Fubang, Wang Demin, Zhu Guangya, Liu Tianquan, Yang Qixun, He Jishan, Zhang Guangdou, Zhang Zongyou, Fan Weitang, Zhao Renkai, Qian Gaoyun, Tang Yinfo, Peng Shilu
Division of Civil, Hydraulic and Construction Engineering (13)	Wang Guangyuan, Wen Fubo, Li Guohao, Zhang Wei, Zhang Jinqiu, Zhou Jing, Zhou Ganzhi, Zheng Zhemin, Qian Qihu, Liang Yingchen, Fu Xinian, Xie Lili, Pan Jiazheng
Division of Agronomy, Textile and Environmental Engineering (12)	Ding Dewen, Wang Tao, Wang Mingxiu, Shi Yuanchun, Lu Liangshu, Liu Gengling, Liu Hongliang, Ji Guobiao, Hu Haitao, Duan Zhenji, Qian Yi, Zhang Jijia

Sources Zhongguo gongchengyuan yuanshi quanti mingdan中国工程院院士全体名单(Full List of Members of Chinese Academy of Engineering). In Zhongguo gongchengyuan中国工程院(Chinese Academy of Engineering) (Eds.) (1995). *Zhongguo gongchengyuan shoujie yuanshi dahui wenjian jinkan*中国工程院首届院士大会文件集刊(Collection of Documents of the First General Assembly of Members of Chinese Academy of Engineering). Beijing: Zhongguo gongchengyuan 中国工程院.50–63

for secretary-general and deputy secretary-general of the conference's secretariat. They adopted an agenda and schedule for the establishment of the conference.[154] At this point, the preparatory work for the CAE was officially complete.

(5) **The Establishment of the CAE and the First General Assembly**

---

[154] 1994 nian dashiji 1994年大事记(Chronicle of Events in 1994). In Zhongguo gongchengyuan中国工程院(Chinese Academy of Engineering) (Eds.) (2004). *Zhongguo gongchengyuan nianjian* (1994–1997) 中国工程院年鉴(1994–1997) (Yearbook of Chinese Academy of Engineering, 1994–1997). Beijing: Zhongguo gongchengyuan中国工程院.156.

From June 3 to 8, 1994, the founding meeting of the CAE and the First General Assembly of CAE Members were held. This marked the establishment of the CAE's membership system. The conference was held at the same time as the Seventh General Assembly of the CAS Members. The opening ceremony was held in Zhongnanhai and was presided over by Zhou Guangzhao. The Members of both academies and party and state leaders Li Peng, Zou Jiahua, Wen Jiabao, Lu Jia-xi, Song Jian, Luo Gan, Qian Zhengying, and others attended.[155] In his speech, Li Peng, on behalf of the Central Committee and the State Council, warmly congratulated the establishment of the CAE, pointing out that "the establishment of the CAE is an important event in the engineering and technology field and is of great significance to China's socialist modernization."[156] Executive Chairman of the Congress Zhu Guangya generally introduced the preparatory work of the CAE and its nature, functions and tasks.[157] On the day of the opening ceremony, Jiang Zemin met with the Members of both academies and made a speech. He said: "The establishment of the CAE will greatly inspire and stimulate the creative spirit of a vast number of engineers and technicians, and will play a positive role in promoting the development of engineering technology and improving its research, design, construction, and operation capabilities."[158]

On June 7, the First General Assembly of CAE Members passed the *Constitution of Chinese Academy of Engineering*. The *Constitution* referred to and drew lessons from the *Constitution for the Academic Division Members of Chinese Academy of*

[155] 1994 nian dashiji 1994年大事记(Chronicle of Events in 1994). In Zhongguo gongchengyuan中国工程院(Chinese Academy of Engineering) (Eds.) (2004). *Zhongguo gongchengyuan nianjian* (1994–1997) 中国工程院年鉴(1994–1997) (Yearbook of Chinese Academy of Engineering, 1994–1997). Beijing: Zhongguo gongchengyuan中国工程院.156.

[156] Li Peng李鹏. Zai Zhongguo gongchengyuan chengli dahui he Zhongguo kexueyuan diqici yuanshi dahui shang de jianghua在中国工程院成立大会和中国科学院第七次院士大会上的讲话(Speech at the Founding Conference of Chinese Academy of Engineering and the Seventh General Assembly of Members of Chinese Academy of Sciences). In Zhongguo gongchengyuan 中国工程院(Chinese Academy of Engineering) (Eds.) (1995). *Zhongguo gongchengyuan shoujie yuanshi dahui wenjian jinkan*中国工程院首届院士大会文件集刊(Collection of Documents of the First General Assembly of Members of Chinese Academy of Engineering). Beijing: Zhongguo gongchengyuan中国工程院.4–6.

[157] Zhu Guangya朱光亚. Zai Zhongguo gongchengyuan chengli dahui shang de jianghua在中国工程院成立大会上的讲话(Speech at the Founding Conference of Chinese Academy of Engineering). In Zhongguo gongchengyuan中国工程院(Chinese Academy of Engineering) (Eds.) (1995). *Zhongguo gongchengyuan shoujie yuanshi dahui wenjian jinkan*中国工程院首届院士大会文件集刊(Collectsion of Documents of the First General Assembly of Members of Chinese Academy of Engineering). Beijing: Zhongguo gongchengyuan中国工程院.9–14.

[158] Zhonggong zhongyang zogshuji guojia zhuxi Jiang Zemin tongzhi zai huijian Zhongguo gongchenguan Zhongguo kexueyuan quanti yuanshi shi de jianghua中共中央总书记、国家主席江泽民同志在会见中国工程院、中国科学院全体院士时的讲话(Speech by Comrade Jiang Zemin, General Secretary of the CPC Central Committee andPresident of the People's Republic of China, Meeting with all the CAE Members and the CAS Members). In Zhongguo gongchengyuan 中国工程院(Chinese Academy of Engineering) (Eds.) (1995). *Zhongguo gongchengyuan shoujie yuanshi dahui wenjian jinkan*中国工程院首届院士大会文件集刊(Collection of Documents of the First General Assembly of Members of Chinese Academy of Engineering). Beijing: Zhongguo gongchengyuan中国工程院.3.

Sciences (Trial) as well as the regulations of some similar foreign academic institutions[159]; it was divided into general provisions and eight chapters. The General Provisions stipulated the nature, functions, and tasks of the CAE. According to the regulations, the CAE was the highest honorary and advisory academic institution in the national engineering science and technology circle. The functions and tasks of the CAE included: (1) to unite the national engineering and technology community, to promote the continuous improvement of China's engineering and technology level and the construction of engineering teams, and to stimulate the growth of outstanding talents; (2) accepting the entrustment of the government to provide consultation on major national engineering and technology decisions, development plans, schemes, and their implementation; (3) to study and discuss the development goals of important engineering and technology fields, and put forward reports and suggestions on the development and application of engineering and technology in combination with national economic and social development plans; (4) organizing academic exchanges and cooperation in the field of engineering technology, representing China's engineering technology community, and participating in counterpart international organizations and relevant international academic activities.

Chapter 1 is titled *Membership*, which stipulates the nature, obligation and rights, standards and conditions, co-optation system and procedures, and the revocation system of the title of Member. According to the regulations, Membership in the CAE is the highest academic title in engineering and technology established by the state and is a lifelong honor. Senior engineers, researchers, professors, or experts with the same title (including Chinese experts residing in Taiwan Province, Hong Kong, Macao, and other countries) who have made significant, creative achievements and contributions in engineering technology, love the motherland, have a decent style of studying, and are of Chinese nationality may be nominated and elected as CAE members. The co-optation of the Members shall be held biennially, and the total number of co-opted candidates shall not exceed one hundred. The number of additional members to be selected in each division shall be decided by the standing leading body of the Assembly through discussion. Candidates for Membership are nominated through two channels: one is the direct nomination of CAE Members; the other is the nomination of relevant domestic engineering research, design, construction, operation institutes, universities, factories, and mines according to organizational systems and through selection procedures; the evaluation of candidates shall be conducted by the Members organized by each division. During the election, the difference will be carried out by secret ballot. The election results shall be reported to all the Members in writing, after being reviewed by the standing leading body of the Assembly.

The second chapter is titled *Election of Foreign Members*, which stipulates the standards, elections, and rights of foreign Members. According to the regulations,

---

[159]Zhu Gaofeng朱高峰. Guanyu qicao Zhongguo gongchengyuan zhangcheng de shuoming关于起草《中国工程院章程》的说明(Explanation for Drafting *The Constitution of Chinese Academy of Engineering*). In Zhongguo gongchengyuan中国工程院(Chinese Academy of Engineering) (Eds.) (1995). *Zhongguo gongchengyuan shoujie yuanshi dahui wenjian jinkan*中国工程院首届院士大会文件集刊(Collection of Documents of the First General Assembly of Members of Chinese Academy of Engineering). Beijing: Zhongguo gongchengyuan中国工程院.35.

foreign scholars who have made important contributions to the development of China's engineering and technology undertakings and have high international reputation may be nominated and elected as foreign Members of the CAE. The election of foreign Members would be held biennially. Candidates for foreign Membership must be nominated by no fewer than five CAE Members; otherwise they will be invalid. Foreign Members are elected by secret ballot at the Assembly. They have the right to make suggestions on the development of China's engineering technology and the work of the CAE. They can be invited to attend relevant meetings organized by the CAE and its academic divisions, and can obtain publications presented by the CAE. Foreign Members do not have the right to nominate candidates for Membership or foreign Membership. Foreign Members do not have the right to vote and be elected at the General Assembly.

Chapter 3 is titled the *General Assembly of CAS Members*, and it defines the nature and functions of the General Assembly. According to regulations, the General Assembly is the highest authority in the CAE. It is held every two years. There are six functions of the General Assembly: (1) reviewing the work report of the CAE; (2) formulating and amending the *Constitution of Chinese Academy of Engineering*; (3) deciding on the setup and adjustment of the division; (4) electing the president, vice-presidents, and some members of the permanent leading bodies; (5) electing foreign Members; (6) carrying out academic activities and discussing major engineering and technical issues; (7) discussing and considering other issues and proposals put forward by the standing leading body of the General Assembly.

Chapter 4 is titled *Standing Leadership Organ*, and it stipulates that the permanent leadership organization is the Presidium of the CAE when the General Assembly is not in session. The Presidium consists of the president, the vice-presidents, directors of each division, and several members elected by the Members. The president is the executive chairman of the Presidium and presides over its work. The meeting of the Presidium is generally held every six months. At the same time, Chapter 4 also stipulates the establishment and responsibilities of the president, vice-president, and secretary-general, as well as the functions of the Presidium.

Chapter 5 is titled *Academic Divisions*, and it stipulates that Members shall participate in the activities of a division according to their specialties. According to the categories of engineering technology, the CAE is temporarily set up as follows: the Division of Mechanical and Transportation Engineering, Division of Information and Electronic Engineering, Division of Chemical Engineering, Division of Metallurgy and Materials Engineering, Division of Energy and Mining Engineering, Division of Civil Engineering, Hydraulic and Construction Engineering, and the Division of Agronomy, Textile, and Environmental Engineering. The plenary meeting of the divisions is held once a year. This chapter also has clear provisions on the Standing Committee of the Academic Divisions and the functions and tasks of the Academic Divisions.

Chapters 6 and 7 respectively stipulate the *Publications* and *Funding and Financial Management* of the CAE. Chapter 8 is a supplementary provision, and it set out that unless there are special provisions the General Assembly, the meeting of the Presidium, the plenary meeting of each division, and the meeting of the Standing

Committee may be decided upon with a quorum of two-thirds. When the resolution is put to the vote, it should be passed by a majority of more than one-half of the people present.[160]

As the basic regulations of the CAE, the *Constitution of Chinese Academy of Engineering* provides a legal basis for the CAE to elect co-opted members and to carry out other work in the future.

On June 6, the first General Assembly of the CAE held a plenary meeting of the Members and elected the first president and vice president by secret ballot. The president was Zhu Guangya, and the vice presidents were Zhu Gaofeng, Shi Changxu, Pan Jiazheng, and Lu Liangshu.[161] On June 7, the members of the six divisions of CAE were elected by secret ballot. The first meeting of the Presidium approved the selection of the standing committees, directors and deputy directors of the divisions. The Standing Committee of the Division of Mechanical and Transportation Engineering consisted of five members: Shen Zhiyun, Yao Fusheng, Gu Songfen, Huang Xuhua, and Tu Shan-cheng. Yao Fusheng was the director of the division and Tu Shan-cheng was the deputy director. The Standing Committee of Information and Electronic Engineering Division consisted of five members: Wang Yue, Chen Junliang, Jin Guofan, Jin Yilian, and Jiang Xinsong. The director of the division is Jin Yilian and the deputy director is Wang Yue. The Standing Committee of the Division of Chemical, Metallurgical and Material Engineering consisted of five members: Wang Dianzuo, Zhou Lian, Xu Gengguang, Xu Chengen and Yin Ruiyu. Yin Ruiyu is the director of the division and Wang Dianzuo was the deputy director. The Standing Committee of the Division of Energy and Mining Engineering consisted of three members: Wang Demin, Fan Weitang, and Qian Gaoyun. The director of the division was Fan Weitang. The Standing Committee of the Division of Civil, Hydraulic and Construction Engineering consisted of three members: Zhou Ganzhi, Qian Qihu, and Liang Yingchen. The director of the division was Zhou Ganzhi. The Standing Committee of the Division of Agronomy, Textile and Environmental Engineering consisted of three members: Wang Tao, Liu Gengling, and Ji Guobiao. The director of the division was Ji Guobiao.[162] On June 8, upon nomination by President

---

[160]Zhongguo gongchengyuan zhangcheng中国工程院章程(*The Constitution of Chinese Academy of Engineering*). In Zhongguo gongchengyuan中国工程院(Chinese Academy of Engineering) (Eds.) (1995). *Zhongguo gongchengyuan shoujie yuanshi dahui wenjian jinkan*中国工程院首届院士大会文件集刊(Collection of Documents of the First General Assembly of Members of Chinese Academy of Engineering). Beijing: Zhongguo gongchengyuan中国工程院.29–34.

[161]Xinwengao新闻稿(Press Release). In Zhongguo gongchengyuan中国工程院(Chinese Aca demy of Engineering) (Eds.) (1995). *Zhongguo gongchengyuan shoujie yuanshi dahui wenjian jinkan*中国工程院首届院士大会文件集刊(Collection of Documents of the First General Assembly of Members of Chinese Academy of Engineering). Beijing: Zhongguo gongchengyuan 中国工程院.44.

[162]Guanyu pizhun Zhongguo gongchengyuan ge xuebu changwu weiyuanhui ji xuebu zhu ren fuzhuren de jueding关于批准中国工程院各学部常务委员会及学部主任、副主任的决定(Decision on Approving the Standing Committee of Each Academic Division of Chinese Academy of Engineering and the Directors and Deputy Directors of the Academic Divisions). In Zhongguo gongchengyuan中国工程院(Chinese Academy of Engineering) (Eds.) (1995). *Zhongguo gongchengyuan shoujie yuanshi dahui wenjian jinkan*中国工程院首届院士大会文件

Zhu Guangya and approval by the Presidium meeting, Ge Nengquan was appointed Secretary General of the CAE.[163] These laid the organizational foundation for the future operation and work of the CAE.

10.3 The Efforts of the CASS to Establish a Membership System

After the plan to establish the membership system ran aground in 1981, the CASS did not give up establishing their own membership system and made many efforts to establish it. In 1983, the year following Hu Qiaomu's election as a member of the Political Bureau of the CPC Central Committee, he gave written instructions on the establishment of the membership system: "Consider proposing that the Chinese membership system include Members in literature and art (there are precedents in foreign countries such as France and the Soviet Union). Therefore, it is necessary to study and prepare the first list, and invite the Literature and Arts Bureau and the Ministry of Culture to consider it together. The list should be strict; put quality before quantity."[164] Judging by the instructions, Hu Qiaomu had hoped that the Literature and Art Bureau and the Ministry of Culture, which were related to the work of literature and art, would prepare a list for the first batch of CASS members with high standards. The principle of "be strict; put quality before quantity" that was put forward by Hu was in line with the 1980 plan of the CAS, CASS, and the State Science and Technology Commission to establish a lifelong membership system of pure academic honor by electing outstanding scientists as Members.

After Hu Qiaomu's instructions, the Publicity Department of the CPC Central Committee proposed that the CASS should take the lead, and that the Ministry of Culture and other relevant units should participate in the study to solve the problem of establishing a membership system. The Scientific Research Office of the CASS was instructed to draft *Some Ideas on Establishing a Membership System (Draft)*. The content of *Some Ideas (Draft)* included the purpose of establishing the membership system, the status of the academic honorary title, the conditions for it, the number of the Members, the method of generating the Members, tasks, and rights,

---

集刊(Collection of Documents of the First General Assembly of Members of Chinese Academy of Engineering). Beijing: Zhongguo gongchengyuan中国工程院.49.

[163]Guanyu renming Zhongguo gongchengyuan mishuzhang de jueding关于任命中国工程院秘书长的决定(Decision on Appointment of Secretary General of Chinese Academy of Engineering). In Zhongguo gongchengyuan中国工程院(Chinese Academy of Engineering) (Eds.) (1995). *Zhongguo gongchengyuan shoujie yuanshi dahui wenjian jinkan*中国工程院首届院士大会文件集刊(Collection of Documents of the First General Assembly of Members of Chinese Academy of Engineering). Beijing: Zhongguo gongchengyuan中国工程院.48.

[164]Wei Lili韦莉莉(2009). Zhongguo shehui kexueyuan xuebu weiyuan zhidu de xiangguan wenti yanjiu中国社会科学院学部委员制度的相关问题研究(Studies on Related Problems of the CASS Membership System). *Shehui kexue guanli yu pinglun*社会科学管理与评论(Management and Review of Social Sciences) (1): 57.

and the revocation of the title of Member. Mei Yi, the Secretary-General of the CASS, instructed the Party members to issue *Some Ideas (Draft)* in order to solicit opinions and proposed amendments for it in September, which were submitted to the Party members for discussion after revision by the Scientific Research Office.[165] However, as the research to solve the problem of establishing a membership system ended, the plan of establishing a membership system in the CASS was shelved.

In 1988, as Hu Keshi and 41 other representatives put forward a proposal to establish a membership system at the first meeting of the Seventh National People's Congress, the State Science and Technology Commission took lead in forming a "Joint Research Group on the Membership System" with the CAS, the CASS, the State Education Commission, and other units and departments to carry out research on the membership system.[166] The research team took several months and changed its draft several times before forming the *Research Report on Implementing Academician System in China*. As a result, the CASS began to pay attention to the establishment of the membership system. However, due to the death of the head of the research team and different views on the establishment of the membership system, the issue of establishing the membership system was shelved again.[167]

Five years later, in 1993, the CASS was commissioned by the Cadre Bureau of the Publicity Department of the CPC Central Committee to carry out research on issues related to the restoration of the Social Sciences division's membership system. The Party Committee of the Academy instructed the personnel and education bureau of the academy to undertake the work and form a research group on "restoring the membership system of the division of social sciences," headed by director Shan Tianlun. After investigation, the research group completed *It is Imperative to Restore the Membership System of the Division of Social Sciences—A Survey Report on the Restoration of the Membership System of the Division of Social Sciences* (hereinafter referred to as the *Survey Report*) in October 1993. The *Survey Report* introduced the history and current situation of the membership system in China and held that "restoring the membership system of the Division of Social Sciences as soon as possible is a strong demand of a vast number of social scientists and is also a requirement of the objective situation—it is imperative." The *Survey Report* suggested that "the Party Central Committee and the State Council formally make a decision to

165 Wei Lili韦莉莉(2009). Zhongguo shehui kexueyuan xuebu weiyuan zhidu de xiangguan wenti yanjiu中国社会科学院学部委员制度的相关问题研究(Studies on Related Problems of the CASS Membership System). *Shehui kexue guanli yu pinglun*社会科学管理与评论(Management and Review of Social Sciences) (1): 57.

166 Wei Lili韦莉莉(2009). Zhongguo shehui kexueyuan xuebu weiyuan zhidu de xiangguan wenti yanjiu中国社会科学院学部委员制度的相关问题研究(Studies on Related Problems of the CASS Membership System). *Shehui kexue guanli yu pinglun*社会科学管理与评论(Management and Review of Social Sciences) (1): 58.

167 Yan Ren严仁(1993). Huifu shehui kexue xuebu weiyuan zhidu shizai bixing: Guanyu huifu shehui kexue xuebu weiyuan zhidu de diaoyan baogao恢复社会科学学部委员制度势在必行——关于恢复社会科学学部委员制度的调研报告(It is Imperative to Restore the Membership System of the Division of Social Sciences—A Survey Report on the Restoration of the Membership System of the Division of Social Sciences). *Shehui kexue guanli*社会科学管理(Management of Social Sciences) (4): 17.

resume the activities of the members of the Division of Socials Sciences, to establish a membership system in the CASS, and to order and instruct the Academy of Social Sciences to plan and preside over the matter and organize its implementation as soon as possible." At the same time, opinions were also put forward on the number, nature, and method of the formation of the academic division members.[168]

Prior to this, both the *Request for Instructions on Issues Concerning the Establishment of Chinese Academy of Engineering,* jointly submitted by the CAS and the State Science and Technology Commission to the Central Committee and the State Council on February 4, 1993, and the revised *Request for Instructions on Issues Concerning the Establishment of Chinese Academy of Engineering,* submitted by the CAS and the State Science and Technology Commission to the State Council and the Central Committee on August 31, 1993, had already put forward the proposal that the academic division members of the CAS should be renamed as CAS Members. In this context, the *Survey Report* proposed the assumption that the CASS would establish a membership system through the conversion of its academic division members into Members, or that the CASS would implement a membership system if the CAS implemented one:

> It is understood that the CAS has recently written a report to the Central Committee requesting the implementation of a membership system and the conversion of all academic division members into Members. Our opinion is that the system of academic division members in the social sciences should be restored first. If the Party Central Committee and the State Council approve this requirement, the criteria for selecting academic division members of the division should be strictly controlled so that, after the national membership system is implemented, all the academic division members can smoothly become the first batch of Members. If the restoration of the membership system of the Division of Social Sciences has not yet been implemented, the Central Committee will approve the implementation of the membership system of the CAS. We hope that the Central Committee can also approve the implementation of the Division of Social Sciences' membership system, in order to unify the national scientific system. Then, the Members will be directly selected from the CASS.[169]

The CASS was established in May 1977 on the basis of the CAS' Division of Philosophy and Social Sciences. "the system of academic division members in the social sciences should be restored" referred to the restoration of the membership system of the Division of Philosophy and Social Sciences, which was originally affiliated to the

[168]Yan Ren严仁(1993). Huifu shehui kexue xuebu weiyuan zhidu shizai bixing: Guanyu huifu shehui kexue xuebu weiyuan zhidu de diaoyan baogao恢复社会科学学部委员制度势在必行——关于恢复社会科学学部委员制度的调研报告(It is Imperative to Restore the Membership System of the Division of Social Sciences—A Survey Report on the Restoration of the Membership System of the Division of Social Sciences). *Shehui kexue guanli*社会科学管理(Management of Social Sciences) (4): 14–22.

[169]Yan Ren严仁(1993). Huifu shehui kexue xuebu weiyuan zhidu shizai bixing: Guanyu huifu shehui kexue xuebu weiyuan zhidu de diaoyan baogao恢复社会科学学部委员制度势在必行——关于恢复社会科学学部委员制度的调研报告(It is Imperative to Restore the Membership System of the Division of Social Sciences—A Survey Report on the Restoration of the Membership System of the Division of Social Sciences). *Shehui kexue guanli*社会科学管理(Management of Social Sciences) (4): 22.

CAS. On the basis of the *Survey Report*, the Cadre Bureau of the Publicity Department of the CPC Central Committee listened to the opinions of various sectors of the social science community, comprehensively demonstrated the urgency and feasibility of implementing the membership system in the CASS, finally formed the *Suggestions on Establishing the Membership System in the CASS,* and submitted it to the relevant leaders.[170] However, the *Suggestions* was not implemented.

In addition to the Personnel and Education Bureau, Ma Hong,[171] former president of the CASS, also made efforts to establish a membership system in the Academy. In 1993, when the State Council discussed the question of changing the title of academic division member of the CAS to Member, he attended the meeting. At that time, he proposed that the CAS and CASS should jointly establish a membership system. Ma Hong recalled that the State Council accepted his opinion and "stated during the meeting that the Division of Social Sciences should also have a membership system, and this issue should be discussed separately in the future. However, there was no further discussion on this matter. "[172]

However, in early 1994, the academic division members of the CAS were renamed as Members and the CAE went under preparations for its membership system, which greatly promoted the CASS' determination to solve the problem of establishing a membership system. From February 21 to 25, the 1994 Annual Work Conference of the CASS was held. In his work report at the meeting, President Hu Sheng listed "accomplishing the preparatory work for the establishment of the membership system well" as one of the ten tasks that should be done well in 1994.[173] In order to implement this work, the Personnel and Education Bureau of the CASS, led by director Shan Tianlun, did much work on the establishment of the membership system of the Academy: (1) on the basis of the *Survey Report*, they completed the work of drafting the documents that were to be submitted to the central government and the State Council, such as the *Request for Instructions on Establishing a Membership System in the CASS*, *Constitution for Membership in the CASS (Draft)*, and *Methods for Selecting the First Batch of CASS Members*; (2) they proposed the principle of allocating a number of candidates for the first batch of Members from the national social science community; (3) they communicated with the relevant departments

---

[170] Wei Lili韦莉莉(2009). Zhongguo shehui kexueyuan xuebu weiyuan zhidu de xiangguan wenti yanjiu中国社会科学院学部委员制度的相关问题研究(Studies on Related Problems of the CASS Membership System). *Shehui kexue guanli yu pinglun*社会科学管理与评论(Management and Review of Social Sciences) (1): 59–60.

[171] Ma Hong waspresident of the CASS from May 1982 to June 1985 and was thendirector of the Development Research Center of the State Council in 1993.

[172] Ma Hong tongzhi fangtanlu马洪同志访谈录(Interview with Comrade Ma Hong). *Yuanshi yanjiu* 院史研究(Research on History of the Chinese Academy of Social Sciences) 2004, Supplement (1): 10–12.

[173] Zhongguo shehui kexueyuan yuanshi yanjiushi中国社会科学院院史研究室(The Research Office of History of the Chinese Academy of Social Sciences) (2007). *Zhongguo shehui kexueyuan biannian jianshi(1977–2007)* 中国社会科学院编年简史(1977–2007) (A Brief Chronological History of the Chinese Academy of Social Sciences, 1977–2007). Beijing: Shehui kexue wenxian chubanshe社会科学文献出版社.286.

in the five major systems of the social science community to solicit opinions and suggestions.[174]

According to the 1994 plan of implementation for the establishment of the membership system in the CASS, the quota of the first batch of Members would be several dozen. The candidates for this batch of Members were members of the Division of Philosophy and Social Sciences of the CAS who were still alive, as well as people who were not elected for the Division of Philosophy and Social Sciences of the CAS in 1955 and 1957 and were recognized by the academic community as uncontroversial and highly respected. This batch of Members should reflect the characteristics of being few and precise and should be approved by the organization. The second batch of Members would be elected independently by the first batch of Members.[175]

After the aforementioned documents were formed, they were discussed and approved by the Party Committee of the CASS. Later, the vice-president of the Academy, Ru Xin, consulted the relevant leaders of the Publicity Department of the CPC Central Committee, the Central Party School, the State Education Commission, and other units. Among them were Zheng Bijian, executive vice-minister of the Publicity Department of the CPC Central Committee, Wang Jialiu, executive vice-president of the Central Party School, Gong Yuzhi, vice-president, and Wei Yu, deputy director of the State Education Commission. The leaders of the consulted units expressed their support for the establishment of a membership system in the CASS and the documents drafted by the Academy on the establishment of the membership system.[176] On December 6, 1994, Ru Xin reported to Liu Qibao, Deputy Secretary-General of the State Council, on the establishment of the membership system in the CASS.[177] However, the plan to establish the membership system was later shelved.

On January 5, 2003, *Opinions of the CPC Central Committee on Further Prospering and Developing Philosophy and Social Sciences* was issued. The *Opinions* expounded the important role of philosophy and social sciences in the modernization of socialism with Chinese characteristics, and clearly put forward the guiding principles, development goals, and main tasks of philosophy and social sciences. It was a programmatic document that guided the research of philosophy and social sciences in China and promoted the cause of philosophy and social sciences with Chinese characteristics. After the *Opinion* was issued, the leading party group of the

---

[174] Wei Lili韦莉莉(2009). Zhongguo shehui kexueyuan xuebu weiyuan zhidu de xiangguan wenti yanjiu中国社会科学院学部委员制度的相关问题研究(Studies on Related Problems of the CASS Membership System). *Shehui kexue guanli yu pinglun*社会科学管理与评论(Management and Review of Social Sciences) (1): 59.

[175] According to the author's interview with Mr. Shan Tianlun on September 11, 2013.

[176] According to the author's interview with Mr. Shan Tianlun on September 11, 2013.

[177] Wei Lili韦莉莉(2009). Zhongguo shehui kexueyuan xuebu weiyuan zhidu de xiangguan wenti yanjiu中国社会科学院学部委员制度的相关问题研究(Studies on Related Problems of the CASS Membership System). *Shehui kexue guanli yu pinglun*社会科学管理与评论(Management and Review of Social Sciences) (1): 59.

CASS attached great importance to it and implemented it.[178] As a result, the issue of establishing a membership system was raised again. He Bingmeng, a member of the National Committee of the Chinese People's Political Consultative Conference and Deputy Secretary-General of the CASS, put forth a proposal to establish a membership system in the CASS at the meeting of the Chinese People's Political Consultative Conference to implement the spirit of the *Opinions*:

> The membership system is an internationally accepted academic system. At present, most of the countries in the world that have established an academician (membership) system has included the field of social sciences. In 1955, the CAS established its academic divisions and the Division of Philosophy and Social Sciences was one of the four major divisions. During the 'Cultural Revolution,' the academic divisions stopped their activities. After the 'Cultural Revolution', the CAS resumed its divisions, but the Division of Philosophy and Social Sciences was not resumed because it had been established independently. In 1990, the State Council clearly pointed out in document No.63 that they approved of the co-optation of CASS members: "On issues related to the academic division members of the CASS, separate investigations and studies will be organized to listen to opinions and propose appropriate measures." In 1993, the system of CAS academic division members was changed into the membership system.[179] The philosophy and social sciences circles strongly called for the approval of the establishment of the CASS membership system as soon as possible. We believe that the conditions for establishing the membership system of the CASS are essentially ripe. If there are still some difficulties in immediately setting up the membership system, we can first restore the system of CASS academic division members to lay the foundation for the future transition to the membership system.[180]

This proposal represented the wishes of the CASS and reflected its urgent desire to establish a membership system.

On May 19, 2005, Chen Kuiyuan, the president of the CASS, reported to the Standing Committee of the Political Bureau of the CPC Central Committee on the implementation of the *Opinions of the CPC Central Committee on Further Prospering and Developing Philosophy and Social Sciences*.[181] In terms of the innovation

---

[178] Zhongguo shehui kexueyuan yuanshi yanjiushi中国社会科学院院史研究室(The Research Office of History of the Chinese Academy of Social Sciences) (2007). *Zhongguo shehui kexueyuan biannian jianshi(1977–2007)* 中国社会科学院编年简史(1977–2007) (A Brief Chronological History of the Chinese Academy of Social Sciences, 1977–2007). Beijing: Shehui kexue wenxian chubanshe社会科学文献出版社.502–503.

[179] The membership system of academic divisions of the CAS was changed to the membership system of the CAS in 1994.

[180] Guanche luoshi yijian jingshen fanrong fazhan zhexue shehui kexue: Quanguo zhengxie weiyuan woyuan fu mishuzhang He Bingmeng zai quanguo zhengxie huiyi shang de fayan贯彻落实《意见》精神，繁荣发展哲学社会科学——全国政协委员、我院副秘书长何秉孟在全国政协会议上的发言(Implementing the Spirit of *Opinions* and Prospering Philosophy and Social Sciences: Speech by He Bingmeng, member of the National Committee of the Chinese People's Political Consultative Conference and Deputy Secretary-General of the CASS, at the Chinese People's Political Consultative Conference). *Zhongguo shehui kexueyuan yuanbao*中国社会科学院院报(CASS Review), 2004-03-09(1).

[181] Zhongguo shehui kexueyuan yuanshi yanjiushi中国社会科学院院史研究室(The Research Office of History of the Chinese Academy of Social Sciences) (2007). *Zhongguo shehui kexueyuan biannian jianshi(1977–2007)* 中国社会科学院编年简史(1977–2007) (A Brief Chronological

of a discipline system, Chen Kuiyuan proposed in the report that the academic division of Social Sciences should be set up on the basis of extent of discipline to further integrate the strength of relevant disciplines and strengthen the organization and the coordination mechanism of major comprehensive research.[182] After discussion, the report was approved by the meeting.[183]

After the meeting, the CASS set up a research group and a drafting group for the constitution of academic division members. On the basis of repeatedly soliciting opinions from all sides, the *Constitution for Academic Division Members of the CASS (Draft)* and the *Selection Methods of the First Batch of Academic Division Members and Honorary Academic Division Members of the CASS* were formed. In October and November 2005, the party group of the CASS passed the two documents and decided to set up five academic divisions; namely the Division of Literature, History, and Philosophy, the Division of Economics, the Division of Social Politics and Law, the Division of International Studies, and the Division of Marxism Research. At the beginning of November, the party group of the CASS decided to set up both a working group and a selection committee for the first batch of academic division members and honorary academic division members.[184]

The selection of the first batch of academic division members and honorary academic division members was important for the CASS to establish the academic division and the system of academic division members. The selection process started on December 6, 2005 and ended on July 30, 2006. It was divided into eight stages: convene the mobilization meeting of the Academy; recommend nominations from all units of the Academy; solicit opinions; propose the primary list of candidates for the first batch of academic division members; propose the primary list of candidates for the first batch of honorary academic division members; discuss and approve the candidate list with the leading party group of the Academy of Social Sciences; vote for the Members through the Selection Committee; publicize and approve of the two lists.[185]

History of the Chinese Academy of Social Sciences, 1977–2007). Beijing: Shehui kexue wenxian chubanshe社会科学文献出版社.536.

[182]Zhongguo shehui kexueyuan中国社会科学院(Chinese Academy of Social Sciences) (Eds.) (2007). *Zhongguo shehui kexueyuan xuebu jiniance*中国社会科学院学部纪念册(Memorial Book of the Academic Divisions of Chinese Academy of Social Sciences). Beijing: Jingji guanli chubanshe经济管理出版社.71.

[183]Zhongguo shehui kexueyuan yuanshi yanjiushi中国社会科学院院史研究室(The Research Office of History of the Chinese Academy of Social Sciences) (2007). *Zhongguo shehui kexueyuan biannian jianshi(1977–2007)* 中国社会科学院编年简史(1977–2007) (A Brief Chronological History of the Chinese Academy of Social Sciences, 1977–2007). Beijing: Shehui kexue wenxian chubanshe社会科学文献出版社.536.

[184]Chen Jiagui陈佳贵(2006). Guanyu shoupi xuebu weiyuan rongyu xuebu weiyuan chansheng guocheng de shuoming关于首批学部委员、荣誉学部委员产生过程的说明(Explanation on the Generation Process of the First Batch of Academic Division Members and Honorary Members). *Zhongguo shehui kexueyuan yuanbao*中国社会科学院院报(CASS Review), 2006-8-8 (4).

[185]Chen Jiagui陈佳贵(2006). Guanyu shoupi xuebu weiyuan rongyu xuebu weiyuan chansheng guocheng de shuoming关于首批学部委员、荣誉学部委员产生过程的说明(Explanation on the

Nominations recommended by various units of the CASS began in mid-December 2005 and ended on December 28, 2005. The first batch of candidates for academic division membership, totaling 496 people, was recommended and nominated by the in-service personnel. The first batch of honorary academic division members was recommended and nominated by retirees with senior professional and technical titles, with a total of 338 recommended.[186] During the nomination process, the selection group started to solicit opinions and held a working meeting on December 19, 2005. The meeting decided to solicit the opinions of senior scholars according to their discipline. In order to facilitate the work efficiently, the selection group was divided into five groups according to discipline, and the work of soliciting opinions was carried out separately. On this basis, a primary list of 69 candidates for the first batch of academic division members and a primary list of 80 candidates for honorary academic division membership were formed. Considering that the recommendation and nomination of honorary academic division members were only made among retirees, and some units only had a had few participants in the meeting, the selection group decided to separately solicit opinions from the leaders of each unit, and adjusted the 80-member list to an 83-member list in the end.[187]

In the process of putting forward a primary list of candidates for the first division members, the selection group held a working meeting on March 22, 2006. According to the opinions of President Chen Kuiyuan, a primary list of 45 candidates (38 formal candidates and seven alternate candidates) was initially drawn up, after repeated discussions and studies. Three former members of the Academic Advisory Committee were over 70 years old and should not be put in the primary list, according to the *Constitution of Academic Advisory Committee of the CASS*. On June 12, the selection group included the three candidates in the primary selection list of honorary academic division members according to the opinions presented in the meeting of presidents. At the same time, according to the recommendations from the meeting of presidents, the selection group adjusted the five alternate candidates into actual candidates. In accordance with Chen Kuiyuan's proposal, Leng Rong, Jiang Lansheng, and Chen Jiagui, all vice-presidents of the academy, were also included in the primary list. Additionally, the majority of the selection group believed that the two researchers mentioned at the dean's meeting could be put on the primary list. In the end, a primary list of 49 people was formed.[188]

Generation Process of the First Batch of Academic Division Members and Honorary Members). *Zhongguo shehui kexueyuan yuanbao*中国社会科学院院报(CASS Review), 2006-8-8 (4).

[186]Chen Jiagui陈佳贵(2006). Guanyu shoupi xuebu weiyuan rongyu xuebu weiyuan chansheng guocheng de shuoming关于首批学部委员、荣誉学部委员产生过程的说明(Explanation on the Generation Process of the First Batch of Academic Division Members and Honorary Members). *Zhongguo shehui kexueyuan yuanbao*中国社会科学院院报(CASS Review), 2006-8-8 (4).

[187]Chen Jiagui陈佳贵(2006). Guanyu shoupi xuebu weiyuan rongyu xuebu weiyuan chansheng guocheng de shuoming关于首批学部委员、荣誉学部委员产生过程的说明(Explanation on the Generation Process of the First Batch of Academic Division Members and Honorary Members). *Zhongguo shehui kexueyuan yuanbao*中国社会科学院院报(CASS Review), 2006-8-8 (4).

[188]Chen Jiagui陈佳贵(2006). Guanyu shoupi xuebu weiyuan rongyu xuebu weiyuan chansheng guocheng de shuoming关于首批学部委员、荣誉学部委员产生过程的说明(Explanation on the

In the process of putting forth the initial selection list of candidates for honorary academic division membership, the Personnel and Education Bureau of the CASS provided the list files of Academy scholars well-known from the 1980s for discussion and reference by the selection group, in accordance with Chen Kuiyuan's request that the work of recommendation should be "full of talents" and that senior experts who had made significant contributions should not be left out. After discussions and additions and deletions to the original 83-member list, the selection group put forth a 93-member primary list on June 14.[189]

Since then, the party group of the CASS held two meetings, on June 20 and July 3, to discuss the primary selection lists of both the candidates for the first batch of academic division members and the candidates for the first batch of honorary academic division members. The June 20 meeting decided to put a scholar of over 90 years old from the previous list onto the list of candidates for the first batch of honorary academic division members. After Yu Guang-yuan and Luo Gengmo, two academic division members of the Philosophy and Social Sciences Division of the CAS, were approved, they were also put in the list. The composition of the Selection Committee was discussed at the meeting on July 3 and a 40-member Selection Committee was formed.[190]

On July 3, the General Assembly of the CASS revised the *Selection Methods of the First Batch of Academic Division Members and Honorary Academic Division Members of the CASS*. The *Selection Methods* stipulated the selection principles and methods, criteria and conditions, and organization methods and procedures. The selection principles and methods are as follows: (1) the selection of the first batch of Academic Division Members must adhere to political and academic standards, and the best of the best must be chosen; (2) the number of members in each division shall be determined according to the size of each discipline and the academic level of the expert team, and an equal distribution shall not be made; (3) the selection of the first batch of academic division members and honorary academic division members should be made by the academic division member selection group on the basis of extensive comments. The selection committee should elect the candidates by secret ballot and the General Assembly should examine and approve of the selection.

The criteria and requirements for selection were: (1) support the leadership of the Communist Party of China, love the socialist motherland, abide by the laws of the People's Republic of China, make great achievements and outstanding contributions in the field of philosophy and social sciences, have an academic level recognized

Generation Process of the First Batch of Academic Division Members and Honorary Members). *Zhongguo shehui kexueyuan yuanbao*中国社会科学院院报(CASS Review), 2006-8-8 (4).

189 Chen Jiagui陈佳贵(2006). Guanyu shoupi xuebu weiyuan rongyu xuebu weiyuan chansheng guocheng de shuoming关于首批学部委员、荣誉学部委员产生过程的说明(Explanation on the Generation Process of the First Batch of Academic Division Members and Honorary Members). *Zhongguo shehui kexueyuan yuanbao*中国社会科学院院报(CASS Review), 2006-8-8 (4).

190 Chen Jiagui陈佳贵(2006). Guanyu shoupi xuebu weiyuan rongyu xuebu weiyuan chansheng guocheng de shuoming关于首批学部委员、荣誉学部委员产生过程的说明(Explanation on the Generation Process of the First Batch of Academic Division Members and Honorary Members). *Zhongguo shehui kexueyuan yuanbao*中国社会科学院院报(CASS Review), 2006-8-8 (4).

by domestic and international academic circles, and have a decent or excellent style of study; (2) be selected from outstanding experts and scholars who held high-level professional and technical positions; (3) be honorary academic division members who enjoy high prestige in the academic community and have made outstanding contributions to social science. They should be produced from retired senior experts and scholars with professional and technical positions; (4) in-service scholars over 80 years of age who meet the requirements of academic division membership and are not in leadership positions may also be awarded the title of honorary academic division member.

The selection organization and procedures are as follows: (1) establish a selection group and a selection committee for the first batch of academic division members, who will be responsible for the selection of the first academic division members and honorary academic division members under the leadership of the General Assembly. The Personnel and Education Bureau will be the administrative body of the selection working group; (2) the Institution (Center) and other relevant units will organize in-service party committee members, academic committee members, experts, and scholars with senior, professional and technical positions to nominate academic division members by secret ballot; consequently organize old, retired experts and scholars with senior, professional and technical positions to recommend honorary academic division members by secret ballot; (3) the academic division member selection group will put forward the primary list of candidates on the basis of comprehensively balancing the votes of the candidate recommendations from all units; to solicit the opinions of famous experts and scholars in various disciplines on the primary selection list of candidates for academic division membership; to solicit the opinions of the members of each unit's leadership team on the primary selection list of the candidates for the honorary academic division members; after the selection group discusses, a list of candidates shall be formed and submitted to the General Assembly for discussion; (4) the candidates shall fill in the *Registration Form for Candidates of the First Batch of Academic Division Members* and the *Registration Form for Candidates of the First Batch of Honorary Academic Division Members* respectively, and provide three to four copies of personal representative works and award certificates after the General Assembly deliberates on the candidates for both academic division membership and honorary academic division membership; (5) the selection committee for academic division members will review and comment on the candidates' representatives and academic standards and vote for the first batch of members and honorary members. The number of votes cast by the members of the Selection Committee shall not be less than two-thirds of the members of the Selection Committee. Only the candidates who have obtained at least or more than two-thirds of the votes can be elected; (6) submit the list of academic division members and honorary academic division members elected by voting to the General Assembly for deliberation; (7) promulgate the first batch of academic division members for a period of fourteen days, and select a working group to deal with any problems raised during publicity. After the promulgation period is over, report to the academy affairs meeting for approval.[191]

It can be seen from the *Selection Methods* that the CASS had quite high academic standards for the first batch of members and honorary members, but the selection was not entirely based on academic standards; it also had to adhere to political standards. The member selection group, selection committee, and the General Assembly of the Academy played an important role in the selection work.

From July 10 to 11, 2006, the selection committee selected the first batch of academic division members and honorary academic division members by vote. Thirty-four members attended the meeting, with attendance exceeding two-thirds of the elected members. By secret ballot, 47 candidates for the first batch of academic division members and 95 candidates for the first batch of honorary academic division members were elected. All the candidates had won more than two thirds of the votes.[192] On July 13, the party group and the General Assembly of the CASS approved the list of honorary members and decided to promulgate the approved list of the first batch of members. The promulgation period was from July 17 to 30. During this period, the academy leaders and involved departments received sixteen pieces of material that reflected the situation; some of these materials raised objections to individuals on the list, others put forward opinions on the selection methods and procedures, and still others recommend people outside the list. These materials did not affect the election results in the end. On August 1, the CASS held a General Assembly and approved the election results of the selection committee.[193]

The average age of the 47 academic division members in the first batch was 65.5 years old, accounting for 7% of the in-service, senior professional and technical personnel in the CASS and 3.1% of the professional and technical personnel above the vice-senior professional title. Of these 47 people, four were from the field of literature, seven from the field of history, eleven from the field of philosophy, twelve were from the field of economics, seven from the field of social politics and law, and six were from the field of international studies. There were three leaders and fifteen directors in office, and eighteen members on the Academic Advisory Committee.[194]

---

[191]Zhongguo shehui kexueyuan中国社会科学院(Chinese Academy of Social Sciences) (Eds.) (2007). *Zhongguo shehui kexueyuan xuebu jiniance*中国社会科学院学部纪念册(Memorial Book of the Academic Divisions of Chinese Academy of Social Sciences). Beijing: Jingji guanli chubanshe经济管理出版社.69–70.

[192]Chen Jiagui陈佳贵(2006). Guanyu shoupi xuebu weiyuan rongyu xuebu weiyuan chansheng guocheng de shuoming关于首批学部委员、荣誉学部委员产生过程的说明(Explanation on the Generation Process of the First Batch of Academic Division Members and Honorary Members). *Zhongguo shehui kexueyuan yuanbao*中国社会科学院院报(CASS Review), 2006-8-8 (4).

[193]Chen Jiagui陈佳贵(2006). Guanyu shoupi xuebu weiyuan rongyu xuebu weiyuan chansheng guocheng de shuoming关于首批学部委员、荣誉学部委员产生过程的说明(Explanation on the Generation Process of the First Batch of Academic Division Members and Honorary Members). *Zhongguo shehui kexueyuan yuanbao*中国社会科学院院报(CASS Review), 2006-8-8 (4).

[194]Chen Jiagui陈佳贵(2006). Guanyu shoupi xuebu weiyuan rongyu xuebu weiyuan chansheng guocheng de shuoming关于首批学部委员、荣誉学部委员产生过程的说明(Explanation on the Generation Process of the First Batch of Academic Division Members and Honorary Members). *Zhongguo shehui kexueyuan yuanbao*中国社会科学院院报(CASS Review), 2006-8-8 (4).

The average age of the 95 members in the first batch of honorary academic division members was 78.7 years old, accounting for 9% of the senior professional and technical retirees and 5.1% of the vice-senior professional and technical retirees. Of these 95 people, eighteen were from the field of literature, twenty-two from the field of history, fifteen from the field of philosophy, seventeen from the field of economics, ten from the field of social politics and law, and thirteen from the field of international studies. Of the 74 members of the CPC, fifty did not hold leadership positions at or above the bureau-level.[195]

On August 3, 2006, the Inaugural Assembly of the Academic Divisions of CASS was held. Vice President Li Shenming read out the *Decision of the CASS on the Establishment of Academic Divisions*. Vice President Jiang Lansheng read out the *Decision of the CASS on the First Batch of Academic Division Members of the CASS* and the *Decision on Granting the Honorary Academic Division Members of the CASS*. Chen Jiagui, a vice President, presented the *Explanation on the Generation Process of the First Batch of Academic Division Members and Honorary Members of the CASS*. President Chen Kuiyuan presented certificates to the first batch of academic division members. Members of the leading party group of the institute presented certificates to the first batch of honorary academic division members.[196] The holding of this conference marked the establishment of the Academic Divisions of the CASS and its system of academic division members.

The CASS held its preparatory meeting of the First General Assembly of Academic Division Members immediately after the founding of its academic divisions. At the meeting, Leng Rong was elected president of the Presidium of the Academic Divisions.[197] On the afternoon of August 3, the First General Assembly of Academic Division Members was held. Leng Rong gave a report on the work of the academic divisions of the CASS, and Chen Jiagui made an explanation on the *Constitution for the Academic Divisions of Chinese Academy of Social Sciences (Trial)*.[198] It stipulated: "the academic divisions of the CASS is an academic guidance, academic consultation, and scientific research coordination organization under the leadership of the General Assembly of the Academy"; "academic division member of the CASS

[195]Chen Jiagui陈佳贵(2006). Guanyu shoupi xuebu weiyuan rongyu xuebu weiyuan chansheng guocheng de shuoming关于首批学部委员、荣誉学部委员产生过程的说明(Explanation on the Generation Process of the First Batch of Academic Division Members and Honorary Members). *Zhongguo shehui kexueyuan yuanbao*中国社会科学院院报(CASS Review), 2006-8-8 (4).

[196]Zhongguo shehui kexueyuan中国社会科学院(Chinese Academy of Social Sciences) (Eds.) (2007). *Zhongguo shehui kexueyuan xuebu jiniance*中国社会科学院学部纪念册(Memorial Book of the Academic Divisions of Chinese Academy of Social Sciences). Beijing: Jingji guanli chubanshe经济管理出版社.74.

[197]Xuebu zhuxituan chansheng学部主席团产生(The Generation of the Presidium of the Academic Divisions). *Zhongguo shehui kexueyuan yuanbao*中国社会科学院院报(CASS Review), 2006-8-8 (1).

[198]Chongfen lüxing xuebu he xuebu weiyuan zhize zhashi zuohao xuebu jianshe kaiju gongzuo充分履行学部和学部委员职责，扎实做好学部建设开局工作(Fulfil the Duties of Academic Division and Its Members, and Do a Solid Job in the Beginning of the Establishment of the Academic Divisions). *Zhongguo shehui kexueyuan yuanbao*中国社会科学院院报(CASS Review), 2006-8-8 (1).

is the highest academic title of the CASS and is a lifelong honor." With regard to the criteria and conditions for membership, the *Constitution for the Academic Divisions of Chinese Academy of Social Sciences (Trial)* stipulate: "In-service experts and scholars of the CASS who love the motherland, support the party's basic line, abide by the constitution and laws of the people's Republic of China, have made outstanding achievements and contributions in the field of philosophy and social sciences, are recognized by domestic or international academic circles as having a decent style of work and a fine style of studying, and have high-level professional and technical positions may be recommended and elected as academic division members of the CASS."[199] This regulation gave consideration to both political and academic standards, and reflects the requirements for the members of the divisions to be politically qualified and academically leading.[200]

The General Assembly also elected directors and deputy directors for each division. The director of the Division of Literature, History, and Philosophy was Jiang Lansheng, and the deputy directors were Zhang Haipeng and Li Jingyuan. The director of the Division of Economics was Chen Jiagui, and the deputy director was Liu Shucheng. The director of the Division of Social Politics and Law was Hao Shiyuan, and the deputy director was Jing Tiankui. The director of the Division of International Studies was Zhang Yunling, and the deputy director was Zhou Hong. The director of the Division of Marxism Research was Cheng Enfu.[201]

As an important achievement in the reform of the management system of the CASS and as an incentive mechanism for scientific research personnel, the establishment of the system of academic division and academic division members was an important step towards the establishment of the membership system in the Academy, enabling personnel within the Academy to enjoy the highest academic title for the rest of their lives. This is of great significance in affirming the achievements and values of those elected as academic division members and honorary academic division members, stimulating and encouraging researchers in the divisions to work actively, work hard, and climb the peak of philosophy and social sciences, and in promoting the development of philosophy and social sciences. The difference between this academic title and the usual academician title is mainly that the former is the highest lifelong academic title of the internal personnel of the CASS, while academician generally refers to the highest lifelong academic title in the country, and candidates are experts and scholars throughout the country.

---

[199]Zhongguo shehui kexueyuan xuebu zhangcheng (shixing)《中国社会科学院学部章程》(试行) (*Constitution for the Academic Divisions of Chinese Academy of Social Sciences* (*Trial*)). *Zhongguo shehui kexueyuan yuanbao*中国社会科学院院报(CASS Review), 2006-8-8 (3).

[200]Chen Jiagui陈佳贵(2006). Guanyu Zhongguo shehui kexueyuan xuebu zhangcheng (shixing) gao de jidian shuoming关于《中国社会科学院学部章程》(试行) 稿的几点说明(Some Explanations on *Constitution for the Academic Divisions of Chinese Academy of Social Sciences* (*Trial*)). *Zhongguo shehui kexueyuan yuanbao*中国社会科学院院报(CASS Review), 20s06-8-8 (3).

[201]Diyici xuebu weiyuan dahui xuanju chansheng ge xuebu zhuren fuzhuren第一次学部委员大会选举产生各学部主任副主任(The First General Assembly of the Academic Division Members Elected the Director and Deputy Directors of the Academic Divisions). *Zhongguo shehui kexueyuan yuanbao*中国社会科学院院报(CASS Review), 2006-8-8 (1).

10.4 Summary

After 1981, the rebuilding process of membership systems in China was quite tortuous and unique. Along with experts and scholars calling for the establishment of the membership system, the Secretariat of the Central Committee decided to establish the membership system in 1984. The CAS discussed the establishment of a membership system several times. The CASS made many efforts to establish one and even formed documents, but they did not succeed. In 2006, the CASS established Academic Divisions and the system of academic division members, which made an important step towards establishing the membership system, and enabled its staff to enjoy the highest academic title in life. However, the first batch of academic division members and honorary academic division members were not elected in strict accordance with democratic procedures; the shortcomings were obvious. It is still unknown when the Academy will establish their membership system.

However, with the approval of the Secretariat of the Central Committee in 1983, academic division member of the CAS became the country's highest academic honorary title in science and technology. The nature of being an academic division member of the CAS is no different from that of being a Member. Compared to the 1980 addition, the 1991 co-optation of the academic division members of the CAS increased the selection criteria and the standard, strictness, and democracy of the selection, thus adding an important bargaining chip to the conversion of academic division members of the CAS into Members. In the end, the CAS established the membership system in 1994 on the basis of the gradual evolution from academic division membership to the Membership, after the academic division members were renamed as Members. In the same year, the CAE also established their membership system. The membership system was, therefore, completely rebuilt in New China.

As an integral part of the reconstruction of the membership system in China, the establishment of the membership system of the CAE was directly due to the *Proposals on the Early Establishment of the Chinese Academy of Engineering and Technological Sciences,* put forward by the six academic division members Zhang Guangdou, Wang Daheng, Shi Changxu, Zhang Wei, Hou Hsiong-Lin, and Luo Peilin in 1992. It also had a profound international and domestic background and was also under the attention and support of the State Council and the Central Committee. The criteria for selecting CAE members focus on the major and creative achievements and contributions of the candidates in the field of engineering technology. This was different from the selection criteria for the members of CAS, which focused on "making systematic and creative achievements and significant contributions in the field of science and technology".[202] In the selection of the first batch of CAE Members, the number of candidates was strictly controlled, the candidates were broadly representative, and the selection results were generally fair. However, there were some shortcomings in this selection, such as the nomination of candidates, which was not fully spread to

---

[202]Yuanshi tiaoli (Zhangcheng) 院士条例(章程) (Statutes for Membership of the Chinese Academy of Sciences). Beijing: Zhongguo kexueyuan danganguan中国科学院档案馆(Archives of Chinese Academy of Sciences), 1994-36-018.

all provinces, cities, and municipalities directly under the central government, and the list of thirty recommended candidates determined by the Presidium of the CAS was not produced by a democratic election in the strictest sense.

The establishment of the CAE and its membership system was of great significance to the implementation of Deng Xiaoping's guiding ideology that "science and technology are primary productive forces," the promotion of the development of China's engineering technology, the improvement of the social status of China's engineering technology community, the mobilization of the enthusiasm of engineering technicians, and the protection and strengthening of exchanges between China and the international engineering technology community. Moreover, since the question of "the appellation of CAE members" was raised in the preparatory process of the CAE and its membership system and was approved by the State Council and the Central Committee, the academic division members were renamed as Members. This has solved the problem of converting academic division members of the CAS into CAS Members, which had pended since the 1980s, and contributed to the final establishment of the membership system of the CAS.

Conclusion

The academician system was born in Europe during the 17th century scientific revolution. By the beginning of the 20th century, the academician system had already been popular in the west, and it was regarded as an important institution that promoted academic development. The establishment of Academia Sinica's academician system led to this system becoming firmly established in China. After it was abandoned, the Chinese Academy of Sciences' and the Chinese Academy of Engineering's academician systems were integral to the rebuilding of this system in China. The establishment and reestablishment of China's academician system is only a small part—in the context of world history—of the systematization of modern science. However, it was an important chapter in 20th century China, worthy of attention because of how it demonstrated the ups and downs of scientific and academic development.

Academia Sinica's academician system involved the drawing up of legislation, nominations and examinations for prospective candidates. It also involved the drawing up of a list announcing candidates, a formal election process, and the General Assembly of Members of Academia Sinica . The legislative work involved the revision of *The Organization Act of Academia Sinica, The Regulations of the Council of Academia Sinica* and *The Regulations for the Election of the Council of Academia Sinica*. It also involved the formulation of new regulations for both the election of members of Academia Sinica and General Assembly of Members of Academia Sinica. *The Regulations for the Election of Members of Academia Sinica* stipulated that the election system of members of Academia Sinica needed to be strict, and involve a democratic election process that was presided over by the Council and independently operated. The requirements for nominees submitting their related works were lowered due to many scholars' works being lost or destroyed during the Anti-Japanese War. The PCE and the council strictly abided by *The Regulations for the Election of Members of Academia Sinica* when both nominating member candidates and formally electing members. They were objective, impartial and academically independent from politics, and so they upheld an attitude of solemnity and prudence. They attached great importance to academic standards and adhered to democratic principles. Their behavior and attitude made the election of the first batch of members

J. Guo, *The Establishment and Reconstruction of the Academician System in China*,
https://doi.org/10.1007/978-981-15-7208-1

of Academia Sinica a success, and it became a model for the independent election of academic elites in modern Chinese academia.

As the leading role on establishing the academician system of Academia Sinica, the Second Council included the elected council members and the acting president, director and director-general of Academia Sinica. The Council brought together outstanding scientists and academic elites from various institutions across the country such as Hu Shih, Fu Ssu-Nien, Chu Chia-Hua , Wong Wen-hao, Adam Pen-Tung Sah, Li Chi, Woo Y. H., Ping Chi, Li Shu-Hua and Mao T. E. During the election of the first batch of members of Academia Sinica, the council respected the existing academic standards and attached great importance to them. They behaved in the same way electing the elected Members of the Second Council by the First Council. They upheld the academic traditions of advocating academic autonomy and respecting academic freedom, which were important in western academies of science and ensured their full development. Almost all of the Members of the Second Council had studied abroad in either Europe or the United States. They adhered to the principles of academic freedom and independence when preparing and holding the election of the first batch of members of Academia Sinica. But at the same time, they integrated nationalism into it.

It was a noble title and an honor to be a member of Academia Sinica, as it was for other academicians around the world. However, members of Academia Sinica were different to conventional academicians in that some were granted the title for reasons other than having made outstanding academic achievements and/or contributions. "Those who have led or presided over academic institutions for more than five years with outstanding achievements" also qualified for the title. This was because the Council had taken into account the model of China's modern, scientific development. Candidates were initially nominated based on their outstanding achievements by universities, independent colleges, specialist societies and research institutes. They could be nominated jointly by five Members of Council from one of three groups, specializing in mathematics and physics, biology or humanities. In other countries, academician candidates were generally recommended by academy members and/or corresponding members. According to *The Organization Act of Academia Sinica* and *The Regulations for the Election of Members of Academia Sinica*,[1] after the very first election, Academia Sinica needed to, once a year, re-elected a maximum of 15 members, five within each of the three groups. However, the number of academicians in each discipline was relatively fixed in other countries. If any academician passed away and there was a vacancy, a by-election would be held. Academia Sinica conducted things differently because the Second Council felt that if international practices were adopted to co-opt academicians, those with outstanding academic achievements/contributions could miss out on being elected just because there were no vacancies. This, they felt, was not conducive to academic progress. Article 6 of

[1]This refers to *The Organization Act of Academia Sinica* amended and promulgated on March 13th, 1947 and *The Regulations for the Election of Members of Academia Sinica* adopted at the Fourth Annual Meeting of the Second Council of Academia Sinica on October 15th, 1947.

The Organization Act of Academia Sinica, amended by the Second Council during its Third Annual Meeting in October 1946, explained this:

> When national academies elect academicians, they mostly stipulate the number of academicians per discipline. In the event of death, a by-election is held. Many people have criticized this system for its corrupt practices. If every discipline has a fixed number of places, it will mean that in the long run, academic progress will be inconsistent. If there is a shortage of people in a certain discipline, the election can be suspended. However, it would be a pity if a new person demonstrated outstanding achievements in a certain discipline, but failed to be selected due to a full quota. This is not the way to promote academic development. Therefore, this article is formulated. The election of the first batch of members of Academia Sinica will be conducted according to the country's current situation. During future elections, more people would be chosen each year to accommodate rising stars. If someone died, the vacancy would not be filled. This way, China could keep pace with academic progress.[2]

The regulations/modes of academician qualification, selection process for academicians and related affairs changed once the academician system prevailing in the West was transplanted into China. Although the changes allowed those who had led or presided over academic institutions for more than five years to be selected as academicians, the council still elected members of Academia Sinica based on internationally accepted qualifications, arguing that "only those who have been recognized by their peers as authorities can be selected".[3] This did not open many doors for those who only met this qualification. Therefore, taking the academician system that had prevailed in the West as a reference point and establishing an academician system in China, was not only in line with the international standards, but also in line with China's national conditions. That was the logic of the Council when they established the academician system of Academia Sinica.

In any case, the establishment of the academician system of Academia Sinica was a milestone for systematizing science in China. It meant that China's modern, scientific system was gradually maturing and beginning to integrate itself into the world's academic system. Its establishment was a small, but historic step and saw China transplanting itself into a modern, western academic system. This process showed that China's academic elite had a high degree of autonomy when constructing a national, academic system under the Republic of China.

---

[2]Zhongyang yanjiuyuan zuzhifa pingyihui tiaoli pingyihui guicheng xianxing fagui zhangcheng mulu deng中央研究院组织法、评议会条例、评议会规程、现行法规章程目录等(The Organization Act, the Regulations of the Council, the Aritcles for the Council, the Catalogue of Current Regulations and Articles of Academia Sinica, etc.). Nanjing: Zhongguo dier lishi danganguan中国第二历史档案馆(The Second Historical Archives of China), Total No.393, File No.2539. The Organization Act of Academia Sinica, Rules of Council, Regulations of Council, Catalogue of Current Regulations and Articles of Association, etc. Nanjing: The Second Historical Archives of China, Total No.393, File No.2539.

[3]Zhongyang yanjiuyuan zuzhifa pingyihui tiaoli pingyihui guicheng xianxing fagui zhangcheng mulu deng中央研究院组织法、评议会条例、评议会规程、现行法规章程目录等(The Organization Act, the Regulations of the Council, the Aritcles for the Council, the Catalogue of Current Regulations and Articles of Academia Sinica, etc.). Nanjing: Zhongguo dier lishi danganguan中国第二历史档案馆(The Second Historical Archives of China), Total No. 393, File No. 2539.

During the Republic of China, the national government, Academia Sinica, and the academic elite, were occasionally subject to contradictions or conflicts.[4] However, these were not so serious as to be irreconcilable. For example, in 1940, after Tsai Yuan-Pei's death, and when during the First Council elected the president of Academia Sinica, Gu Mengyu was not elected because of resisting a letter from Chiang Kai-shek. This caused a storm between Academia Sinica and the national government. However, Chiang Kai-shek finally came to respect the election results of the Council. In 1941, Chen Lifu suggested that Chiang Kai-shek merge Academia Sinica into the Ministry of Education or the Examination Yuan, after the latter had been later unblocked by Academia Sinica. This proposal was "dismissed". During the process of establishing Academia Sinica's academician system, the national government gave their support to the institution having legislation, and there was no political or administrative interference in the election of the first batch of members of Academia Sinica. This was an important guarantee that the Council would maintain their academic freedom and academic independence. When the First General Assembly of Members of Academia Sinica was held, Chiang Kai-shek, who rarely visited Academia Sinica, attended the opening ceremony and delivered a speech. This further demonstrated the national government's goodwill towards the establishment of the academician system of Academia Sinica. These reflect that the government maintained a basic balanced relationship with the national academy of sciences and the academic elite. It did not excessively infiltrate state power into the public, academic realm, especially after the Anti-Japanese War, during the establishment of Academia Sinica's academician system.

After the Kuomintang regime stabilized in Taiwan, the academician system of Academia Sinica was restored. It was reestablished in Taiwan in 1958, and continues to exist to the present day. The Council system has also been preserved. Currently, Academia Sinica still requires that its members qualify with one or more of the two requirements stipulated in *The Organization Act of Academia Sinica*, which was revised and promulgated on 13 March, 1947. *The Regulations for the Election of Members of Academia Sinica* were passed during the Fourth Annual Meeting of the Second Council in 1947, and by 2009, they had been amended 13 times. There were a large number of amendments made to the provisions for the "election", but the provisions for "nominations" and the "examinations for qualifying academician candidates" did not change much. The tasks of the academician assembly, such as electing academicians, honorary academicians, and Council members, also did not change. They stayed as they were stipulated in *The Regulations for the General Assembly of Members of the Academia Sinica*, adopted during the Fifth Annual Meeting of the Second Council in 1948, requiring that attendees read important academic papers at General Assemblys. Therefore, some important rules/regulations formulated during the establishment of Academia Sinica's academician system were not eliminated either by history, or the changing times in Taiwan. They still have an impact on the academic development of Academia Sinica, and Taiwan.

[4]Chen Shiwei (1998). Government and Academy in Republican China: History ofs Academia Sinica, 1927–1949. Cambridge: Doctoral Thesis of Harvard University. 171–220.

After 1949, Academia Sinica's academician system was completely abandoned in New China, following the change in regime. In 1955, the CAS took lessons from the Soviet Union, and set up academic division memberships. It did not, however, set up an academician system. This showed that China's modern, scientific system had changed under the leadership of the new regime. In 1955, the Chinese Academy of Sciences did not draw up relevant laws or regulations in advance when selecting and appointing members of academic divisions. The processes were somewhat disorderly, and continuously being explored. Such was the mode of selecting scientists under the Communist Party of China's centralized leadership system. It was a completely different process to the previous method of electing elected council members and electing members of Academia Sinica. The selection and appointment of CASAD members was directly led by the Publicity Department of the CPC Central Committee and was interfered by the government. At the same time, it had strong political ideologies. This reflects that the government's basic balanced relationship with the national academy of sciences and the academic elite before 1949 was broken. However, after the regime changed and the thought reform, China's academic elite went from being completely autonomous to part of the Republic of China's national, academic system. The selection process showed that their participation involved being under the leadership and subject to the intervention of the government.

Academic division members were different to academicians . As the academician system had become popular internationally, the establishment of an academician system was the general trend in New China. Therefore, after the Chinese Academy of Sciences set up a system of academic division members in 1955, it tried to establish an academician system that was compatible with it. But despite several discussions, the CAS failed. After 1980, the Chinese Academy of Sciences and the Chinese Academy of Social Sciences tried multiple times to establish an academician system, but they failed every time. In the end, the Chinese Academy of Sciences established an academician system by gradually evolving its membership system into an academician system and recognizing all academic division members as academicians. Compared to the methods used by Academia Sinica and other National Academies of Science around the world, this was a unique way of establishing an academician system. Since academic division members basically became academicians, there was a reasonable basis for establishing an academician system. However, because there were many division members and their academic levels differed, this inevitably meant that the overall level of Chinese academicians was limited.

The establishment of the Chinese Academy of Engineering and its academician system received the attention and support of the State Council and the Central Committee. Both of these departments were vigorous about implementing Deng Xiaoping's guiding ideology that "science and technology are the first productive forces". This showed that there was a strong relationship between the state, and science and technology. In general, things went smoothly with the establishment of the Chinese Academy of Engineering and its academician system. But there were a few hiccups. The State Council intervened to move the preparatory work from the hands of the Chinese Academy of Sciences , to the State Science and Technology

Commission. Both the Department of Mathematics and Physics and the Department of Technology at the Chinese Academy of Sciences experienced difficulties preparing their initial lists of candidates for the first batch of academicians at the Chinese Academy of Engineering . There were also some deficiencies with regards the selection process of the first batch of academicians of Chinese Academy of Engineering. However, the establishment of the Chinese Academy of Engineering and its academician system was extremely significant and institutionalizing engineering and technology was an important achievement for China.

After the 1994 academician system was rebuilt in China, the Chinese Academy of Engineering's and the Chinese Academy of Sciences' membership system respectively underwent stages of improvement. In 1995, once a round of members had been co-elected at the Chinese Academy of Engineering, the nominations for new member candidates were carried out across the country. This time, the evaluation procedure was much more rigorous and candidates needed to pass two rounds. The first round was organized by the standing committees/joint standing committee of various departments, and it was conducted by members whose professional field corresponded to that of respective, valid candidates. The members reviewed material, exchanged opinions, held discussions, and then prepared to vote. After they had voted, they selected candidates for a preliminary list. They included more than twice the number of names that would make the official list.[5] The second round of evaluation was then conducted in the same way as the first round of evaluation; it was within the scope of all academicians to discuss each candidate's conditions and review their material, before holding group discussions and evaluations. The academicians would introduce and discuss relevant academic divisions, and then go to a secret ballot.[6] Since the 2007 academician co-elections, the Chinese Academy of Engineering has raised the threshold for candidates to be elected. They can now only be elected if they receive votes from two-thirds of the number of academicians voting. Previously they needed a half.[7]

---

[5]Zhongguo gongchengyuan 1995 nian zengxuan yuanshi houxuanren de timing linxuan pingshen he xuanjubanfa中国工程院1995年增选院士候选人的提名、遴选、评审和选举办法(Methods for Nomination, Selection, Evaluation and Election of the 1995 Co-optation of Member Candidates of the Chinese Academy of Engineering). In Zhongguo gongchengyuan中国工程院(Chinese Academy of Engineering) (Eds.) (2004). *Zhongguo gongchengyuan nianjian* (1994–1997) 中国工程院年鉴(1994–1997) (Yearbook of Chinese Academy of Engineering, 1994–1997). Beijing: Zhongguo gongchengyuan中国工程院.99–101.

[6]Zhongguo gongchengyuan 1995 nian yuanshi zengxuan dierlun pingshen he xuanju banfa 中国工程院1995年院士增选第二轮评审和选举办法(Methods for Second Round Evaluation and Election of the 1995 Co-optation of Member of the Chinese Academy of Engineering) . In Zhongguo gongchengyuan中国工程院(Chinese Academy of Engineering) (Eds.) (2004). *Zhongguo gongchengyuan nianjian* (1994–1997) 中国工程院年鉴(1994–1997) (Yearbook of Chinese Academy of Engineering, 1994–1997). Beijing: Zhongguo gongchengyuan中国工程院. 242–245.

[7]Zhongguo gongchengyuan 2007 nian yuanshi zengxuan jieguo jiexiao中国工程院2007年院士增选结果揭晓(The Chinese Academy of Engineering Announces the Results of the Co-optation of Members in 2007). *Zhongguo gongcheng kexue*中国工程科学(Strategic Study of CAE) 2008, 10 (1): 4.

After the 1997 co-election of academicians and prior to the first round of review meetings, the Chinese Academy of Sciences stated that all relevant material recommending valid academician candidates (under the column "Major scientific and technological achievements, contributions and moral styles of study") should be published by grass-roots units where the candidates are located. In this way, there was a system in place so that academics could seek and listen to further opinions. This increased the transparency of the academician co-optation process and showed that an improved effort was being made to make candidates' recommended material much more reliable. During the evaluation stage, academician candidates had a relatively straightforward procedure involving a pre-conference "preliminary examination", a round of external "preliminary examinations" and "reviews", and another round of "final examinations". The academic department also took steps to improve the handling of complaint letters by formulating the Interim Measures for Handling Complaint Letters.[8] On 14 February, 1997, "in order to strengthen the construction of spiritual civilization in the work of academician co-optation", it was stated that the department needed to "ensure the quality of recommendation, evaluation and election [of candidates], and maintain the reputation of the academician department". During the first plenary meeting of the Chinese Academy of Sciences' scientific ethics committee, a "Code of Conduct for Co-optation of Members of Chinese Academy of Sciences" was formulated. It stipulated a code of conduct for academicians when recommending, evaluating and introducing candidates.[9] In 2006, during the meeting of the Chinese Academy of Sciences presidium, changes were also made to the proportion of academicians who could be elected. The "election of academicians" section of the Chinese Academy of Sciences constitution was changed, stating that a candidate needed more than two-thirds of the total votes to be elected (previously, they needed half).[10]

Disputes about the co-optation of academicians are common in China, often due to either unhealthy practices or improper behavior during the selection process. In 2012, the Chinese Academy of Engineering's leading party group made a confession of sorts in an article published in the *Qiushi* magazine. The article, entitled "Strengthening and Improving Academician Ranks" stated that "various unhealthy social

[8]Guanyu 1997 nian Zhongguo kexueyuan yuanshi zengxuan gongzuo de baogao关于1997年中国科学院院士增选工作的报告(Report on the Co-optation of Members of the Chinese Academy of Sciences in 1997). In Zhongguo kexueyuan bangongting中国科学院办公厅(General Office of the Chinese Academy of Sciences) (Eds.). *Zhongguo kexueyuan nianbao*中国科学院年报(The Annals of the Chinese Academy of Sciences) 1997: 153–154.

[9]Zhongguo kexueyuan yuanshi zengxuan gongzuo zhong yuanshi xingwei guifan中国科学院院士增选工作中院士行为规范(code of conduct for co-optation of members of Chinese Academy of Sciences). In Zhongguo gongchengyuan中国工程院(Chinese Academy of Engineering) (Eds.) (2004). Zhongguo gongchengyuan nianjian (1994–1997) 中国工程院年鉴(1994–1997) (Yearbook of Chinese Academy of Engineering, 1994–1997). Beijing: Zhongguo gongchengyuan中国工程院. 641–642.

[10]Zhongguo kexueyuan dishisanci yuanshi dahui wenjian huibian中国科学院第十三次院士大会文件汇编(Compilation of Documents on the 13th General Assembly of Members of Chinese Academy of Sciences). Beijing: Zhongguo kexueyuan danganguan中国科学院档案馆(Archives of Chinese Academy of Sciences), 2006-36-001.

practices have impacted the co-optation process of academicians at the academy". Interferences had manifested in three ways: via "campaign aid", "canvassing" activities, and a phenomenon of "integrating candidates as a package", making it difficult for individual academicians to be co-selected. The environment for discussing the co-optation of academicians was made somewhat uneasy.[11] People were also worried, because academicians are granted special rights in China; they don't need to work to retirement, and they can enjoying vice-ministerial levels of medical treatment and similar benefits. Academicians are recognized as authoritative and having administrative power. Being an "academician" and having the highest honorary, academic title, has actually come to be more powerful than it was originally meant to be. Academicians have become a special class in China's scientific and technological circles.[12] As the General Assembly of CAS Members is an advisory body and not the highest decision-making body internationally, the decision-making power of the CAS's member group is still relatively limited. The academician system, therefore, still needs to be continuously improved and perfected in China.

[11]Zhongguo gongchengyuan dangzu中国工程院党组(Party Group of Chinese Academy of Engineering) (2012). Jiaqiang he gaijin yuanshi duiwu jianshe加强和改进院士队伍建设(Strengthen and Improve the Construction of Members). *Qiushi*求是(Seek Truth) (3): 59.

[12]Huang Yannan黄延楠(2013). Yuanshi zhidu hequ hecong院士制度何去何从?(Academician System: Where to Go?). *Zhongguo keji jiangli*中国科技奖励(China Awards for Science and Technology) (3): 21; Wang Yangzong王扬宗(2005). Zhonsgguo yuanshi zhidu de jianli jiqi wenti中国院士制度的建立及其问题(The Academician System in China and Its Defect). *Kexue wenhua pinglun*科学文化评论(Science & Culture Review) 2 (6): 15–16.

Glossary

Academia Sinica 中央研究院
Academia Sinica Member Records《国立中央研究院院士录》
Academic Division 学部
Academic Review Committee of the Ministry of Education 教育部学术审议委员会
Academician 院士
Académie des Sciences 法国皇家科学院
Adam Pen-Tung Sah 萨本栋
Ai Siqi 艾思奇
All-China Federation of Specialized Societies in Natural Science 中华全国自然科学专门学会联合会
Amoy University 厦门大学
Anhui University 安徽大学
Bao Erhan 鲍尔汉
Bao Guobao 鲍国宝
Bao Zhenghu 鲍正鹄
Bi Dexian 毕德显
Bo Shang 伯商
Botanical Society of China 中国植物学会
Cai Yi 蔡仪
Catholic University of Peking 辅仁大学
Cao Fu 曹孚
Cen Qixiang 岑麒祥***Central Daily***《中央日报》
Central Weather Bureau 中央气象局
Chang C. F. 张忠绂
Chang Chin-yueh 张景钺
Chang Hsiang-tung 张香桐
Chang-Lin Tien 田长霖
Chang Poling 张伯苓
Chang T. S. 张宗燧
Chang Tsinglien 张青莲

J. Guo, *The Establishment and Reconstruction of the Academician System in China*,
https://doi.org/10.1007/978-981-15-7208-1

Chang Tung 常迥
Chang Wen-yu 张文裕
Chang Yinfo 常印佛
Chang Yu-Che 张钰哲
Chao Chung-Yoa 赵忠尧
Chao Yuen Ren 赵元任
Chang Hsiao-chien 张孝骞
Cheeloo University 齐鲁大学
Chen Boda 陈伯达
Chen Kang 陈康
Chen Fei 陈菲
Chen Fengtong 陈凤桐
Chen Han-seng 陈翰笙
Chen Heqin 陈鹤琴
Chen Hsin-tao 陈心陶
Chen Jiagui 陈佳贵
Chen Junliang 陈俊亮
Chen Kangbai 陈康白
Chen Kien Kwong 陈建功
Chen Ko Kuei 陈克恢
Chen Kuiyuan 陈奎元
Chen Lifu 陈立夫
Chen Liwei 陈力为
Chen Meng-chia 陈梦家
Chen-Ning Yang 杨振宁
Chen Rong 陈嵘
Chen Shisan C. 陈桢
Chen Shixiang 陈世骧
Chen Shouyi 陈受颐
Chen Su-ching 陈序经
Chen Ta 陈达
Chen Wangdao 陈望道
Chen Wengui 陈文贵
Chen Yi 陈毅
Chen Yin-k'o 陈寅恪
Chen Yuan 陈垣
Chen Yun 陈云
Chen Zhiqian 陈志潜
Chen Zong 陈总
Cheng Dan'an 承澹盦
Cheng Enfu 程恩富
Cheng Ganyun 程干云
Cheng Min-Teh 程民德
Cheng Xiaogang 程孝刚
Cheng Y. C. 程裕淇

Chi Yuoh-Fong 纪育沣
Chia-Chiao Lin 林家翘
Chia Ching-teh 贾景德
Chiang Kai-shek 蒋介石
Chiang Li-Fu 姜立夫
Chiang Ming-chien 蒋明谦
Chiang Ting-fu 蒋廷黻
Chien Jen-yuan 钱人元
Chien Pao-kung 钱保功
Chien-Shiung Wu 吴健雄
Chien Shung-shu 钱崇澍
Chien Wei-zang 钱伟长
Chien Mu 钱穆
Ch'in Chung Mien 岑仲勉
Chin Yuch-Lin 金岳霖
China Society 中国学会
China Zoological Society 中国动物学会
China's Muslim Youth Service Agency for Building Country 中国回教青年建国服务社
Chinese Academy of Engineering 中国工程院
Chinese Academy of Sciences 中国科学院
Chinese Academy of Social Sciences 中国社会科学院
Chinese Association of Frontier Political Affairs 中国边政学会
Chinese Astronomical Society 中国天文学会
Chinese Chemical Society 中国化学会
Chinese Frontier Research Association 中国边疆学术研究会
Chinese Mathematical Society 中国数学会
Chinese Meteorological Society 中国气象学会
Chinese Physical Society 中国物理学会
Chinese Physiological Society 中国生理学会
Ching Ren Chang 秦仁昌
Chou Keng-Shang 周鲠生
Chou Pei-yuan 周培源
Chou T. Q. 赵承嘏
Chow Hung-ching 周鸿经
Chow Jen 周仁
Chow Tung-ching 周同庆
Chu Chia-Hua 朱家骅
Chu Coching 竺可桢
Chu F. T. 诸福棠
Chu Hung-fu 朱弘复
Chu Kun-ching 朱公瑾
Chu Tze-tsin 朱子清
Chu Yinghuang 褚应璜
Chuang Chang-Kong 庄长恭

Chun W. Y. 陈焕镛
Chung H. L. 钟惠澜
Chung S. M. 钟士模
Commercial Press 商务印书馆
Communist Party of China 中国共产党
Constitution for the Academic Divisions of Chinese Academy of Social Sciences (Trial)《中国社会科学院学部章程》(试行)
Constitution for the Academic Division Members of Chinese Academy of Sciences (Trial)《中国科学院学部委员章程(试行)》
Constitution for the Members of Chinese Academy of Sciences《中国科学院院士章程》
Council 评议会
Cui Zhilan 崔之兰
Dai Songen 戴松恩
Dai Wensai 戴文赛
Daye Iron Mine 大冶铁矿
Deng Jiaxian 邓稼先
Deng Tuo 邓拓
Deng Xiaoping 邓小平
Deng Yi-zhe 邓以蛰
Di Chaobai 狄超白
Ding Daoheng 丁道衡
Ding Dewen 丁德文
Ding Henggao 丁衡高
Ding Shengshu 丁声树
Ding Shunnian 丁舜年
Ding Ying 丁颖
Ding Zan 丁瓒
Director 所长
Director-General 总干事
Dong Chenglang 董承琅
Dong Weichuan 董渭川
Draft Constitution of the Chinese Academy of Sciences《中国科学院章程(草案)》
Du Dianying 杜殿英
Du Guoxiang 杜国庠
Duan Zhenji 段镇基
Examination Yuan 考试院
Executive Yuan 行政院
Fan Memorial Institute of Biology 静生生物调查所
Fan Weitang 范维唐
Fan Wenlan 范文澜
Fang Guangtao 方光焘
Fang Jun 方俊
Fang Yi 方毅

Fei Xiaotong 费孝通
Feng Binglin 酆炳林
Feng Ding 冯定
Feng Jinglan 冯景兰
Feng Lan-chou 冯兰洲
Feng Ping-chuan 冯秉铨
Feng Teh-Pei 冯德培
Feng You-lan 冯友兰
Feng Zefang 冯泽芳
Feng Zhi 冯至
First Draft of Draft Constitution of the Chinese Academy of Sciences《中国科学院章程草案初稿》
Fong H. D. 方显廷
Fu C. Y. 傅承义
Fu Maoji 傅懋勣
Fu Ssu-Nien 傅斯年
Fu Xinian 傅熹年
Fu Ying 傅鹰
Fudan University 复旦大学
Gao Ji-yu 高济宇
Ge Helin 葛和林
Ge Nengquan 葛能全
Ge Tingsui 葛庭燧
Geng Yili 耿以礼
Geographical Society of China 中国地理学会
Geological Society of China 中国地质学会
Geological Survey of Kwangtung and Kwangsi 两广地质调查所
Ginling College 金陵女子大学
Gong Yuzhi 龚育之
Gu Gong-xu 顾功叙
Gu Mengyu 顾孟余
Gu Songfen 顾诵芬
Guan De 管德
Guan Dongbo 关东伯
Guan Qiao 关桥
Guan Songtao 关颂韬
Guangming Daily《光明日报》
Guangxi University 广西大学
Guiyang Medical College 贵阳医学院
Guo Bingkuan 郭秉宽
Guo Dali 郭大力
Guo Konghui 郭孔辉
Guo Tanxian 过探先
Guo Xingxian 过兴先
Hao Shiyuan 郝时远

He Bingmeng 何秉孟
He Dequan 何德全
He Ganzhi 何干之
He Jie 何杰
He Jishan 何继善
He Qifang 何其芳
Hiong King-lai 熊庆来
Ho L. 何廉
Ho Tso-lin 何作霖
Ho Ying-chin 何应钦
Ho Zah-wei 何泽慧
Hoff Lu 卢鹤绂
Honorary Member of Academia Sinica 中央研究院名誉院士
Honorary Secretary of the Council 评议会秘书
Hou Guangjiong 侯光炯
Hou Hsiong-Lin 侯祥麟
Hou T. P. 侯德榜
Hou Teh-feng 侯德封
Hou Wai-lu 侯外庐
Hou Zonglian 侯宗濂
Hsia Chien-pei 夏坚白
Hsiao Chu 萧蘧
Hsiao Kung-chuan 萧公权
Hsieh C. K. 谢志光
Hsieh Chia-yung 谢家荣
Hsu L. C. 徐利治
Hsu Pao-Lu 许宝騄
Hsu Shuhsi 徐淑希
Hu C. H. 胡正祥
Hu Haitao 胡海涛
Hu Hsen-hsu 胡先骕
Hu Ji-min 胡济民
Hu Jingfu 胡经甫
Hu K. F. 胡刚复
Hu Keshi 胡克实
Hu M. T. 胡明复
Hu Ning 胡宁
Hu Qiheng 胡启恒
Hu Qiaomu 胡乔木
Hu Sheng 胡绳
Hu Shih(Hu Shizhi) 胡适
Hu Xikui 胡锡奎
Hu Yaobang 胡耀邦
Hu Yitong 胡祎同
Hua Loo-Keng 华罗庚

Huai Guomo 怀国模
Huang Bingwei 黄秉维
Huang C. S. 黄家驷
Huang Changgu 黄昌穀
Huang Kewei 黄克维
Huang Kun 黄昆
Huang Minlon 黄鸣龙
Huang Peiyun 黄培云
Huang Ruilun 黄瑞纶
Huang Shulin 黄书麟
Huang Songling 黄松龄
Huang T. C. 黄子卿
Huang T. K. 黄汲清
Huang Wenxi 黄文熙
Huang Xuhua 黄旭华
Huang Yuanbin 黄元彬
Huang Zongzhen 黄宗甄
Hunan University 湖南大学
Hung S. L. 洪式闾
Institute for the Prevention and Cure of Tuberculosis in Nanjing 南京结核病防治院
James K. Shen 沈克非
Jiang Zemin 江泽民
Jao Chin-chih 饶钦止
Jaw Jeou-Jang 赵九章
Ji Chaoding 冀朝鼎
Ji Guobiao 季国标
Ji Wenfu 嵇文甫
Ji Xian-lin 季羡林
Jian Bozan 翦伯赞
Jiang Lansheng 江蓝生
Jiang Nanxiang 蒋南翔
Jiangsu Medical College 江苏医学院
Jiang Xinsong 蒋新松
Jin Guofan 金国藩
Jin Shanbao 金善宝
Jin Shuliang 靳树梁
Jin Yilian 金怡濂
Jin Yufu 金毓黻
Jin Zhonghua 金仲华
Jing Tiankui 景天魁
John C. H. Wu 吴经熊
John Lee 吕炯
Kang Xirong 康熙荣
Kao C. H. 高崇熙

Kiang Tsai-han 江泽涵
King P. C. 金邦正
Ko Chao 柯召
Ku Chieh-kang 顾颉刚
Ku Yu-hsiu 顾毓琇
Kuomintang 国民党
Kuo Mo-jo 郭沫若
Kuo Y. H. 郭永怀
Kuo Yun-kuan 郭云观
Kuo Zing-Yang 郭任远
Kwei C. T. 桂质廷
Lai Lian 赖琏
Le Tianyu 乐天宇
Lee Ching-kwei 李庆逵
Lee J. S. 李四光
Lee S. P. 李善邦
Legislative Yuan 立法院
Lei Tianjue 雷天觉
Leng Rong 冷溶
Leroy L. Chang 张立纲
Li Bingquan 李炳权
Li Chenggan 李承干
Li Chi 李济
Li Chunyu 李春昱
Li Da 李达
Li Dadong 李大东
Li Deren 李德仁
Li Erzhong 李尔重
Li F. H. 李方训
Li Fanfu 李凡夫
Li Fang Kuei 李方桂
Li Guohao 李国豪
Li Guoping 李国平
Li Haopei 李浩培
Li Hen 李珩
Li Hongzhi 李鸿志
Li Hsien-wen 李先闻
Li Jishi 李吉士
Li Jiannong 李剑农
Li Jinxi 黎锦熙
Li Jingyuan 李景源
Li Lianjie 李连捷
Li Ju-chi 李汝祺
Li Qiang 李强
Li Shenming 李慎明

Li Shizeng 李石曾
Li Shu-Hua 李书华
Li Shutian 李书田
Li Siwei 李四维
Li Tongbao 李同保
Li Tsi-tung 李继侗
Li Wencai 李文采
Li Xie 李协
Li Xun 李薰
Li Yan 李俨
Li Yanong 李亚农
Li Zongen 李宗恩
Liang Boqiang 梁伯强
Liang Qichao 梁启超
Liang Shuquan 梁树权
Liang Ssu-cheng 梁思成
Liang Ssu-yung 梁思永
Liang Xi 梁希
Liang Yingchen 梁应辰
Liao Shicheng 廖世承
Lieu D. K. 刘大钧
Lin Chuanguang 林传光
Lin Hanxiong 林汉雄
Lin Hua 林华
Lin Qiaozhi 林巧稚
Lin Rong 林镕
Lin Wei-gan 林为干
Lin Wenbing 林文炳
Ling Chunsheng 凌纯声
Ling Hung-hsun 凌鸿勋
Liou Tchen-Ngo 刘慎谔
Liu C. C. 刘承钊
Liu Chi-ying 柳支英
Liu Chun 刘春
Liu Chung-lo 刘崇乐
Liu Danian 刘大年
Liu Daosheng 刘导生
Liu Ding 刘鼎
Liu Dunzhen 刘敦桢
Liu Gengling 刘更另
Liu Guiwu 刘桂五
Liu Hongliang 刘鸿亮
Liu Hui-xian 刘恢先
Liu Hsian-chou 刘仙洲
Liu Jibin 刘积斌

Liu Naicheng 刘乃诚
Liu Qibao 刘奇葆
Liu Ruiheng 刘瑞恒
Liu S. H. 刘士豪
Liu Shaoqi 刘少奇
Liu Shucheng 刘树成
Liu Sichi 刘思炽
Liu Simu 刘思慕
Liu Ta-kong 柳大纲
Liu Tianquan 刘天泉
Liu W. T. 刘文典
Liu Weitao 刘为涛
Liu Weitong 刘纬通
Liu Xianlin 刘先林
Liu Yei-Tsen 柳诒徵
Liu Yongtan 刘永坦
Liu Zhe 刘哲
Lo C. C. 罗忠忱
Loo Tsung-Le 罗宗洛
Lou Chenghou 娄成后
Lu Dingyi 陆定一
Lu Gwei-djen 鲁桂珍
Lu Jia-xi 卢嘉锡
Lu Liangshu 卢良恕
Lu S. S. 陆学善
Lu Yongxiang 路甬祥
Lu Yuanjiu 陆元九
Lu Yuanlei 陆渊雷
Lu Zhijun 鲁之俊
Lu Zhiwei 陆志韦
Luo Changpei 罗常培
Luo Gan 罗干
Luo Gengmo 洛耕漠
Luo Peilin 罗沛霖
Luo Xibei 罗西北
Lü Cheng 吕澂
Lü Shuxiang 吕叔湘
Lü Zhenyu 吕振羽
Ma Fubang 马福邦
Ma Heng 马衡
Ma Hong 马洪
Ma Jung-chih 马溶之
Ma Te 马特
Ma Wen-chao 马文昭
Ma Xulun 马叙伦

Ma Yifu 马一浮
Ma Yin-chu 马寅初
Maa Dah-you 马大猷
Mao Dun 茅盾
Mao Henian 毛鹤年
Mao T. E. 茅以升
Mao Tse-tung 毛泽东
Mei Ju-Ao 梅汝璈
Mei Yi-chi 梅贻琦
Member of Academia Sinica 中央研究院院士
Member of Academic Division 学部委员
Member of the CAS 中国科学院院士
Member of the Council 评议员
Meng Chao-ying 孟昭英
Meng H. M. 孟宪民
Meng Yongqian 孟用潜
Min Enze 闵恩泽
Min Guirong 闵桂荣
Min Szu-hoa 闵嗣鹤
Mineral Survey Office 矿产测勘处
Ming Nai-Ta 闵乃大
Ministry of Industry 实业部
Nan Y. T. 南延宗
Nankai University 南开大学
National Academy of Peiping 北平研究院
National Agricultural Research Bureau 中央农业实验所
National Central University 中央大学
National Chengchi University 政治大学
National Geological Survey of China 中央地质调查所
National Health Experimental Institute 中央卫生实验院
National Institute for Compilation and Translation 国立编译馆
National Jinan University 暨南大学
National Jiao Tong University 交通大学
National Northwest Engineering College 西北工学院
National Northwest University 西北大学
National Pei-yang University 北洋大学
National Social Education College 国立社会教育学院
National Southwest Associated University 西南联合大学
National Sun Yatsen University 中山大学
National Tongji University 同济大学
National Tsing Hua University 清华大学
National University of Peking 北京大学
National Wu-han University 武汉大学
Natural Science Society of China 中华自然科学社
Ni Guangnan 倪光南

Nie Rongzhen 聂荣臻
North China Daily《华北日报》
Northeastern University 东北大学
Northwest Agricultural College 西北农学院
Ny Tsai-ze 严济慈
Ou Pao-san 巫宝三
Ouyang Yuqian 欧阳予倩
Ouyang Zhu 欧阳翥
Pai Sitsan 贝时璋
Pan Beilei 潘蓓蕾
Pan Jiazheng 潘家铮
Pan Kuang-tan 潘光旦
Pan Shu 潘菽
Pan Zinian 潘梓年
Pei Chien 裴鉴
Pei Lisheng 裴丽生
Pei W. C. 裴文中
Peiping Union Medical College 北平协和医学院
Peng Huan-wu 彭桓武
Peng Jiqun 彭济群
Peng Shilu 彭士禄
People's Daily《人民日报》
Peter P. T. Sah 萨本铁
Ping Chi 秉志
Preparatory Committee for the Election of the First Batch of Members 第一次院士选举筹备委员会
Preparatory Committee for the Election of the Second Council Members 第二届评议员选举筹备委员会
President of the Council 议长
Publicity Department of the CPC Central Committee 中共中央宣传部
Qian Baocong 钱宝琮
Qi Shounan 戚寿南
Qi Yuanjing 戚元靖
Qian Changzuo 钱昌祚
Qian Gaoyun 钱皋韵
Qian Jiaju 千家驹
Qian Junrui 钱俊瑞
Qian Ling-xi 钱令希
Qian Qihu 钱七虎
Qian Xu 钱煦
Qian Yi 钱易
Qian Zhengying 钱正英
Qian Zhidao 钱志道
Qian Zhongshu 钱锺书
Qiantang River Engineering Office 钱塘江工程处

Qiao Guanhua 乔冠华
Qin Lisheng 秦力生
Qin Zhongda 秦仲达
Qingfeng Cotton Mill 庆丰纱厂
Ren Hongjun 任鸿隽
Ren Mei-e 任美锷
Ren Shangwu 任尚武
Research Assistant 助理研究员
Research Associate 副研究员
Research Fellow 研究员
Resources Committee 资源委员会
Revised Organizational Regulations for Academia Sinica《修正国立中央研究院组织条例》
Robert K. S. Lim 林可胜
Rong Dushan 荣独山
Rong Geng 容庚
Royal Society of London for Improving of Natural Knowledge 英国皇家学会
Ru Xin 汝信
Ruan Xueyu 阮雪榆
Samuel C. C. Ting 丁肇中
Science and Art Society of China 中华学艺社
Science Society of China 中国科学社
Selection Committee 推选委员会
Shandong University 山东大学
Shang Yue 尚钺
Shanghai Business School 上海商学院
Shanghai Medical College 上海医学院
Shao Xianghua 邵象华
Shao Xunke 邵循恪
Shen Chia-jui 沈嘉瑞
Shen Jianshi 沈兼士
Shen Qizhen 沈其震
Shen Shizhang 沈诗章
Shen Yi 沈怡
Shen Yu-cheng 申又枨
Shen Zhijie 沈之介
Shen Zhiyuan 沈志远
Shen Zhiyun 沈志云
Shen Zonghan 沈宗翰
Sheng Shuren 盛树仁
Sheng Tongsheng 盛彤笙
Shi Changxu 师昌绪
Shi Jiayang 施嘉炀
Shi Yafeng 施雅风
Shi Yuanchun 石元春

Shi Zhiquan 石志泉
Shi Zhiren 石志仁
Shiah Nae 夏鼐
Shih J. W. 施汝为
Shiing-Shen Chern 陈省身
Shing-Tung Yau 丘成桐
Si Xingjian 斯行健
Sichuan University 四川大学
Sino-British Cultural Association 中英文化协会
Sino-French Industrial College 中法工学院
Song Jian 宋健
Song Tao 宋涛
State Scientific and Technological Commission 国家科学技术委员会
Su Buchin 苏步青
Su Junhong 苏君红
Sun C. C. 孙健初
Sun Cheng-e 孙承谔
Sun Dan 孙鏞
Sun Dehe 孙德和
Sun Jingzhi 孙敬之
Sun Keji 孙克基
Sun Keming 孙克明
Sun Pen-wen 孙本文
Sun Xiaocun 孙晓邨
Sun Xuewu 孙学悟
Sun Y. C. 孙云铸
Sun Yat-sen 孙中山
Supervision Yuan 监察院
Synthesis and Summary 《综合与摘报》
T. Ko 柯俊
Ta Kung Pao 《大公报》
Tai A. P. 戴安邦
Tai Fon-Land 戴芳澜
Taiwan University 台湾大学
Tan Chia-chen 谈家桢
Tang Au-chin 唐敖庆
Tang Bingyuan 唐炳源
Tang Changru 唐长孺
Tang F. F. 汤飞凡
Tang Lan 唐兰
Tang P. S. 汤佩松
Tang Yueh 唐钺
Tang Yung-tung 汤用彤
Tangshan Engineering College 唐山工学院
Tao Hengxian 陶亨咸

Tao L. K. 陶孟和
Tao Yuanming 陶渊明
Tchang Si 张玺
Teng Chih-I 邓植仪
Teng S. C. 邓叔群
Tentative Draft Constitution of the Chinese Academy of Sciences《中国科学院试行章程(草案)》
The Organization Act of Academia Sinica《国立中央研究院组织法》
The Regulations for the Election of Members of Academia Sinica《国立中央研究院院士选举规程》
The Regulations for the Election of the Council of Academia Sinica《国立中央研究院评议会选举规程》
The Regulations for the General Assembly of Members of Academia Sinica《国立中央研究院院士会议规程》
The Regulations of Handling Affairs for the Council of Academia Sinica《国立中央研究院评议会处务规程》
The Regulations of Procedure for the Council of Academia Sinica《国立中央研究院评议会议事规程》
The Regulations of the Council of Academia Sinica《国立中央研究院评议会条例》
Thought Reform Movement 思想改造运动
Tien C. C. 田奇瑰
Ting S. L. 丁燮林
Tong Jun 童寯
Tsai Burg 蔡堡
Tsai Chiao 蔡翘
Tsai Chin-tao 蔡金涛
Tsai Fang-yin 蔡方荫
Tsai Liu-sheng 蔡镏生
Tsai Pang-hwa 蔡邦华
Tsai Yuan-Pei 蔡元培
Tsao Pen-hsi 曹本熹
Tseng C. K. 曾呈奎
Tseng Chao-lun 曾昭抡
Tseng Y. Y. 曾远荣
Tsien H. S. 钱学森
Tsien L. C. 钱临照
Tsien San-Tsiang 钱三强
Tsien T. S. 钱端升
Tsung-Dao Lee 李政道
Tu Chang-wang 涂长望
Tu Shan-cheng 屠善澄
Tu Zhi 涂治
Tuan Hsio-fu 段学复
Tung Ti-Cho 童第周

Tung Tso-pin 董作宾
University of Chekiang 浙江大学
University of Nanking 金陵大学
V. K. Ting 丁文江
Wang C. C. 王竹泉
Wang C. Y. 王宠佑
Wang Chengwei 汪成为
Wang Chia-Chi 王家楫
Wang Ching-wei 汪精卫
Wang Chung Hwei 王宠惠
Wang Daheng 王大珩
Wang Demin 王德民
Wang Dexi 汪德熙
Wang De-zhao 汪德昭
Wang Dezi 王德滋
Wang Dianzuo 王淀佐
Wang Ganyu 王赣愚
Wang Ganchang 王淦昌
Wang Ging-Hsi 汪敬熙
Wang Guangyuan 王光远
Wang H. C. 汪胡桢
Wang Hefu 王和夫
Wang Hengsheng 王恒升
Wang Hong-zhen 王鸿桢
Wang Jialiu 汪家镠
Wang Juqian 汪菊潜
Wang Jwu-Shi 王竹溪
Wang Li 王力
Wang Mingxiu 王明庥
Wang Pao-jen 王葆仁
Wang Pu 王普
Wang S. C. 王守竞
Wang Shifeng 王世丰
Wang Shijun 王世浚
Wang Shih-Chieh 王世杰
Wang Shou 王绶
Wang Shu-min 王叔岷
Wang Tieh-yai 王铁涯
Wang Tao 王涛
Wang Xianghao 王湘浩
Wang Xuan 王选
Wang Xuewen 王学文
Wang Yanan 王亚南
Wang Yangzong 王扬宗
Wang Ying-Lai 王应睐

Wang Yongzhi 王永志
Wang You 汪猷
Wang Yousheng 汪槱生
Wang Yue 王越
Wang Yuelun 王曰伦
Wang Zhibao 王志宝
Wang Zhihua 汪志华
Wang Zhixi 王之玺
Wang Zhizhuo 王之卓
Wei Jiangong 魏建功
Wei Jingchao 魏景超
Wei Xi 魏曦
Wei Yu 韦钰
Wen Fubo 文伏波
Weng Dujian 翁独健
Weng Wen-bo 翁文波
Wong Wen-hao 翁文灏
Woo Ting-liang 吴定良
Woo Tsin-Hang 吴敬恒
Woo Sho-Chow 吴学周
Woo Y. H. 吴有训
Wu C. K. 吴征铠
Wu Chi-Yuen 伍启元
Wu Ching-chao 吴景超
Wu Chung-hua 吴仲华
Wu Guang 吴光
Wu Han 吴晗
Wu Heng 武衡
Wu Hsien 吴宪
Wu Hsien-Wen 伍献文
Wu Ta-You 吴大猷
Wu Ting-chang 吴鼎昌
Wu Wen-Tsün 吴文俊
Wu Xuelin 吴学蔺
Wu Y. K. 吴英恺
Wu Yi-fang 吴贻芳
Wu Yousheng 吴有生
Wu Yuzhang 吴玉章
Wu Zheng-yi 吴征镒
Wu Zhongwei 吴中伟
Xia Kangnong 夏康农
Xiang Da 向达
Xiangya Medical College 湘雅医学院
Xiao Guanying 萧冠英
Xiao Longyou 萧龙友

Xie Jiasheng 谢家声
Xie Lili 谢礼立
Xie Shaowen 谢少文
Xie Youbai 谢友柏
Xiehe Agency 协和行
Xing Qi-yi 邢其毅
Xiong Daren 熊大任
Xiong Shili 熊十力
Xiong Yi 熊毅
Xu Bingchang 徐炳昶
Xu Chengen 徐承恩
Xu Dixin 许涤新
Xu Fengyan 徐丰彦
Xu Gengguang 徐更光
Xu Hongbao 徐鸿宝
Xu Jie 许杰
Xu Kai 须恺
Xu Keqin 徐克勤
Xu Shigao 徐士高
Xu Yitang 徐益棠
Xu Zhijian 徐志坚
Xu Zhongshu 徐中舒
Xue Muqiao 薛暮桥
Yan Chunhui 严春辉
Yan Dongsheng 严东生
Yan Enyu 严恩棫
Yan Kai 严恺
Yang Duanliu 杨端六
Yang Hsi-meng 杨西孟
Yang Jeng-tsong 杨承宗
Yang K. C. 杨克纯
Yang Kaiqu 杨开渠
Yang Liangui 杨连贵
Yang Qixun 杨奇逊
Yang Quan 杨铨
Yang Shixian 杨石先
Yang Shu-Tah 杨树达
Yang Tingbao 杨廷宝
Yang We-I 杨惟义
Yang Xianzhen 杨献珍
Yang Xiangkui 杨向奎
Yang Xiufeng 杨秀峰
Yang Yunkui 杨允奎
Yao Fusheng 姚福生
Yao Y. T. 饶毓泰

Yap Chu-Phay 叶渚沛
Ye Du-zheng 叶笃正
Ye Juquan 叶橘泉
Ye Liangfu 叶良辅
Yeh Chi-Sun 叶企孙
Yellow River Water Conservancy Committee 黄河水利委员会
Yenching University 燕京大学
Yen Shu-Tang 燕树棠
Yin Da 尹达
Yin Hung-chang 殷宏章
Yin Ruiyu 殷瑞钰
Yin T. H. 尹赞勋
Yingshi University 英士大学
Yongli Chemical Industry Company 永利化学工业公司
Yongli Soda Plant 永利制碱厂
You Guo-en 游国恩
Young C. T. 杨澄中
Young Chung-Chien 杨钟健
Yu C. C. 俞建章
Yu C. S. 余青松
Yu Cha-Shih 余嘉锡
Yu Fuchun 虞福春
Yu Guanying 余冠英
Yu Guang-yuan 于光远
Yu Hongzheng 虞宏正
Yu Ruihuang 余瑞璜
Yu T. F. 俞大绂
Yu T. Y. 喻德渊
Yu Ta-wei 俞大维
Yuan-Cheng B. Fung 冯元桢
Yuan Hanqing 袁翰青
Yuan I-Chin 袁贻瑾
Yue Senxun 乐森玥
Yun Zhen 恽震
Yun Ziqiang 恽子强
Yunnan University 云南大学
Zang Yugan 臧玉淦
Zeng Qi 曾琦
Zhang Bingxi 张炳熹
Zhang Changshao 张昌绍
Zhang Cunhao 张存浩
Zhang Dayu 张大煜
Zhang Deqing 张德庆
Zhang Fengxiang 张凤祥
Zhang Guangdou 张光斗

Zhang Haipeng 张海鹏
Zhang Hongyuan 张洪沅
Zhang Hongzhao 章鸿钊
Zhang Jijia 章基嘉
Zhang Jiafu 张稼夫
Zhang Jiangshu 张江树
Zhang Jin 张锦
Zhang Jinqiu 张锦秋
Zhang Jinfu 张劲夫
Zhang Jingjiang 张静江
Zhang Mingtao 章名涛
Zhang Qiyun 张其昀
Zhang Ruxin 张如心
Zhang Wei 张维
Zhang Wensong 张文松
Zhang Wen-you 张文佑
Zhang Xijun 张锡钧
Zhang Xiruo 张奚若
Zhang Xianlin 张先林
Zhang Yi 张毅
Zhang Yuanji 张元济
Zhang Yuanshan 章元善
Zhang Yun 张云
Zhang Yungang 张云岗
Zhang Yunling 张蕴岭
Zhang Zhaoqian 张肇骞
Zhang Zhirang 张志让
Zhang Zhizhong 张直中
Zhang Zongyou 张宗祐
Zhao Feike 赵飞克
Zhao Guangzeng 赵广增
Zhao Hongzhang 赵洪璋
Zhao Lianfang 赵连芳
Zhao Renkai 赵仁恺
Zhao Yibing 赵以炳
Zhao Zongyu 赵宗燠
Zheng Baojun 郑宝钧
Zheng Bijian 郑必坚
Zheng Guangdi 郑光迪
Zheng Wanjun 郑万钧
Zheng Zhemin 郑哲敏
Zheng Zhenduo 郑振铎
Zhi Bingyuan 支秉渊
Zhong Jinglin 钟敬林
Zhong Lin 钟林

Zhongxing Daily《中兴日报》
Zhongzheng Medical College 中正医学院
Zhou En-lai 周恩来
Zhou Ganzhi 周干峙
Zhou Guangzhao 周光召
Zhou Hong 周弘
Zhou Jing 周镜
Zhou Lian 周廉
Zhou Mingzang 周明牂
Zhou Shilu 周拾禄
Zhou Yang 周扬
Zhou Yiliang 周一良
Zhou Zezhao 周泽昭
Zhou Zhihong 周志宏
Zhou Zhongyi 周仲义
Zhu Bingnan 朱炳南
Zhu Erming 朱尔明
Zhu Gaofeng 朱高峰
Zhu Guangya 朱光亚
Zhu Kaixuan 朱开轩
Zhu Lilan 朱丽兰
Zhu Qifeng 朱起凤
Zhu Ruhua 朱汝华
Zhu Sen 朱森
Zhu Wuhua 朱物华
Zhu Xi 朱洗
Zhu Yuanding 朱元鼎
Zhuang Qianding 庄前鼎
Zhushao Railway Administration 株韶铁路管理局
Zong Bai-hua 宗白华
Zou Jing 邹竞

Select Bibliography

Archival Sources

Beiping yanjiuyuan deng yanjiu jiguan suoti de yuanshi houxuanren mingdan ji youguan hanjian北平研究院等研究机关所提的院士候选人名单及有关函件(The Nomination Lists of Member Candidates Sent by the National Academy of Peiping and Other Research Institutions and Related Correspondence). Nanjing: Zhongguo dier lishi danganguan中国第二历史档案馆(The Second Historical Archives of China), Total No. 393, File No. 1614.

Chen Yi fuzongli 1955 nian 6 yue zai kexueyuan xuebu lianxihui shang de jianghua陈毅副总理一九五五年六月在科学院学部联席会上的讲话(Vice Premier Chen Yi's Speech at the Joint Meeting of the Academic Divisions of Chinese Academy of Sciences in June 1955. Beijing: Zhongguo kexueyuan danganguan中国科学院档案馆(Archives of Chinese Academy of Sciences), 1955-01-010.

Dangzu huiyi jiyao dangzu bangong huiyi jilu党组会议纪要、党组办公会议记录(Summaries of the Party Group Meetings, Minutes of the Party Group Office Meetings). Beijing: Zhongguo kexueyuan danganguan中国科学院档案馆(Archives of Chinese Academy of Sciences), 1956-01-008.

Dahui choubei gongzuo huiyi jianbao (1–6) ji dahui jianbao (1–20) 大会筹备工作会议简报(1–6)及大会简报(1–20) (Bulletins on the Meetings of the Preparatory Work for the Assembly, Issue 1–6, and Bulletins on the Assembly, Issue 1–20). Beijing: Zhongguo kexueyuan danganguan中国科学院档案馆(Archives of Chinese Academy of Sciences), 1981-02-111.

Disici xuebu weiyuan dahui jianbao(21–40 qi) 第四次学部委员大会简报(21–40期) (Bulletins on the Fourth General Assembly of Members of the Academic Divisions, Issue 21–40). Beijing: Zhongguo kexueyuan danganguan中国科学院档案馆(Archives of Chinese Academy of Sciences), 1981-02-112.

Diyici yuanshi xuanju第一次院士选举(The Election of the First Batch of Members). Nanjing: The Second Historical Archives of China, Total No. 393, File No. 494.

Dongbei daxue deng yuanxiao hansong yuanshi houxuanren timingdan ji youguan hanjian东北大学等院校函送院士候选人提名单及有关函件(The Nomination Lists of Member Candidates Sent by Northeastern University and Other Universities and Colleges and Related Correspondence). Nanjing: Zhongguo dier lishi danganguan中国第二历史档案馆(The Second Historical Archives of China), Total No. 393, File No. 1613.

Dui dierci xuebu weiyuanhui qingkuang jianbao ji gebie weiyuan fayan zhailu对第二次学部委员会情况简报及个别委员发言摘录(Bulletins on the Second General Assembly of Members of Academic Divisions and Excerpts from Speeches of Individual Members). Beijing: Zhongguo kexueyuan danganguan中国科学院档案馆(Archives of Chinese Academy of Sciences), 1957-02-021.

J. Guo, *The Establishment and Reconstruction of the Academician System in China*,
https://doi.org/10.1007/978-981-15-7208-1

Gefang dui diyici yuanshi xuanju dangxuan yuanshi houxuanren zhi piping ji yijian各方对第一次院士选举当选院士候选人之批评及意见(Criticisms and Opinions of All Sides on the Candidates for the Election of the First Batch of Members of Academia Sinica). Nanjing: Zhongguo dier lishi danganguan中国第二历史档案馆(The Second Historical Archives of China), Total No. 393, File No. 646.

Gefang hansuo diyici yuanshi houxuanren mingdan各方函索第一次院士候选人名单(People of All Sides Asked for the List of Candidates for the Election of the First Batch of Members of Academia Sinica). Nanjing: Zhongguo dier lishi danganguan中国第二历史档案馆(The Second Historical Archives of China), Total No. 393, File No. 1623.

Guanyu fangsu daibiaotuan xiang zhongyang de gongzuo baogao关于访苏代表团向中央的工作报告(Report to the Central Committee on the Work of the Delegation to the Soviet Union). Beijing: Zhongguo kexueyuan danganguan中国科学院档案馆(Archives of Chinese Academy of Sciences), 1953-02-026.

Guanyu sulian guwen Kovda dui zuzhi quanguoxing kexue yanjiu gongzuo de yijian ji yu Zhang fuyuanzhang tanhua jiyao关于苏联顾问柯夫达对组织全国性科学研究工作的意见及与张副院长谈话纪要(About Soviet Consultant Koffida's Opinions on Organizing National Scientific Research Work and Summary of His Conversation with Vice President Zhang). Beijing: Zhongguo kexueyuan danganguan中国科学院档案馆(Archives of Chinese Academy of Sciences), 1955-01-036.

Guanyu tuijian he linxuan Zhongguo kexueyuan xuebu weiyuan houxuanren de tongzhi he jige juti wenti de yijian关于推荐和遴选中国科学院学部委员候选人的通知和几个具体问题的意见(Notice on Recommending and Selecting Candidates for Academic Division Members of the Chinese Academy of Sciences and Opinions on Several Specific Issues). Beijing: Zhongguo kexueyuan danganguan中国科学院档案馆(Archives of Chinese Academy of Sciences), 1980-36-008.

Guanyu zengpin xuebu weiyuan de yijian关于增聘学部委员的意见(Opinions on the Supplement of Academic Division Members). Beijing: Zhongguo kexueyuan danganguan中国科学院档案馆(Archives of Chinese Academy of Sciences),1957-15-002.

Guojia kewei kexueyuan shehui kexueyuan gunayu jianli yuanshi zhidu wenti de qingshi yuanshi tiaoli (caoan) ji guowuyuan changwu huiyi taolun jiyao国家科委、科学院、社会科学院关于建立院士制度问题的请示、院士条例(草案)及国务院常务会议讨论纪要(Request for Instructions by the State Scientific and Technological Commission, the Chinese Academy of Sciences and the Chinese Academy of Social Sciences on the Establishment of Academician System, *Academician Regulations (Draft)* and Summary of Discussion at the Executive Meeting of the State Council). Beijing: Zhongguo kexueyuan danganguan中国科学院档案馆(Archives of Chinese Academy of Sciences), 1980-02-024.

Guoli zhongyang yanjiuyuan banli diyici yuanshi xuanju jingguo qingxing jielue国立中央研究院办理第一次院士选举经过情形节略(Summary of Academia Sinica's Election of Its First Batch of Members). Nanjing: Zhongguo dier lishi danganguan中国第二历史档案馆(The Second Historical Archives of China), Total No. 393, File No. 1085.

Guoli Zhongyang yanjiuyuan diyici yuanshi xuanju houxuanren timingce国立中央研究院第一次院士选举候选人提名册(Nomination Lists of the Member Candidates for the Election of the First Batch of Members of Academia Sinica). Nanjing: Zhongguo dier lishi danganguan中国第二历史档案馆(The Second Historical Archives of China), Total No. 393, File No. 1597.

Guowuyuan zhongkeyuan xuebuban youguan pizhuan zengxuan de qingshi zengxuan tongzhi tuijian qingkuang pingshen banfa xinwen fabuhui deng国务院、中科院、学部办有关批转增选的请示、增选通知、推荐情况、评审办法、新闻发布会等(Requests for Instructions on Supplement Approved and Transferred, Supplement Notices, Recommendation Information, Evaluation Methods and Press Conference, etc. Handled by the State Council, the Chinese Academy of Sciences and Its Office of Academic Divisions Respectively). Beijing: Zhongguo kexueyuan danganguan中国科学院档案馆(Archives of Chinese Academy of Sciences), 1991-36-001.

Hu Shih Loo Tsung-Le dengren suo ti de yuanshi houxuanren mingdan ji youguan hanjian胡适、罗宗洛等人所提的院士候选人名单及有关函件(The Nomination Lists of Member Candidates

Proposed by Hu Shih, Loo Tsung-Le, etc. and Related Correspondence). Nanjing: Zhongguo dier lishi danganguan中国第二历史档案馆(The Second Historical Archives of China), Total No. 393, File No. 1615.

Jishu kexuebu changwei huiyi de youguan wenjian (1–15 ci) 技术科学部常委会议的有关文件(1–15次) (Relevant Documents of the Standing Committee Meetings of the Division of Technological Sciences (No. 1–15). Beijing: Zhongguo kexueyuan danganguan中国科学院档案馆(Archives of Chinese Academy of Sciences), 1956-17-001.

Jishu kexuebu dishiwu zhi shibaci changwei hui wenjian技术科学部第15–18次常委会文件(Documents of the 15th–18th Meetings of Standing Committee of the Division of Technological Sciences). Beijing: Zhongguo kexueyuan danganguan中国科学院档案馆(Archives of Chinese Academy of Sciences), 1956-17-003.

Jishu kexuebu wei zhongkeyuan dierci xuebu dahui zhunbei de huiyi wenjian ji ti'an de chuli yijian技术科学部为中科院第二次学部大会准备的会议文件及提案的处理意见(The Documents and Processing Opinions on the Proposals Prepared by the Division of Technological Sciences for the Second General Assembly of Members of the Academic Divisions of the Chinese Academy of Sciences). Beijing: Zhongguo kexueyuan danganguan中国科学院档案馆(Archives of Chinese Academy of Sciences), 1957-17-001.

Kexueyuan dangzu shuji Li Chang gei Fang Yi deng yuanlingdao tonghzi de xin科学院党组书记李昌同志给方毅等院领导同志的信(Comrade Li Chang, Secretary of the Party Group of the Chinese Academy of Sciences, Addressed a Letter to Fang Yi and Other Leaders of the Academy). Beijing: Zhongguo kexueyuan danganguan中国科学院档案馆(Archives of Chinese Academy of Sciences), 1981-02-035.

Pingyihui erjie zai yu pingyiyuan tanhuahui评议会二届在渝评议员谈话会(The Symposium of Members of the Second Council in Chongqing). Nanjing: Zhongguo dier lishi danganguan中国第二历史档案馆(The Second Historical Archives of China), Total No. 393, File No. 577.

Woyuan dui quanguo renda liujie erci huiyi daibiao jianyi piping yijian banli qingkuang我院对全国人大六届二次会议代表建议、批评、意见办理情况(Our Academy's Handling of Suggestions, Criticisms and Opinions of Deputies at Two Sessions of the Sixth National People's Congress). Beijing: Zhongguo kexueyuan danganguan中国科学院档案馆(Archives of Chinese Academy of Sciences), 1984-02-020.

Shengwu dixue bu changweihui changwei kuoda huiyi yiji xukezu huodong wenjian生物地学部常委会、常委扩大会议以及学科组活动文件(Documents of the Standing Committee Meetings and Enlarged Meetings of the Division of Biology and Earth Sciences, and of the Activities of the Discipline Teams). Beijing: Zhongguo kexueyuan danganguan中国科学院档案馆(Archives of Chinese Academy of Sciences), 1957-16-006.

Shengwu dixue bu dishici zhi shisici changwu kuoda huiyi jiyao生物地学部第十次至十四次常务扩大会议纪要(Summaries of the 10th to 14th Expanded Meetings of the Standing Committee of the Division of Biology and Earth Sciences). Beijing: Zhongguo kexueyuan danganguan中国科学院档案馆(Archives of Chinese Academy of Sciences),1956-16-003.

Shengwu xuebu changwei kuodahui tanlun zengbu dierci xuebu dahui weiyuan mingdan ji tuijian lunwen生物学部常委扩大会讨论增补第二次学部大会委员名单及推荐论文(Discussion at the Enlarged Meeting of the Standing Committee of the Biology Division for List of Supplementing Academic Division Members and Recommending Papers for the Second General Assembly of the Academic Divisions). Beijing: Zhongguo kexueyuan danganguan中国科学院档案馆(Archives of Chinese Academy of Sciences),1957-16-004.

Shulihua xuebu zhaokai diyi er wu ci xuebu changwei huiyi jilu ji youguan cailiao数理化学部召开第一、二、五次学部常委会议记录及有关材料(Minutes of the First, Second and Fifth Standing Committee Meetings held by the Division of Mathematics, Physics and Chemistry and Related Materials). Beijing: Zhongguo kexueyuan danganguan中国科学院档案馆(Archives of Chinese Academy of Sciences), 1956-15-005.

Xingzhengyuan jiaoyubu guanyu boshi xuewei kaoshi ji kexue yanjiu huodong deng wenti yu Zhongyang yanjiuyuan pingyihui laiwang wenshu行政院、教育部关于博士学位考试及科学研究活动等问题与中央研究院评议会来往文书(Correspondence between Executive Yuan and

Ministry of Education on Doctoral Degree Examination and Scientific Research Activities with the Council of Academia Sinica). Nanjing: Zhongguo dier lishi danganguan中国第二历史档案馆(The Second Historical Archives of China), Total No. 393, File No. 1546.

Xuebuban youguan yuanshi zhangcheng renzhi (Wang Chen Qian Yingjie Ai Xinyuan) 学部办有关《院士章程》、任职(王晨、钱莹洁、艾新源) (Relevant Statutes for Membership of the Chinese Academy of Sciences by the Office of CASAD, Position Held (Wang Chen, Qian Yingjie, Ai Xinyuan). Beijing: Zhongguo kexueyuan danganguan中国科学院档案馆(Archives of Chinese Academy of Sciences), 1994-36-007.

Xuebu changwei dui zengxuan gongzuo xize tichu de xiugai buchong yijian学部常委对"增选工作细则"提出的修改、补充意见(Revision and Supplemental Opinions of the Standing Committee of the Academic Divisions on the *Detailed Rules for Co-optation*). Beijing: Zhongguo kexueyuan danganguan中国科学院档案馆(Archives of Chinese Academy of Sciences), 1994-36-016.

Xuebu zhuxituan huiyi cailiao学部主席团会议材料(Meeting Materials of the Presidium of the Academic Divisions). Beijing: Zhongguo kexueyuan danganguan中国科学院档案馆(Archives of Chinese Academy of Sciences), 1994-36-017.

1957 nian xuebu changwei huiyi ji xueke xiaozu huiyi youguan wenjian一九五七年学部常委会议及学科小组会议有关文件(Relevant Documents of the 1957 Meetings of the Standing Committees of the Academic Divisions and the Meetings of the Discipline Teams). Beijing: Zhongguo kexueyuan danganguan中国科学院档案馆(Archives of Chinese Academy of Sciences), 1957-15-001.

1981 nian dangzu changwei huiyi zhenggongkou pengtou huiyi jilu一九八一年党组常委会议、政工口碰头会议记录(Minutes of the 1981 Meetings of the Standing Committee of the Party Group and Brief Meetings of the Political Work Section). Beijing: Zhongguo kexueyuan danganguan中国科学院档案馆(Archives of Chinese Academy of Sciences), 1981-01-005.

1981 nian dangzu huiyi jiluben一九八一年党组会议记录本(Notebooks of the Party Group Meetings in 1981). Beijing: Zhongguo kexueyuan danganguan中国科学院档案馆(Archives of Chinese Academy of Sciences), 1981-01-004.

1991 nian xuebu weiyuan zengxuan cailiao1991年学部委员增选材料(Materials on Co-optation of Academic Division Members in 1991). Beijing: Zhongguo kexueyuan danganguan中国科学院档案馆(Archives of Chinese Academy of Sciences), 1991-36-035.

Youguan yuanshi zhidu cankao ziliao guanyu xuewei xuexian tiaoli deng wenti de baogao有关院士制度参考资料,关于学位、学衔条例等问题的报告(Reference Materials on Academician System, Reports on Regulations of Academic Degrees, Academic Titles, etc). Beijing: Zhongguo kexueyuan danganguan中国科学院档案馆(Archives of Chinese Academy of Sciences), 1964-02-041.

Yu Wen tongzhi fu hang xiang Li fuyuanzhang huibao gongzuo qingkuang ji Li fuyuanzhang de yijian郁文同志赴杭向李副院长汇报工作情况及李副院长的意见(Comrade Yu Wen Went to Hangzhou to Report to Vice-president Li on his Work and Vice-president Li's Opinions). Beijing: Zhongguo kexueyuan danganguan中国科学院档案馆(Archives of Chinese Academy of Sciences), 1957-01-019.

Yuanshi Tanhuahui院士谈话会(Symposium of Members of Academia Sinica). Nanjing: Zhongguo dier lishi danganguan中国第二历史档案馆(The Second Historical Archives of China), Total No. 393, File No. 569.

Yuan sici xuebu dahui zhuxituan huiyi jiyao院四次学部大会主席团会议纪要(Summaries of the Presidium Meeting of the Fourth General Assembly of Members of the Academic Divisions of Chinese Academy of Sciences). Beijing: Zhongguo kexueyuan danganguan中国科学院档案馆(Archives of Chinese Academy of Sciences), 1981-2-98.

Zhang Jiafu Wu Heng tongzhi zai zhongyang zhaokai de zhishi fenzi wenti huiyi shang fayan ji kexueyuan xiaozu huibao张稼夫、武衡同志在中央召开的知识分子问题会议上发言及科学院小组汇报(Speeches made by Comrade Zhang Jiafu and Comrade Wu Heng at the Meeting on Intellectual Issues held by the Central Committee and the Report of the Group of Chinese Academy of Sciences). Beijing: Zhongguo kexueyuan danganguan中国科学院档案馆(Archives of Chinese Academy of Sciences),1956-1-010.

Zhongguo kexueyuan shixing zhangcheng中国科学院试行章程(The Tentative Constitution of the Chinese Academy of Sciences). Beijing: Zhongguo kexueyuan danganguan中国科学院档案馆(Archives of Chinese Academy of Sciences), 1981-02-092.

Zhongguo kexueyuan xuebu weiyuan quanti huiyi(1980 nian 3 yue 27 ri zhi 4 yue 2 ri) (1) (Li Chang Qin Lisheng deng tongzhi zai bimuhui shang jianghua)中国科学院学部委员全体会议(1980.3.27–4.2) (1) (李昌、秦力生等同志在闭幕会上讲话) (The Plenary Meeting of Academic Division Members of Chinese Academy of Sciences (March 27–April 2, 1980) (1) (Speeches made by Comrades Li Chang, Qin Lisheng and Others at the Closing Meeting). Beijing: Zhongguo kexueyuan danganguan中国科学院档案馆(Archives of Chinese Academy of Sciences), 1980-36-005.

Zhongguo kexueyuan zhaokai dishici zhi dishiliuci yuanwu changwu huiyi de tongzhi jilu jiyao ji youguan cailiao中国科学院召开第十次至第十六次院务常务会议的通知、记录、纪要及有关材料(Notices, Minutes, Summaries and Related Materials of the 10th to 16th Executive Meetings of the Chinese Academy of Sciences). Beijing: Zhongguo kexueyuan danganguan中国科学院档案馆(Archives of Chinese Academy of Sciences), 1957-02-032.

Zhongkeyuan bangongting guanyu jianli Zhongguo gongchengyuan timing Qian Zhengying tongzhi wei 1995 nian qiaola fasi ziran kexuejiang houxuanren Zhou Guangzhao yuanzhang zai Zhongguo kexuebao gongzuo huiyi shang de jianghua tongzhi qingshi中科院、办公厅关于建立中国工程院、提名钱正英同志为1995年乔拉法斯自然科学奖候选人、周光召院长在中国科学报工作会议上的讲话、通知、请示(Remarks, Notices and Requests of Chinese Academy of Sciences and Its General Office on Establishing the Chinese Academy of Engineering, Nominating Comrade Qian Zhengying as Candidate for the 1995 Chorafas Natural Sciences Award, and President Zhou Guangzhao at the Work Conference of China Science Daily). Beijing: Zhongguo kexueyuan danganguan中国科学院档案馆(Archives of Chinese Academy of Sciences), 1993-02-055.

Zhongkeyuan guanyu tuixuan xuebu weiyuan jingguo qingkuang xiang zhongyang de baogao中科院关于推选学部委员经过情况向中央的报告(Report of the Chinese Academy of Sciences to the Central Committee on the Selection of Academic Division Members). Beijing: Zhongguo kexueyuan danganguan中国科学院档案馆(Archives of Chinese Academy of Sciences), 1955-01-005.

Zhongkeyuan xuebu chengli dahui choubei weiyuanhui ji dahui zhuxituan huiyi jilu中科院学部成立大会筹备委员会及大会主席团会议记录(Meeting minutes of the Preparatory Committee and the Presidium of the Inaugural Assembly of the Academic Divisions of Chinese Academy of Sciences). Beijing: Zhongguo kexueyuan danganguan中国科学院档案馆(Archives of Chinese Academy of Sciences), 1955-02-029.

Zhongkeyuan xuebu chengli dahui qingkuang jianbao中科院学部成立大会情况简报(Bulletins on the Inaugural Assembly of the Academic Divisions of Chinese Academy of Sciences). Beijing: Zhongguo kexueyuan danganguan中国科学院档案馆(Archives of Chinese Academy of Sciences), 1955-02-038.

Zhongkeyuan xuebu chengli dahui richeng ji canjia renyuan mingdan中科院学部成立大会日程及参加人员名单(The agenda of the Inaugural Assembly of the Academic Divisions of the Chinese Academy of Sciences and the List of Participants). Beijing: Zhongguo kexueyuan danganguan中国科学院档案馆(Archives of Chinese Academy of Sciences), 1955-02-017.

Zhongguo kexueyuan xuebu chengli dahui zongjie bing xiang guowuyuan de baogao中国科学院学部成立大会总结并向国务院的报告(Summary of the Inaugural Assembly of the Academic Divisions of Chinese Academy of Sciences and Its Report to the State Council). Beijing: Zhongguo kexueyuan danganguan中国科学院档案馆(Archives of Chinese Academy of Sciences), 1955-02-037.

Zhongkeyuan xueshu mishuchu 1954 nian chuwu huiyi jiyao gongzuo jihua (yi zhi ershijiu ci) 中科院学术秘书处一九五四年处务会议纪要、工作计划(一至二十九次) (Office Meeting Summaries and Work Plans (1st–29th) of the Academic Secretariat of the Chinese Academy of Sciences in 1954). Beijing: Zhongguo kexueyuan danganguan中国科学院档案馆(Archives of Chinese Academy of Sciences), 1954-02-022.

Zhongkeyuan 1957 nian zhaokai dierci xuebu dahui shulihua xuebu xiaozu de huiyi jilu中科院1957年召开第二次学部大会数理化学部小组的会议记录(Minutes of the Meetings of Group of the Division of Mathematics, Physics and Chemistry at the Second General Assembly of Members of Academic Divisions of the Chinese Academy of Sciences held in 1957). Beijing: Zhongguo kexueyuan danganguan中国科学院档案馆(Archives of Chinese Academy of Sciences), 1957-2-018.

Zhongkeyuan 1953 nian zhaokai dishiyi zhi sanshici yuanwu changwu huiyi jilu ji youguan wenjian中科院一九五三年召开第十一至三十次院务常务会议记录及有关文件(Minutes and Related Documents of the 11th to 30th Executive Meetings of the Chinese Academy of Sciences Held in 1953). Beijing: Zhongguo kexueyuan danganguan中国科学院档案馆(Archives of Chinese Academy of Sciences), 1953-02-003.

Zhongkeyuan youguan dui xuebu weiyuan gongzuo de yijian ji xuebu weiyuan mingdan zhuanchang jianli he dui mingdan de yijian中科院有关对学部委员工作的意见及学部委员名单、专长、简历和对名单的意见(The Opinions of the Chinese Academy of Sciences on the Work of Academic Division Members and the Lists of Academic Division Members and Their Specialities and Curriculum Vitae, and Opinions on the Lists). Beijing: Zhongguo kexueyuan danganguan中国科学院档案馆(Archives of Chinese Academy of Sciences), 1955-02-036.

Zhongkeyuan zhaokai dishiwu zhi ershijiu ci yuanwu changwu huiyi de tongzhi jiyao ji youguan cailiao中科院召开第15–29次院务常务会议的通知、纪要及有关材料(Notices, Summaries and Related Materials of the 15th-29th Executive Meetings of the Chinese Academy of Sciences). Beijing: Zhongguo kexueyuan danganguan中国科学院档案馆(Archives of Chinese Academy of Sciences), 1954-02-004.

Zhongguo kexueyuan zhaokai diwuci zhi dibaci yuanwu changwu huiyi de tongzhi jilu jiyao ji youguan cailiao中科院召开第5次至第8次院务常务会议的通知、记录、纪要及有关材料(Notices, Minutes, Summaries and Related Materials of the 5th to 8th Executive Meetings of the Chinese Academy of Sciences). Beijing: Zhongguo kexueyuan danganguan中国科学院档案馆(Archives of Chinese Academy of Sciences), 1958-02-005.

Zhongkeyuan zhaokai diyi zhi shisi ci yuanwu changwu huiyi de tongzhi jiyao ji youguan cailiao中科院召开第1–14次院务常务会议的通知、纪要及有关材料(Notices, Summaries and Related Materials of the 1st–14th Executive Meetings of the Chinese Academy of Sciences). Beijing: Zhongguo kexueyuan danganguan中国科学院档案馆(Archives of Chinese Academy of Sciences), 1954-02-003.

Zhongkeyuan zhaokai diwuci zhi dijiuci yuanwu changwu huiyi de tongji jilu jiyao ji youguan cailiao中科院召开第五次至第九次院务常务会议的通知、记录、纪要及有关材料(Notices, Minutes, Summaries and Related Materials of the Fifth to Ninth Executive Meetings of the Chinese Academy of Sciences). Beijing: Zhongguo kexueyuan danganguan中国科学院档案馆(Archives of Chinese Academy of Sciences), 1957-02-031.

Zhongyang lingdao dui Zhongguo kexueyuan youguan wenti de pishi中央领导对中国科学院有关问题的批示(Instructions of the Central Leaders on Relevant Issues of Chinese Academy of Sciences), Beijing: Zhongguo kexueyuan danganguan中国科学院档案馆(Archives of Chinese Academy of Sciences), 1985-02-072.

Zhongyang yanjiuyuan dierjie pingyihui linshi huiyi jilu中央研究院第二届评议会临时会议纪录(Minutes of Provisional Meetings of the Second Council of Academia Sinica). Nanjing: Zhongguo dier lishi danganguan中国第二历史档案馆(The Second Historical Archives of China), Total No. 393, File No. 1559.

Zhongyang yanjiuyuan dierjie pingyiyuan juexuan xuanjupiao中央研究院第二届评议员决选选举票(Tickets of the Final Election of the Second Council Members of Academia Sinica). Nanjing: Zhongguo dier lishi danganguan中国第二历史档案馆(The Second Historical Archives of China), Total No. 393, File No. 1551.

Zhongyang yanjiuyuan diyici yuanshi huiyi yubeihui cankao ziliao ji huiyi tian中央研究院第一次院士会议预备会参考资料及会议提案(Reference Materials for the Preliminary Meeting and Proposals of the First General Assembly of Members of Academia Sinica). Nanjing: Zhongguo

dier lishi danganguan中国第二历史档案馆(The Second Historical Archives of China), Total No. 393, File No. 1625.
Zhongyang yanjiuyuan diyici yuanshi xuanju houxuanren fenzu taolun toupiao qingkuangbiao 中央研究院第一次院士选举候选人分组讨论投票情况表(Voting Tables of Groups Discussion on Candidates for the Election of the First Batch of Members of Academia Sinica). Nanjing: Zhongguo dier lishi danganguan中国第二历史档案馆(The Second Historical Archives of China), Total No. 393, File No. 1619.
Zhongyang yanjiuyuan diyici yuanshi xuanju timing biaoge中央研究院第一次院士选举提名表格(Nomination Forms for the Election of the First Batch of Members of Academia Sinica). Nanjing: Zhongguo dier lishi danganguan中国第二历史档案馆(The Second Historical Archives of China), Total No. 393, File No. 2600 (4).
Zhongyang yanjiuyuan diyici yuanshi xuanju timing jigou tongjibiao中央研究院第一次院士选举提名机构统计表(Statistical Table of Nominating Institutes during the Election of the First Batch of Members of Academia Sinica). Nanjing: Zhongguo dier lishi danganguan中国第二历史档案馆(The Second Historical Archives of China), Total No. 393, File No. 1599.
Zhongyang yanjiuyuan diyijie yuanshi houxuanren mingdan中央研究院第一届院士候选人名单(List of Member Candidates for the First Batch of Members of Academia Sinica). Nanjing: Zhongguo dier lishi danganguan中国第二历史档案馆(The Second Historical Archives of China), Total No. 393, File No. 2596.
Zhongyang yanjiuyuan guanyu sheji yuanshi dangxuan zhengzhang shiyang de wenshu中央研究院关于设计院士当选证章式样的文书(Academia Sinica's Documents on Designing Certificates for Elected Members). Nanjing: Zhongguo dier lishi danganguan中国第二历史档案馆(The Second Historical Archives of China), Total No. 393, File No. 1629.
Zhongyang yanjiuyuan guanyu yuanshi houxuanren timing wenti de tonggao中央研究院关于院士候选人提名问题的通告(The Announcement of Nomination problems of Member Candidates of Academia Sinica). Nanjing: Zhongguo dier lishi danganguan中国第二历史档案馆(The Second Historical Archives of China), Total No. 393, File No. 1617.
Zhongyang yanjiuyuan gonggao yuanshi houxuanren ji xuanding yuanshi mingdan中央研究院公告院士候选人及选定院士名单(The Announcement of Member Candidates and the List of Members of Academia Sinica). Nanjing: Zhongguo dier lishi danganguan中国第二历史档案馆(The Second Historical Archives of China), Total No. 393, File No. 2558.
Zhongyang yanjiuyuan hanxun ge danwei zhengji diyici yuanshi houxuanren nianji zili deng xiang ziliao中央研究院函询各单位征集第一次院士候选人年籍资历等项资料(Academia Sinica Sent Letters to Ask Each Unit for Information on the Ages, Hometowns and Qualifications of Candidates for the First Batch of Members). Nanjing: Zhongguo dier lishi danganguan中国第二历史档案馆(The Second Historical Archives of China), Total No. 393, File No. 1611.
Zhongyang yanjiuyuan pingyihui dierjie dierci nianhui jilu中央研究院评议会第二届第二次年会纪录(Minutes of the Second Annual Meeting of the Second Council of Academia Sinica). Nanjing: Zhongguo dier lishi danganguan中国第二历史档案馆(The Second Historical Archives of China), Total No. 393, File No. 1556.
Zhongyang yanjiuyuan pingyihui dierjie disanci nianhui jilu中央研究院评议会第二届第三次年会纪录(Minutes of the Third Annual Meeting of the Second Council of Academia Sinica). Nanjing: Zhongguo dier lishi danganguan中国第二历史档案馆(The Second Historical Archives of China), Total No. 393, File No. 1557.
Zhongyang yanjiuyuan pingyihui dierjie disanci nianhui ti'an中央研究院评议会第二届第三次年会提案(Proposals of the Third Annual Meeting of the Second Council of Academia Sinica). Nanjing: Zhongguo dier lishi danganguan中国第二历史档案馆(The Second Historical Archives of China), Total No. 393, File No. 1565.
Zhongyang yanjiuyuan pingyihui dierjie disici nianhui jilu中央研究院评议会第二届第四次年会纪录(Minutes of the Fourth Annual Meeting of the Second Council of Academia Sinica). Nanjing: Zhongguo dier lishi danganguan中国第二历史档案馆(The Second Historical Archives of China), Total No. 393, File No. 1558.

Zhongyang yanjiuyuan dierjie diyi san si wuci nianhui jilu中央研究院评议会第二届第一、三、四、五次年会纪录(Minutes of the First, Third, Fourth and Fifth Annual Meetings of the Second Council of Academia Sinica). Nanjing: Zhongguo dier lishi danganguan中国第二历史档案馆(The Second Historical Archives of China), Total No. 393, File No. 2551.

Zhongyang yanjiuyuan dierjie pingyihui tiaoli jilu ti'an ji baogao中央研究院第二届评议会条例、记录、提案及报告(Regulations, Records, Proposals and Reports of the Second Council of Academia Sinica). Nanjing: Zhongguo dier lishi danganguan中国第二历史档案馆(The Second Historical Archives of China), Total No. 393, File No. 2926.

Zhongyang yanjiuyuan juban diyici yuanshi xuanju tuijian houxanren mingdan ji youguan guizhang banfa he laiwang wenshu中央研究院举办第一次院士选举推荐候选人名单及有关规章办法和来往文书(List of Recommended Candidates for the Election of the First Batch of Members of Academia Sinica and Relevant Regulations, Methods and Correspondence). Beijing: Qinghua daxue danganguan清华大学档案馆(Archives of Tsinghua University), Total No. 1, Cat. No. 4–2, File No. 191.

Zhongyang yanjiuyuan pingyihui dierjie pingyiyuan xuanju choubei weiyuanhui baogao中央研究院评议会第二届评议员选举筹备委员会报告(Report of the Preparatory Committee for the Election of the Second Council Members of Academia Sinica). Nanjing: Zhongguo dier lishi danganguan中国第二历史档案馆(The Second Historical Archives of China), Total No. 393, File No. 1549.

Zhongyang yanjiuyuan pingyihui diyici yuanshi xuanju choubei weiyuanhui zuzhi ji huiyi jilu中央研究院评议会第一次院士选举筹备委员会组织及会议记录(The Organization and Minutes of the Preparatory Committee for the Election of the First Batch of Members of Academia Sinica). Nanjing: Zhongguo dier lishi danganguan中国第二历史档案馆(The Second Historical Archives of China), Total No. 393, File No. 134.

Zhongyang yanjiuyuan diyijie dierjie diyici nianhui ti'an ji youguan wenshu中央研究院评议会第一届第二届第一次年会提案及有关文书(Proposals and Related Documents of the First Annual Meeting of the First and the Second Council of Academia Sinica). Nanjing: Zhongguo dier lishi danganguan中国第二历史档案馆(The Second Historical Archives of China), Total No. 393, File No. 1563.

Zhongyang yanjiuyuan pingyihui pingyiyuan xuanju wendian ji houxuanren cankao mingdan中央研究院评议会评议员选举文电及候选人参考名单(The Telegram and the Candidate Reference List for the Election of Council Members of the Council of Academia Sinica). Nanjing: Zhongguo dier lishi danganguan中国第二历史档案馆(The Second Historical Archives of China), Total No. 393, File No. 1547.

Zhongyang yanjiuyuan pingyiyuan xuanju mingdan huiyi jilu deng wenjian中央研究院评议员选举名单、会议纪录等文件(Documents such as the Selection Lists and Minutes of Council Members of Academia Sinica, etc.). Nanjing: Zhongguo dier lishi danganguan中国第二历史档案馆(The Second Historical Archives of China), Total No. 393, File No. 2591.

Zhongyang yanjiuyuan shulizu yuanshi houxuanren timingbiao (2) 中央研究院数理组院士候选人提名表(二) (The Nomination Forms of Member Candidates of the Mathematics and Physics Group of Academia Sinica(II)). Nanjing: Zhongguo dier lishi danganguan中国第二历史档案馆(The Second Historical Archives of China), Total No. 393, File No. 1602.

Zhongyang yanjiuyuan 1945 niandu 5 yue 8 yue 12 yue pingyiyuan tanhuahui jilu中央研究院一九四五年度五月八月十二月评议员谈话会纪录(*Minutes of the Symposiums of the Council Members of Academia Sinica in May, August and December, 1945*). Nanjing: Zhongguo dier lishi danganguan中国第二历史档案馆(The Second Historical Archives of China), Total No. 393, File No. 647.

Zhongyang yanjiuyuan yuanshi houxuanren mingdan ji xuanjupiao中央研究院院士候选人名单及选举票(The List of Candidates of Members of Academia Sinica and the Ballots). Nanjing: Zhongguo dier lishi danganguan中国第二历史档案馆(The Second Historical Archives of China), Total No. 393, File No. 1620.

Zhongyang yanjiuyuan yuanshi huiyi tuiding lunwen weiyuanhui deng weiyuan mingdan中央研究院院士会议推定论文委员会等委员名单(Lists of Members of the Thesis Committee and

Other Committees Selected at the First General Assembly of Members of Academia Sinica). Nanjing: Zhongguo dier lishi danganguan中国第二历史档案馆(The Second Historical Archives of China), Total No. 393, File No. 1630.

Zhongyang yanjiuyuan yuanshi xuanju guicheng ji jinghu pingyiyuan tanhuahui jilu中央研究院院士选举规程及京沪评议员谈话会纪录(*The Regulations for the Election of Members of Academia Sinica* and Minutes of the Symposiums of the Council Members in Nanjing and Shanghai). Nanjing: Zhongguo dier lishi danganguan中国第二历史档案馆(The Second Historical Archives of China), Total No. 393, File No. 1034.

Zhongyang yanjiuyuan yuanshi xuanju guicheng ji yuanshi houxuanren timingbiao中央研究院院士选举规程及院士候选人提名表(The Regulations for the Election of Members of Academia Sinica and Nomination Form for Member Candidates). Beijing: Beijing daxue danganguan北京大学档案馆(Archives of Peking University), File No. BD1947058.

Zhongyang yanjiuyuan zai yu pingyiyuan tanhuahui jilu中央研究院在渝评议员谈话会纪录(Minutes of the Symposium of the Council Members of Academia Sinica in Chongqing). Nanjing: Zhongguo dier lishi danganguan中国第二历史档案馆(The Second Historical Archives of China), Total No. 393, File No. 546.

Zhongyang yanjiuyuan zuzhi banfa yuanshi huiyi jilu jiqi xuanju guicheng deng youguan wenshu 中央研究院组织办法、院士会议纪录及其选举规程等有关文书(Relevant Documents such as the Organization Methods of Academia Sinica, Minutes of the General Assembly of Members and the Regulations for Election). Nanjing: Zhongguo dier lishi danganguan中国第二历史档案馆(The Second Historical Archives of China), Total No. 393, File No. 2824.

Zhongyang yanjiuyuan zuzhifa pingyihui tiaoli pingyihui guicheng xianxing fagui zhangcheng mulu deng中央研究院组织法、评议会条例、评议会规程、现行法规章程目录等(The Organization Act, the Regulations of the Council, the Aritcles for the Council, the Catalogue of Current Regulations and Articles of Academia Sinica, etc.). Nanjing: Zhongguo dier lishi danganguan中国第二历史档案馆(The Second Historical Archives of China), Total No. 393, File No. 2539.

Articles and Books

Chen Shiwei (1998). Government and Academy in Republican China: History of Academia Sinica, 1927–1949. Cambridge: Doctoral Thesis of Harvard University.

Chu Coching竺可桢(2005). *Chu Coching quanji*竺可桢全集(The Complete Works of Coching Chu), Vol. 7. Shanghai: Shanghai keji jiaoyu chubanshe上海科技教育出版社.

—(2006a) *Chu Coching quanji*竺可桢全集(The Complete Works of Coching Chu), Vol. 8. Shanghai: Shanghai keji jiaoyu chubanshe上海科技教育出版社.

—(2006b) *Chu Coching quanji*竺可桢全集(The Complete Works of Coching Chu), Vol. 10. Shanghai: Shanghai keji jiaoyu chubanshe上海科技教育出版社.

—(2006c) *Chu Coching quanji*竺可桢全集(The Complete Works of Coching Chu), Vol. 11. Shanghai: Shanghai keji jiaoyu chubanshe上海科技教育出版社.

—(2007) *Chu Coching quanji*竺可桢全集(The Complete Works of Coching Chu), Vol. 13. Shanghai: Shanghai keji jiaoyu chubanshe上海科技教育出版社.

—(2008) *Chu Coching quanji*竺可桢全集(The Complete Works of Coching Chu), Vol. 14. Shanghai: Shanghai keji jiaoyu chubanshe上海科技教育出版社.

Fan Hongye樊洪业(Eds.) (1999). *Zhongguo kexueyuan biannianshi: 1949–1999*中国科学院编年史:1949–1999 (Chronicle of Chinese Academy of Sciences, 1949–1999) . Shanghai: Shanghai keji jiaoyu chubanshe上海科技教育出版社.

Hu Songping胡颂平(1990). *Hu Shizhi xiansheng nianpu changbian chugao*胡适之先生年谱长编初稿(The First Draft of Mr. Hu Shizhi's Chronology), Vol. 6. Taipei: Lianjing chuban shiye gongsi联经出版事业公司.

Li Jishi李吉士(2012). Diwuci xuebu weiyuan dahui buxuanchu zhuxituan wuwei zhixing zhuxi hou wuwei zhixing zhuxi zhuyao huodong jishi第五次学部委员大会补选出主席团五位执行主席后,五位执行主席主要活动纪事(The Chronicle of Main Activities of the Five Executive Chairmen of the Presidium after They Were Re-elected at the Fifth General Assembly of Members of the Academic Divisions of Chinese Academy of Sciences). *Yuanshi ziliao yu yanjiu*院史资料与研究(Historical Materials and Research of the Chinese Academy of Sciences) (2): 1–15.

Li Lairong李来容(2010). Yuanshi zhidu yu minguo xueshu: 1948 nian yuanshi zhidu de queli yu yunzuo院士制度与民国学术—1948年院士制度的确立与运作(Academician System in Republic of China: Establishment and Operation of the Academician System in 1948). Tianjin: Doctoral Dissertation of Nankai University.

Liu Guiyun刘桂云 and Sun Chengrui孙承蕊(Eds.) (2008).*Guojia tushuguan cang guoli zhongyang yanjiuyuan shiliao congbian*国家图书馆藏国立中央研究院史料丛编(Collection of Historical Materials of Academia Sinica Collected by the National Library), Vol.1, 2, 7. Beijing: Guojia tuchuguan chubanshe国家图书馆出版社.

Lyons, Henry (1944). *The Royal Society, 1660–1940: A History of Its Administration under Its Charters*. Cambridge: Cambridge University Press.

Vucinich, Alexander (1984). *Empire of Knowledge: The Academy of Sciences of the USSR (1917–1970)*. Berkeley and Los Angeles: University of California Press.

—(1956) *The Soviet Academy of Sciences*. Stanford: Stanford University Press.

Wang Yangzong王扬宗(2005). Zhongguo yuanshi zhidu de jianli jiqi wenti中国院士制度的建立及其问题(The Academician System in China and Its Defect).*Kexue wenhua pinglun*科学文化评论(Science & Culture Review)2 (6): 5–22.

—Cong yuanshi dao xuebu weiyuan: Zhongguo kexueyuan xueshu tizhi jianli de kunjing从院士到学部委员——中国科学院学术体制建立的困境(From Academicians to Academic Division Members: Difficulties in Establishing Academic System of the CAS). In Yu Miin-ling余敏玲(Ed.) (2012). *Liangan fenzhi: Xueshu jianzhi tuxiang xuanchuan yu zuqun zhengzhi(1945–2000)* 两岸分治:学术建制、图像宣传与族群政治(1945–2000) (Divided Rule across the Taiwan Straits: Educational Reorganization, Visual Propaganda, and Ethnic Politics, 1945–2000). Taipei: Institute of Modern History, Academia Sinica. 65–135.

Wang Yujun王聿均 and Sun Bin孙斌(Eds.) (1977). *Chu Chia-Hua xiansheng yanlun ji*朱家骅先生言论集(Speech by Mr. Chu Chia-Hua). Taipei: Zhongyang yanjiuyuan jinaishi yanjiusuo中央研究院近代史研究所(Institute of Modern History of Academia Sinica) .

Shiah Nae夏鼐(2011). *Shiah Nae riji*夏鼐日记(Shiah Nae's Diary), Vol. 4. Shanghai: Huadong shifan daxue chubanshe华东师范大学出版社.

Xue Pangao 薛攀皋(2011). *Keyuan qianchen wangshi*科苑前尘往事(Past Events of the Scientific Centre). Beijing: Kexue chubanshe科学出版社.

Yang Shangkun杨尚昆(2001). *Yang Shangkun riji* (I) 杨尚昆日记(上册) (Yang Shangkun's Diary), Vol. I. Beijing: Zhongyang wenxian chubanshe中央文献出版社.

Young Chung-Chien杨钟健(1983). *Young Chung-Chien huiyilu*杨钟健回忆录(Memoirs of Young Chung-Chien). Beijing: Dizhi chubanshe地质出版社.

Zhang Jiafu张稼夫(1984). *Gengshen yishi*庚申忆逝(Recalling the Passed in 1980). Taiyuan: Shanxi renmin chubanshe山西人民出版社.

Zhongguo gongchengyuan中国工程院(Chinese Academy of Engineering) (Eds.) (2004). *Zhongguo gongchengyuan nianjian* (1994–1997) 中国工程院年鉴(1994–1997) (Yearbook of Chinese Academy of Engineering, 1994–1997). Beijing: Zhongguo gongchengyuan中国工程院.

Zhongguo gongchengyuan中国工程院(Chinese Academy of Engineering) (Eds.) (1995). *Zhongguo gongchengyuan shoujie yuanshi dahui wenjian jinkan*中国工程院首届院士大会文件集刊(Collection of Documents of the First General Assembly of Members of Chinese Academy of Engineering). Beijing: Zhongguo gongchengyuan中国工程院.

Index

A

Académie des Sciences, 1–3, 16, 49

Adam Pen-Tung Sah, 11, 31, 44–48, 50, 52, 56, 58–61, 66, 68, 72–77, 80, 81, 83, 84, 86, 87, 89, 91, 93, 94, 96, 101, 112, 113, 115, 116, 119, 121, 122, 124, 126, 134–138, 144, 146–150, 152, 154, 156, 161–163, 166–169, 171, 360

C

Chang Chin-yueh, 69, 89, 117, 121, 122, 127, 139, 154, 155, 205, 232

Chang Hsia Ch'ien, 200

Chao Chung-yao, 9, 68

Chao Yuen Ren, 4, 13, 22, 32, 33, 70, 90, 95, 112, 118, 121, 122, 127, 129, 139, 152, 156, 171

Chen Ko Kuei, 11, 29, 69, 79, 91, 92, 117, 122, 127, 129, 130, 139, 153, 154, 171

Chen Kuiyuan, 349–352, 355

Chen Lifu, 38, 362

Chen Shisan, 27, 32, 40, 87, 89, 94, 112, 117, 121, 122, 124, 127, 139, 153–156, 195, 205

Chen Ta, 70, 90, 95, 110, 118, 122, 127, 139, 144, 146, 153–157, 159, 161–163, 207

Chen Yi, 93, 173, 188, 189, 207, 209, 210, 212, 215, 261

Ch'en Yin-k'o, 12, 22, 29, 30, 32, 33, 70, 90, 94, 112, 118, 121, 122, 127, 139, 153–157, 197, 206

Chen Yuan, 12, 22, 32, 33, 44, 45, 70, 87, 90, 94, 112, 118, 122, 124, 127–129, 146, 149, 152, 156, 165, 206

Chen Yun, 180

Chiang Kai-shek, 6, 7, 13, 14, 19, 31, 38, 41, 133, 139–141, 168, 171–173, 362

Chiang Li-Fu, 21, 30, 32, 33, 41, 42, 66, 68, 79, 81, 89, 94, 95, 112, 116, 121, 122, 126, 130, 139, 146–148, 152, 154, 156, 159, 162, 163, 173, 200, 207, 220, 221

Chien Shung-shu, 44, 47, 63, 66, 69, 87, 89, 94, 99, 112, 117, 120, 122, 124, 125, 127–129, 134, 139, 149, 153–156, 158, 160, 163, 165, 205

Chinese Academy of Engineering, 285, 304–306, 309–312, 314–328, 330–334, 337–344, 346, 357, 363–366

Chin Yuch-Lin, 70, 90, 94, 117, 118, 122, 127, 139, 153, 154, 198, 202, 206, 214

Chi Yuoh-Fong, 91, 94, 100, 101, 116, 122, 132, 200, 205, 206

Chou Pei-Yuan, 9, 68, 75, 76, 86, 89, 94, 116, 122, 125, 148, 150, 152, 154, 194, 205, 262, 265

Chou T. Q., 21, 29, 69, 79, 89, 91, 194, 205

Chou Keng-Shang, 22, 24, 32–35, 44, 45, 47, 70, 87, 90, 92, 95, 112, 118, 120, 122–124, 127, 130, 134, 136, 139, 146–149, 152, 156, 157, 161, 165, 199, 228, 229, 234–236

Chow Jen, 4, 23, 30, 40, 44, 66, 68, 80, 81, 87, 89, 94, 112, 115, 116, 121, 122, 124–126, 130, 134, 136, 139, 146, 149, 152, 154, 156, 162, 163, 173, 202, 205

J. Guo, *The Establishment and Reconstruction of the Academician System in China*, https://doi.org/10.1007/978-981-15-7208-1

Chu Chia-hua, 14, 15, 23, 25, 30–33, 37, 39–44, 47, 48, 50, 58, 61–63, 66, 68, 71, 86, 87, 89, 92–94, 101, 112, 113, 115, 116, 119, 122, 124, 126, 130, 134, 138, 139, 141–144, 146, 149–152, 154, 156, 160, 162, 167, 168, 170–172, 174–176, 360
Chu Coching, 10, 19, 23, 30, 31, 40, 42, 44, 45, 68, 89, 93, 94, 115, 116, 122, 126, 139, 144, 146, 147, 151, 152, 156, 159, 161, 162, 167, 173, 183, 189, 191
Chuang Chang-Kong, 10, 23, 32, 40–42, 68, 87, 89, 94, 112, 115, 116, 120, 122–124, 126, 134, 139, 146, 149, 152, 156, 162, 194, 195, 199, 205, 207
Council, 8, 13–15, 17–35, 37–50, 52–67, 71–75, 77–79, 82–88, 91–104, 106, 107, 110–115, 117, 119–121, 123–126, 128, 130, 132–136, 140–142, 144–152, 154–159, 161, 163, 165, 166, 176, 181, 182, 188, 189, 204, 207, 209, 210, 212, 216, 220, 234, 237, 242, 244–247, 249, 251, 254, 255, 257–271, 273, 274, 281, 283, 286, 287, 292–295, 297–302, 304, 305, 310, 311, 313, 315–318, 321–323, 325–330, 335, 338, 340, 346–349, 357–363

D
Deng Xiaoping, 242, 243, 260, 313, 358, 363
Director, 17, 19, 21–23, 26, 27, 34, 43, 47, 48, 51, 74, 78–80, 88, 91, 107, 131, 142, 145, 151, 159, 169, 173, 174, 184, 187, 189–191, 193–196, 198–200, 207, 208, 221, 224, 227, 230, 232–234, 243, 245, 248, 249, 251, 252, 257–259, 265, 274, 285, 288, 290–292, 294, 296–298, 300, 303, 315, 321, 323, 336, 342–345, 347, 348, 354, 356, 360
Director-General, 17, 19, 22, 25, 44, 47, 48, 52, 57, 58, 66, 92, 94, 121, 135, 136, 138, 145, 151, 159, 161, 162, 169, 172, 360

F
Fang Yi, 257, 259, 264, 271, 273, 279, 281, 282, 288
Feng Teh-Pei, 12, 69, 94, 117, 127, 134, 139, 146, 148, 152, 156, 157, 162, 174, 196, 262
Feng You-lan, 12, 70, 90, 94, 118, 122, 127, 139, 146, 150–152, 156, 161, 162, 164–166, 198, 202, 206, 214
Fu Ssu Nien, 23, 25, 30, 31, 37, 39, 40, 42–45, 47, 50–57, 60, 61, 64, 66, 70, 79, 90, 94, 97, 101–112, 114, 118, 122, 127, 128, 131, 132, 139, 146–149, 151, 153–156, 158–160, 171–173, 360

G
Ge Tingsui, 194, 205, 285, 286

H
He Bingmeng, 349
Hou Hsiong-Lin, 200, 205, 206, 310, 313, 319, 322, 323, 326, 327, 329, 334, 335, 337, 339, 357
Hou T. P., 11, 31–33, 66, 68, 79, 89, 94, 112, 115, 116, 122, 126, 139, 152, 154–157, 197, 205, 227
Hsiao Kung-chuan, 31, 70, 90, 95, 118, 122, 123, 127, 134, 139, 150, 153–155, 162, 171
Hsieh C. Y., 27, 29, 32, 33, 40, 42, 44, 50, 61, 66, 68, 87, 89, 93, 94, 112, 116, 122, 126, 139, 147, 149, 152, 154, 155, 157, 159, 161, 162, 167, 195, 202, 205, 241
Hsu Pao-Lu, 67, 79, 89, 94, 103, 109, 116, 122, 123, 126, 128, 139, 152, 154, 194, 205, 221
Hu Keshi, 274, 275, 279, 280, 290, 345
Hu Qiaomu, 206, 257–259, 264, 274, 275, 281, 282, 288, 289, 344
Hu Shih, 12, 22–24, 29–33, 37, 44, 45, 47, 50, 52, 53, 60, 66, 70, 73–75, 77, 78, 85–88, 90–94, 101, 103, 112, 115, 118, 119, 121, 122, 124, 125, 127, 134–136, 139, 143, 144, 146–152, 156, 161, 164, 165, 171, 172, 176, 360
Hu Yaobang, 242, 288, 289
Hua Loo-Keng, 9, 14, 68, 79, 89, 94, 109, 116, 122, 126, 128–130, 139, 152, 154, 175, 189, 194, 199, 205, 207, 208, 262

Huang T. K., 10, 68, 89, 94, 116, 121, 122, 126, 134, 139, 152, 154, 195, 202, 205, 208, 262
Huang Zongzhen, 173, 180–182
Hu Hsen-hsu, 11, 30, 32, 44, 47, 69, 87, 89, 94, 99, 112, 117, 120, 122, 124, 127, 139, 152, 156, 161, 167, 173, 195, 202, 203, 207, 208, 224, 225, 230, 232

J
Jaw Jeou-Jang, 50, 66, 68, 76, 79, 87, 99, 112, 115, 124, 125, 149, 156, 196, 205, 238

K
Ku Chieh-kang, 27, 70, 90, 95, 118, 122, 124, 126, 127, 129, 139, 153–155, 207, 228, 229
Kuo Mo-jo, 29, 70, 90, 92, 93, 95, 96, 118, 122, 123, 127, 139, 153, 154, 174, 183, 186, 188, 189, 192, 193, 197, 199, 204, 206–208, 210–212, 217, 237, 260

L
Lee J. S., 10, 19, 23, 30, 40–42, 68, 89, 94, 112, 115, 116, 121, 122, 126, 139, 152, 154, 156, 173–175, 183, 195, 196, 205, 239
Li Chi, 13, 22, 24, 30, 32, 33, 40, 50, 52, 60, 61, 66, 70, 87, 90, 92, 95, 101, 112, 118, 122, 124, 127, 134, 136–138, 144, 146–149, 152, 156, 159, 162, 171, 360
Li Fang Kuei, 27, 29, 70, 90, 95, 118, 122, 127, 129, 139
Li Hsien-wen, 67, 90, 94, 99, 117, 121–123, 127, 134, 138, 144, 146, 153, 154, 171, 172
Ling Hung-hsun, 30–33, 44, 50, 61, 66–68, 87, 89, 94, 112, 115, 116, 122, 124–126, 129, 130, 134, 139, 146, 149, 152, 154–157, 171, 172
Li Shizeng, 5–7
Li Shu-Hua, 21, 24, 30, 32, 33, 40–42, 44, 45, 63, 66, 68, 87–89, 91, 93–95, 112, 115, 116, 119, 120, 122–124, 126, 130, 134, 136, 138, 144, 146–149, 152, 156, 160, 162, 163, 171, 360
Li Zongen, 69, 89, 94, 98, 99, 106, 107, 117, 121, 122, 127, 132, 138, 144, 146, 152, 156, 159, 161, 207
Liang Qichao, 13
Liang Ssu-cheng, 13, 27, 29, 70, 90, 95, 118, 122, 127, 136, 137, 139, 146, 153–156, 161, 162, 205
Liang Ssu-yung, 13, 29, 67, 79, 90, 95, 118, 122, 127, 139, 153, 154, 206
Liu W. T., 90, 94, 103, 104, 106, 112, 118, 122
Liu Yei-Tsen, 70, 90, 94, 118, 122, 123, 126–129, 139, 146, 153–155, 207
Loo Tsung-Le, 44, 50, 63, 66, 69, 70, 74, 77, 79, 83, 87, 89, 91, 94, 99, 112, 117, 122, 124, 127, 134, 139, 147, 148, 153, 154, 156, 165, 173, 205, 209
Lu Dingyi, 180, 182, 217
Luo Peilin, 226, 310, 313, 319, 322, 326, 327, 329, 337, 339, 357

M
Mao T. E., 30, 32, 40, 42, 44, 50, 52, 60, 66, 68, 71, 73, 82, 87, 89, 93, 94, 112, 115, 116, 120, 122–124, 126, 139, 146–150, 152, 156, 162, 199, 205, 208, 227, 262, 360
Mao Tse-tung, 180, 187, 197, 216, 257
Ma Yin-chu, 29, 70, 88, 90, 95, 118, 122, 127, 134, 139, 153–155, 197, 202, 206, 214
Mei Yi-chi, 34, 35, 74–77, 96, 150, 152, 154

N
Ny Tsai-ze, 27, 29, 68, 89, 94, 116, 122, 126, 130, 139, 152, 154, 160, 162, 164, 189, 191, 194, 199, 205, 208, 221, 227, 252, 254, 262, 265, 279, 280, 288, 291

P
Pai Sitsan, 68, 89, 94, 103, 117, 122, 123, 127, 138, 146, 153, 154, 159, 189, 195, 205, 262
Peter P. T. Sah, 27, 68, 76, 89, 91
Ping Chi, 11, 21, 30, 32, 44, 45, 66, 68, 77, 83, 87–89, 91, 93, 94, 112, 115, 117, 119, 122, 124, 127, 139, 144, 146–149, 151, 152, 156, 165, 173, 195, 205, 360

R

Robert K. S. Lim, 11, 21, 30, 32, 44, 66, 69, 87, 90, 94, 99, 106, 112, 117, 121, 122, 124, 127, 139, 146, 149, 152, 156, 157

Royal Society of London for Improving of Natural Knowledge, 1, 2, 6

S

Shiah Nae, 86, 88, 92, 93, 104, 105, 107–109, 112, 126, 130, 131, 143, 172, 206, 266, 267, 269, 270, 283

Shi Changxu, 310, 313–315, 319, 322, 323, 326, 327, 329, 336, 337, 339, 343, 357

Shiing-Shen Chern, 9, 68, 74–76, 80, 86, 89, 94, 109, 116, 122, 126, 128, 134, 139, 146, 150–152, 156–159, 162, 163, 171, 307, 380

Song Jian, 288, 290, 291, 315–317, 322, 323, 326–329, 337–340

Su Buchin, 27, 68, 89, 94, 116, 122, 123, 126, 130, 139, 152, 154–156, 162, 194, 205, 221, 262, 276, 277

T

Tai Fon-Land, 27, 32, 40, 42, 66, 69, 89, 94, 99, 112, 117, 122, 127, 129, 139, 149, 153, 154, 157, 205

Tang F. F., 69, 89, 94, 98, 99, 107, 117, 121, 122, 125, 132, 200, 202, 203, 224, 225, 230, 232, 235, 236, 239, 241

Tang P. S., 69, 90, 94, 117, 122, 127, 139, 153–156, 159, 160, 162, 164, 198, 202, 205

Tang Yung-tung, 70, 90, 94, 118, 122, 127, 139, 146, 152, 156, 159, 161, 197, 198, 206

Tao L. K., 23, 25, 30, 40, 42, 66, 70, 79, 80, 87, 90, 93, 95, 112, 115, 118, 122, 127, 128, 139, 146, 149, 151, 153–156, 159, 162, 163, 173, 174, 183, 202, 206, 214

Teng S. C., 70, 90, 94, 117, 122, 127, 134, 139, 153, 154, 159, 195, 205

Tsai Chiao, 11, 12, 69, 90, 94, 117, 122, 127, 134, 139, 144, 151, 153, 154, 196, 205

Tsai Yuan-Pei, 5–8, 18–20, 23, 27, 30, 34, 38, 101, 140, 362

Tseng Chao-lun, 27, 29, 32, 40, 68, 89, 94, 112, 116, 122, 123, 126, 139, 152, 154, 157, 194, 205, 221, 241

Tsien H. S., 77, 89, 91, 92, 221–223, 226, 231, 235, 236, 239, 241, 262, 288, 335–337, 339

Tsien San-Tsiang, 180–182, 184, 186, 189, 190, 192, 194, 205, 206, 248, 259, 260, 262–267, 269, 271, 274, 291, 292, 294

Tsien T. S., 70, 90, 95, 118, 122, 127, 139, 153–156, 199, 207, 228, 229, 234–236

Tung Ti-Cho, 68, 89, 94, 117, 121–123, 127, 139, 153, 154, 195, 205, 208, 224

Tung Tso-pin, 13, 27, 70, 90, 95, 118, 122, 127, 129, 139, 153, 154, 171

V

V. K. Ting, 4, 17, 19–24, 34, 35

W

Wang Chia-Chi, 23, 30, 40, 42, 50, 66, 77, 79, 83, 87, 89, 94, 112, 117, 122, 124, 127, 134, 139, 153, 154, 156, 165, 173, 195, 205

Wang Chung Hwei, 70, 88, 90, 95, 118, 122, 127, 128, 134, 138, 144, 146, 152, 156, 171

Wang Daheng, 205, 262, 310, 313, 319, 322, 323, 326, 327, 329, 337, 339, 357

Wang Ganchang, 68, 194, 205, 221, 262

Wang Ging-Hsi, 23, 30, 40–42, 44, 45, 66, 69, 79, 90, 94, 112, 117, 122, 127, 139, 147, 153, 154, 156, 171

Wang Shih-Chieh, 19, 22, 25, 30, 32, 33, 39, 44, 66, 70, 88, 90, 95, 112, 113, 118, 122, 127, 139, 147, 148, 153–157, 171

Wang Shu-min, 103, 104

Wong Wen-hao, 4, 8, 10, 22, 24, 25, 27, 30–32, 34, 37–42, 44–47, 50, 52, 56, 58, 61–64, 66, 68, 72, 73, 83, 87, 89, 94, 101, 112, 116, 119, 122, 124, 126, 130, 134–136, 138, 142, 143, 146, 149, 152, 156, 157, 160, 161, 166, 167, 172, 173, 207, 360

Woo Sho-Chow, 10, 27, 40, 42, 44, 50, 66, 68, 79, 83, 87, 89, 94, 112, 115, 116, 121, 122, 124, 126, 134, 138, 146, 149, 152, 154, 156, 159, 162, 163, 173, 194, 205

Woo Ting-liang, 22, 30, 32, 33, 40, 42, 44, 61, 66, 69, 87, 90, 94, 112, 117, 122, 124, 127, 134, 138, 146, 149, 153–155, 157, 207, 228
Woo Tsin-Hang, 70, 90, 94, 103, 117, 118, 122, 125–129, 139, 153, 154, 171
Woo Y. H., 8, 27, 29, 32, 40, 42, 44, 45, 47, 50, 52, 60, 66, 68, 83, 87, 89, 93, 94, 112, 115, 116, 121, 122, 124, 126, 139, 152, 156, 173, 189, 191, 194, 199, 205, 207, 208, 237, 240, 360
Wu Heng, 184, 187, 189, 202, 205, 206, 217, 259, 260, 262, 264, 265, 267, 269, 288, 290
Wu Hsien, 10, 21, 29, 31, 68, 69, 89, 91, 94, 99, 115, 116, 122, 126, 139, 152, 154, 171
Wu Hsien-Wen, 29, 31, 89, 94, 117, 121, 127, 134, 138, 144, 146, 153–157, 159, 161, 167, 195, 205
Wu Ta-You, 9, 68, 89, 94, 116, 122, 126, 130, 139, 152, 154, 171

Y

Yang Shu-Tah, 70, 90, 94, 103, 118, 121, 122, 127, 128, 139, 153, 154, 161, 166, 206
Yao Y. T., 9, 68, 89, 94, 116, 122, 126, 139, 152, 154, 160, 162, 163, 194, 205
Yeh Chi-sun, 21, 25, 29–31, 68, 89, 94, 95, 106, 116, 122, 126, 130, 152, 156, 162, 163, 194, 205
Yin Hung-chang, 69, 89, 94, 117, 121, 122, 127, 139, 146, 150, 151, 153–156, 161–163, 195, 205
Young Chung-chien, 10, 29, 68, 86, 87, 89, 94, 116, 122, 126, 131, 134, 139, 152, 154, 155, 161, 167, 195, 205, 208
Yu Cha-Shih, 70, 90, 94, 103, 118, 122, 125, 126, 128, 129, 138, 144, 146, 151, 153, 154, 206
Yuan I-chin, 69, 89, 94, 106, 107, 121, 122, 127, 130, 134, 139, 146, 150, 153–155, 171
Yu T. F., 70, 90, 94, 98, 99, 117, 121–123, 127, 132, 139, 146, 152, 156, 157, 205
Yun Ziqiang, 180, 199, 202, 205–207

Z

Zhang Guangdou, 198, 205, 227, 262, 276, 277, 310, 313, 319, 322–327, 330, 337, 339, 357
Zhang Jiafu, 184, 186–189, 197, 202, 206, 213, 217
Zhang Wei, 205, 227, 310, 313, 319, 322, 323, 326, 327, 329, 337, 339, 357
Zhang Yuanji, 70, 90, 94, 103, 118, 122, 127–130, 139, 143, 144, 146, 153, 154, 171, 207
Zhou En-lai, 174, 181, 188, 204, 207, 216, 219
Zhou Guangzhao, 294, 305, 315–317, 319, 322, 323, 325–327, 330, 335, 336, 338, 340
Zhu Guangya, 310, 311, 322, 326, 327, 330, 337, 339, 340, 343

The manufacturer's authorised representative in the EU is Springer Nature Customer Service Centre GmbH, Europaplatz 3, 69115 Heidelberg, Germany. If you have any concerns regarding our products, please contact ProductSafety@springernature.com

Printed and bound by CPI Group (UK) Ltd, Croydon, CR0 4YY
06/10/2026
02225252-0013